ENVIRONMENTAL CONTAMINANTS, ECOSYSTEMS AND HUMAN HEALTH

The Pennsylvania Academy of Science Publications
Books and Journal
Editor: Shyamal K. Majumdar
Professor and Head of Biology, Lafayette College
Easton, Pennsylvania 18042
(610) 250-5464

1. *Energy, Environment, and the Economy,* 1981. ISBN: 0-9606670-0-8. Editor: S.K. Majumdar.
2. *Pennsylvania Coal: Resources, Technology and Utilization,* 1983. ISBN: 0-9606670-1-6. Editors: S.K. Majumdar and E.W. Miller.
3. *Hazardous and Toxic Wastes: Technology, Management and Health Effects,* 1984. ISBN: 0-9606670-2-4. Editors: S.K. Majumdar and E.W. Miller.
4. *Solid and Liquid Wastes: Management, Methods and Socioeconomic Considerations,* 1984. ISBN: 0-9606670-3-2. Editors: S.K. Majumdar and E.W. Miller.
5. *Management of Radioactive Materials and Wastes: Issues and Progress,* 1985. ISBN: 0-9606670-4-0. Editors: S.K. Majumdar and E.W. Miller.
6. *Endangered and Threatened Species Programs in Pennsylvania and Other States: Causes, Issues and Management,* 1986. ISBN: 0-9606670-5-9. Editors: S.K. Majumdar, F.J. Brenner and A.F. Rhoads.
7. *Environmental Consequences of Energy Production: Problems and Prospects,* 1987. ISBN: 0-9606670-6-7. Editors: S.K. Majumdar, F.J. Brenner and E.W. Miller.
8. *Contaminant Problems and Management of Living Chesapeake Bay Resources,* 1987. ISBN: 0-9606670-7-5. Editors: S.K. Majumdar, L.W. Hall, Jr. and H.M. Austin.
9. *Ecology and Restoration of The Delaware River Basin,* 1988. ISBN: 0-9606670-8-3. Editors: S.K. Majumdar, E.W. Miller and L.E. Sage.
10. *Management of Hazardous Materials and Wastes: Treatment, Minimization and Environmental Impacts,* 1989. ISBN: 0-9606670-9-1. Editors: S.K. Majumdar, E.W. Miller and R.F. Schmalz.
11. *Wetlands Ecology and Conservation: Emphasis in Pennsylvania,* 1989. ISBN: 0-945809-01-8. Editors: S.K. Majumdar, R.P. Brooks, F.J. Brenner and R.W. Tiner, Jr.
12. *Water Resources in Pennsylvania: Availability, Quality and Management,* 1990. ISBN: 0-945809-02-6. Editors: S.K. Majumdar, E.W. Miller and R.R. Parizek.
13. *Environmental Radon: Occurrence, Control and Health Hazards,* 1990. ISBN: 0-945809-03-4. Editors: S.K. Majumdar, R.F. Schmalz and E.W. Miller.
14. *Science Education in the United States: Issues, Crises, and Priorities,* 1991. ISBN: 0-945809-04-2. Editors: S.K. Majumdar, L.M. Rosenfeld, P.A. Rubba, E.W. Miller and R.F. Schmalz.
15. *Air Pollution: Environmental Issues and Health Effects,* 1991. ISBN: 0-945809-05-0. Editors: S.K. Majumdar, E.W. Miller and J.J. Cahir.
16. *Natural and Technological Disasters: Causes, Effects and Preventive Measures,* 1992. ISBN: 0-945809-06-9. Editors: S.K. Majumdar, G.S. Forbes, E.W. Miller and R.F. Schmalz.
17. *Global Climate Change: Implications, Challenges and Mitigation Measures,* 1992. ISBN: 0-945809-07-7. Editors: S.K. Majumdar, L.S. Kalkstein, B. Yarnal, E.W. Miller and L.M. Rosenfeld.
18. *Conservation and Resource Management,* 1993. ISBN: 0-945809-08-5. Editors: S.K. Majumdar, E.W. Miller, D.E. Baker, E.K. Brown, J.R. Pratt and R.F. Schmalz.
19. *Biological Diversity: Problems and Challenges,* 1994. ISBN: 0-945809-09-3. Editors: S.K. Majumdar, F.J. Brenner, J.E. Lovich, J.F. Schalles and E.W. Miller.
20. *The Oceans: Physical-Chemical Dynamics and Human Impact,* 1994. ISBN: 0-945809-10-7. Editors: S.K. Majumdar, E.W. Miller, G.S. Forbes, R.F. Schmalz and A.A. Panah.
21. *Medicine and Health Care Into the 21st Century,* 1995. ISBN: 0-945809-11-5. Editors: S.K. Majumdar, L.M. Rosenfeld, D.B. Nash and A.M. Audet.
22. *Environmental Contaminants, Ecosystems and Human Health,* 1995. ISBN: 0-945809-12-3. Editors: S.K. Majumdar, E.W. Miller and Fred J. Brenner.

ENVIRONMENTAL CONTAMINANTS, ECOSYSTEMS AND HUMAN HEALTH

EDITED BY

S.K. MAJUMDAR, Ph.D.
Marshall R. Metzgar Professor
and Head of Biology
Department of Biology
Lafayette College
Easton, PA 18042

E.W. MILLER, Ph.D.
Professor and Associate Dean (Emeritus)
The Pennsylvania State University
University Park, PA 16802

FRED J. BRENNER, Ph.D.
Professor of Biology
Department of Biology
Grove City College
Grove City, PA 16127

Founded on April 18, 1924

A Publication of
The Pennsylvania Academy of Science

Library of Congress Cataloging in Publication Data

Bibliography
Index
Majumdar, Shyamal K. 1938-, ed.

Library of Congress Catalog Card No.: 95-74813

ISBN-0-945809-12-3

A Publication of
The Pennsylvania Academy of Science, Easton, PA 18042

Printed in the United States of America by

Typehouse of Easton
Easton, PA 18042

ACKNOWLEDGEMENT

The publication of this book was aided by contributions from the E. Willard and Ruby S. Miller book publication fund.

PREFACE

Ever since the industrial and agricultural revolution, humans have been discharging contaminants into natural ecosystems which not only adversely impacting these systems but also human health as well. Following the second world war, the use of chemicals to control insect pests, plant diseases, and vegetation has resulted in increased crop yields. This widespread use of agricultural and other chemicals has raised public health concerns because of their potential as mutagenic, carcinogenic and teratogenic agents, as well as their harmful effects on non-target species and bioaccumulation in aquatic and terrestrial food chains. Chemicals, pesticides, radioactive materials, and their by-products are important components of environmental contaminants. In addition, the excavation and utilization of fossil fuels to meet our ever increasing demand for electric power generation along with the disposal of the by-products of our industrial society pose concerns for ecosystems and human health in all regions of the world. Many of these issues are major public policy frontiers in the 1990's. The solution of these problems will center around critical decision processes in advanced and the Third World nations.

In the current volume, the authors cover a variety of concerns regarding the adverse impacts of contaminants on ecosystems and human health. The twelve chapters in the first section of the text address the impact of contaminants on ecosystem function, and ten of the remaining twenty-two chapters are devoted to the effects of contaminants on human health. Part three presents eight case studies in humans, while the final four chapters of the text provide the reader with an assessment of environmental problems and analyses.

The editors express their appreciation to the authors for the quality of their work and for excellent cooperation as the volume developed. We trust that this volume will be a valuable asset to a wide audience. Special gratitude is extended to Lafayette College, the Pennsylvania State University, Grove City College and the Pennsylvania Academy of Science for their support during the development of the book. The editors extend heartfelt thanks to their spouses for their encouragement and support during the preparation of this volume.

S.K. Majumdar
E. Willard Miller
Fred J. Brenner
Editors
November, 1995

Environmental Contaminants, Ecosystems and Human Health

Table of Contents

Environmental Contaminants, Ecosystems and Human Health. Edited by S.K. Majumdar, E.W. Miller, and Fred J. Brenner. © 1995, The Pennsylvania Academy of Science.

Chapter One

ENVIRONMENTAL HAZARDS: Survey And Perspective

E. WILLARD MILLER
Professor of Geography and Associate Dean, Emeritus
The College of Earth and Mineral Sciences
The Pennsylvania State University
University Park, PA

INTRODUCTION

Since the earliest civilization there has been some contamination of the environment. The quantities of contaminants were, in general, small and localized in time and space. It was generally felt that the world was so vast that it could not be permanently affected by human activities. With the development of a worldwide industrial society, pollution of the earth has become a reality. It is now recognized that issues of environmental contamination of the total ecosystem are looming as major public policy frontiers.

Because of the vastness of the earth's environment, the problems have been created over decades, and it is difficult, if not impossible, to detect short-range alterations. As a result, policy decisions must be made on a long-range basis.

HISTORICAL ENVIRONMENTAL CONTAMINATION

There is evidence in caves occupied by early man that smoke filled their living quarters. In the Bronze and Iron Ages, industrial villages thrived and smoke and particulates filled the atmosphere. Clay was baked and glazed as early as 4000 BC. The early industries relied on charcoal, and the conversion of wood to charcoal created a smoky environment. Coal was used before 1000 AD and began to be converted into coke about 1600.

Over the centuries, there have been references in literature to the health problems of smoke pollution in cities. As early as AD 61 the Roman philosopher Seneca wrote about air pollution in Rome. The problem of air pollution in London has been reported for centuries. Conditions became so bad in 1306 that Edward I issued a royal proclamation forbidding the use of "sea coal" in the city and Queen Elizabeth I banned the burning of coal in London when Parliament was in session. (Miller and Miller, 1989).

With the coming of the Industrial Revolution, which began in England in the middle of the 18th century, air pollution increased tremendously. The most important fuel of the Revolution was coal, with petroleum and natural gas becoming major sources in the 19th century.

The quality of the water for human consumption has been a problem since the earliest settlements of civilization. Without a sewage system water contamination was inevitable. The common method of disposing of raw sewage was to deposit it outside the home on the street. As a response fevers were common occurrences, particularly during the warm season. Because there appeared to be no solution, the problem of water pollution was long ignored. The first legislation was enacted in 1647 when the Massachusetts colony legislature passed a law designed to prevent the pollution of the Boston harbor. This Act was not to control disease but to improve navigation. It was not until the 19th century that polluted water was recognized as a potential source of disease.

By the middle of the 19th century before pathogenic microorganisms had been isolated, John Snow, a physician in England, demonstrated the causal relationship between a polluted water supply and disease transmission. By empirical evidence he showed that people who drank polluted water from the Southwash and Vandhall Water companies had a higher incidence of cholera than other Londoners. At the same period, physicians in the United States and Germany began campaigns for clean water supplies.

Of all environmentally dispersed diseases plagues have been the most destructive to human populations. These infectious diseases caused by Pasteurelle pestis and Bacillas pestis are a response to the contamination of the environment by rats and fleas. The plague takes three forms: bubonic swelling of the lymph glands, pneumonic swelling of the lungs, and septumic blood distribution due to invasion of Pasteurelle pestis.

The earliest recorded plague was 320 BC in the Philistines population and the last recorded plague in the United States was reported in 1955. Plagues are a worldwide phenomenon. They usually existed a few years and then disappeared. The most destructive one was in Europe in the 14th century known as the "Black Death". There was an estimated 25,000,000 deaths which was two-thirds to three-quarters of the total population.

The plagues have occurred where sanitation was essentially lacking. There have been no satisfactory explanations why plagues devastated the population and then disappeared quickly. In more recent times it is recognized that sanitation, cleanliness, and hospitalization have lessened the problem of plagues. As a response, plagues have nearly disappeared in the 20th century.

AIR CONTAMINATION 1930 -

By the 1930s air pollution had increased in all industrial nations. Further, there was a growing recognition that this situation was becoming intolerable. Large-scale surveys of cities and regions throughout the world highlighted the problems. Although the technical foundation was well established to control air pollution, no comprehensive plans had been devised or legislation enacted. After 1945 with the potential increase in radioactivity due to explosions of atomic bombs and some catastrophic air incidents such as at Donora, Pennsylvania, provided the catalysts for initial action to control air pollution.

Radioactive Contamination

The potential for radioactivity contaminating the atmosphere is a modern problem. The potential arises from two sources-atomic explosions and nuclear plant accidents. The explosion of an atomic bomb releases a massive amount of radioactivity into the atmosphere (Turco, 1983). After World War II the Soviet Union, United Kingdom, and France continued to test atomic bombs in the atmosphere. By the mid-1950s radioactive contamination had spread over the world. Great concern was developing as to the health effects that would occur. In 1956 President Eisenhower was able to achieve a moratorium on testing bombs in the atmosphere. The Soviet Union, United Kingdom, and the United States discontinued this practice although France has exploded a few bombs in the atmosphere since then. It is now known that the radioactive contaminants remain in the soils of such islands as Bikini in the Pacific Ocean and it will be many decades before these areas can once again become habitable.

Urban Air Pollution

There are a great number of atmospheric pollutants created by human activities in the cities. The basic pollutants are carbon dioxide, carbon monoxide, sulfur dioxide (SO_2), and nitric oxide (NO_x). The major sources of these gases in the atmosphere is the burning of fossil fuel (Elsom, 1987).

When SO_2 and NO_x combine with water, sulfur and nitric acids are formed. In addition in the presence of oxygen and sunlight, the nitric oxide (NO_x) combines with other organic compounds, such as the hydrocarbon from unburned gardens, to produce a whitish haze, sometimes tinged with a yellow brown color. This is commonly known as smog. In the process a large number of new hydrocarbons and oxyhydrocarbons are produced. These secondary hydrocarbon products may compose as much as 95 percent of the total organic matter in a severe smog episode.

The term oxidant refers to chemicals that oxidize substances not easily oxidized by oxygen. Photochemical oxidants consist primarily of ozone (O_3), and nitrogen dioxide (NO_3) as well as small quantities of peroxyacetyl (PAN) and other peroxy compounds. Heavy concentrations of ozone can be produced only when hydrocarbons, aldehydes and other reactive gases are converted to peroxy radicals by photochemical processes. Of these oxidants, ozone is the most important.

Air pollutants have a number of health effects. Ozone irritates the mucous membranes of the respiratory system causing coughing, choking and impaired lung function. Nitrogen oxide can increase susceptibility to vital infections such as influenza. Carbon monoxide interferes with the blood's ability to absorb oxygen, and many of the toxic emissions in smog are suspected or known to cause cancer and birth defects (Miller and Miller, 1989).

The atmospheric movement of SO_2 and NO_2 pollutants is a complex and politically charged issue, and scientific studies since the early 1950s have revealed a number of findings. Local emissions of pollutants will cause some acid rain to fall in that area, but in addition, the pollutants in many areas are transported hundreds, or even thousands of miles downwind across national and international borders.

There is growing evidence that acid rain proves a threat to human health. The New York State Department of Health has linked acid rain to an outbreak of gastroenteritis in the Adirondack Mountains-the outbreak was caused by acid resistant strains of bacteria found in wells fed by acidified groundwater. In another report the department found that water from acidified lakes contained toxic chemicals that were potentially dangerous to health and recommended that the residents of affected areas flush their water pipes thoroughly before drinking the water.

Studies have shown that sulfur and nitric oxides pose serious problems for people with respiratory diseases and the presence of sulphate particulates in acid rain can lead to serious illness and even death. The U.S. Council on Environmental Quality has estimated that the health related problems directly attributable to acid precipitation are costing the country $2 billion annually. In a Canadian study undertaken at Brookhaven National Laboratory in New York, it was estimated that 5,000 Canadians may be dying each year of diseases directly related to acid rain. Many cases of chronic bronchitis and emphysema leading to chronic heart disease can be traced to the effects of acidic particulate matter.

Control of Air Pollution

The control of air pollution to protect public health and welfare developed slowly in the United States There was the general belief that air pollution control was the responsibility of local and state governments and it was not until the late 50s that it began to be recognized as a national problem.

One of the earliest local air pollution control programs in the United States began in southwestern Pennsylvania. Pittsburgh in the industrial valley of Western Pennsylvania was known as the "Smoky City." In order to revitalize the city, laws were passed in the late 1940s prohibiting the use of coal in factories, railroads, and residences without abatement controls.

California in 1950 passed the first state law to control pollutants from automobiles. The legislation stated: "that the control and elimination of those pollutants is of primary importance for the protection and preservation of the public health and well-being and for the prevention of irritation to the senses, interference with visibility, and damage to vegetation and property." The pollution control legislation pertaining to motor vehicles was implemented by placing emission control devices on the fuel system.

By the 1950s it became apparent that air pollution was not a local or even a regional phenomenon but a national problem. The first federal attempts in 1955, 1960, 1963, 1965 and 1967 to provide a framework for the control of air pollution were tentative, but these early acts gradually began to address the problem more directly. By 1970 the philosophical basis for federal controls had been well established.

The 1970 Clean Air Act inaugurated the modern era of air pollution control. The Act established a federal program, which had as a goal the prevention, control, and abatement of air pollution from stationary and mobile sources, and the administration of the program was removed from the Department of Health, Education and Welfare and placed in the newly created Environmental Protection Agency (EPA). The original objective was to achieve clean air throughout the nation by July 1975.

The major provisions of the Act included:

1. Expanded research funds totaling $360 million to be spent over a three-year period on fuels and stationary sources. Special emphasis was to be placed on (1) the means to clean fuels before combustion, (2) the development of new and synthetic fuels, (3) epidemiological studies of the effects of air pollutants on mortality and morbidity, and (4) studying the immunological, biochemical, physiological, and toxicological effects of air pollutants.
2. An assessment of the causes and effects of noise pollution, with an Office of Noise Pollution and Abatement to be established under the Environmental Protection Agency.
3. Establishment of primary (health) and secondary (welfare) ambient air quality standards.
4. Emission limitations enforceable by both state and federal government. Enforcement to be through criminal penalties and injunctions as well as provisions for citizen suits.

5. Transportation plans must include the "land use and transportation controls" necessary to attain air quality standards.

The key issues in the establishment of ambient air-quality standards concerned the SO_2 and NO_x portions of the standards. In 1971 the EPA established a uniform system known as the New Source Performance Standards (NSPS) for fossil-fired utility boilers

1. Sulfur dioxide (SO_2) 1/2 lb/mill, Btu heat input emission limit and a three-hour stack test to show compliance.

2. Nitric oxide (NO_2) 0.7 lb/mill. Btu emission limit on coal burned.

3. Particulate 0.10 lb/mill. Btu heat input with a normal stack test for compliance.

The goals of clean air through the nation have been more difficult to achieve than was originally visualized. The legislation and then implementation have been in a constant state of change in order to achieve established goals. For example, the 1977 Clean Air Amendments revised the plans to provide for a reasonable further progress, and the 1987 Amendments recognized that air pollution was an international problem and that the U.S. president was to institute negotiations with Canada and Mexico to control transboundary air pollution. All of the amendments have stipulated strict regulations for mobile sources of emission. After nearly a quarter century of legislation many complex issues remain unresolved. These include such questions as: What is the economic cost of clean air? If clean air is not achieved, what will be the health cost to the individual and the nation? How will the traditional patterns of life be changed? These are only a few of the questions that must be answered, if clean air is to be achieved.

CHEMICAL CONTAMINATION 1960 -

In the 1950s there was much concern expressed about the contamination of the environment due to the atmospheric testing of nuclear weapons. With the termination of atmospheric nuclear testing other environmental problems persisted. The widespread use of pesticides and other contaminants took the place of radioactive fallout as an environmental issue. In the early 1960s, under the stimulation of President John F. Kennedy, many Americans were seeking ways to create a better world. The pollution of the environment gave them the cause they were looking for.

Pesticide Controls

Within this framework for environmental change, Rachel Carson, a nature lover, wrote *Silent Spring*, a book that set the stage for the new crusade: the case against pesticide spraying. She wrote:

> The most alarming of all man's assaults upon the environment is the contamination of air, earth, rivers, and sea with dangerous and even lethal materials. The pollution for the most part is irreversible, the chain of evil it initiates not only in the world that must support life, but in living tissues is for the most part irreversible. In this now universal contamination of the environment, chemicals are the sinister and little recognized partners of radiation in changing the very nature of the world-the very nature of its life.

Of the use of pesticides, Carson continued:

> Chemicals sprayed on croplands and forests or gardens lie long in soil, entering into living organisms, passing from one another in a chain of poisoning and death. Or they pass mysteriously by underground streams until they emerge and through the alchemy of air and sunlight, combine into new forms that kill vegetation, sicken cattle, and work unknown harm on those who drink from once pure wells . . . The chemicals to which life is asked to make its adjustment are . . . the synthetic creations of man's inventive mind.
>
> These sprays, dusts, and aerosols are now applied almost universally to farms, gardens, forests, and homes-nonselective chemicals that have the power to kill every insect, the "good" and the "bad", to still the songs of birds and the leaping of the fish in the streams, to coat the leaves with a deadly film and to linger on in soil-all this though the intended target may be only a few weeds or insects. Can anyone believe it is possible to lay down such a barrage of poisons on the surface of the earth without making it unfit for all life?

Although *Silent Spring* contained some scientific inaccuracies and some distortions of scientific facts, it served as a catalyst, sounding the alarm and reminding people that a clean environment is a precious commodity that must be protected from human carelessness and irresponsibility. The stage was now set for the first major Federal environmental legislation on hazardous materials and toxic wastes. The public entered directly into the environmental discussions through such organizations as the Sierra Club, the Environmental Defense Fund, and the National Audubon Society. The environmental movement now had focus—one that persists to the present day.

During the 1950s and 1960s a number of programs evolved to control the use of pesticides. In 1957 the U.S. Congress passed the Fire Ant Eradication Act. The fire ants delivered a painful sting and were a greater nuisance because they built high mounds of dust. Also in 1957 the U.S. Department of Agriculture began spraying trees to eradicate the gypsy moth. While these programs were only moderately successful they brought attention to the need for federal control of pesticide programs.

These early programs set the stage in the 1960s for the campaign to ban DDT. This single pesticide became the starting point for the control of unrestricted spraying. Although DDT had been used to control malaria for over 200 million people in the tropics its control dominated the campaign. There were three reasons for this. First, it was the least expensive insecticide available. Second, it exhibited an exceptional stability and persistence for years after spraying, and third, DDT combined high toxicity for insects with low toxicity for other forms of life.

In spite of many favorable scientific reports, the crusade against DDT continued Rachel Carson wrote:

> Once it (DDT) has entered the body it is stored largely in organs rich in fatty substance (DDT itself is fat-soluble) such as the adrenals, testes, or thyroid. Relatively large amounts are deposited in the liver, kidneys and the fat of the large, protective mesenteries that enfold the intestines . . . Because these small amounts of pesticides cumulatively are stored and only slowly excreted, the threat of chronic poisoning and degenerative changes of the liver and other organs is very real.

In 1970 the U.S. Department of Agriculture developed a plan to gradually phase out the use of DDT. This did not please environmentalists and in 1972 the Environmental Protection Agency banned all use of DDT in the United States. An alliance had now been created between the environmentalists and a major government agency.

Herbicide Control of Vegetation

In the late 1960s the attack on the use of herbicides was added to pesticides by the environmentalists. Of the herbicides, the phenoxy acid compounds are most used, some of which are contaminated with dioxin, a deadly substance. There are about 75 types of phenoxy acid compounds, of which the most popular is 2,4-dichlorophenoxy acetic acid, or 2-4D.

The spraying of 2,4-D or 2,4,5-T beginning in 1962 to destroy vast areas of vegetation in Vietnam, drew the attention of the general public to the use of herbicides to a greater degree than any other event. Because the use of the herbicides received so much adverse publicity it was withdrawn from spraying in Vietnam in 1970.

In 1976 the first Vietnam veteran reported that his cancer was due to his exposure to Agent Orange (2,4-D). The U.S. Veterans Administration by March 1,

1983 had collected names of 16,821 victims who had filed claims for suspected Agent Orange damage. As a response a court trail against the chemical companies was set for May 7, 1984. Before the trial began, an out-of-court settlement was reached totaling about $180 million.

While questions remain about 2,4-D being a carcinogen, studies are not conclusive. Obviously this remains a concern, but the fact that 2,4-D is a doomsday menace is based less on medical evidence than on some kind of psychological phenomenon.

Pesticide and Herbicide Controls

In 1972 Congress passed the Federal Pesticide Act. This new law with later amendments emphasized the need for health safety in the use of pesticides and herbicides.

The Environmental Protection Agency was given the formidable task of reviewing more than 50,000 pesticides. The task has been too formidable. Less than 1,000 chemicals had been tested by 1992. At the current rate, the final assessment will not be completed until well into the 21st century. The slowness of evaluation has been due to two problems. First, there has been a failure in enforcement of the legislation. The review procedure has been remarkably slow and minimal. Because pesticide control was not given a high priority the staff from 1980 to 1986 was reduced from 829 to 591 positions. The second factor has its tradition in the 1950s when it was believed that pesticides were essential to agricultural productivity and quality. No alternatives have taken the place of pesticides in producing quality products demanded by the public.

HAZARDOUS AND TOXIC WASTE SITES

As the environmental studies broadened it became evident that the United States contained thousands of hazardous chemical and biological by-product sites that were contaminating the land. A number of these such as Times Beach in Missouri, Valley of the Drum in Kentucky, Saltville in Virginia, and especially the Love Canal area of Niagara Falls, New York had attained national publicity. Because it was believed that these sites were health hazards there was a growing demand to initiate cleanup programs.

By 1980 it was recognized that a national program was needed to focus on the cleanup of these hazardous waste sites. The passage of the Comprehensive Environmental Response, Compensation and Liability Act (CERCLA), known as the Superfund began the national efforts. These efforts began with high expectations for they had strong public and congressional support and a growing network of state, local, and private authorities. The initial appropriation was $1.6 billion.

Regrettably, the implementation of the program has been far more complex than was originally visualized. At the beginning there was no proven technology on how to manage waste sites. The EPA had to devise ad hoc control strategies to deal with hundreds of unique problems. In reality, the public had too high expectations for great results of this program.

By 1985 more than 31,000 hazardous sites had been identified with about 2,000 new sites added each year. Because this number was overwhelming, the EPA identified those sites most hazardous for early cleanup. Because remediation of the sites usually requires years of work the EPA in 1985 stated:

> As a consequence, the nation is now beginning to confront the real dilemma: as to how to reduce environmental risks from a growing list of sites presenting ever more complexities, in a situation characterized by incomplete knowledge, immature technology, and relentless pressure on a limited pool of resources.

In order for the Superfund to be ultimately successful, the EPA in 1989 proposed the following strategies:

1. Control acute hazardous sites immediately.
2. Treat the worst sites first.
3. Carefully monitor and maintain sites over the long term.
4. Emphasize enforcement to induce private party cleanup.
5. Develop and use new technologies for more effective cleanups.
6. Improve efficiency of program operations.
7. Encourage full participation by communities.
8. Foster cooperation with state agencies and with the National Resource Trustees and the Agency for Toxic Substance and Disease Registry (ATSDR).

The exact number of waste sites in the United States is still unknown, and estimates within the 50 states vary greatly. The number in each state varies from a few to over 6,600 presently identified in California. The number of sites reported by a state is largely a function of the level of the state's program. Since the passage of the Comprehensive Environmental Response Compensation and Liability Act of 1980 partnership has developed between federal and state governments in the cleanup of hazardous waste sites. Coordinated efforts are cleaning up numerous sites on the National Priorities List. At the present time 39 states have funded enforcement authorities. The remaining 11 states have limited capabilities.

The progress to remove hazardous waste sites has been amazingly slow. The number has averaged between 25 and 35 sites annually. This does not appear to be an impressive record but evaluating the success of the Superfund by tallying completed sites is inherently misleading. Each site requires not only years of initial cleanup but EPA monitors sites long after land quality standards have been achieved so that the area remains free of pollution.

Success of the Superfund is more appropriately measured in terms of the successive interim steps that quickly provide a margin of safety to local residents. For example, the EPA has conducted emergency removals to attack the most immediate sources of toxic exposure at more than 1,500 sites in communities across the nation. In addition, long-term cleanup work is currently under way to neutralize the sources, contain the spread, and systematically reduce or eliminate toxic pollution at more than 400 sites under Superfund's remedial program.

In order for the Superfund to be ultimately successful, the Environmental Protection Agency must clarify, communicate, and give broad acceptance for its fundamental strategies. Such an approach makes it possible to solve the riskiest problems first and then turn to remaining sources of long-term risk on an established priority basis.

PERSPECTIVE

Since the beginning of the Industrial Revolution the contamination of the environment has steadily increased. In spite of this, the human population has not only survived but has grown greatly in numbers, and life expectancy has increased. This raises a fundamental question: What is the purpose of managing environmental contamination? Is it to control the environment or is it to determine how humans use the environment? There are those who believe that our society is wasteful, morally divided, and sustained only by pollution, resource-depleting, high technology. Their conclusion is that, no matter what the benefits are, the long-term risks are too great and amount to a basic violation of human rights. In contrast, there are those who argue that the problems of environmental contamination can be resolved by modern scientific technology, and that the future is most promising.

The problem thus is not only technological but also ethical. It is undeniable that regulatory standards and institutionalized structures are necessary for the protection of some measure of human health. However, what constitutes a hazard depends upon values on a scale of real possibilities. In essence, hazards are a response to value standards. Further hazards are not absolute. Consequently, many have adopted the uncritical assumption that risk is a normative concept certifying that everything can be classified either good or bad. As a result, a false antithesis has been established between risks and benefits-as if there were a way to have one without the other. In reality, in the decision process what is actually "at risk" is the possibility that the intended benefits from risk taking may not occur. If harm occurs it is an unwanted and unintended side effect. Thus risks and benefits are inseparable. To illustrate, if we are to have nature produce a perfect apple, there must be a control of insects. In controlling insects in order to achieve the perfect fruit, there may be some risk. Thus, each qualitative benefit has a corresponding level of risk.

It is usually not recognized that there are degrees of harm, extending from trivial to disastrous. If a harm is trivial and prevents a greater harm, there is ethical justification for the trivial harm. In contrast, if the harm is so severe that human disaster results, then there is no ethical justification for this act. One of the greatest problems in determining the magnitude of harm to a human being is the incremental risk of exposure to a toxic substance or radiation over time. A single exposure or several exposures may cause little or no harm, while continuous exposure may result in great risk to human health. Under this condition there must be restructuring of the prior patterns of benefits and risks. Only systematic risk accounting can provide accurate data for this type of situation. To illustrate, a single contact with a toxic chemical may result in no health risk, but continual contact may result in death.

There is an additional problem in determining benefits and risks, in that there are many naturally occurring toxics. In the creation of any policy for the use of toxic materials, the policymakers must initially take into account the wide variation of personal exposure to naturally occurring background sources. Human tolerance for natural toxicant demonstrates that increments from human-made applications can be kept well within the range of variance without inflicting unjustifiable harm on the general population or depriving humankind ofpossible benefits.

There is an ethical need for the formulation of principles essential for providing a sound basis for the proper management of hazardous material and toxic waste, management that will protect both the environment and the health of individuals. As a fundamental premise it must be recognized that such a policy cannot be guided by the philosophy that no risk is valid. Any policy will entail risks of some harm to the environment and individuals at some future time. A policy for the protection of health must be based on a method of assessing systematic risks so as to prevent the environment and human beings from experiencing basic harm deemed unjustifiable.

This policy is not thus based on the absence of risk but rather on how much risk is acceptable. There is an urgent need to reach a consensus on this fundamental question. Without an answer to this problem the basic rights to health protection and a quality environment may be lost in the obsession with hypothetical health effects from only a few toxic substances. This distorted viewpoint averts public attention from a host of more pressing problems such as preventable causes of malnutrition and diseases. Thus, a policy of social justice requires management of the potential sources of basic harm, that is, management that is proportional to actual, identifiable basic harms that human effort and time can reduce.

To implement such a policy, we need a full evaluation of the technology available for the reduction of each potential toxic hazard. Only then can standards be established to provide the greatest benefits for the most people. It must be remembered that risk assessment is not a utilitarian tool for placing a monetary value on human welfare. In reality, risk assessment maximizes the value placed on human welfare so that a finite amount of money can be used to maximize the greatest benefits to human welfare and the quality of the environment.

SELECTED REFERENCES

Armstrong, Neal E. and Akita Kudo, eds. 1983. *Toxic Materials: Methods for Control*, Water Resources Symposium, No. 10, Austin, TX: Center for Research and Water Resources, University of Texas, 380 pp.

Carson, Rachel. 1962. *Silent Spring*, Boston: Houghton Mifflin, 368 pp.

Chatterji, Manaseu, ed. 1987. *Hazardous Materials Disposal: Siting and Management*, Brookfield, VT: Gower Publishing Company, 331 pp.

Cooper, M.G., ed. 1985. *Risk: Man-Made Hazards to Man*, New York: Oxford University Press, 141 pp.

Elsom, Derek. 1987. *Atmospheric Pollution: Causes, Effects, and Control Policies*, New York: Basil Blackwell, 319 pp.

Epstein, Samuel S., Lester O. Brown and Carl Pope. 1982. *Hazardous Waste in America*, San Francisco: Sierra Club Books, 593 pp.

French, Hilary. 1990. *Changing the Air: A Global Agenda*. Worldwatch Paper 94, Washington, DC: Worldwatch Institute 94 pp.

Gallaway, et al. 1984. *Acid Precipitation*, "Natural Versus Anthropogenic Components", *Science*, 226 (No. 4676): 829-831.

Gibson, Mary, ed. 1985. *To Breathe Freely: Risk, Consent, and Air*, Maryland Studies in Public Philosophy, Totowa, NJ: Rowman and Allanheld, 294 pp.

Graves, Philip E. and Ronald J. Krumm. 1981. *Health and Air Quality: Evaluating the Effects of Policy*, Studies in Economic Policy, AEI Studies 322, Washington, DC: American Enterprise Institute for Public Policy Research, 156 pp.

Greenberg, Michael R., ed. 1987. *Public Health and the Environment: The United States Experience,* New York: Guilford Press, 395 pp.

Hynes, H. Patricia. 1989. *The Recurring Silent Spring*, The Athene Series, New York: Pergamon Press, 225 pp.

Key, David and Harold Jackobson, eds. 1983. *Environmental Protection: The International Dimension*, Totowa, NJ: Rowman and Littlefield, 352 pp.

Majumdar, Shyamal K. and E. Willard Miller, eds. 1984. *Hazardous and Toxic Wastes: Technology, Management and Health Effects*, Easton, PA: Pennsylvania Academy of Science, 442 pp.

Majumdar, Shyamal K., E. Willard Miller and John J. Cahir. 1991. *Air Pollution: Environmental Issues and Health Effects*, Easton, PA: Pennsylvania Academy of Science, 496 pp.

Miller, E. Willard. 1991. "Air Pollution: A Perspective," in Majumdar, S.K. and E.W. Miller and J.J. Cahir, eds., *Air Pollution: Environmental Issues and Health Effects*, pp. 1-16.

Miller, E. Willard and Ruby M. Miller. 1989. *Environmental Hazards: Air Pollution*, Santa Barbara, CA: ABC/CLIO, 250 pp.

______ and ______. 1990. *Environmental Hazards: Radioactive Materials and Wastes*, Santa Barbara, CA: ABC/CLIO, 298 pp.

______ and ______. 1991. *Environmental Hazards, Toxic Waste and Hazardous*

Material, Santa Barbara, CA: ABC/CLIO, 259 pp.

Piasecki, Bruce, ed. 1984. *Beyond Dumping: New Strategies for Controlling Toxic Contamination*, Westport, CT: Quorum Books, 235 pp.

Sax, N. Irving and Richard J. Lewis, Sr. 1989. *Dangerous Properties of Industrial Materials,* 7th ed., New York: Van Nostrand Reinhold, 3 vols.

Turco, R., et al. 1983. Nuclear Winter: Global Consequences of Multiple Nuclear Explosions. *Science* 222 (No. 4630):1283-1292.

Environmental Contaminants, Ecosystems and Human Health. Edited by S.K. Majumdar, E.W. Miller, and Fred J. Brenner.

Chapter Two

IMPACT OF STRATOSPHERIC OZONE DEPLETION ON TERRESTRIAL ECOSYSTEMS

C.K. VARSHNEY and ARUN K. ATTRI
School of Environmental Sciences
Jawaharlal Nehru University
New Delhi - 110 067 India

INTRODUCTION

As we are entering into the twenty first century, the problem of ozone depletion has acquired global dimension. There is overwhelming evidence to show that chemically innocuous chlorofluorocarbons (CFCs), the pall bearer of man's industrial prowess, believed to be safe till recently, have chemically scissored the stratospheric ozone which prevents the deadly solar ultraviolet radiation from reaching the surface of the earth (Manhan, 1990). The concentration of CFCs in the stratosphere is increasing at a rate of 5 to 8 percent per year, even if their industrial production remains limited in accordance with 1987 Montreal Protocol, the abundance of Arctic chlorine is expected to triple by the year 2050 (Brasseur and Hitchman, 1988).

OZONE IN THE STRATOSPHERE

The ozone, an allotropic form of oxygen is a minor, forming $1.0x10^{-6}$ part by volume, but highly important constituent of the atmosphere. It occurs at all heights but mostly concentrated in the stratosphere between 20 and 30 km above the earth's surface (Ramade, 1988). Origin of stratospheric ozone is linked with photosynthesis which began with the advent of marine autotroph about 3 billion years ago. Steady buildup of oxygen and its interaction with UV radiation promoted ozone formation in the stratosphere (Chaisson, 1988). UV radiation through photodissociation produces oxygen atoms which in turn combine with oxygen molecules to form ozone. Stratospheric ozone production is largely over tropics from where it is subsequently transported to the mid and upper latitudes (Mani, 1993; Subbaraya, 1993). It took two to three billion years to form an effective ozone layer which is since maintained in a dynamic equilibrium through ozone formation and breakdown.

The UV region of solar spectrum for convenience is divided into three parts, UV-A, UV-B and UV-C, which represents the wavelength range from 320 to 400 nm, 280 to 320 nm and 200 to 280 nm, respectively. The absorption cross section of ozone for UV is a function of wavelength. In the wavelength region, 245 to 300 nm, the absorption cross section is lowered by 32 times (Molina and Molina, 1986). The long wave part of UV, mostly UV-A, is transmitted further through the atmosphere. Albeit, some attenuation of UV-A occurs due to suspended particulates, water vapor and pollutants present in the lower atmosphere. Approximately 600 million BP, the earth experienced a rapid torrent in number and diversity of complex life forms, which paved the way for the evolution of *Homo sapiens* (Chaisson, 1988). It is implicit that any crack or dilution of stratospheric ozone will proportionately enhance the UV-B flux on the earth surface.

CURRENT STATUS OF STRATOSPHERIC OZONE LAYER

In 1974 Molina and Rowland (1974a,b) , predicted the depletion of stratospheric ozone due to CFCs on the basis of model calculations. The problem is alarming because ozone scavenging by CFCs is a catalytic process mediated by halogen atoms. It is estimated that a single atom of Cl is capable of destroying 100,000 molecules of ozone (Steinfeld et al., 1989).

Predictions based on two dimensional model of the atmosphere (Brasseur and Hitchman, 1988) suggest that 3.3 fold increase in active chlorine causes a net ozone depletion varying from 2.6 percent at equator to 5.5 percent at poles. Ozone depletion trend between 1970-1980 for middle and upper stratosphere (30-40 km) was 2 to 3 percent, representing an annual loss of 0.25 percent. In 1985, a British Antarctic team, while working over Halley Bay startled the world scientific community by reporting 40 percent reduction in ozone between 1977 to 1984 (Stolarski, 1988). The area of annual spring time ozone depletion of approximately 6 percent, was larger than the area of Antarctica. The boundary of ozone-void region over Antarctica got pushed upward from 12 km to 24 km from the earth surface. Such three dimensional space devoid of ozone was termed as, "ozone hole" (Stolarski, 1988).

The amount of UV-B radiation is a function of latitudinal gradient (Caldwell,

1981). The daily UV-B flux at equator is almost twice relative to its flux at 40° latitude (Cladwell et al., 1986). The examination of ozone depletion trend over poles, mid and high latitudes and tropics, make the inherent latitudinal variations distinct.

The reduction of ozone over Antarctic has been more dramatic as compared to its depletion over mid and high latitudes. Column ozone values of 110 DU were recorded in October 1991, which represents 60 percent reduction since 1970. The ozone hole over south pole region during Sep-Oct 1992 covered an area of 14.5×10^6 km^2, which was merely 4.8×10^6 km^2 in 1984. In October 1993 ozone over Antarctic reduced to an all time low of 91 DU, which represents its lowest value recorded till date (Rodenhuis et al., 1993; Niu et al., 1992).

Ozone data for Arctic is rather sparse. The first expedition undertaken in 1989 to monitor ozone found that the Arctic stratosphere was full of destructive chlorine species akin to the situation which is believed to be the cause of ozone depletion over Antarctica. More disturbing findings came to the fore in January 1992 when the NASA expedition reported chlorine monoxide (ClO) concentration of 1.5 ppb (Zurer, 1993), which is significantly higher over the value of ClO recorded from Antarctic. Such high concentration of ClO suggests that ozone depletion over Arctic may occur at a phenomenal pace of 1 to 2 percent per day (WMO, 1991; NOAA, 1992; Stolarski et al., 1992; Frederick, 1993; Rodenhuis et al., 1993; Zurer, 1993).

As evident from Table 1 there has been a significant reduction in total ozone over northern and southern latitudes, but no significant change has occurred over tropics (Mani, 1993; Subbaraya, 1993; Nagabhushana, 1993).

The total column ozone values over 50°S-50°N during 1989 to late 1993 decreased from 350 DU to 265 DU (NOAA, 1993). In January 1993, the ozone level over northern mid-latitudes was 14 percent lower than the calculated value (Zurer, 1993). The average ozone depletion trend for the last twelve years, excluding the polar regions, is 6 to 8 percent over high latitudes and 3 to 5 percent over mid-latitudes. The ozone depletion trend was found to be independent of seasonal, semi-annual and annual fluctuations.

In the next 15 to 20 years the ozone reduction is expected to be at least as

TABLE 1

Percent decadel loss of total column ozone.

Season	45°S [TOMS] 1979-91	Equator [TOMS] 1979-91	45°N [TOMS] 1979-91	26°-64°N Ground Based 1979-91	26°-64°N Ground Based 1970-91
Dec-Mar	-5.2±1.5	+0.3±4.5	-5.6±3.5	-4.7±0.9	-2.7±0.7
May-Aug	-6.2±3.0	+0.1±5.2	-2.9±2.1	-3.3±1.2	-1.3±0.4
Sept-Nov	-4.4±3.2	+0.3±5.0	-1.7±1.9	-1.2±1.6	-1.2±0.6

Source: WMO Global Ozone Research and Monitoring Report No. 25 1991.

much as witnessed during the recent years, barring year to year fluctuations. Ozone depletion will result in corresponding increase in UV-B flux. The terrestrial ecosystems will undergo long term and cumulative change on account of adverse impact of enhanced UV-B flux on global climate, living organisms, ecosystem dynamics and biogeochemical cycles.

EFFECT OF UV-B ON BIOLOGICAL SYSTEMS

The UV-B part of solar radiation is potentially dangerous to biological systems ranging from biomolecules to ecosystems. For long UV-B is known to be a potent mutagenic agent. The UV-B induced mutations are stochastic in nature. The genetic changes are due to the molecular alteration in DNA which is the prime hereditary material (Setlow, 1968). In 1960, Beukers and Berends, for the first time showed UV-B induced dimer formation in thymine. Subsequently, intrastrand dimer formation between adjacent pyrimidine bases was observed in DNA (Setlow, 1974). Recently, UV has been also shown to induce interstrand pyrimidine dimer formation in Z-DNA (Attri and Minton, 1986; Love et al.,1986).

The nature of UV-B interaction with biomolecules and the accompanied changes are of fundamental importance. The energy of UV-B photons elevates the molecules to excited state (Phillips, 1983). Molecules in excited state are relatively unstable and tend to lose energy by; i) returning to ground state, ii) reacting chemically, iii) molecular decomposition, iv) photosensitization of adjacent molecules.

In general any biological response to UV radiation is linked with action spectra i.e., a relationship between degree of biological response and the range of wavelength invoking such response. For instance, DNA absorption curve lies between 240 to 290 nm with maxima at 260 nm. The action spectra for skin erythema and photokeratitis is at 290 nm (Diffey, 1983). The erythema spectral curve range coincides with the absorption spectra of most of the nuclear and serum proteins (Van der Leun, 1984). Proteins are the largest UV-B absorbing component in living cells having absorption maxima at 280 nm. The UV-B flux between 280 to 320 nm causes irreversible changes in lipids, steroids, melanin and urocanic acid (Emmett, 1986). The schematic representation of UV-B effects on living cells is shown in Figure 1.

Actinic effects of UV-B on human health are mainly from its absorption by skin affecting DNA and protein synthesis (Hawk and Parrish, 1982; Quevedo et al., 1985). Most of the adverse effects attributed to UV-B are skin cancer, actinic elastosis, equamous skin cancer and pre-mature aging. Indirect effects of UV-B include genetic disorders, pellagra, kwashiorkor, suppression of immune response, metabolic disorders and eye diseases (Fredrick, 1986).

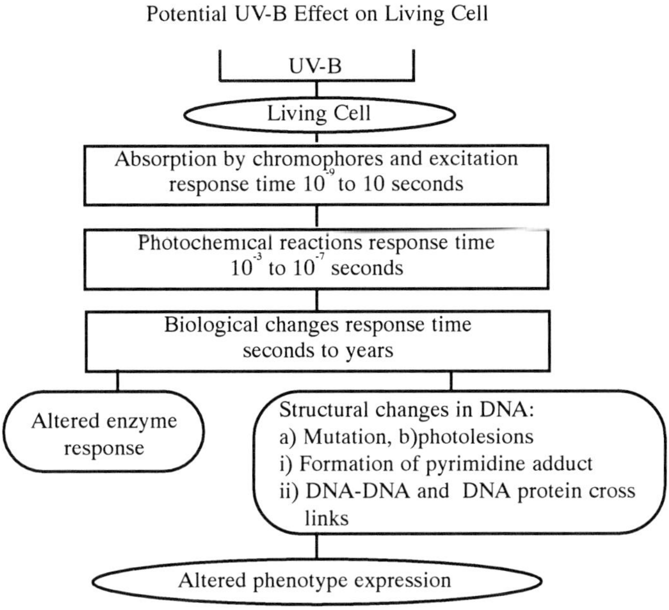

FIGURE 1.

EFFECT OF UV ON PLANTS

Enhanced UV-B radiation can cause significant alteration in plants. Damage to plant cells can arise from adverse effects of UV-B on DNA, membranes and photosynthesis. Reduction in leaf area, stem elongation, enhanced shoot branching and increase in leaf number are common manifestations of UV-B stress observed in widely different species representing both monocots and dicots.

Leaves are the main photoreceptors and UV-B affects the light relations of leaves significantly. Penetration of UV-B widely varies in leaves of different plant species. A study involving 22 species of plants has revealed that UV-B transmittance of leaf epidermal layer was maximum in herbaceous dicots, which ranged between 18 to 41 percent, and it was found to be minimum in needle leaves of pine. UV-B penetration in the leaves of woody dicots and grasses was in between herbaceous dicots and conifers (Teramura et al., 1991).

Enhanced UV B radiation significantly changes internal light regime of plant leaves. The leaf thickness of *Brassica campestris* plants exposed to UV-B radiation increased by 45 percent and synthesis of UV-B absorbing pigments in leaves increased by 21 percent over control (Bornman and Vogelnann, 1991). UV-B induced changes in leaves affect plant growth adversely.

Higher plants are likely to be affected mainly by two mechanisms, namely suppression of photosynthesis and reduced leaf expansion. The photosynthetic suppression is primarily on account of the adverse effects of UV-B on photosystem II, albeit the site of action is not very clear (Caldwell, 1981; Teramura, 1983). Reduction in leaf expansion is on account of inhibited or delayed cell division, in addition to the diminshed supply of photosynthate (Dickson and Caldwell, 1978). In general, enhanced UV-B radiation has been shown to affect the following plant parameters and processes (Table 2).

UV-B has been shown to suppress flowering. Plants grown in UV-B, excluded of sunlight, exhibited more profuse flowering as compared to plants grown in sunlight with normal UV-B component. Under experimental conditions UV-B has been shown to reduce the photoperiodic flower induction in a long day plant (Rau et al., 1988). The UV-B effect on flowering has been confirmed by the studies of Hart et al. (1975). Photoperiodic response of plants to UV-B are not similar in all species, but in those species in which the photoperiodic behavior is affected will have serious consequences. For example, alteration in flowering time may result either in mis-match between the time of anthesis and availability of pollinators or may lead to excessive flower damage by insects, birds or other herbivores, which remain absent at the normal time of flowering (Tevini and Teramura, 1989). Any shift in the onset of flowering can lead to reproductive failure in plants.

Pollen grains are highly sensitive to UV-B radiation. *In vitro* studies have shown that pollen germination and pollen tube growth are adversely affected by UV-B radiation in a large number of important crop and ornamental plants (Flint and Caldwell, 1984). How far the results of *in vitro* studies can be extrapolated to *in vivo* situation is not clear? But UV-B suppression of pollen germination and pollen tube growth, if valid, under field conditions, will have serious implications for agricultural crops. Pollen grains represent the most vulnerable stage in the life cycle of plants where potential for mutations or chromosomal aberrations in germinal cells is very high (Caldwell, 1981).

EFFECT OF UV-B ON TERRESTRIAL ECOSYSTEMS

The stratospheric ozone depletion leading to increased UV-B irradiance on

TABLE 2.
Plant parameters and processes affected by enhanced UV-B.

Serial No.	Parameter	Reduced (-)	Increased (+)
1	Cell Division	-	
2	Cell Elongation	-	
3	Seedling Elongation	-	
4	Stem Elongation	-	
5	Stem Branching		+
6	Leaf Area(Length/Size)	-	
7	Leaf Thickness		+
8	Stomatal Closure	-	
9	Pigmentation		+
10	Wax, (Cuticular)		+
11	Photosystem II	-	
12	Flowering	-	
13	Pollen Germination	-	
14	Pollen Tube Growth	-	

earth will have long term and cumulative impact on terrestrial ecosystems comprising virtually pristine, mutilated and intensively man managed systems tailored to produce food, fibre and to meet other basic human needs. Many of these production systems already suffer from over exploitation, soil erosion, air, water and soil pollution due to growing human pressure (Varshney, 1994). Enhanced UV-B flux from ozone depletion will be an additional stress on the terrestrial ecosystems. Photosynthetic fixation of solar energy is the fundamental tenet of all ecosystems. Chronic depression of photosynthesis due to enhanced UV B flux raises critical questions about indirect effects of ozone depletion on the structure and function of terrestrial ecosystems (Teramura et al., 1990). Changes in floristic composition will, in turn alter ecosystem physiology, particularly carbon sequestering, evapotranspiration and nutrient cycling.

Long term studies on the effect of UV-B at ecosystem level are lacking. Hence our present understanding of the UV-B induced ecosystem level changes is rather sketchy. The current state of knowledge pertaining to the effect of thinning of stratospheric ozone layer is based upon the studies carried out with isolated cell organelles, plant organs and with individual plants under laboratory conditions. Results of such studies may not be able to explain the interactions and feedback responses at population, community and ecosystem levels. However, physiological and isolated single species responses provide the first sign posts of ecosystem level changes. There is a pressing need to develop predictive capability to ascertain the quantum and direction of such changes in relation to prime agronomic species, keystone species, natural and managed forests, pastures, range lands an wetlands. The implicit UV-B effect on agricultural and non-agricultural ecosystems are distinct and are discussed separately (Gold and Caldwell, 1983).

EFFECT OF UV-B ON CULTIVATED ECOSYSTEMS

Enhanced UV-B is going to initiate complex changes in cultivated ecosystems. Increased exposure to UV-B reduces leaf area, photosynthesis, water use efficiency, and yield. UV-B has been shown to reduce leaf size and length of internodes affecting canopy organization and light interception by plants. UV-B promotes development of daughter tillers in monocots, and increases leaf thickness, number of leaves and affects allocation of photosynthate to different plant organs. The consequential morphological changes will affect light relations and net photosynthesis at ecosystem level.

Significant differences exist in the sensitivity of plants to UV-B (Teramura, 1983). Screening of more than 200 different plant species, most of them crops, has revealed that 70 percent were sensitive to UV-B. Variations in the response of different plant species to UV-B has serious implications for crop-weed interaction by altering competitive balance (Fox and Caldwell, 1978). Increase in UV-B can be expected to alter species composition. The implications of UV-B induced changes will have different implications for intra-specific and inter-specific interactions. Studies on the effect of UV-B carried out on single species stand of wheat, wild oat and jointed goat grass, grown under UV stress did not show any significant impact (Gold and Caldwell, 1983). However, the effect of UV-B under conditions of interspecific competition representing mixed stands of two or more species, as seen in mixed crop agriculture and non-agricultural ecosystems, will change the competitive balance drastically due to differential morphological response of competing species (Gold and Caldwell, 1983). Variation in photosynthetic response among different species may shift the competitive advantage to otherwise less aggressive species. Relatively better performance of UV-B tolerant species can alter species composition of plant communities over time.

Soybean, a versatile and protein rich crop, is particularly susceptible. Teramura (1983), found that a simulated loss of 25 percent ozone reduced the yield of one important soybean species by as much as 25 percent. UV-B also reduced protein and lipid contents of the seeds by 5 and 3-5 percent respectively (Teramura, 1983). Agro-ecosystems of irrigated and moister rich areas are likely to suffer more relative to drought stressed agricultural communities. *Glycine max* grown under well irrigated conditions clearly shows adverse effect of UV-B on photosynthesis, transpiration, plant growth and number of pods. Whereas UV-B had no significant effect on these parameters in drought stressed plants (Murali and Teramura, 1986). High phosphorus content in soil enhances plant sensitivity to UV-B implying that heavily fertilized cultivated ecosystems may be more vulnerable to enhanced UV-B.

The ozone depletion can stimulate some plants to evolve UV-B protective mechanisms such as UV-B stimulated synthesis of UV-B absorbing phenolic compounds, flavonoid and flavones in leaf epidermis, to enhance their UV-B tolerance (Caldwell, 1981). It may appear attractive to breed plants for UV tolerance to overcome the impending stress of enhanced UV-B. The UV-B resistant cultivars may be able to thrive under enhanced UV-B climate but such plants may not possess the desirable economic and nutritive characteristics. The UV-B resistant plants are likely to be inefficient because adaptation to any stress is achieved at considerable expense of energy (Odum, 1971).

EFFECT ON NON-CULTIVATED ECOSYSTEMS

About 20 tree species have been tested experimentally; only eight have been tested in field situation; half were found to be sensitive. No information is available on tropical plants which constitutes almost 50 percent of the total global biodiversity. Vegetation will change due to the loss of UV-B sensitive species and from the selection of species which are relatively resistant or tolerant of UV-B radiation. Over time, sensitive species will substitute for resistant species and some niches may be opened for lack of suitable substitutes. Herbaceous vegetation and early successional communities which thrive in open under fully exposed conditions will suffer relatively more from UV-B stress.

UV-B induced changes in terrestrial ecosystems will have serious implications for rare endangered, and endemic species as they are highly specialized and exacting in their environmental requirements. They have narrow genetic base and their geographical distribution is highly restricted. Changes in species composition and competitive balance will prove highly inimical to such specialized floral and faunal elements of terrestrial ecosystems.

Currently, there is a mounting concern about the rapid and accelerating loss of biological diversity. A crucial question is how the changes prompted by stratospheric ozone depletion will alter the terrestrial ecosystems which embrace wide species diversity and multitude of species interactions? Mayers (1988a), in context of global biodiversity conservation has identified 18 hot spots. They deserve urgent attention for the conservation of global biodiversity, which is a common heritage of mankind. In Figure 2 an attempt has been made to examine the distribution of the 18 hot spots and their vulnerability in relation to the enhanced UV-B radiation due to progressive stratospheric ozone depletion. A perusal of the Figure 2 shows that maximum biodiversity is concentrated between 40°N and 40°S latitude. The UV-B enhancement in absolute terms over the prevailing base value will be significantly high in this belt. The likely genetic changes between 40°N and 40°S will be highly prejudicial to the rare and endangered species, whose list is growing every day. In practice the problem is very much complicated because most of the tropical ecosystems, which harbor over 50 percent of the total global biodiversity (Myers, 1988b), are already under great stress. Large areas have been cleared due to deforestation, over grazing and suffered from the impact of galloping urbanization. Vast areas in tropics, currently support successional communities which are open and directly exposed to sun rays in contrast to natural ecosystems, where overlapping tree canopies shield many plants from direct sunlight. The open areas formed from the degradation of natural vegetation will be most severely impacted by enhanced UV-B flux and will be a major deterrent to ecological succession and ecosystem development.

Terrestrial ecosystems which are specially susceptible to increased UV-B include; (a) those already receiving high inputs of solar radiation (arid and semi arid environment, high altitudes); (b) those which are likely to receive greater amounts of UV-B radiation in the future (latitude and altitude dependence); and (c) those in which large amounts of litter is exposed to direct sunlight for prolonged periods (e.g. open canopy forest, grasslands, deserts, etc.). In tropics where UV-B load is already high (Ilyas, 1989) forest degradation is creating such conditions over wide areas.

How far plants can acclimate to enhanced UV-B flux will determine the degree of selection pressure and the extent of ecosystem level changes.

EFFECT OF OZONE DEPLETION ON GLOBAL CLIMATE AND BIOGEOCHEMICAL CYCLES

The radiation balance of the earth is, primarily regulated by water molecules,

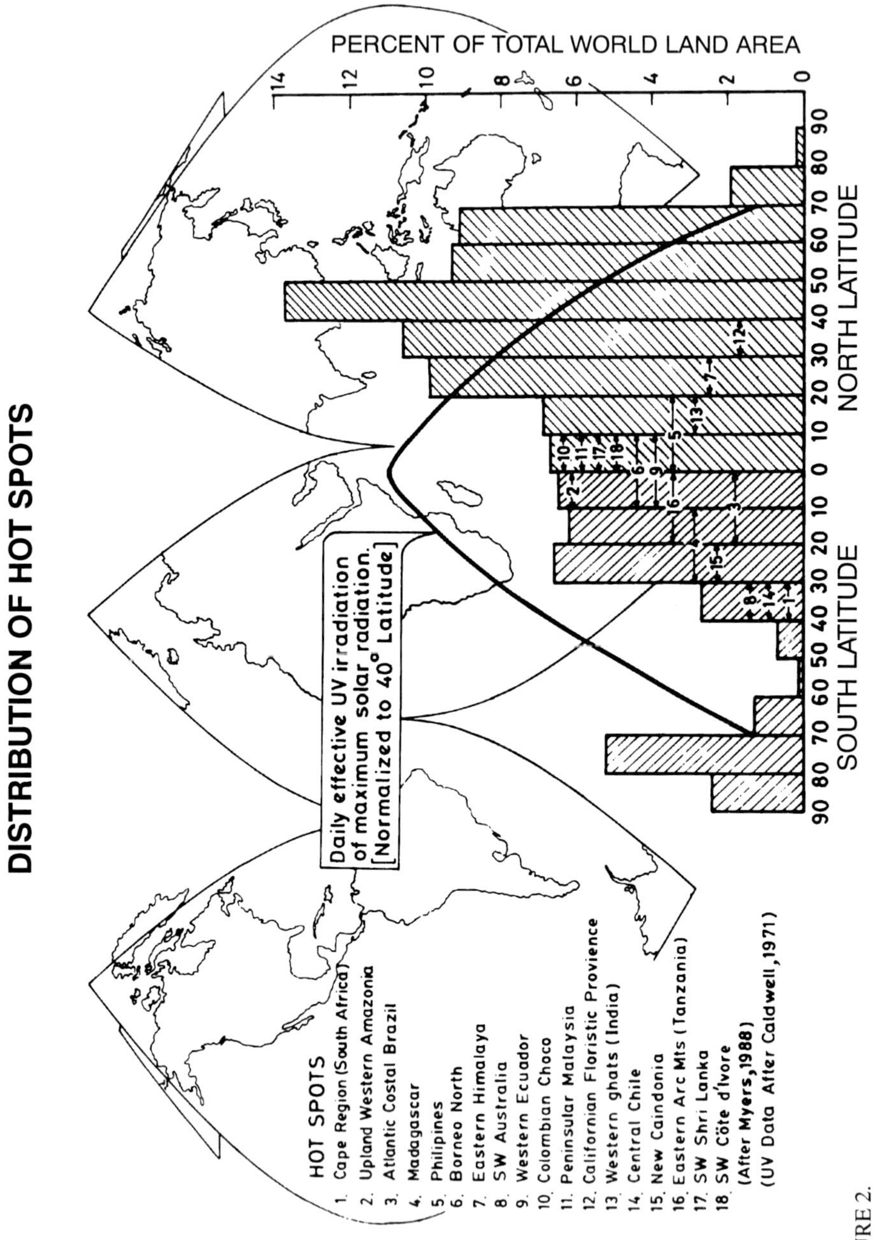

DISTRIBUTION OF HOT SPOTS
PERCENT OF TOTAL WORLD LAND AREA
0
2
4
6
8
10
12
14
90 80 70 60 50 40 30 20 10 0 10 20 30 40 50 60 70 80 90
SOUTH LATITUDE
NORTH LATITUDE
Daily effective UV irradiation of maximum solar radiation. [Normalized to 40° Latitude]
HOT SPOTS
1. Cape Region (South Africa)
2. Upland Western Amazonia
3. Atlantic Costal Brazil
4. Madagascar
5. Philipines
6. Borneo North
7. Eastern Himalaya
8. SW Australia
9. Western Ecuador
10. Colombian Choco
11. Peninsular Malaysia
12. Californian Floristic Provience
13. Western ghats (India)
14. Central Chile
15. New Caindonia
16. Eastern Arc Mts (Tanzania)
17. SW Shri Lanka
18. SW Côte d'Ivore
(After Myers, 1988)
(UV Data After Caldwell, 1971)

FIGURE 2.

carbon dioxide and ozone (Manahan, 1990). This is accomplished by selective absorption of incoming and outgoing infrared (IR) radiation regulating the stratospheric thermal ambience. The stratospheric thermal structure in conjunction with tropospheric radiative balance is responsible for maintaining the mean surface temperature of earth at about 15°C (Lovelock, 1987).

Stratospheric ozone depletion will enhance transmission of solar radiation which will prompt photochemical ozone synthesis in troposphere involving CO, NO_x, CH_4 and volatile hydrocarbons as feedstock, whose emission is steadily increasing (Varshney and Aggarwal, 1992, 1993). Tropospheric ozone buildup will further enhance radiative forcing. It has been estimated that 50 percent increase in tropospheric ozone, over the present level, will increase the surface temperature by 0.3°K (Krause et al., 1989).

The stratospheric ozone depletion will have cooling effect and the same has been corroborated by temperature records of lower stratosphere (WMO, 1991). Lowering of stratospheric temperature is likely to affect the gaseous chemistry affecting ozone formation *per se*. Eclectic nature of stratospheric chemistry and the dynamics of CFC buildup are the major constraints in comprehending the impact of ozone depletion in terms of UV-B flux at earth surface and atmospheric temperature gradient. Directly and indirectly consequences of these complexities will have strong bearing on global climate.

Stratospheric ozone depletion will promote progressive changes in biogeochemical cycles of major elements. The dynamic equilibrium prevailing between the fluxes and active pools of nutrients, which got established through a long process of evolution, is likely to shift unpredictably in subtle manner over time. Detailed studies of these aspects in relation to ozone depletion are lacking. However, a strong correlation between ozone depletion and the corresponding increase in UV-B flux can be used to assess the impact in broad terms (Green et al., 1980). UV-B is an extremely potent suppressor of photosynthesis. Consequently, primary production, food chain and food web relationships will suffer adversely. The adverse effects of UV-B on net photosynthesis will suppress carbon sequestering on global scale leading to carbon dioxide buildup in the atmosphere. Direct exposure of dead organic matter to UV-B releases volatile organic compounds and CO_2 abiotically which will have additive effects on atmospheric CO_2 buildup (Valentine and Zepp, 1993). Suppression of photosynthesis by UV-B in marine phytoplankton will be an additional factor in bolstering atmospheric CO_2 buildup (Smith and Baker, 1989). Recent evidence shows that UV-B affects nitrifying bacteria which implies that nitrogen cycle will be affected more directly (Lean, 1993). Moreover, changes in the cycling of carbon will adversely affect biogeochemical cycling of the major nutrients.

The complex organization of ecosystems defies a proper assessment of the extent and pace of changes in biogeochemical cycles, but there is sufficient evidence to show that the effect of stratospheric ozone depletion will be crucial. It is important to model the potential response of terrestrial ecosystems in relation to changing UV-B scenario, mirroring stratospheric ozone depletion. There is an urgent need to undertake field experiments on whole ecosystem to monitor the effect of enhanced UV-B flux on ecosystem level linkages and feedback including plant-atmosphere, soil-atmosphere and plant-microbe interactions.

REFERENCES

Attri, A.K. and K.W. Minton. 1986. An automated method for determining buoyant density of nucleic acids using a preparative ultracentrifuge. *Anal. Biochem.* 159:88-95.

Beukers, R. and W. Berends. 1960. Isolation and identification of the irradiation product of thymine. *Boiochim. Biophys. Acta.* 41:550-551.

Bornman, J.F. and T.C. Vogelmann. 1991. Effect of UV-B radiation on leaf optical properties measured with fibre optics. *J. Exp. Bot.* 42:547-554.

Braseur, G. and M.H. Hitchman. 1988. Stratospheric response to trace gas perturbation: Changes in Ozone and temperature distribution. *Science.* 240:634-637.

Caldwell, M.M. 1971. Solar UV irradiation and the growth and development of higher plants. In: A.C. Giese (Ed.) *Photophysiology*, Vol. 6, Academic Press, New York. pp. 131.

Caldwell, M.M. 1981. Plant response to solar UV radiation. pp. 169-197. In: O.L. Lange, P.S. Nobel, C.B. Osmond and H. Ziegler (Ed.) *Physiological Plant Ecology. I. Response to Physical Environment.* Springer-Verlag, Berlin.

Caldwell, M.M., Camp, L.B., Warner, C.W. and S.D. Flint. 1986. Action spectra and their role in assessing biological consequences of solar UV-B radiation change, pp. 87-112. In: R.C. Worrest and M.M. Caldwell (Ed.) *Stratospheric Ozone Reduction Solar Ultraviolet Radiation and Plant Life.* Springer-Verlag, Berlin.

Caldwell, M.M., Teramura, A.H. and M. Tevini. 1989. The changing solar ultraviolet climate and the ecological consequences for higher plants. *Trends in Ecology and Evolution.* 4:363-366.

Chaisson, Eric. 1988. Universe And Evolutionary Approach To Astronomy. Prentice Hall, Englewood Cliffs, New Jersey.

Diffey, B.L. 1986. Possible Errors involved in dosimetry of solar UV-B radiation. pp. 63-74. In: R.C. Worrest and M.M. Caldwell (Ed.) *Stratospheric Ozone Reduction Solar Ultraviolet Radiation and Plant Life.* Springer-Verlag, Berlin.

Dickeson, J.G. and M.M. Caldwell. 1978. Leaf development of Rumex patientia L. (polygonaceae) exposed to UV radiation (280-320nm). *Am. J. Bot.* 65:857-863.

Emmett, E.A. 1986. Health effects of ultraviolet radiation. pp. 129-145. In James G. Titus (Ed.) Effects of Changes in Stratospheric Ozone and Global Climate Vol. I: Overview. UNEP. USEPA.

Flint, D.S. and M.M. Caldwell. 1984. Partial inhibition of in vitro pollen germination by simulated solar UV-B radiation. Ecology. 65:792-795.

Fox, F.M. and M.M. Caldwell. 1978. Competitive interaction in plant population exposed to supplementary ultraviolet-B radiation. Oecologia. 36:173-190.

Frederick, J.E. 1986. The ultraviolet radiation environment of biosphere. pp. 121-128. In: James G. Titus (Ed.) Effects of Changes in Stratospheric Ozone and Climate Vol. I: Overview, UNEP, USEPA.

Frederick, J.E. 1993. Ultraviolet sunlight reaching the earth's surface: A review of recent research. *Photochemistry and Photobiology.* 57, 175-178.

Gold, W.G. and M.M. Caldwell. 1983. The effects of ultraviolet-B radiation on plant competition in terrestrial ecosystems. *Physiol. Plant.* 58:435-444.

Green, A.E.S., Cross, K.R. and L.A. Smith. 1980. Improved analytic characterization of ultraviolet skylight. *Photochem. Photobiol.* 31:59-65.

Hart, R.H., Carlson, G.E., Klueter, H.H. and H.R. Carns. 1975. Response of economically valuable species to ultraviolet radiation. In: D.S. Nacthtwey, M.M. Caldwell, R.H. Biggs (Ed.). *Impacts of climatic change on biosphere, part I, ultraviolet radiation effects, monograph 5.* Climatic Impact Assessment Program, US Dept. Transportation Report No. DOT-TST-75-55, NTIS, Springfield, Virginia. (4)263.

Hawk, J.L.M. and J.A. Parrish. 1982. Responses of normal skin to ultraviolet radiation. pp. 216-260. In: J.D. Regan and J.A. Parrish (Ed.) *The Science of photomedicine.* Plenum Press, New York.

Ilyas, M. 1989. The danger of ozone depletion in the tropics. *Search.* 20:148-149.

Krause, F., Wilfrid, B. and Jon Koomey. 1989. Energy policy in the greenhouse Vol. I *From Warming Fate to Warming Limit: Benchmark For Global Climate Convention.* IPSEP. European Environmental Bureau, California.

Lean, D. 1993. Ecosystem effects. In: *Workshop On The Effects Of Increased UV-B Radiation.* Environment Canada. Toronto.

Love, J.D., Nguyen, H.T., Attri, A.K. and K.W. Minton. 1986. UV induced interstrand cross-linking of d(GT)n.d(CA)n is facilitated by structural transition. *J. Biol. Chem.* 261:100051-100057.

Lovelock, J.E. 1987. *GAIA A New Look At Life On Earth.* Oxford University Press, Oxford.

Manahan, Stanley. E. 1990. *Environmental Chemistry.* Lewis Inc., Florida.

Mani, A. 1993. Ozone in tropics. *Current Science.* 64:335-339.

Mayers, N. 1988a. Theoretical biotas: hot spots in tropical forests. *Environmentalist.* 8:11-20.

Mayers. N. 1988b. Tropical forest and their species going going . . .? *Biodiversity.* National Academy Press, Washington, D.C.

Molina, M.J. and F.S. Rowland. 1974a. Stratospheric sink for chlorofluromethanes: Chlorine atom catalyzed destruction of ozone. *Nature.* 249:810-812.

Molina, M.J. and F.S. Rowland. 1974b. Predicted present stratospheric abundances of chlorine species from photodissociation of carbon tetrachloride. *Geophys. Res. Lett.* 1:309-312.

Molina, L.T. an M.J. Molina. 1986. Absolute absorption cross section of ozone in the 185 to 350 nm wavelength range. *J. Geophys. Res.* 91:145501-14508.

Murali, M.S. and A.H. Teramura. 1986. Effectiveness of ultraviolet radiation on the growth and physiology of field grown soybean modified by water stress. *Photochem. Photobiology*. 44:215-217.

Nagabhushana, R.K. 1993. A Statistical analysis of total ozone data at four Indian stations. *Current Science*. 64:344-348.

Niu, X., Frederick, J.E., Stein, M. and G.C. Tiao. 1992. Trends in column ozone based on TOMS data: dependance on month, latitude and longitude. *J. Geophys. Res*. 97, 14661-14669.

NOAA. 1992. In: M.S. Halpert and C.F. Ropelwski (Ed.) *Fourth Annual Climate Assessment.*

Odum, H.T. 1971. *Environment, Power and Society*. John Wiley and Sons. New York. pp. 331.

Phillips, R. 1983. Photochemistry and photopolymerization, pp. 56-106. In: *Sources and Appliations of Ultraviolet Radiation*. Academic Press. New York.

Quevedo, W.C., Fitzpatrick, T.B. and K. Jimbow. 1985. Human skin color: origin, variation and significance. *J. Human. Evolution*. 14:43-56.

Ramade, F. 1988. Ecology Of Natural Resources. John Wiley and Sons. New York.

Raue, W., Hoffman, H., Huber-Willer, A., Mitzk-Schnabel and E. Shrott. 1988. Die wirkung von UV-B anf photoreguregulierte Entwicklungsvorganga bei Pflanzen. Strahlen-und Unweltforschung mbH.

Rodenhuis, David R., O'Lenic, Edward and Alvin J. Miller. 1993. *Climate Review-1993* NOAA. Washington, D.C.

Setlow, R.B. 1968. Thc photochemistry, photobiology, and repair of polynucleotides. *Prog. Nucleic. Acid. Molecular. Biology*. 8:257-294.

Setlow, R.B. 1974. The wavelengths in sunlight effective in producing cancer: A theoretical analysis. *Proceed. Natl. Acad. Sciences*, USA. 9:3363-3366.

Smith, R.C. and A.S. Baker. 1989. Stratospheric ozone, middle ultraviolet radiation and phytoplankton prouctivity. *Oceanogr. Mag*. 2:4.

Steinfeld, J.I., Francisco, J.S. and W.L. Hase. 1989. *Chemical Kinetics and Dynamics*. Prentice Hall. New Jersey.

Stolarski, R.S. 1988. The antarctic ozone hole. *Scientific. American*. 258:20-26.

Stolarski, R., Bojkov, R., Bishop, L., Zerefos, C., Staehelin, J. and J. Zawodny. 1992. Measured trends in stratospheric ozone. *Science*. 256:342-349.

Subbaraya, B.H. 1993. Ozone distribution in tropics. *Current. Science*. 64: 339-344.

Teramura, A.H. 1983. Effect of ultraviolet-B radiation on the growth and yield of crop plants. *Physiologia Plantarum*. 58:415-427.

Teramura, A.H., Sullivan, J.H. and J. Lydon. 1990. Effects of UV-B radiation on soybean yield and quality: A six year field study. *Physiology. Plant*. 80:5-11.

Teramura, A.H., Tevini, M., Bornman, J.F., Caldwell, M.M., Kulandaivelug, G. and L.O. Bjorn. 1991. Terrestrial plants. In: *Environmental Effects of Ozone Depletion*: 1991 Update UNEP, Nairobi, Kenya.

Tevini, M. and A.H. Teramura. 1989. Ultraviolet effects on terrestrial plants.

Photochemistry Photobiology. 50:479-487.

Valentine, R. and R.G. Zepp. 1993. Formation of carbon monoxide from photodegradation of terrestrial organic matter. *Environ. Sci. Technology*. 27: 409-412.

Van der Leun, J.C. 1984. UV carcinogenesis. *Photochem. Photobiol*. 39:861-868.

Varshney, C.K. 1994. Physical aspects of environment and global problems. In: *Source Book On Environmental Education for Elementary Educators*. UNEP-UNESCO.

Varshney, C.K. and M. Aggarwal. 1993. Vertical ozone variation in the lower troposphere of Delhi. *Environ. Monitor*. and *Assess*. 25:41-49.

Varshney, C.K. and M. Aggarwal. 1991. Ozone levels in urban atmosphere of Delhi. *Atmosphere. Environment*. Part B. 26:291-294.

WMO. 1991. Scientific Assessment of Ozone Depletion. WMO Report 25. Geneva.

Zurer, P.S. 1993. Ozone depletion's recurring surprises challenge atmospheric scientists. *C & EN,* May, 8-18.

Environmental Contaminants, Ecosystems and Human Health. Edited by S.K. Majumdar, E.W. Miller, and Fred J. Brenner. © 1995, The Pennsylvania Academy of Science.

Chapter Three

CHEMICAL CONTAMINANTS IN PUBLIC DRINKING WATER WELLS IN CALIFORNIA

HELMUT KLOOS

Department of Epidemiology and Biostatistics
School of Medicine
University of California
San Francisco, CA 94143-0560

INTRODUCTION

The most severe groundwater contamination in the United States is associated with synthetic pesticides, industrial chemicals, nitrate and seawater intrusion (1). The National Drinking Water Survey carried out by the Environmental Protection Agency (EPA) recently estimated that 10.4% of all community system wells and 4.2% of the rural domestic wells contained one or more detectable pesticides. Nitrate detections were projected in 57.0% of the community water systems and in 52.1% of the domestic water systems (2). This EPA study also reported that the number of pesticides detected in ground water in the USA increased from 12 in 18 States in 1984 to 46 in 26 States in 1988 (3). This trend must be expected to have significant health implications for rural Americans, 90% of whom depend on groundwater for drinking and cooking (4), and for periurban areas surrounded by farmland.

In California, the rapid growth of the population, agriculture and industry and the relative scarcity of water resources in much of this State have made the supply, delivery and regulation of drinking water an important and sensitive issue. Numerous agricultural pesticides have been applied in large amounts on California's highly productive farmland and many industrial chemicals discharged or have escaped into the environment. An increasing number of contaminants has been identified in drinking water wells throughout the State. Increased pumpage in excess of natural recharge since the accelerated development of irrigated agriculture and urbanization has resulted in hundreds of feet of head decline in the San Joaquin Valley and other groundwater basins since the 1940s, a trend which has recently been reversed in some areas with increased interbasin transfer of water (5). The impact of these changes on well water contamination is poorly understood, largely because few studies have fully examined the spatial distribution, mobility and persistence of chemical contaminants in the soil and groundwater environment (6-8). Most chemical contamination of groundwater in California is non-point source pollution (9), the control of which is particularly difficult and expensive (10).

Few State-wide studies of chemical contamination have been carried out because of constraining technical and financial factors. There is also little nformation on the public health significance of contaminants in well water; drinking water standards are periodically revised as new toxicological data become available, but the extrapolation of animal study results to humans remains problematic (11-13).

The objective of this paper is (a) to briefly describe California's drinking water regulations and public well water supply and monitoring programs and (b) to examine the spatial distribution of the major agricultural, industrial and naturally occurring chemical contaminants in public drinking water wells in California.

We reviewed the literature on well water contamination in California and on related health risks, State regulatory legislation and well water monitoring. The chemical test results for the major contaminants found in the 14,000 public wells monitored by the California Department of Health Services between 1984-1992 (14, 15) were examined and remapped.

THE SUPPLY, REGULATION AND USE OF WELL WATER IN CALIFORNIA

More than 10,000 public drinking water systems, including about 450 surface sources and more than 39,000 wells, supply drinking water to over 30 million consumers in California. About 300,000 residents obtain their water from private wells. Fewer than 10% of the medium and large systems public water systems serve more than 95% of the state's population. Approximately 70% of the drinking water in California comes from surface sources and the remainder from wells, with significant regional variations (14). Most communities in the densely populated groundwater basins, particularly the Central Valley and Santa Ana or Los Angeles Basin, obtain much or all of their drinking water from groundwater sources (Figure 1).

Overall, California's drinking water is of good quality, and its drinking water standards are higher than those of most other States. Well standards and laboratory detection capabilities are improving, and a better understanding of health risks have created an increased awareness of the public health risk of certain contaminants. This awareness has resulted in increased public demand for safe drinking water and in related laws and regulations. Nevertheless, consumer confidence in public water supplies has declined in recent years, largely due to excessive media publicity or perceived physical problems (taste, odor and color) that represent no health risk. Microbial contamination due to waste disposal in surface waters has not been a widespread problem in California because of strict State drinking water regulation. About 50% of the residents of many urban areas are now using bottled water or home treatment devices, and Californians spend annually an estimated one billion dollars for bottled water (14).

Several regulatory, planning and economic problems are threatening the safety of California's water supplies. Regulation of drinking water is presently fragmented between state and local governments, rendering the implementation of regulations and programs difficult. The counties regulate small systems with fewer than 200 service connections and the state regulates systems serving more than 200 connections. Many of the approximately 9,000 small systems have relatively low quality water supplies and low levels of compliance, largely due to the financial and technical difficulties they face in meeting water standards and requirements. There is also a lack of coordinated planning to ensure reliable

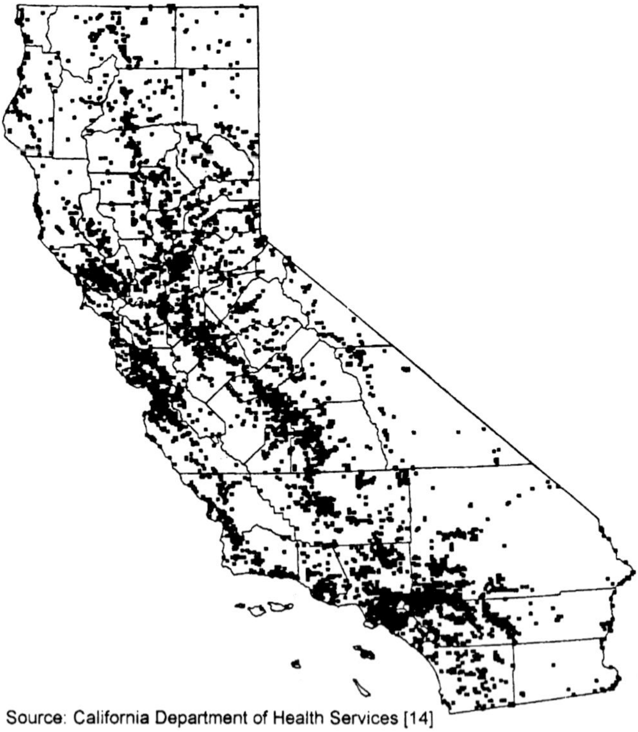

Source: California Department of Health Services [14]

FIGURE 1. Public Drinking Water Wells in California.

drinking water supplies at the local and regional levels. Of particular concern are certain types of land use associated with agricultural and industrial development and urbanization (14).

GROUNDWATER MONITORING PROGRAMS

The California Department of Water Resources periodically monitors all drinking water sources in the state and several other agencies conduct special sampling surveys. All public suppliers operating large systems with more than 10,000 service connections are required by State law to develop their own monitoring plan. The number of regulated contaminants has gradually increased in recent years as a reflection of mounting public health concerns; however, MCLs are established for only about half of all commonly used organic chemicals, as well as radon and the inorganic chemical arsenic. All public wells are required to routinely sample and test at least once every three years for general mineral, physical, inorganic chemical and organic pesticides and industrial substances, and every four years for radioactivity. In areas of known chemical contamination test intervals for each well are two years. If chemical contaminants are detected, monthly or quarterly sampling and testing may be required for a period up to three years depending on the level of contamination. If the average concentration exceeds the MCL the affected wells must be fitted with GAC (granulated activated carbon) filters, nitrate removal systems, air strippers, other wellhead treatment methods, or taken out of service (16). GAC is one of the most widely used and effective methods for removing synthetic organic chemicals from drinking water supplies. Field testing achieved around 98% to 100% removal of DBCP (dibromochloropropane) from well water (17). DBCP is a pesticide used in the 1970s to control nematodes in agricultural crops.

The California Department of Health Services carried out a separate, statewide drinking water testing program between 1984 and 1989 covering over 380 different industrial, agricultural and naturally occurring chemicals. Assembly Bill 21, enacted in 1989, directed the Department of Health Services to undertake a follow-up comprehensive assessment of drinking water in the State. Normally this department samples and tests 82 chemicals quarterly (14, 15).

The major deficiencies of the State, County and City water quality database relate to the absence of longitudinal data for most major contaminants before the middle and late 1980s, the limited information on and the absence of spatially and temporally synchronized testing programs. Contaminants in well water vary not only daily and seasonally but also vertically within wells and with fluctuations in pumpage (6, 18, 19).

HEALTH RISK ASSESSMENT

Although health risk assessment of chemical contaminants in drinking water is beset with many conceptual and methodological problems, it can provide information essential for the establishment of drinking water standards. Nevertheless, calculated risk should only be treated as an indication of the relative risk of a substance rather than a precise estimate of the actual risk (14). The theoretical lifetime risk of 10 major contaminants for which data were available are summarized in Table 1. The relatively high MCLs of DBCP (dibromochloropropane) nematocide, an animal carcinogen and testicular toxin in humans (20, 21), and radon reflect their widespread distribution and the technical and economic constraints of lowering contamination levels in the environment. The most detailed health risk analysis of contaminated well water in California was done in Santa Clara County, where high levels of TCE (trichloroethylene) due to a leaking tank were associated with spontaneous abortion (22). Carcinogenic, reproductive and acute health risks of these and other chemical drinking water

TABLE 1
Estimated Lifetime Health Risks of Various Contaminants in California's Public Drinking Water Wells

	MCL	Risk (no. of excess cancers)*
Organic chemicals		
Pesticides:		
DBCP	0.2 ppb***	1 per 10,000 population
EDB	0.02 ppb***	1 per 1 million population
VOCs** and chlorinated solvents		
PCE	5 ppb***	1 per 400,000 population
TCE	5 ppb***	1 per 1 million population
Vinyl chloride	0.5 ppb	2.5 per 100,000 population
Benzene	1 ppb	1 per 1 million population
Inorganic chemicals:		
Nitrate	45 ppm	No cancer risk but methemoglobinemia in infants at concentrations above 45 ppm
Arsenic	50 ppb***	2 per 100: mostly nonlethal skin cancer
Fluoride	1.4-2.4 ppm	No cancer risk but dental and skeletal fluorosis risk at concentrations above MCL
Chlorinated disinfection byproducts:		
Trihalomethanes	100 ppb	1.7-3 per 10,000-100,000 pop.
Radiochemicals:		
Radon	300 pCi/L***	9-39 per 100,000 population
Combined radium 226 and 228	5 pCi/L	5 per 100,000 population

*The theoretical risk of contracting cancer if exposed to a carcinogenic substance through drinking water at a concentration equal to MCL, assuming continuous lifetime exposure for 70 years at 2 liters per day.
**Volatile organic chemicals: drinking water standards proposed by the US Environmental Protection Agency (11).
***These MCLs may soon be lowered due to recent studies (14).
Source: California Department of Health Services (14).

contaminants were reviewed by Golub (22) and the EPA (11). The Santa Clara County incident had far reaching implications for drinking water regulation as it led to the establishment of new EPA and California MCLs for organic chemicals and the monitoring of public small water systems (14).

One epidemiological study in Fresno County found a correlation between DBCP contamination of community drinking water wells and cancer mortality (stomach cancer and lymphatic leukemia) (21). Another study found no relationship between these variables (23). The results of these two studies are inconclusive due to their failure to control for confounding factors, and to associate exposure risk with source wells at the census tract level.

WELL WATER CONTAMINATION IN CALIFORNIA

Slightly over 5% of the 14,000 public drinking water sources monitored between 1984 and 1989 exceeded one or more MCLs. Most sources with values above MCL were in the category of 10,000 and more service connections (2.9%), and the smallest number was in the 200-299 service connections category (0.5%) (14, 15).

Synthetic Contaminants

The group of contaminants most commonly detected was the chlorinated solvents. They are carcinogens which are widely used as cleaners and degreasing agents by industry and commercial cleaners. Five percent of all wells, mostly in urban and industrial areas, had detectable concentrations and 2.3% exceeded one or more MCL. TCE and PCE (tetrachloroethylene) were the most common chlorinated solvents, exceeding their MCLs in 0.89% and 0.72% of all wells, respectively (Figure 2A). Another eight chlorinated solvents exceeded MCLs in 63 (0.45%) of the wells. They were 1,1,2,2,-tetrachloroethane, 1,1-dichloroethene, 1,1-dichloroethane, 1,2-dichloroethane, carbon tetrachloride, cis-1,2-dichloroethene, trans-1,2-dichloroethene, and vinyl chloride (Figure 2B). The great majority of wells containing chlorinated solvent residues was in larger urban areas and industrial, transportation or military centers (Figure 2A).

A total of 650 wells (4.6%) were found to contain organic pesticides and 1.4% exceeded MCLs. DBCP represents the most significant of all pesticide detections. It was found in 430 wells and exceeded the MCL in 190 well sources (1.4%), mostly in the San Joaquin Valley and Los Angeles Basin (2C). DBCP is underreported here since it is more prevalent in small systems and private wells in rural areas, few of which were sampled (6, 24). Much of the San Joaquin Valley is particularly susceptible to groundwater contamination. A total of 28.5 million kilograms of pesticides were applied in this valley in 1986. This represents 52% of all pesticides applied in California and an estimated 10% applied in the USA (9). The only other pesticides reported by the Office of Drinking Water database to be in excess of MCL were EDB (ethylene dibromide), cis-1,2-

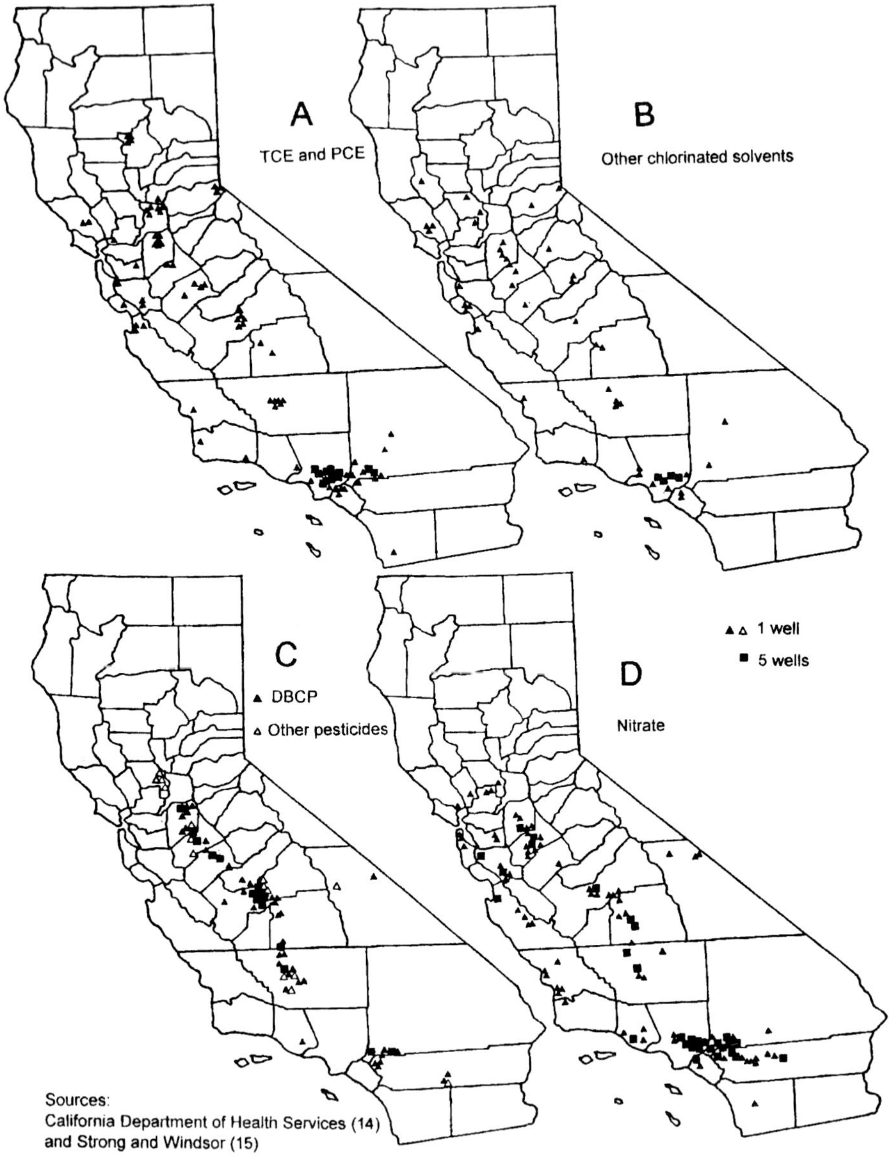

FIGURE 2. Chlorinated Solvents (A and B), Organic Pesticides (C) and Nitrate (D) Above MCL in Public Wells, 1984-1992.

dichloropropane and cis-1,3-dichloropropene (Figure 2C). EDB and DBCP are particularly susceptible to leaching into groundwater because of their high mobility in soil, high solubility and long half-life (26). Their production and use was prohibited after 1977 (DBCP) and 1983 (EDB) due to their toxicity in humans. Eleven additional pesticides and herbicides (atrazine, bromacil, 2,4-DB, diazinon, 1,2-dibromoethane, dicamba, diuron, prometon, promotryn, propazine and simazine) were detected by the California Department of Agriculture between 1982 and 1987 (9, 25). Although these pesticides were applied fairly uniformly throughout the San Joaquin Valley, nearly all contaminated ground water was confined to the highly leacheable and deeply irrigated soils of the alluvial fans near the Sierra Nevada. The fine-grained soils in the western and southern valley bottom lands inhibit pesticide leaching (9).

Nitrate was the most common inorganic contaminant in this sample, with 2.1% of all wells tested above MCL. But nitrate exceeded the MCL in 9% of 121,000 analyses from over 38,000 wells (mostly private wells) made by the California State Water Resources Board indicating poor compliance by private well owners with drinking water regulations. More than half of the wells closed due to high nitrate levels were in the San Gabriel and Santa Ana basins and another 25% in the eastern San Joaquin Valley (Figure 2D). Farming, dairying, wineries and other agricultural industries, and sewage are the major sources of nitrate contamination in California (14, 27). The ingestion of nitrate at levels above the 45 ppm MCL may cause methemoglobinemia in infants, and several other health effects, including birth defects, have recently been implicated (28).

Natural Contaminants

Fluoride, selenium, arsenic, intruding seawater, and the radiochemicals total alpha and uranium were found to be common natural contaminants in different parts of California.

Fluoride levels above the 1.4-2.4 ppm MCL were recorded in 123 well sources in central and southern California (Figure 3A). In spite of the large number of wells with fluoride above MCL, the risk of dental or skeletal fluorosis as reported from developing countries in Africa and South Asia (29) is apparently small due to the strong chemical taste of fluoride-contaminated water and the widespread availability and use of alternative drinking water sources in California.

The selenium MCL was exceeded in 49 wells, mostly in localities in the central and southern coastal mountains having marine deposits high in selenium (Figure 3B). An essential nutrient, selenium causes deformities in humans and wildlife and cancer at high concentrations. In some areas, including western Fresno County, selenium has not percolated into the groundwater although it occurs in high concentrations in ponds formed by the rising water table on irrigated land, where it has caused major damage to the local avifauna (30). Arsenic, another carcinogen, was found in relatively few wells, mostly in the Central Valley and in southern California (Figure 3C). Nevertheless, recent studies by the

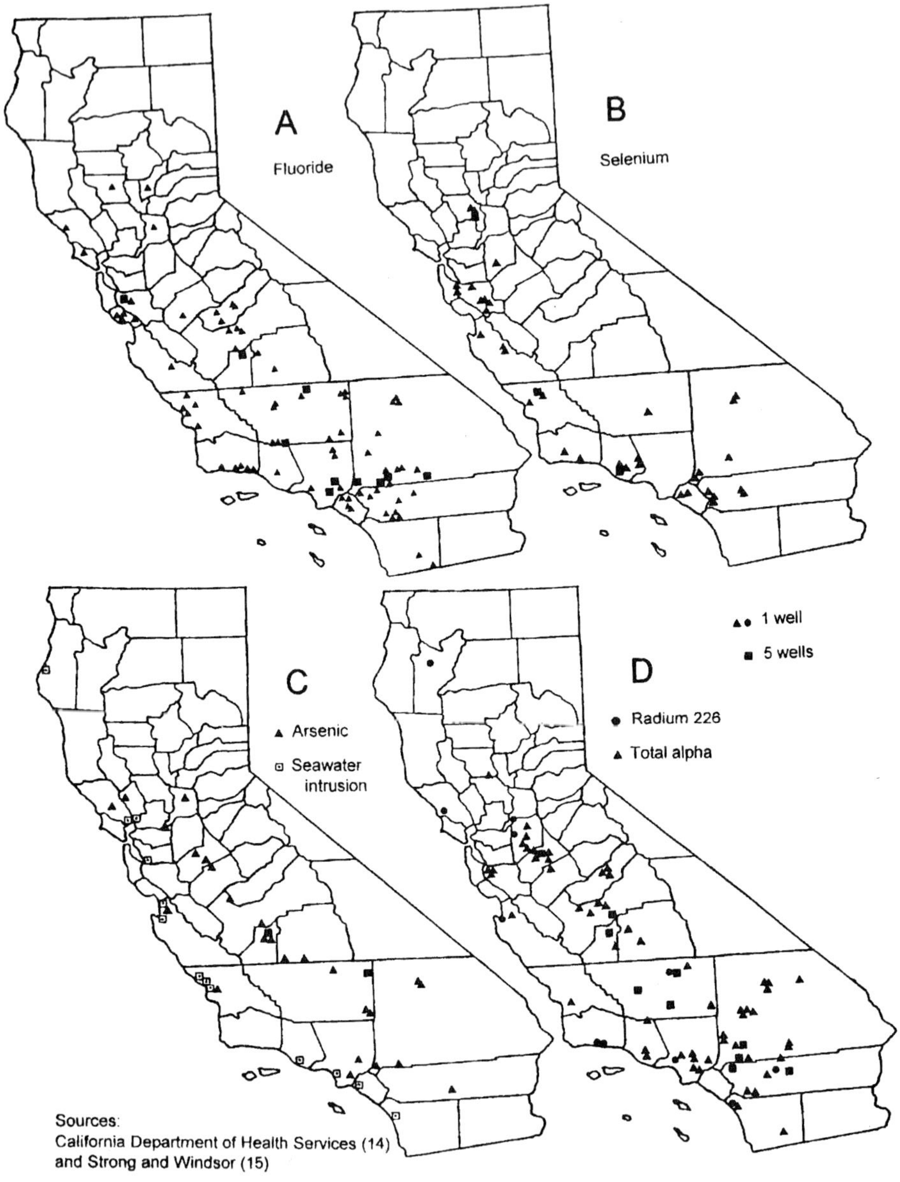

FIGURE 3. Inorganic Chemicals (A, B and C), Seawater Intrusion (C) and Selected Radiochemicals (D) Above MCL in Public Wells, 1984-1992.

California Office of Drinking Water consider arsenic as one of the most serious health hazards in drinking water (15).

Seawater intrusion has increasingly degraded groundwater sources in basins near the California coast due to continued pumping of groundwater and the recent drought (Figure 3C). Artificial recharge remains the major remedial measure. A serious health problem in seawater intrusion is the formation of carcinogenic trihalomethanes in areas of high organic soil content, as in the Sacramento-San Joaquin Delta (14).

Over 37% of the 1,884 wells tested by the Office of Drinking Water for total radioactive contamination had detectable concentrations and many exceeded the MCLs for uranium, radium, total alpha and total beta (Figure 3D) (15). Recent studies identified radon as a potentially serious health hazard in much of the United States (31). Radon will probably exceed the proposed MCL of 300 pCi/L in about half of all California large water systems (14). In separate surveys in the central and southern Sierra Nevada and in the southern portion of the San Joaquin Valley, 72 out of 96 (75%) wells had radon levels above 300 pCi/L. The highest radon concentrations, between 12,000 and 39,730 pCi/L, were in wells in granitic rock (32-34).

This discussion shows that the overall distributional pattern of all the above-MCL wells for most organic and inorganic contaminants tested for is characterized by two clusters, one in the Los Angeles Basin and the other in eastern Fresno and Tulare Counties. The Los Angeles cluster is of particular significance because of the large population exposed there. Smaller clusters exist in the southern and northern parts of the Central Valley and along the central and southern coast.

Studies of temporal changes in well water contamination have been made only for DBCP and nitrate, because these are the only chemicals for which long-term data are available. Between 1982 and 1992, DBCP concentrations in domestic and community wells in eastern Fresno County decreased about 4-fold (8, 35) and 3-fold in Fresno City (36). But levels increased in one-third of the wells due to intrusion of DBCP (8). When combining the natural decline and the effect of well closure and GAC filtration, mean DBCP levels declined 9 fold in 35 highly contaminated large public wells in Fresno City and 22 fold in 8 wells in adjacent Clovis City during that period (36). The long-term decline in depth of water and vertical hydraulic head gradient due to accelerated pumpage appears to have contributed to these variations (18). Nitrate concentrations have remained relatively stable in different communities in California but further studies using standardized data and depth sampling are required to determine long-term changes (36-38).

The popularity of GAC adsorption vessels for the control of organic chemicals and plans to lower the MCLs for DBCP from 0.2 ppb to 0.02 ppb, of EDB from 0.02 to 0.01 ppb, TCE from 5 ppb to 2.5 ppb and of PCE from 5 ppb to 0.7 ppb may reduce health risks further. However, the high cost of GAC filtration systems makes them costly for many small rural public systems and private wells, where they are needed most. Additional well closures and monitoring and filtration requirements that will become necessary with the proposed lowering of the various MCLs may constitute the major constraint in the provision of safe public drinking throughout California (16). The overwhelming approval by the US Senate in May 1994 of a drinking water bill that eases regulations for the nation's drinking water system (39) reflects the growing concern nation-wide over the financial burden imposed by drinking water legislation.

CONCLUSION

The spatial distribution and environmental parameters of well water contamination in California have become better known during the last decade and both preventive and remedial measures taken at all levels, from State legislature to the individual consumer, are beginning to reduce health risks. Assessment of the health risk of drinking water contaminants will have to include studies of their changing distribution over time and evaluation of control measures. Modeling the coupled system of distribution, degradation, vertical and lateral movement of organic and inorganic contaminants in different soil, climatic, hydrological and agricultural environments, water use patterns and morbidity/mortality may assist in the development of control strategies. Such holistic studies are necessary in view of the changing occurrence of many agricultural and industrial contaminants over time, the promulgation of new drinking water regulations, changes in pesticide use patterns and increasing consumer concerns over drinking water quality and cost. This review also points to the need for more comprehensive health risk and cost-benefit analyses that include all contaminants found in community and domestic water systems. The decline of some major contaminants and the appearance of new chemicals, which tends to render the evaluation of their health effects in relation to the general carcinogenic background of the urban and rural environment more difficult, will have to receive due attention in modeling efforts and health risk analyses.

ACKNOWLEDGMENTS

The author is grateful to David Strong and Cynthia Forbes of the California Office of Drinking Water, Martin McIntyre and Bob Dunn of the Fresno City Water Division, Lisa Koehn of the City of Clovis Department of Public Works, Gunther Redlin and Mark Reitz of Boyle Engineering, Gary Shanks, Arvey Swanson and Dennis Williams of the California Department of Water Resources, Kenneth Schmidt, and George Hanna of the Engineering Research Institute of California State University Fresno for making various data available and for valuable discussions.

REFERENCES

1. Hosseinipour, E.Z. 1991. Groundwater models and aquifer restoration: An Overview. *ASCE Int. Symp.Groundwater in Practice.* Nashville, Tennessee, July 12-14, 1991, pp.132-141.
2. Cohen, S. 1992. Results of the National Drinking Water Survey: Pesticides, Nitrates, and Well Characteristics. *Water Well* 46:35-42.
3. Weinberg, A.C. 1990. Low Input Agriculture Reduces Non-point Source Pollution. J. Soil Water Conserv. 45:48-50.
4. Murphy, David. 1990. Working Toward Groundwater Policies for the 1990s (Commentary). *J. Soil Water Conserv.* 45:278-280.
5. Bertoldi, G.L. et al. 1991. Groundwater in the Central Valley, California-a Summary Report. USGS Professional Paper, No. 1401.
6. Kloos, Helmut. 1983. DBCP Pesticide in Drinking Water Wells in Fresno and Other Communities in the Central Valley of California. *Ecology of Disease* 2:353-367.
7. Loague, K. 1991. Statistical and graphical methods for evaluating solute transport models: review and application. *J. Contam. Hydrol.* 7:51-62.
8. Shanks, W. Gary. 1994. Changes in Dibromochloropropane Concentrations in Domestic and Municipal Wells in the Kingsburg-Parlier-Dinuba and Livingston-Winton Areas Between 1980 and 1992. State of California Department of Water Resources, Fresno.
9. Domagalski, Joseph L. and Neil M. Dubrovsky. 1992. Pesticide Residues in Groundwater of the San Joaquin Valley, California. *J. Hydrol.* 130:299-338.

10 Griffen, R. Jr. Introducing NPS Water Pollution. *EPA Journal* 17:56-65.

11. Environmental Protection Agency. 1992. National Primary Drinking Water Regulations; Synthetic Organic Chemicals and Inorganic Chemicals; Final rule. *Federal Register* 57, No. 138: July 17, 1992. See also *Federal Register* 56, No. 20: January 30, 1991.
12. Huff, James. 1993. Chemicals and Cancer in Humans: First Evidence in Experimental Animals. *Envir. Hlth. Persp.* 100:201-210.
13. Ames, Bruce N., Renae Magaw and Lois Swirsky Gold. 1987. Ranking Possible Carcinogenic Hazard. *Science* 23:271-280.
14. California Department of Health Services. 1993. *Drinking Water into the 21st Century.* Office of Drinking Water, Sacramento.
15. Strong, David L. and Jim Windsor. 1993. Drinking Water Quality in California. California Department of Health Services, Division of Drinking Water and Environmental Management, Sacramento.
16. California Department of Health Services. 1992. *Domestic Water Quality and Monitoring Regulations.* Office of Drinking Water, Sacramento.
17. Reitz, Mark P. and Martin McIntyre. 1993. Design and Management Considerations for Full Wellhead Treatment Using Activated Carbon for DBCP Removal. Paper presented at the Groundwater Treatment Conference, California-Nevada Section of the American Water Works

Association, Sacramento, May 18, 1993.

18. Schmidt, Kenneth D. 1984. The Distribution of DBCP in Groundwater in Southeast Fresno. Paper prepared for the City of Fresno Department of Public Works, Fresno.
19. Whorton, M. Donald et al. 1988. Problems Associated with Collecting Drinking Water Data for Community Studies: A Case Example, Fresno County, California. *Am. J. Publ. Hlth.* 78:47-51.
20. Kluwe, William M. 1981. Acute toxicity of 1,2-dibromochloropropane in the F 344 male rat. *Toxicol. Appl. Pharmacol.* 59:71-95.
21. Jackson, Richard J. et al. 1982. Literature Review on the Toxicological Aspects of DBCP and an Epidemiological Comparison of Patterns of DBCP Drinking Water Contamination with Mortality Rates from Selected Cancers in Fresno County, California 1970-1979. California Department of Health Services, Berkeley.
22. Golub, Mari S. 1992. Reproductive Toxicology of Water Contaminants Detected by Routine Water Quality Testing. *Epidemiology* 3:125-129..
23. Wong, O. et al. 1989. Ecological Analyses and Case-Control Studies of Gastric Cancer and Leukemia in Relation to DBCP in Drinking Water in Fresno County, California. *Brit. J. Industr. Med.* 46:521-528.
24. California Department of Health Services. 1990. Organic Chemical Contamination of Small Public Water Systems in California. Office of Drinking Water, Sacramento.
25. California Department of Food and Agriculture. 1982. Pesticide Movement to Ground Water. Vol. 1. Survey of Ground Water Basins for DBCP, EDB, Simazine and Carbofuron. Sacramento
26. Wagenet, R.J., J.L. Hulton and J.W. Biggar. 1989. Simulating the Fate of a Volatile Pesticide in Unsaturated Soil: A Case Study of DBCP. *J. Envir. Qual.* 18:78-84.
27. Redlin, Gunther A., Clapp, Jonathan W. and Kurt Walden. 1992. Water Quality Monitoring Plan for Wells, Wellhead Treatment Facilities, and Treatment Distribution System and Water Quality Standards. Report prepared for the City of Fresno Water Department, Fresno.
28. Bouchard, D.C. 1992. Nitrate Contamination of Groundwater: Sources and Potential Health Effects. *J. Am. Water Works Assoc.* 84:85-93.
29. Kloos, Helmut and Redda Tekle Haimanot. 1993. Fluorosis, pp. 455-461. In: Kloos, Helmut and Z.A. Zein (Ed.) *The Ecology of Health and Disease in Ethiopia.* Westview Press, Boulder.
30. Tokunga, T.K. and S.M. Benson. 1992. Selenium in Kesterson Reservoir Ephemeral Pools Formed by Groundwater Rise. 1. A Field Study. *J. Envir. Qual.* 21:246-251.
31. Owen, L. 1993. Radon-a New Environmental Health Hazard. *Geography* 78:194-198.
32. Shanks, Gary W. 1994. Radioactivity in Groundwater in the Joaquin Valley

Portion of Kings and Tulare Counties. California Department of Water Resources, Fresno.
33. Shanks, Gary W. 1993. Radioactivity in Groundwater in the San Joaquin Valley Portion of Kern County. Memorandum Report. California Department of Water Resources, Fresno.
34. Shanks, Gary W. 1990. Natural Radioactivity in Groundwater of the Western Sierra Nevada. District Report, California Department of Water Resources, Fresno.
35. Schmidt, Kenneth W. 1994. Surface Water and Groundwater Aspects of the Kings River DBCP Evaluation. Paper presented at the Technical Seminar of the Engineering Research Institute, California State University Fresno, Jan. 12, 1994.
36. Kloos, Helmut. 1995. DBCP and EDB in Well Water in the Fresno/Clovis Metropolitan Area, California. *Arch. Envir. Hlth.* in press.
37. California Department of Water Resources. 1985. Follow-up Investigation of Groundwater Contamination by Dibromochloropropane in the Kingsburg Area. San Joaquin District, Fresno.
38. Schmidt, Kenneth D. 1977. Variations for Pumping Wells. *Ground Water* 15:130-137.
39. Editorial. 1994. A Drinking Water Bill was Approved. *The Wallstreet Journal* May 20, p. 1.

Environmental Contaminants, Ecosystems and Human Health. Edited by S.K. Majumdar, E.W. Miller, and Fred J. Brenner.

Chapter Four

GUIDELINES AND STANDARDS FOR THE REHABILITATION OF SOURCES OF ASBESTOS POLLUTION IN SOUTH AFRICA

J.J.P. VAN WYK

Director
Research Institute for Reclamation Ecology
Potchefstroom University for Christian Higher Education
Private Bag X6001
Potchefstroom 2520
South Africa

INTRODUCTION

A need exists in South Africa (Hart 1988, Sluts-Creamce 1970, 1991) as well as internationally for a scientific approach for techniques for the rehabilitation of mine-related sources of asbestos-pollution. Since 1984 the Research Institute for Reclamation Ecology of the Potchefstroom University for Christian Higher Education in South Africa has been involved in the rehabilitation of mine-related asbestos-pollution in South Africa (Van Der Walt 1993, Van Wyk and Hart 1990).

The guide-lines set forth in this chapter describes the recommendations for mine-related asbestos fibre material may be rehabilitated. It must be stressed, however, that these recommendations are rehabilitation **guide-lines** and that **each and every rehabilitation-plan must be site-specific** as local conditions must be taken into consideration.

The chemical properties of the three types of asbestos are summarized in Table 1.

The Pollution of the Environment by Asbestos Fibre

Asbestos-pollution in the natural environment can be represented schematically by Figure 1.

The aim of the rehabilitation of mining waste materials containing asbestos fibre, is to prevent the suspension of fibres in the atmosphere as the floating fibres can be inhaled by humans. Fibres, which measure between 3 and 100 micron, can penetrate the alveoli in the lungs and can cause several pathogenic disorders, some even fatal.

The rehabilitation of tailings containing fibre must conform to the following requirements:

1. Fibre must be isolated from any mechanism (wind, water, animals and man) capable of spreading it into the environment.
2. All relevant legal requirements must be addressed.
3. The site must be as stabilized and vegetation established as self-sustaining as possible in order to require minimum maintenance.
4. It must satisfy the conditions of BATNEEC (Best Available Technology Not Entailing Excessive Cost).
5. The final product must, as far as possible, blend visually with the local or surrounding environment.

GUIDELINES FOR THE REHABILITATION OF MINE-RELATED ASBESTOS-POLLUTION

The following terminology is used to define asbestos related mine tailings:

1. **Asbestos-pollution**: The distribution of asbestos fibres from any mine-related source of pollution to the environment.
2. **Source of Asbestos-pollution**: Any material in which unbound asbestos fibres occur. Two main types may be distinguished:
   A. Primary sources of pollution: sources such as tailings dumps and processing plants/mills of asbestos mines that release fibre into the environment.
   B. Secondary sources of pollution: These occur when the asbestos-containing

TABLE 1.
General properties of the different types of asbestos occurring in South Africa.

Mineral	Type of Fibre	Non-Fibrous	General Chemical Formula	Color
Serpentine	Chrysotile	Antigorite Lizardite	$Mg_6[(OH)_4Si_2O5]_2$	White
Amphibole	Amosite	Grunerite Commingtonite	$Fe_7(OH)_2Si_8O_{22}$ $Mg_7(OH)_2Si_8O_{22}$	Brown and/or Gray
	Crocidolite	Riebeckite	$Na_2Fe_5[(OH)Si_4O_{22}]_2$	Blue

material or fibres have already spread into the environment from the primary sources e.g. transported there by wind, water or humans. This term also refers to sources of pollution that have been created by human activity e.g. through the unsafe use of asbestos fibres in building material, as well as for agricultural- and filling-purposes and in the building of roads.

3. **Shrouding material**: A mixture of alluvial soil, gravel and rock, free of any asbestos fibre, to encapsulate the fibre-containing material.

In order to obtain effective isolation of asbestos fibre-containing material, and to conform to the requisites for rehabilitation, the following methods for the ecological rehabilitation of sources of asbestos-pollution are recommended:

1. Identify the source of pollution.
2. An environmental and terrain analyses of the polluted area must be done to determine the extent of the pollution.
3. Do a complete survey and draw a subsequent plan/design for the rehabilitation of the area. This rehabilitation-plan is concerned with the effectiveness, the feasibility and the economic attainability of the means of rehabilitation. It must be determined what must be done, how it will be done, what the cost-implications will be, which alternatives may be considered and what final methods and techniques will be applied. It is imperative that counselling with the relevant parties or departments concerned with financing the rehabilitation, take place in order to reach a consensus of opinion concerning the cost and logistic matters.
4. Determine the practicability of the plan by visiting the terrain with all the relevant parties, e.g. the government/state, property owners, consultants, etc. on which occasion the plans will be submitted for comment.

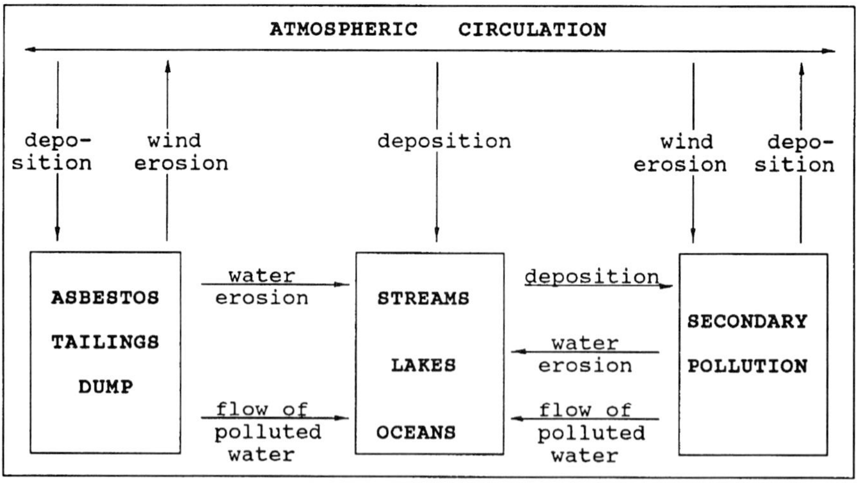

FIGURE 1. The distribution of asbestos fibre in the natural environment.

5. Invite contractors to tender for the practical execution of the rehabilitation. The best available technology not entailing excessive cost (BATNEEC) must be employed in order to obtain optimal results on a long-term basis.
6. After levelling or burying of the dump, it must be covered with a fibre-free mixture of soil, gravel and rock. The physical and chemical properties of a suitable shrouding material, as well as the optimal thickness of the shrouding, should be described.
7. Provision must be made for run-off water and storm-water control by means of weirs, contours, toe-dams and drainage channels on and around the rehabilitated locality (Figure 2).

APPENDIX 1.
GRAPHICAL REPRESENTATION OF DIFFERENT REHABILITATION TECHNIQUES

Different methods of rehabilitation, applicable to different types of sources of asbestos-pollution, are illustrated by means of diagrams on the following pages.

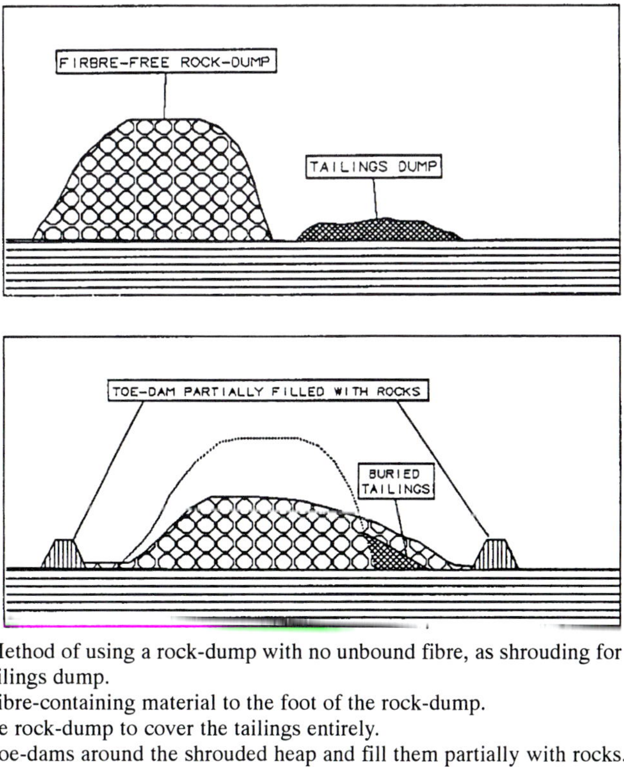

FIGURE 2. Method of using a rock-dump with no unbound fibre, as shrouding for the rehabilitation of a tailings dump.
1. Move the fibre-containing material to the foot of the rock-dump.
2. Reshape the rock-dump to cover the tailings entirely.
3. Construct toe-dams around the shrouded heap and fill them partially with rocks.

8. Revegetate the entire area with self-sustaining indigenous plant species (Figures 3, 4).
9. An environmental audit plan must be followed in order to detect flaws in the rehabilitation-process and bring about improvements. Thus the projects are re-evaluated in terms of probable unsuspected circumstances which may have occurred during the implementation phase. During the auditing process special attention must be paid to the following aspects:
   - The quality of the air.
     Monitoring of the concentration of fibre in the atmosphere.
   - Vegetation
     Basal and crown cover, species-diversity and vitality of plant growth.
   - Soil
     Monitoring of erosion.
   - Water
     Monitoring of the concentration of fibre in the run-off water (where applicable).

   The duration of the auditing program must be extended over a period determined by representatives of all concerned and affected parties.

10. The responsibility for the maintenance of the rehabilitated areas must be clear to all the parties concerned and must be put in writing.

STANDARDS FOR REHABILITATION

Introduction

The rehabilitation of sources of asbestos-pollution must conform to certain minimum standards. These standards must be kept in mind during the drafting of the rehabilitation-plans to prevent planning problems from surfacing during the implementation-phase. The nature of the terrain on which the tailings dump is situated, must be considered during the planning of the rehabilitation. Some examples of different techniques of rehabilitation that may be applied on different terrains are illustrated in Figures 2-6. The rate of the degrading of the rehabilitated source of pollution, will be ascertained during the maintenance phase.

Maximum Levels of Pollution Permitted

In South Africa the maximum fibre-concentration allowed in rural areas in the atmosphere is 0.02 f.ml-1 (fibres per millilitre air). Rehabilitation-planning must at least aim at measuring up to this standard, and to maintain it in the long term. There are as yet no standards regarding the concentration of fibre in water. Naturally it has to be as low as possible as the fibre is distributed by the water into

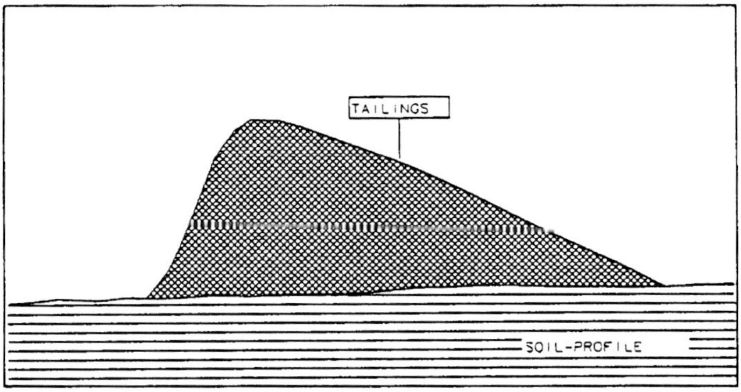

Representation of the original profile

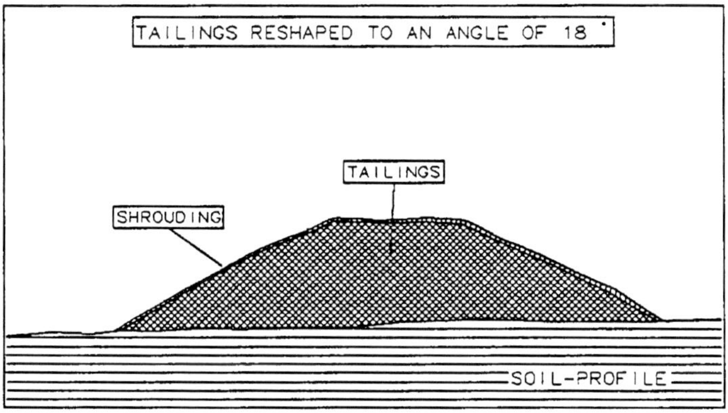

Profile of tailings dump reshaped to 18 and shrouded

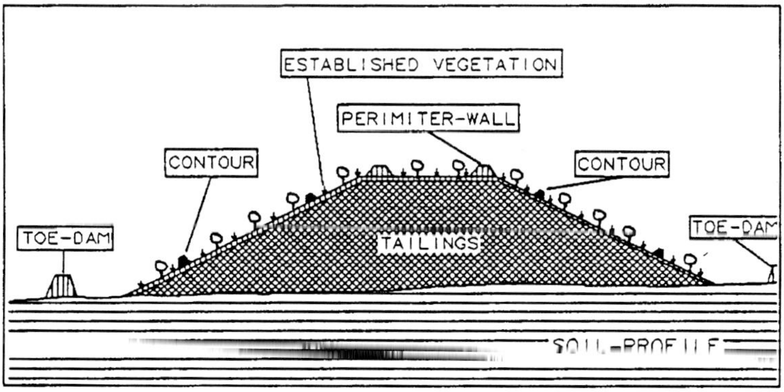

Rehabilitated area showing toe-dams, berms and plant-cover

FIGURE 3. Representation of the rehabilitation of a source of asbestos-pollution situated on a flat surface.

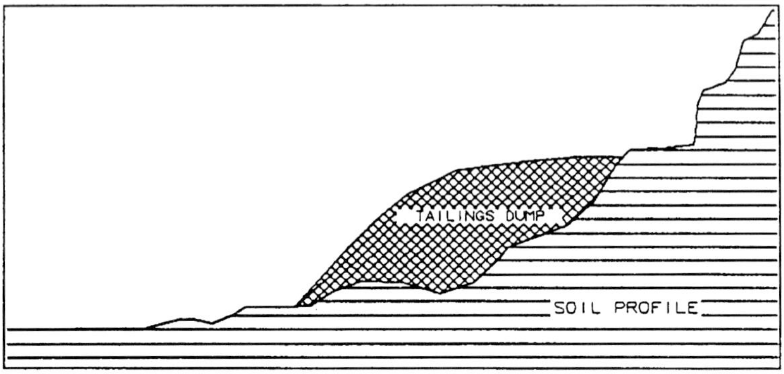

Representation of a source of asbestos-pollution

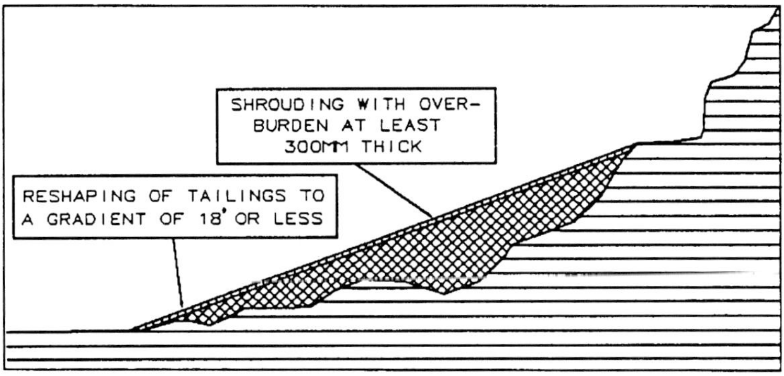

Source of asbestos pollution after reshaping and shrouding

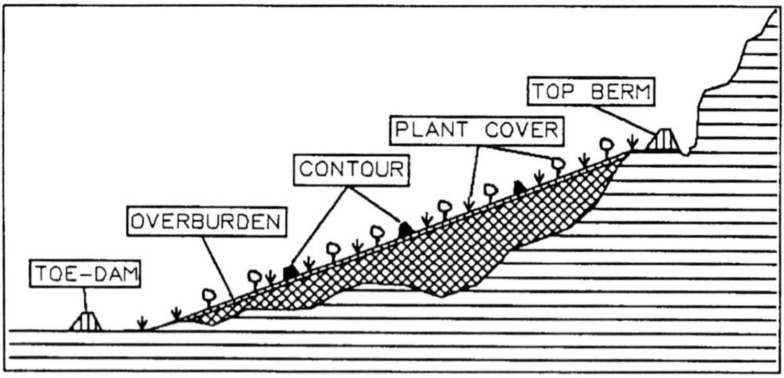

Profile of rehabilitated source of asbestos-pollution showing berms, contours, overburden, plant-cover and toe-dams

FIGURE 4. Rehabilitation by levelling and shrouding a source of asbestos pollition situated at the base of a hill-site.

the environment. The surface area of the source of pollution is gradually enlarged as a result of the disseminating influence of water and wind, which in turn can lead to a higher concentration of fibre in the atmosphere. In practice this means that rehabilitation must occur in such a way that a fibre-free environment is established.

Survey Standards

Wherever possible, surveys must be based on the national system (Lo.). The scale for contours must not be less than 1:1000 and contour intervals (preferably of 1 metre) should be used. Normal standards for topographical survey must be followed at all times.

Pedological Standards

1. After reshaping of the tailings, the tailings must be shrouded with a clean erosion-resistant cover, (even though the tailings can be completely bound or sealed chemically) to prevent its dispersal by wind and water.
2. The prerequisites for a suitable shrouding material are the following:
   • The material must be able to sustain a permanent, self-maintaining vegetation, and
   • Provide the maximum resistance against erosion.
3. After reshaping, all exposed fibre-containing surfaces must be covered with a layer of alluvial soil-gravel-rock (Table 2) at least 300 mm thick. This mixture of soil-gravel-rock should be "borrowed" from a suitable quarry. (The thickness of the shrouding has been determined experimentally. The minimum thickness that provides protection against erosion in the long term is 300-350 mm). This shrouding material must be fibre-free. Horizontal surfaces may be covered with normal clean topsoil if there is a shortage of coarser material, but it is imperative to use coarse material as a cover on surface with an angle of more than 5°, to prevent erosion. The following variations in particle-distribution composition, suitable for shrouding on different slopes, are recommended:
   • 0-10° slope: material of which a minimum of 40% of the particles have an average diameter of more than 2 mm.
   • 10-15° slope: material of which a minimum of 55% of the particles have an average diameter of more than 2 mm.
   • 15-20° slope: material of which a minimum of 70% of the particles have an average diameter of more than 2 mm.

   The particle-size distribution of the overburden used at the Penge mine in the northern Transvaal, and which gave satisfactory results, is illustrated in Table 2.

   In exceptional cases material from different quarries may have to be mixed in order to obtain the desired particle-distribution.

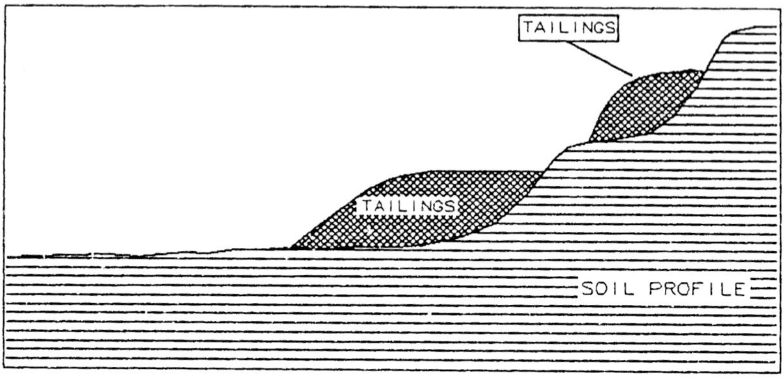

Representation of the original profile

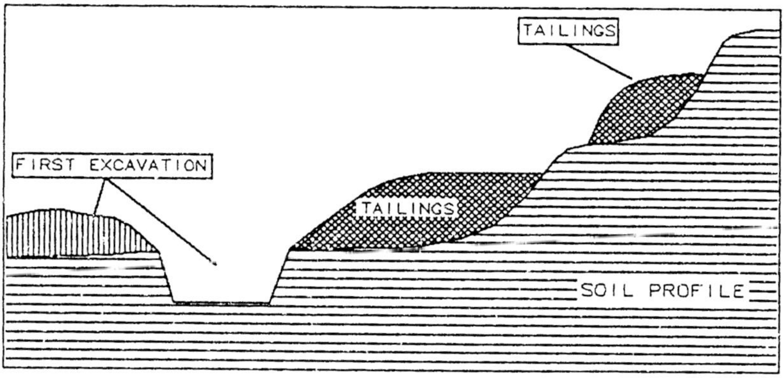

First excavation

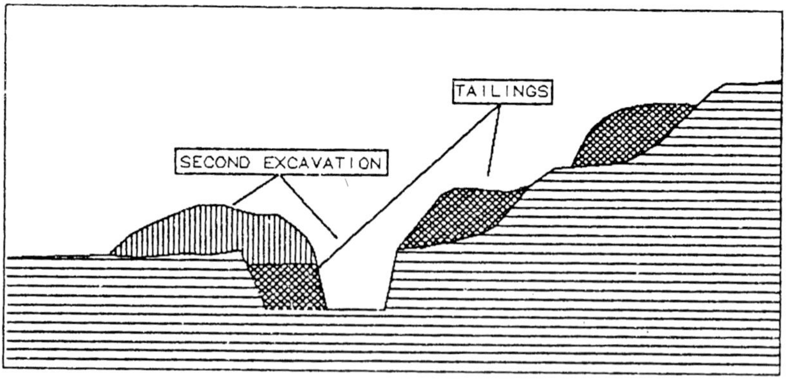

Filling the tailings into the excavation

FIGURE 5. Rehabilitation of a source of asbestos-pollution situated on a steep hill-side by excavating and filling.

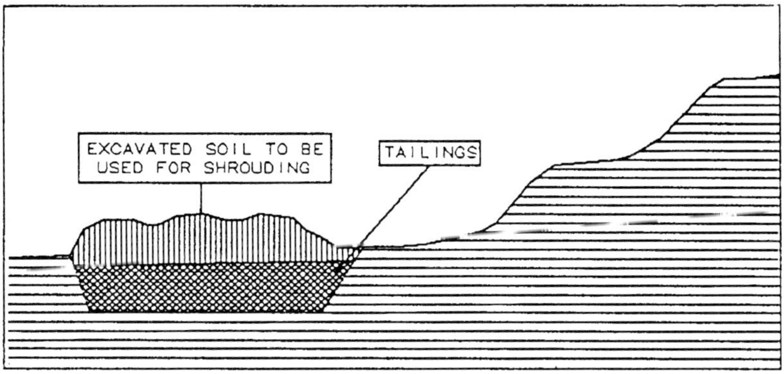

Completed excavation with tailings from the hill-side pushed down into the hole

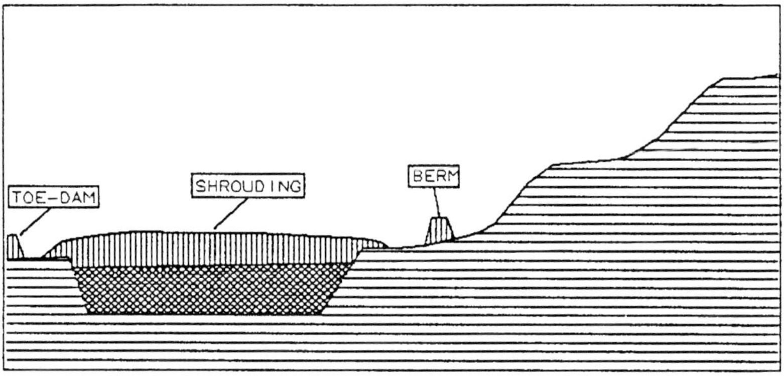

Rehabilitated area before establishment of vegetation. (The construction of berms and contours is preferable, but may be reduced as the shrouding will normally be very thick (1 - 2 metres)

FIGURE 5. *Continued.*

TABLE 2
Composition of Overburden

Particle-size (average diameter in mm)		Percentage (of mass)
<2	-	±30%
2 - 5	-	±10%
5 - 10	-	[illegible]
10 - 20	-	±10%
20 - 50	-	±10%
50 - 100	-	±10%
>100	-	±30%

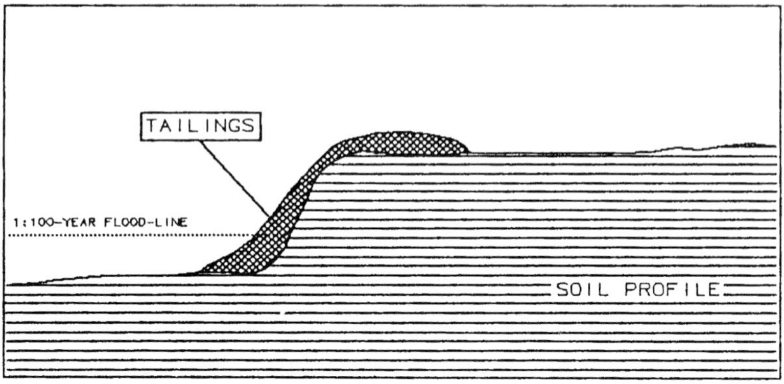

Representation of the original profile

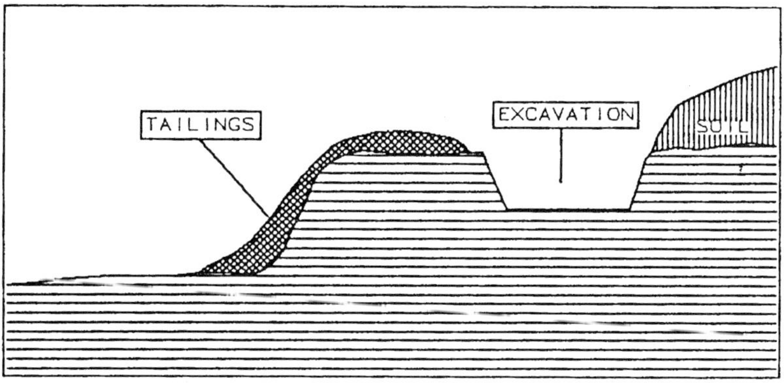

Excavation above the 1:100-year flood-line

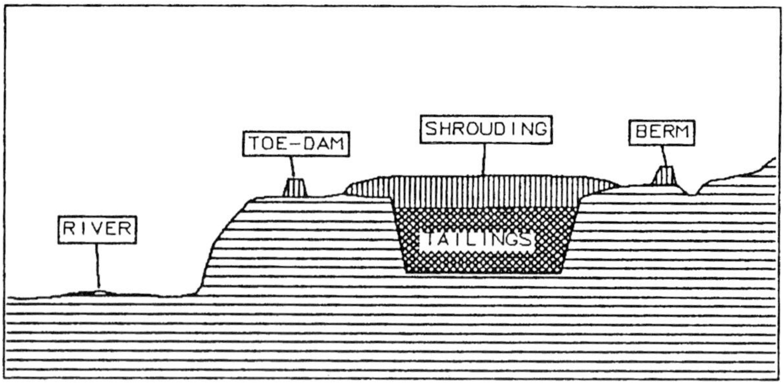

Completed rehabilitation showing the source of asbestos-pollution buried and covered in a new position before establishment of vegetation.

FIGURE 6. Rehabilitation of a source of asbestos-pollution situated in a river-valley.

At the same time care must be taken not to obtain an overburden with a texture that is too coarse, as plants must be able to establish on it. Deviations from the recommended textures can be considered depending on:

- Slope-angle;
- Length of slope (top to base);
- Geographical location (rainfall);
- Access-potential (animals and humans);
- Potential utilization of the rehabilitated terrain.

4. Requirements for the rectification of the tailings to allow root-development:

   Chrysotile waste is the only type of asbestos tailings which, in some cases, may be unsuitable for the establishment of vegetation. The most important soil-chemical problem regarding chrysotile waste, is the high concentration of soluble magnesium (Mg) which causes an imbalance in the nutrient elements. With the high concentration of Mg, the pH of the material is abnormally high which consequently affects the solubility of the nutrient elements. High concentrations of nickel (Ni) and boron (B) are present in many tailings dumps. The above-mentioned tendencies occur to different degrees in practically all chrysotile tailings. Natural vegettion however invades some dumps after a lapse of time. It is therefore essential to do a thorough analysis of the tailings on which rehabilitation must be implemented in order to determine the appropriate rectification steps that must be taken.

   Rectification may, separately or collectively, include the following steps:

   - Extensive leaching, after flattening, of the tailings;
   - Addition of the required nutrient elements;
   - Addition of organic material.

5. When recommending the types and quantities of fertilizers to be used, one must bear in mind that the after-use (ultimate land-use) following rehabilitation of the terrain, is not optimal agricultural utilization, but merely to establish an effective erosion-preventative vegetation on the rehabilitated surfaces.

GRADIENT (SLOPE) STANDARDS

The natural angle of repose of asbestos tailings is about 35-45°. It is therefore essential to reshape tailings dumps in order to implement effective storm-water control and to prevent erosion. Research on asbestos tailings dumps in the North Western Cape, proved that material containing fibre must be reshaped to a gradient of *18° or less* in order to obtain long-lasting erosion-control on the rehabilitated areas. The results of the research were confirmed by the success achieved since 1985 by the Research Institute for Reclamation Ecology in the rehabilitation of numerous sources of asbestos pollution in South Africa. The slope gradient should not exceed 20°, as the slope gradient has a direct influence on the erosion-potential. A lower slope gradient is also preferable in order to ensure optimum penetration of rainwater for vegetation to survive on the rehabilitated areas.

STANDARDS FOR STORM-WATER CONTROL AND RUN-OFF FROM REHABILITATED AREAS

1. *Flat Surfaces on Top of Rehabilitated Tailings Dumps*
   A. A perimeter wall 1.5 metre high and 1.5 metre wide at the base consisting of clean erosion resistant material should be constructed around the perimeter of the flat surfaces. The slopes will be graded to the natural angle of repose of the material used. The texture of the material should correspond relatively well to the analyses specified in Table 2.
   B. A grid-pattern of contours, 0.45 m should be implemented in paddocks of 0.5 hectare over the total flat surface.

2. *Sides of Rehabilitated Tailings Dumps*
   A. In South Africa the rainfall east of 26°EL (Eastern Longitude) is between 600-1400 mm and less than west of the 26°EL below 600 mm, hence the erosion-potential east of 26°EL is consequently greater than west of the 26°EL line. It is recommended therefore that the contours on the sides of dumps east of 26°EL should be constructed at closer intervals than those west of the 26°EL. The distance between the perimeter-wall/berm at the top and the toe-dams at the base of the dump will determine the intervals of contours that should be constructed.

   As a broad guide-line it is recommended that 1:200 contours of 0.75 m be constructed on angles of natural repose at maximum intervals as follows:

   • spaced every 40 m west of 26°EL and spaced every 30 m east of 26°EL (15°-20° slope-angles), and
   • spaced every 50 m west of 26°EL and spaced every 40 m east of 26°EL. (10-15° slope-angles).

The general rule for constructing contours, is to space them as evenly as possible between the upper perimeter-wall and the toe-dams, with shorter intervals east and longer intervals west of the 26° EL.

If possible, as a general rule, it is better to construct one contour more rather than one less, especially east of the 26°EL.

The composition of the material (particule-distribution) must be the same proportion as that of the selected quarry. In other words, care must be taken not to separate the fine and course fractions of the material during transport and contour construction of the contours. This applies chiefly to cases where the material is not placed directly into position on the contours, but moved mechanically to the pre-determined position.

Where, for topographical reasons, the contour comes to an end at the side of a dump, the contour must be lengthened obliquely upwards at the end to prevent damage as a result of increased run-off of water at that point.

Contours are constructed in such a way that water is diverted to protected canals. These canals must debouch into the nearest natural drainage canal. Material used for the contours must contain enough fine particles to prevent tunnelling.

(Originally South African agricultural standards were used as guide-lines for all norms. The norms presented, however, have been adapted by the Institute on the strength of research results obtained during the execution of asbestos-waste rehabilitation, research and implementation since 1985).

STANDARDS FOR STORM-WATER CONTROL

All water running off from adjacent higher-lying areas, must effectively be kept away from rehabilitated areas and deviated in such a way that it does not flow over it. Maximum precipitation as well as the size of the relevant catchment area must be taken into consideration when constructing canals and/or contours for the canalization of storm-water. When a waste-dump is situated near a river, the rehabilitation-plan must be designed in such a way that the waste dump an toe-dams are situated entirely above the 1:100 year flood-line level after completion of the rehabilitation (Figure 5).

STANDARDS FOR TOE-DAMS

Toe-dams can be constructed by excavating, construction of walls, or a combination of the two. Excavating can be advantageous as supplementary cover-material can be obtained through this method. The capacity of the Toe-dams are constructed must be such to accommodate the capacity of the run-off of 1:100 year flood and still have a free-board of 0.5 m.

STANDARDS FOR CONSTRUCTION

Wherever there is a possibility that rehabilitated sources of pollution may become exposed through erosion, or that storm-water or run-off control may cause excessive erosion, techniques such as gabions, Reno-mattresses, Hyson-cells or armed concrete should be considered.

1. Botanical Standards
   A. It is generally accepted that vegetation is the best long term strategy for stabilization and erosion-control on any rehabilitated source of asbestos-pollution.

   Vegetation is established as soon as the earth-works of the rehabilitation-process are completed. Factors to be taken into consideration for the selection and establishment of vegetation are the following:

- Suitability of the overburden as a growth-medium to sustain vegetation;
- species-composition;
- seeding density (kg/ha);
- the quantity of seed already present in the overburden;
- reproduction and survival;
- life-span of the species;
- inter-species competition (for moisture and nutrients)

and

- in some cases the selected species may also have to be hand-planted by means of cuttings.

B. Sufficient quantities of plant nutrients and micro-organisms are present to ensure the establishment of a satisfactory vegetation cover. The following standards are suggested for the establishment of plants on rehabilitated areas:

a. A self-sustaining, dynamic and permanent plant-cover, consisting mainly of local species, or their genetic variations.
b. Where grazing cannot be controlled, plants should be used that are not preferred food species of domestic live-stock.
c. The plants must cover the surface effectively (strive at obtaining a basal- and crown-cover equivalent to, or as near as possible to the surrounding area).
d. The plants bordering the rehabilitated area must be used as a hedge and must preferably bear thorns or be impenetrable in order to render its access difficult, thereby discouraging entrance and trampling onto the area.
e. The plants should be resistant to fire and/or able to regenerated quickly.
f. All plants should be drought tolerant (Kruckberg 1979).
g. Plants should be selected that will not invade surrounding natural habitats and compete with natural vegetation.
h. To reduce human activity, the plants on rehabilitated sites should nave no or limited economic potential (e.g. fruit, firewood, timber, etc.).

STANDARDS FOR EXCAVATIONS AND FILLS

In some cases it is possible to bury an entire waste-dump in trenches (Figure 5 and 6). This type of rehabilitation must, however, correspond to those of tailings dumps that are flattened and shrouded. The tailings must be well compacted in order to obtain long-term geomorphological stability. If tailings dumps are buried in their entirety in excavated trenches and are covered in a manner that the topography corresponds to the original natural state, standards for run-off control can be implemented selectively, as such a surface will then be in balance with the local landscape-forming elements.

STANDARDS FOR ROCK-DUMPS

When rock-dumps do not contain loose or unbound asbestos fibres, there is no need to apply the same high and costly standards for rehabilitation as in the case of tailings with fibres. In some cases, it may be possible to use rock-dumps as overburden for material with loose fibre. This practice must only be used in exceptional cases as no vegetation can be established on it. BEFORE A ROCK-DUMP CAN BE TREATED IN SUCH A MANNER IN SOUTH AFRICA, SPECIAL PERMISSION MUST BE GRANTED BY THE DEPARTMENT OF MINERAL AND ENERGY AFFAIRS.

Toe-dams must in any case be built around the dump according to the same standards as for fibre-containing dumps. These toe-dams should be constructed with 30-60% coarse rock material so that any fibres washed from the dump will settle between the coarse rocks to prevent dispersal by the wind and/or water (Figure 2).

CHEMICAL STABILIZATION

Many companies throughout the world are developing new and better binding and adhesive materials with the aim to bind loose displaceable fibre. In the future, these materials may be used in certain situations for the rehabilitation of fibre-containing mine waste, but their use as an alternative for a plant-cover may not become common practice before obtaining permission from the respective regulatory agencies of the nation involved.

STANDARDS FOR QUARRIES/BORROW-PITS

All quarries must be rehabilitated to the same stable condition as the dumps.

ACKNOWLEDGEMENTS

DeSousa Correia, R.I.
Kuit, H.B.
LeGrange, J.
Prinsloo, H.B.
Van der Walt, Dr. I.J.
Van Wyk, S.

LITERATURE CITED

Botha, J.C., Irwig, L.M. and P.M. Strebel. 1986. Excess mortality from stomach cancer, lung cancer and asbestoses and/or mesothelioma in crocidolite mining districts in South Africa. American Journal of Epidemiology, 123(1):30-40.

DuToit, R.S.J. 1969. Dust in South African asbestos mines and fiberizing plants. Proceedings of the International Conference on Pneumoconiosis:13-17. Johannesburg.

DuToit, R.S.J. 1987. Die gevare en beheer van asbesstof. (Referaat gelewer op die gesondheidskongres: Instituut vir Openbare Gesondheid, Pretoria: 26-29 Oktober 1987).

DuToit, R.S.J. 1991. An estimate of the rate at which crocidolite asbestos fibres are cleared from the lung. Annals of Occupational Hygiene, 35(4):433-438.

DuToit, R.S.J. 1992. Research and prevention of asbestos related diseases in South Africa. (In International Asbestos Medical Research - Butterworth).

Edwards, D. 1989. Asbestos and health. South African Mining: Coal, Gold and Base Minerals: 13-15, Aug.

Felix, M.A., Rendall, R.E.G., Steinberg, M. Roodt, L. and Z. Mabiletja. 1988. Abatement of environmental asbestos pollution in Mafefe in the North Eastern Transvaal. (Report of the National Centre for Occupational Health.) Johannesburg. (Unpubished).

Hart, H.P. 1988. Asbestos in South Africa. South African Institute of Mining and Metallurgy, 88(6):185-198, Jun.

Kruckeberg, A.R. 1979. Plants that grow on serpentine - a hard life. Davidsonia, 10:21-29.

Rendall, R.E.G. 1974. Airborne fibres in the Johannesburg atmosphere. Probe, 74:5-7.

Rendall, R.E.G., Felix, M.A., Mogomotsi, E. and Z. Mabiletja. 1989. Asbestos in the air around abandoned mines in North Eastern Transvaal. (National Centre for Occupational Health.) Johannesburg. (Unpublished).

Schreier, H. 1989. Asbestos in the natural environment. Amsterdam: Elsevier.

Sluis-Cremer, G.K. 1965. Asbestos in South Africa - certain geographical and environmental considerations. Annals of the New York Academy of Sciences, 132:215-234.

Sluis-Cremer, G.K. 1970. Asbestos in South African asbestos mines. Environmental Research, 3:310-319.

Van Wyk, J.J.P. and H.P. Hart. 1990. Erosion control at asbestos mine spoil rehabilitation programmes in South Africa. Potchefstroom: PU for CHE.

Webster, I. 1973. Asbestos and malignancy. South African Medical Journal, 4:165-171.

Webster, I. 1986. Asbestos related diseases. Continuing Medical Education, 4:71-78.

Environmental Contaminants, Ecosystems and Human Health. Edited by S.K. Majumdar, E.W. Miller, and Fred J. Brenner. © 1995, The Pennsylvania Academy of Science.

Chapter Five

THE USE OF NITROGENOUS FERTILIZERS AND ITS EFFECTS ON THE ENVIRONMENT

APURBA SARKAR
Senior Scientist
Head of System Management
Project Directorate for Cropping Systems Research
(PDCSR)
Modipuram, Meerut, (U.P.) 250110 India

INTRODUCTION

About 35% of the earth's 13389 million hectares of total geographical area is under agriculture, of which one-third is under arable and permanent crops and the rest is under permanent meadows and pastures. This small chunk of land feeds an ever-increasing population expected to be 6 billion by the turn of this century. This necessitates the use of fertilizer, an essential input because available soil N is inadequate for crop production. Considering the total of nitrogenous (N), phosphate (P_2O_5) and potassic (K_2O) fertilizers, global consumption comes to about 133 million metric tonne (mt), consumption per hectare of land under arable and permanent crops is 92 kg, per hectare of total agricultural land is 30 kg/ha and per capita consumption is 28 kg/ha. Because of its good response, availability, economic pricing and freedom of choice, nitrogenous type has a unique place among the fertilizers. It comprises more than half of total fertilizers consumed. In this regard, China ranks first, followed by U.S.A. Based on the nitrogenous material dispatched, 40% comes through urea, 11% through ammonium nitrate, 5% through ammonium sulphate and 10% through others.

After application, nitrogenous fertilizer has a well defined fate: a part goes for plants' uptake, a fraction is taken up by microbes called microbial immobilization, a part lost in leaching (most in NO_3^- - form), a part is fixed by soil (mainly as NH_{+4}) - form), a part lost through ammonia volatilization under alkaline condition. Under anaerobic condition, nitrogen is lost through denitrification as different gases like dinitrogen (N_2), nitric oxide (NO), nitrogen dioxide (NO_2), nitrous oxide (N_2O). Some nitrogen is exuded by plant roots as amino acid. There may be gaseous loss of nitrogen through plant leaves. Such use of N fertilizer is not also an unmixed blessing. This large input has secondary environmental consequences. A main concern is harmful environmental impact due to excessive agricultural nitrogen loss. Nitrate leaching contaminates groundwater considered health-threatening and accelerates eutrophication of such water, whereas emission of nitrous oxide causes photochemical destruction of stratospheric ozone layer that protects the earth from ultraviolet rays. Also the emission of nitric oxide and nitrogen dioxide has been implicated in acid rain formation and photochemical smog. An effort has been made here to analyze these consequences in view of the protection of this environment.

N FERTILIZER BEHAVIOR IN SOIL

Under aerobic conditions the soil-applied ammonium sulphate behaves like this: $(NH_4)_2\ SO_4 \rightarrow NH_{+4} + SO_{=4}$, $NH_{+4} \rightarrow NO_{-2} \rightarrow NO_{-3}$ and this NO_{-3} can be uptaken by plants or a part may be leached down if followed by irrigation. NH_{+4} as such may be uptaken in exchange of H^+-ion of plant root or it may be held by the cation exchange site of soil micelle and subsequently may be uptaken by the crop in exchange of H^+ ion of plant root, and may also be nitrified to NO_{-3}. If applied under unaerobic submerged condition, the picture is quite different. In the oxidized layer of the submerged soil, NH_{+4} is converted to NO_{-3} and goes down to the reduced zone where it is denitrified, leading to gaseous loss in the form of N_2, NO, NO_2 and N_2O.

Applied urea is quickly converted to CO $(NH_2)_2 + 2H_2O +$ urease $\rightarrow (NH_4)_2CO_3 \rightarrow 2NH_3 + \rightarrow +H_2CO_3 \rightarrow 2NN_3 + H_2O + CO_2$. Ammonia is thus lost to the atmosphere. Such loss is greater with increasing alkalinity and even greater if applied on bare soil surface. This happens mostly on surface applied urea. Normally, with the addition of a proton, $NH_3 + H^+ \rightarrow NH_{+4}$ and this NH_{+4} has the same fate as noted above for ammonium sulphate, depending upon the aerobic and anaerobic condition. Foliar applied urea has similar transformation because of the presence of urease enzyme in the foliage. Thus, it is rapidly and directly absorbed through leaf epidermis.

Ammonium nitrate is highly hygroscopic and upon application, it becomes: $NH_4NO_3 \rightarrow NH_{+4} + NO_{-3}$. Again the NH_{+4} part has the same fate as that of the ammonium sulphate. The nitrate part also has the same fate as the nitrified NH_{+4} part of ammonium sulphate, as previously cited.

Other nitrogen fertilizers include anhydrous ammonia. Since it is a gas at atmospheric pressure, some part is lost during and after application. After application, it behaves like any other ammoniacal fertilizer. Aqua ammonia and nitrogen solution require a special tank and equipment for application. Low pressure material can be applied directly on foliage as a small stream. Ammonium chloride, ammonium phosphate, and urea-phosphate, behave as any other ammoniacal fertilizer. The NH_{+4} in it has the same fate as the nitrified NH_{4+} part of ammonium sulfate. Other nitrate fertilizers, like sodium nitrate, potassium nitrate, calcium nitrate, hydrolyze into NO_{-3} and the corresponding metal cation. And the NO_{-3} behaves similarly as that of the other nitrate fertilizers like ammonium nitrate.

In slow release nitrogenous fertilizer the nitrogen is gradually and slowly released in the soil and is ammonified. The NH_3 or NH_{+4} released is adsorbed by colliodal fraction of the soil and subsequently nitrified. Once in NH_{+4} form, it behaves like any other ammoniacal source of nitrogen. Examples are ureoform, oxamide ($NH_2COCONH_2$), crotonylidene diurea, dicyandiamide (NH_2C=NHN-HCN), thiourea (NH_2 -CS-NH_2), isobutyric diurea (IBDU).

AMMONIA VOLATILIZATION

Under alkaline conditions or on bare surfaces there may be gaseous loss of NH_3 from applied ammoniacal fertilizers like ammonium sulphate, urea, ammonium chloride, and ammonium nitrate. $NH_4 + H_2O + OH \rightarrow NH_3 + 2H_2O$.

Magnitude of Such Loss

Carter et al. (1961) reported negligible loss in calcium carbonate-treated soil to 3.5% under low level and 36% under higher N level. Watson (1992) reported 29% loss in surface-applied urea in acid soil.

Factors Affecting NH_3 - Volatilization

(a) *pH*: Ammonia is highly soluble, very reactive, and its volatilization increases with increasing pH. (b) *Temperature*: As the temperature increases, ammonia diffusion in soil increases but its solubility decreases, and more ammonia is lost as gas. (c) *Method of fertilizer application*: Surface application of amoniacal fertilizer increases the volatilization especially under alkaline conditions. (d) *Soil moisture*: For anhydrous ammonia, the loss is greater with dry soil, and such loss is lower as the soil moisture content of soil increases. This is because of conversion of NH_3 to NH_{+4} and subsequent retention by soil. (e) *Soil texture*. (f) *Type of soil and type of NH_{+4} salt applied.*

Effects of Ammonia Volatilization on Environments

(a) If seeds are planted too close to the zone of ammonia (aqua ammonia, anhydrous ammonia) application, seedlings and plants will be injured in burning action. (b) The rapid hydrolysis of urea to ammonia causes ammonia injury to seedling if large quantities are placed too close to seeds. (c) It is a loss of agricultural N from soil.

Measures to Reduce Ammonia Volatilization

(a) *Placing the nitrogen material* several inches under the soil surface reduces NH_3 loss. (b) *Burning of sod* (or other available crop residue) before urea application destroys the enzyme urease in plant tissue and reduces the loss (Jackson, et al. 1962). (c) *Ammonia volatilization is directly proportional to rate of evaporation*: Fertilization should be done during low evaporation in the evening or at night when temperature and wind speed are minimal. (d) *To avoid seedling injury* by anhydrous ammonia, banding placement of ammonia 4 to 6 inches under; and 6 inches to the side of seed or plants is recommended. (e) In calcarious soil, apply NH_4NO_3 salt. Here $Ca(NO_3)_2$ is formed and it is highly soluble so the formation of $(NH_4)_2CO_3$ is reduced and probability of NH_3 volatilization is reduced. (f) Application of nitrogen closely timed with crop requirement is most efficient in reducing inorganic N in soil profile. (g) Foliar application of urea reduced ammonia loss. (h) Recently, Watson et al. (1992) reported that use of urease inhibitor may improve the N efficiency by delaying urea hydrolysis and by reducing ammonia loss. The phosphoric triamide analog of NBPT may be a promising one. (i) Proper method of preparation and storage of organic fertilizer: The traditional method of preparation and storing of farm yard manure is faulty in many of the developing and under-developed countries. Gypsum or super-phosphate can be used as preservatives to prevent escape of ammonia from farm yard manure. Likewise, the farm waste, crop residue, etc., can be used as a bedding for cattle urine absorption and subsequently can be carried with dung, wood ash to a pit and plastered to make compost. Such well-decomposed manure should be spread in the field before sowing. Partially decomposed manure should be applied generally three to four weeks before sowing of crop.

DENITRIFICATION

Under water-logged condition, the oxygen is excluded, leading to anaerobic condition. Some anaerobic organisms of the Genera Pseudomonas, Micrococcus, Archromobacter and Bacillus are able to obtain oxygen from nitrate and nitrites with the release of NO, NO_2, N_2O, N_2 gases. Emission of these gases is also possible through other processes. Let us deal one gas at a time.

Nitrous Oxide (N_2O) Emission

It is formed during chemodenitrification, i.e., interaction of nitrite with soil organic matter and pH, during nitrification of NH_{+4} under aerobic condition, denitrification of nitrate under anaerobic condition.

1. *Factors affecting its emission*: (a) *pH*: It occurs mostly at 4.9 to 5.6 of soil pH during denitrification. As days pass after water logging, aerobic conditions arise. Under these conditions in an alkaline poorly buffered soil, the pH may increase. Nitrate oxidizers' (*Nitrobacter*) growth is restricted at higher pH, but that of ammonium oxidizers' (*Nitrosomonas*) is not affected, resulting in accumulation of nitrite with the possible emission of N_2O through chemodenitrification. (b) *Soil moisture*: At more than 70% soil pore space filled with water, emission through denitrification occurs. Under water logged condition, it is the exclusion of oxygen rather than the presence of water that matters. Aulakh et al. (1992) reported that the higher the water content, the lower the relative proportion of N_2O in denitrification. (c) *Partial pressure of oxygen in soil air*: As the partial pressure of oxygen decreases, emission increases. (d) *Amount of organic matter in soil:* Soil organic matter provides the energy for this reaction, (Linn & Doran, 1984). Increased organic matter is directly proportional to increased emission. (e) *Microsite or hot spot in the soil*: Christensen et al. (1990) reported that in acid soil even at field capacity (aerobic condition) denitrification occurs in some microsite (hot spot) which comprises 4-8% of soil volume. Such hot spots contain more organic matter. The enhanced microbial activity used up the already available oxygen and made the hot spot more anaerobic, thereby more conducive for N_2O loss in denitrification.
2. *Magnitude of nitrous oxide emission*: In general, cropland loss of N_2O ranges from 0.1 to 3kg/ha/annum (Smith et al. 1990). Aulakh et al. (1992) reported that the tentative estimate of global N_2O emission is 9.7 to 12 mmt/year for natural terrestrial ecosystems, 2.3 to 3.7 mmt/year for cultivated land and 0.2 to 2.1 mmt/year for fertilized soil (mmt=million metric tonne).
3. *Effects of N_2O emission on environment*: (a) *Stratosphere contains few ppm of ozone (O_3)*. This is a 40 km layer between 10-50 km from the earth's surface. It is formed by ultraviolet rays on O_2 breaking into $O^{=}+O^{=}$ and then $O^{=}+O_2=O_3$. It forms a shield against ultraviolet rays. Nitrous oxide is implicated in the photochemical destruction of ozone layer leading to the hazards for the ecosystem. The atmosphere contains approximately 300 ppb of N_2O. Aulakh et al. (1992) estimated that a doubling of atmospheric N_2O would cause a 10% decrease in ozone layer, causing a 20% increase in ultraviolet radiation reaching the earth resulting in increased incidence of skin cancer, leucoderma and other hazards. Shea (1988) estimated that for each 1% drop in ozone layer there is 4 to 6% increase in skin cancer. Current measurement of atmospheric N_2O indicates an annual concentration increase of 0.25% (Crutzen, 1981). (b) *Greenhouse effect*: The glass in the greenhouse allows the solar radiation to enter it but does not allow it to radiate back totally because a part of it is absorbed by the glass. As a result, the temperature inside a greenhouse increases. Some gases have similar character. These are called greenhouse gas, e.g., CO_2, N_2O). N_2O also acts

like an energy trap. Increased concentration of these gases causes global warming. Melting of polar ice, rising in sea water level, flooding and submerging of part of the earth's land mass are the anticipated effects of global warming. It is estimated that global temperatures will rise @ 0.3° and sea level @ 6 cm per decade. Aulakh et al. (1992) estimated that an increase of 0.2 to 0.3% in N_2O concentration in the atmosphere contributes about 5% to the supposed greenhouse warming.

Emission of Nitric Oxide

It is formed during denitrification, self decomposition of nitrous acid: $3NHO_{2-} \rightarrow 2NO + HNO_3 + H_2O$, chemodenitrification and nitrification.

1. *Factors affecting the emission of nitric oxide (NO)*: (a) *Water logging*: It induces denitrification and in turn the emission of NO. (b) *Type of land.* (c) *Type of ecosystem*: If the ecosystem is disturbed by burning, there is a high rate of mineralization and nitrification and more emission of NO. (d) Rate of nitrification.
2. *Effect of nitric oxide on environment*: Nitric oxide has been implicated in acid rain formation and generation of photochemical smog, a mixture of smoke and fog. Exposed to an atmosphere containing NO or NO_2, water vapor absorbs nitrogen forming nitric acid. Along with rain or other forms of precipitation this acid comes down. The pH of acid rain is less than 4. $2NO + O_2 = 2NO_2$, $4NO_2 + O_2 + 2H_2O = 4HNO_3$, $HNO_3 \rightarrow H^+ + NO_{-3}$. It acidifies the surface and subsurface water and has detrimental effects on the ecology of affected catchments like, fish population, forests, exteriors of buildings, statues, foliar injury and reduction in plant growth.

Emission of Nitrogen Dioxide

Nitrogen dioxide is formed during nitrification reaction, chemodenitrification, self-decomposition, denitrification reaction.

1. *Factor affecting its emission*: (a) Soil surface temperature: In the morning the emission is minimum and in the early afternoon it is maximum (Slemr and Seiler, 1984; Delamy et al. 1986).
2. *Effect of Nitrogen dioxide on environment*: Same as of NO.

Dinitrogen Emission

It is formed during denitrification of nitrate under submerged condition. Under well-drained acid soil, there may be decomposition of $NH_4NO_2 \rightarrow 2H_2O + N_2$; Van Slyke reaction as familiar to organic chemists: $RNH_2 + HNO_2 \rightarrow ROH + H_2O + N_2$.

1. *Factors affecting dinitrogen emission*: (a) *pH*: at pH value 7.3 to 7.9, the loss is largely in the form of N_2. With denitrification, pH increases. At higher

pH, nitrate oxidizers (*Nitrobacter*) are suppressed, apparently due to ammonia toxicity, while growth of ammonia oxidizers (*Nitrosomonas*) is not suppressed. As a result, nitrites (NO_2^-) accumulate with their possible chemodenitrification. (b) *As the soil N_2 decreases,* loss increases through denitrification. (c) *Organic matter*: The greater the organic matter, the greater the loss. (d) The N_2 loss increases with increasing input of nitrogen fertilizer. (e) *Thickness of saturated soil layer:* The thicker the layer, the greater the emission (Steenvoorden and Costerom, 1972).

2. *Magnitude of dinitrogen emission*: In general, annual cropland loss ranges between 5-25 kg/ha (Smith et al. 1990).
3. *Effect of dinitrogen emission on environment*: (a) Truly speaking, by the process of denitrification, dinitrogen is recycled to its original source, the atmosphere from where it was removed by biological or chemical fixation. But to the grower it is an economic loss of available input of nitrogen, the biological yield.

Measures to Reduce Denitrification

1. *Method of N fertilizer application*: (a) Ammoniacal or nitrate fertilizer should not be top dressed on standing water in crops like rice. As previously cited, ammonium sulphate or urea will be nitrified at oxidized layer and will go down to reduced layer and will be lost in denitrification. Mix urea with soil and make it into small dough. Keep it for two to three days to allow fixation to take place. Now place the dough a few inches below the soil surface between the hills of rice plants. In other crops, like wheat and maize, do not apply such fertilizer immediately after irrigation to avoid N loss; in denitrification pockets of soil. (b) *Split application* of N fertilizer timed closely to crop requirements and especially under submerged condition. (c) *Foliar application*: Under a submerged condition, it is often difficult to enter the field where foliar-applied urea decreases N loss.
2. *Type of N fertilizer*: (a) Use of slow release N fertilizer: Nitrogen from such fertilizers is gradually released with time, preventing the possibility of immediate nitrification and subsequent denitrification, e.g., IBDU, urea supergranules, lac coated urea, wax coated urea, tar coated urea, gypsum coated urea. (b) *Nitrification inhibitor coated urea*, capsulated urea, etc. Such materials are useful in areas of heavy rainfall or submerged soil conditions, etc. These chemicals temporarily inhibit the nitrification and thus delay the release of NO_3^- which might have gone down in reduced zone immediately after application in submerged soil. A compound, 2-chloro-6-(trichloromethyl) pyridine can delay nitrification for at least 6 weeks applied @ 5% of urea weight. Thus the N is gradually available to plants for an extended period. Various chemicals are found to have such inhibiting property e.g., neem (*Melia azadirachta*) cake due to the presence of an alkaloid, neembidin. (c) If the farmer has urea and calcium ammonium

nitrate at his disposal in a well-drained soil he should prefer urea because the denitrification loss is only one third of that from CAN.

3. *Time of N fertilizer application*: Powlson et al. (1988) reported that in the wheat crop, denitrification goes well over the temperature range of 5-10°C in temperate regions which often contain much nitrate in spring following application of fertilizer N in autumn. In such places somewhat delayed application may reduce such loss. A similar sharp increase in denitrification occurred in the sub-tropical soil but over the higher temperature range of 10-20°C. Here preplant application for winter wheat should be made in autumn, when soil temperature are lower, rather than in late summer.

LEACHING OF NITROGEN

Leaching of nitrogen occurs mainly as nitrate. A small fraction may leach in ammonium form in some over-percolating sandy soils, but its possibility is reduced by adsorption to the soils cation exchange complex. Leaching of soil organic nitrogen is not major issue. Sherwood and Fanning (1985) found that 25% of applied urea may also be lost in surface run-off rain water.

Nitrate Leaching

Under aerobic conditions, the applied ammoniacal fertilizers are nitrified in two steps by nitrifying bacteria $2NH_{+4} + 3O_2 \rightarrow 2NO_{-2} + 2H_2O + 4H^+$ (*Nitrosomonas*), $NO_{-2}+O2 \rightarrow 2NO_{-3}$ (*Nitrobacter*). This NO_{-3} form of N is highly soluble and completely mobile in soil water. Unlike ammonium, nitrate ion is not retained by soil negatively charged clay micelle. Under conditions of heavy rain it is leached out of the upper horizons of the soil and mixes with the groundwater. This is called nitrate leaching. Of course, during extremely dry weather allowing capillary movement of water, nitrate salts may move with the upward movement of water and be deposited on the surface. Nitrate fertilizer like NH_4NO_3; KNO_3 etc., are readily soluble without going for nitrification steps and have the same fate as above.

1. *Factors affecting nitrate leaching* (a) *Supply of nitrate nitrogen*: The greater the supply, the greater the leaching. This depends on the factors influencing nitrification of added fertilizer material e.g., C:N ratio of soil, availability of nitrifying organism, soil reaction, soil aeration, soil moisture, temperature: winter temperatures are generally low enough to prevent nitrification. But it goes well between 16 to 30°C. (b) *Type of soil*: Kolenbrander (1982) reported that leaching losses decrease in heavier soil. This is due to lower volume of macrospore in heavier soil slowing down water movement in the profile. (c) *Type of vegetation*: Leaching is less in grassland than in arable crop land because of no fallow period in grassland. Vitousek et al. (1982) showed that forest ecosystem has potential for NO_{-3}

leaching when the amount of litter fall is above 60 kg N/ha/year. (d) *Amount of drainage water.* (e) *Amount of fertilizer level applied.* (f) *Atmospheric N deposition*: High atmospheric N deposition in excess of vegetation demand in forest ecosystem causes nitrate leaching in the long run (Gundersen and Rasmussel, 1993). (g) *Season of N fertilizer application*: Smith and Cassel (1990) reported that fall-applied N leached to 180 cm depth, unlike up to 30 cm from spring application.

2. *Magnitude of nitrate nitrogen leaching*: Kolenbrander (1982) showed that in West-European countries, 80 to 300 kg N/ha/year leached with 100 to 1000 kg N/ha/year of applied N when the drainage water was 300 mm/year and groundwater was more than 1 m below the surface. On the other hand, in a review of N losses in tile effluent from corn belts of Iowa, Minnesota and Ohio of USA, Logan et al. (1980) found an annual nitrate-N loss of less than 30 kg N/ha in general, but increased with N application above crop needs.
3. *Effect of nitrate leaching on environment*: (1) *Soil acidification*: Nitrification is a proton-producing reaction: - $NH_4^+ + 2O_2 \rightarrow NO_3^- + 2H^+ + H_2O$. This causes acidity of the soil solution. Likens et al. (1970) did an extensive study at the Hubbard Brook Experimental Forest, USA. A watershed was clear-cut and treated with herbicides to prevent regrowth. Since plant N uptake was totally cut off, the NO_3^- leaching was of the order 142 Kg N/ha/yr. Since for every 14 kg N as NO_3^-, 1 K mol of H^+ is produced, this corresponds; to 10 k mol of H+/yr. which decreases the pH in runoff from 5.1 to 4.3 (2) *Loss of nutrients and fertility of forest soil*: Soil tries to buffer the acidity by releasing bases like Ca, Mg, Na, K at pH 4. Thus apart from N, increased leaching of these bases, essential plant nutrients, may in the long run reduce soil fertility. (3) *Aluminum toxicity in forest soils*: As has already been narrated, nitrate leaching causes soil acidity. At pH <4.2 aluminum ions come into the picture for buffering such enhanced acidity. It was up to 2 millimole/L at Geritsfles in the Netherlands (Mulder, 1988). (4) *Nitrate contamination to ground and surface water used for drinking.* Apart from direct economic loss to the grower, nitrate leaching causes a direct threat to the quality of our drinking water. It contaminates the ground and surface drinking water, causing health hazards. (a) *Human health hazards*: Nitrate is not directly metabolized to other compounds in humans, but bacteria present in stomach and saliva reduces it to nitrite. Nitrate reacts with hemoglobin (this transports oxygen to tissues) in erythrocytes of blood to form methemoglobin which is incapable of transporting oxygen, leading to asphyxia in 3 month old or less infants. (Parks et al. 1981). Normal levels of methemoglobin in human range from 1-2%. A level more than 3% is called methemoglobinemia. The Committee on water quality criteria of the National Academy of Sciences of USA recommends that nitrate-nitrogen concentration in public water supply should not exceed 10 mg/L and nitrite-nitrogen 1 mg/L (Annon, 1972).

World Health Organization (Kolenbrander, 1982), on the other hand, has set the limit 11-22 mg of NO_3-N/L of drinking water. Accepting WHO norm of 11 mg/L under condition of 300 mm of drainage water per year on an average a leaching loss of 33 kg N/ha/year is corresponding to maximum permissible level of applied mineral N,O, 100, 320, 500 Kg/ha/year for arable sandy soil, arable clay soil, grassland sandy soil and grassland clay soil, respectively. (b) *Animal health hazards*: Unlike humans, the immediate toxic effect of nitrite (converted from nitrate) in some species like horses is due to nitrite-induced vasodilation resulting in cardiovascular collapse and shock (Anon, 1985). Other studies show that nitrate in stomach can react with secondary amines and other amine substances to form N-nitroso compounds that may be oncogenic (Ohshima and Bartsch, 1981). (5) *Eutrophication of surface water*: Leaching of nitrates, phosphates and other plant nutrients to an aging aquatic ecosystem like a lake, gradually makes it rich in nutrients. This results in excessive growth of algae and microscopic organism. Water becomes turbid, preventing light penetration. As a result of increased deposition of dead algae, there is extreme competition for oxygen absorption necessary for underwater life. This kills animal life. In extreme cases, spawn of some fish species developing near the bottom can be endangered. (Anon, 1989 a,b,d).

4. *Measures to reduce nitrate leaching*: (1) *Applying optimum level of N fertilizer*: This can be done by determining the crop N requirement and soil available N. Crop N requirement is the crop N uptake at near maximum yield. For corn it is 22 Kg N Mg^{-1} (Meisinger, 1984). Tests for residual nitrate are useful for soils suspected of having an appreciable accumulation of nitrate in the profile. In Illinois (Anonymous, 1989 C), an electronic sensor mounted on a moving soil shank measures nitrate directly and transmits a signal to a micro-computer that controls the fertilizer N application. (2) *To determine when water is likely to percolate below root zone*: Irrigation is frequently implicated as the cause of nitrate leaching. For example, in Chicago (Schepers 1988), precipitation exceeds evapotranspiration between April 1 and June 15. Irrigation should be adjusted accordingly. (3) *Proper method of fertilizer application*: (a) *Placement of fertilizer*: Placing the N fertilizer in band increases its efficiency and decreases leaching. (b) *Split application*: As has already been discussed, crops require nitrogen at some critical physiological growth stages, splitting the N level at those stages. (c) *Foliar application of urea*: Urea sprayed on foliage directly goes in the plant system incorporated into allantonin via allantonic acid, the latter being synthesized from condensation of glyoxylate and urea (Mothes, 1961). Thus the possibility of leaching is minimum, later spray may not increase the yield because plant may not be N-deficient at later part of its life. (d) *Fertigation*: It means applying fertilizer along with irrigation water. This can be timed to coincide with the time of greatest N requirement. (4) *Using proper source of N fertilizers.* (a) *Slow release N*

fertilizer: N from such materials is gradually and slowly available to plants as and when required. This increases its uptake and reduces leaching. (b) *Nitrification inhibitor coated urea*: These materials restrict the activity of the microbes, *Nitrosomonas* doing the first step of nitrification, as a result the ammoniacal fertilizer undergoes delayed nitrification. But such inhibitor-treated aqueous urea should be placed below soil surface to avoid volatilization. (5) *Timely application or incorporation of organic material*: In fertile soil, mineralized N can account for a great chunk of crop N requirement and if not used can contribute to nitrate leaching. The slow and perpetual mineralization and immobilization should be synchronized with crop N requirement when large amounts of crops residue are added to soil. Residues with 1-2% N normally immobilize N reducing potential nitrate leaching. On the other hand, legume green manure and animal manure promote mineralization shortly after addition. Their timely incorporation reduces the possibility of leaching. (6) *Conservation tillage*: The reduced tillage, in general reduces soil erosion, increases available soil moisture and thus increases uptake of N. (7) *Following proper cropping system.* Instead of monoculture, proper crop rotation may reduce N leaching. Papendick et al. (1987) reported that leaching is less in alfalfa-corn system than in corn monoculture. Winter cover crops can be effective in absorbing both nitrate and available water during fall, winter and spring, thereby reducing the N leaching potential, e.g., vetches, clovers, peas, small grains. Scavenging of excess available N from within the rooting zone can be done by crops like rye. For scavenging nitrate accumulation below normal rooting depth, deep rooted crop like alfalfa (having rooting depth more than 5 m) can be used (Mathers et al. 1975). Vegetative filter strip for the riparian zone acted like a filter for particulate waste nitrate which was converted to organic form and thus reduces leaching (Lowrance et al. 1983). (8) *Using terraces, and impoundments*: Such structures provide means for trapping soluble nitrate from run-off. Olsen et al. (1973) noted that nitrate content of inflow to three Nebraska reservoirs averages 1 mg L^{-1}, whereas that of outflow was only 0.05 mg L^{-1}. (9) *Method of irrigation*: Although irrigation is beneficial, it increases the nitrate leaching sometimes. A careful method should be adopted. For saline area, the amount should be just sufficient to maintain an appropriate salt balance. For crops like vegetables, potatoes, furow irrigation is suitable. Run-off water from furrows is only a bit more enriched in nitrate as it passes across the field than its concentration at source. Properly timed N application in small amounts through sprinklers reduce leaching (Rhoads and Stanley, 1981). For high value crops like orchards, vineyards, strawberries, drip or trickle irrigation is probably the most water and nutrient-efficient methods of irrigation. With low water holding capacity, coarse-textured soils are more prone to nitrate leaching. Rate and time of irrigation thus should be determined for each soil for the type of crop to be planted. Late season fertilization may not be effective because crop may have assimilated most of the N by that time. (10) *Protection of well*: Proper sealing and cementing of the casing of boreholes should be ensured to avoid entry of run-off water and seepage of groundwater. As far as possible, the crop field, farmstead, feedlot, septic tank drainage, loading, unloading and mixing of fertilizer should be directed away from the well. (11) *Protection of forest land*: Deforestation results in an overload of atmospheric N deposits (NH_3, NO_2, NO, N_2), reduced uptake of N by trees, enhanced nitrification in such aerobic condition under increased soil temperature followed by leaching (Gundersen and Rasmussel 1993). (12) *Leaf removal from street surface*: Leaf removal in the fall will markedly decrease the effect. Leaf burning on street surface should be avoided because most of the nutrients are in the ash and are transported by rains. (13) *Adopting remedial measures against eutrophication of lake, ponds, etc.* (a) Minimize run-off from animal concentration and manure. (b) Removal of organic substances can be done by mechanical and biological means as is done in Switzerland. (c) Pipeline around lake border collects the waste water from catchment areas in Germany.

DISCUSSION AND CONCLUSION

It shows that an increase in food production through an increased input of N fertilizer can affect to a great extent our environment, i.e., ground and surface water, atmospheric temperature and probably also the ozone layer around the earth. To ameliorate such effects, attention should be given to reduce N loss in denitrification and ammonia volatilization to increase the efficiency of applied N.

Higher loss of N in denitrification was observed (Addiscott and Powlson, 1992) from higher level of applied N (ammonium nitrate, KNO_3) than from lower rate under spring applied N to winter wheat. Due to more applied N there may be enhanced diffusion of NO_{-3}-N to the soil microsite (hot spot) and more denitrification of the applied KNO_3. On the other hand, there may be more nitrification of the applied NH_4NO_3 in the hot spots having more of organic matter. As a result, the microsite atmosphere might be more anaerobic in nature, causing more denitrification. Another possibility may be that due to higher application rate, there may be enhanced production of overall plant biomass (root, stubble) increasing the organic matter content even of the hot spot microsites leading to more activity of microbes causing depletion of oxygen, anaerobic condition and ultimately more denitrification. To prove this, research is needed to know two components of denitrification process taking soil microsite under consideration: (a) diffusion mechanism of NO_{-3}-N to the hot spots, (b) monitoring the organic matter accumulation during the plant growth in the soil especially to the hot spots.

With the help of urease enzyme present in soil and on leaf surface, applied urea hydrolyzes to NH_{+4} and HCO_{-3} ions resulting in a localized increase in pH at the granule microsite which promotes NH_3 volatilization. Such loss can be reduced by using urease inhibitors to delay hydrolysis. N- (n-butyl thiophosphoric triamide) i.e., NBPT was found to be the most effective urease inhibitor. But it is unlikely to reduce urease activity on leaf surface. So there must be more research on other phosphoric triamide analogs which may be more effective, even on leaf surfaces. Studies should be continued to improve N use efficiency synchronizing N availability and crop requirement through integrated nutrient management involving organic and inorganic sources of N, appropriate cropping system and better methods of storage and application of organic fertilizer, etc. Work should be continued on nitrification inhibitor, slow release nitrogenous fertilizer by testing new and newer synthetic and natural indigenous material. High yielding varieties generally require large amounts of N promoting nitrate contamination of groundwater. Efforts should be sustained to select genetically superior N efficient cultivors using genetic engineering and molecular biology techniques. Incorporating appropriate Geographical Information Systems (GIS), N fertilizer management research should be site-specific for uniform soil area instead of whole of the field.

Efforts to reduce N leaching should also be continued. Future work in this line should derive more information on the effect of soil, vegetation and receiving area on the use of buffer strips. More attention is required to the role of soil fauna e.g., arthropods, protozoa, earthworms, on soil N behavior.

Studies should also include N pollution of potable water in respect of nitrate and bacterial (e.g., coliform) contents because infantile methemoglobinemia may cause enteritis even at negligible nitrate intake.

Research on treatment technologies to remove nitrates from drinking water may be continued to find better and cheaper ways.

Lastly, modeling research should also be directed to develop a dynamic model for predicting nitrate movement as affected by different factors.

Since the environment belongs to all of us, it is our common responsibility to reduce such anthropogenic environmental deterioration from the use of N fertilizer. Although political, economical, social and hygenic factors may come in the way, scientific temper should continue towards those positive directions indicated above.

REFERENCES

1. Addiscott, T.M. and D.S. Powlson. 1992. Partioning losses of nitrogen fertilizer between leaching and denitrification. *J. Agril. Sc. Camb. 118*:101-107.
2. Anonymous. 1972. National Academy of Sciences. Water quality criteria. EPA R 3-73-033. *National Academy Press, Washington, D.C.*
3. Anonymous. 1985. Health effects criteria document for nitrate/nitrite. *USEPA criteria and Standard Division, Office of Drinking Water, Washington, D.C.*

4. Anonymous. 1989. a. *Encyclopaedia Britannica, Inc. Chicago. 4.* 611.
5. Anonymous. 1989. b. *Encyclopaedia Britannica, Inc. Chicago. 1*:295.
6. Anonymous. 1989. c. Nonstop nitrogen tests. *Farm J. 113*(6):18.
7. Anonymous. 1989. d. *Encyclopaedia Britannica. 16*:505.
8. Aulakh, M.S., Doran, J.W. and A.R. Mosier. 1992. Soil denitrification, significance, measurement and effects of manuring. *Adv. Soil. Sc. 18*:1-42.
9. Cater, J.N. and F.E. Allison. 1961. The effect of rates of application of ammonium sulphate on gaseous losses of nitrogen from soils. *Soil Sc. Soc. Am. Proc. 25*:404.
10. Christensen, S.S., Simkins, S. and J.M. Tiedje. 1990. Spatial variation in denitrification: dependency of activity centres on the soil environment. *Soil Sc. Soc. Am. J. 54*:1608-1613.
11. Crutzen, P.J. 1981. Atmospheric chemical processes of the oxides of N including N_2O. In: C.C. Delwiche (Ed). *Denitrification and Atmospheric Nitrous Oxide Wiley, N.Y.* pp. 17-44.
12. Delany, A.C., Fitzjarrald, D.R., Lenschow, D.H., Pearson, R. Jr., Weindel, G.J. and B. Woodruff. 1986. Direct measurements of nitrogen oxides and ozone fluxes over grassland. *J. Atm. Chem. 4*:429-444.
13. Gundersen, P. and L. Rasmussel. 1993. Nitrification in forest soils: Effect from N deposition on soil acidification and aluminum release. *Rev. Env. Contamination and Toxicology 113*:1-46.
14. Jackson, J.E. and G.W. Burton. 1962. Influence of sod treatment and nitrogen placement on the utilization of urea nitrogen by Coastal Burmuda grass. *Agron. J. 54*:47.
15. Kolenbrander, G.J. 1982. Fertilizers and pollution. *Trans. 12th Internatl. Cong. Soil Sc.* pp. 248-266.
16. Logan, T.J., Randell, G.W. and D.R. Timmons. 1980. Nutrient content of tile drainage from cropland in the North central region. *Res. Bull. Ohio Agric. Res. Dev. Center, Wooster, OH. No. 1119.*
17. Lowranee, R.R., Todd, R.L. and L.E. Asmussen. 1983. Waterborne nutrient budgets for the riparian zone of an agricultural watershed. *Agric. Ecosystem and Environ. 10*:371-384.
18. Likens, G.E., Bormann, F.H., Johnson, N.M., Fisher, D.W. and R.S. Pierce. 1970. Effects of forest cuttings and herbicide treatment on nutrient budgets in the hubbard brook watershed ecosystem. *Ecol. Monograph. 40*(1): 23-47.
19. Linn, D.M. and J.W. Doran. 1984. Effect of water-filled pore space on carbon dioxide and nitrous oxide productions in tilled and non-tilled soils. *Soil Sc. Soc. Am. J. 48*:1267-1272.
20. Mathers, A.C., Stewart, B.A. and B. Blair. 1975. Nitrate-nitrogen removal from soil profiles by alfalfa. *J. Environ. Qual. 4*:403-405.
21. Meisinger, J.J. 1989. Evaluating plant-available nitrogen in soil-crop systems. In R.D. Hauck (Ed). Nitrogen in Crop Production pp. 391-441. *Amer. Soc. Agron., Madison, WI.*

22. Mothes, K. 1961. The metabolism of urea and ureides, *Can. J. Bot. 39*:1785-1807.
23. Mulder, J. 1988. *Impact of Acid Atmospheric Deposition on Soils: Field Monitoring and Aluminum Chemistry. Doctoral thesis. Agricultural University, Wageningen, The Netherlands.*
24. Olsen, R.A., Seim, E.C. and J. Muir. 1973. Influence of agricultural practices on water quality in Nebraska: A survey of stream, groundwaters and precipitation *Water Resources Bull. 9*:301-311.
25. Ohshima, H. and H. Bartsch. 1981. Quantitative estimation of endogenous nitrosation in humans by monitoring N-nitrosoproline excreted in the urine. *Cancer Res. 41*:3658-3662.
26. Papendick, R.I., Elliott, L.F. and J.F. Power. 1987. Alternative production systems to reduce nitrate in groundwater. *Amer. J. Alternative Agric. 2*: 19-28.
27. Parks, N.J., Krohn, K.A., Mathis, C.A., Chasko, J.H., Geiger, K.R., Gregor, M.E. and N.F. Peek. 1981. Nitrogen-13-labelled nitrate and nitrite: Distribution and metabolism after intratracheal administration. *Science 212*:58-61.
28. Powlson, D.S., Saffigna, P.G. and M.K. Cottaar. 1985. Denitrification at sub-optimal temperatures in soils from different climatic zones. *Soil Biol. Biochem. 20:*(5):719-723.
29. Rhoads, F.M. and R.L. Stanley, Jr. 1981. Fertilizer scheduling, yield and nutrient uptake of irrigated corn. *Agron. J. 73*:971-974.
30. Schepers, J.S. 1988. Role of cropping systems in environmental quality: Groundwater nitrogen. In W.L. Hargrove (Ed). Cropping Strategies for Efficient Use of Water and Nitrogen. pp. 167-178. *Am. Soc. Agron. Spec. Pub. No. 51.*
31. Shea, C.P. 1988. Protecting the life on the earth: Step to save ozone layer. *World Watch paper 87. World Watch Institute. Washington, D.C.*
32. Sherwood, M. and A. Fanning. 1985. Run-off of fertilizer. Soils and grassland production. *Res. Rep. An Foras Taluntais.* pp. 10-11.
33. Slemr, F. and W. Seiler. 1984. Field measurement of NO and NO_2 emissions from fertilized and unfertilized soils. *J. Atmospheric Chem. 2*:1-24.
34. Smith, S.J. and D.K. Cassel. 1990. Estimating nitrate leaching in soil materials. In Follett, R.F. (Ed.) *Managing Nitrogen for Ground Water Quality and Farm Profitability. Soil Sc. Soc. Amer. Pub. Madison, WI.*
35. Steenvoorden, J.H., and A.M. and H.P. Oosterom. 1972. N:Cl ratio in drainage and groundwater. *Cutuurtechnisch Tijdschrift 12*:231.
36. Vitousek, P.M., Gosz, J.R., Grier, C.C., Melillo, Jim and W.A. Reiner. 1982. A comparative analysis of potential nitrification and nitrate mobility in forest ecosystems. *Ecol. Monographs. 52*(2):155-177.
37. Watson, C.J., Stevens, R.J., Laughlin, R.J. and P. Poland. 1992. Volatilization of ammonia from solid and liquid urea surface-applied to perennial ryegrass. *J. Agril. Sc. Camb. 119*:223-226.

Environmental Contaminants, Ecosystems and Human Health. Edited by S.K. Majumdar, E.W. Miller, and Fred J. Brenner. © 1995, The Pennsylvania Academy of Science.

Chapter Six

EFFECTS OF POWERLINE CORRIDORS ON FOREST ECOSYSTEM INTEGRITY

CHARLES E. WILLIAMS
Department of Biology
Clarion University of Pennsylvania
Clarion, PA 16214-1232

INTRODUCTION

High-voltage electric transmission lines and their associated rights-of-way (hereafter referred to as powerlines and powerline corridors, respectively) are conspicuous and often controversial components of human-dominated landscapes. In the United States, construction of new high-voltage powerlines, particularly in urban or scenic areas, often receives heavy public opposition, based primarily on potential environmental health effects and (or) aesthetic concerns (22). For example, electromagnetic fields emanating from powerlines have been implicated as causal agents of some human cancers including brain cancer and childhood leukemia (5, 22, 56), audible noise generated by powerlines can be an annoyance (33), and improper use of herbicides in powerline corridor maintenance can threaten human health and cause unacceptable environmental damage (10). Likewise, powerline corridors can detract from scenic views and recreational experiences (28), and may result in decreased property values or revenues (22). Thus for a variety of anthropocentric reasons, powerlines and powerline corridors are generally perceived as negative entities.

Human values aside, what are the impacts of powerlines and powerline corridors on the natural ecosystems they traverse? To date, there is little evidence to suggest that local electromagnetic fields generated by high-voltage powerlines have any major negative effect on native plants or animals (32, 43 and references therein), although the current database is somewhat limited in ecological and taxonomic scope. In contrast, construction and maintenance of powerline corridors can have significant impacts on ecosystem structure and function by direct habitat destruction or by alteration of the biotic and (or) abiotic environments of natural systems through subdivision, fragmentation, and associated edge effects (3, 19, 21, 28).

In this short review, I focus on the effects of powerline corridors on forest ecosystems, emphasizing their potential impacts on the ecological integrity of forests in the eastern United States. Ecological or biological integrity [defined as "the capability of supporting and maintaining a balanced, integrated, adaptive community of organisms having a species composition and functional organization comparable to that of natural habitat of the region" (26, 27)] has recently become a major theme in ecosystem management (53). Ecological integrity has been suggested as an alternative to biological diversity [defined as the variety of organisms considered at all levels from variation at the genetic level to ecosystems and landscapes (52)] as an ecosystem management goal because it is a holistic concept which includes biological diversity and considers functional relationships within the system (4). Since powerline corridors can affect both the structure and function of natural systems (19, 21), ecological integrity is a desirable measure with which to gauge potential impacts to forest ecosystems.

POWERLINE CORRIDORS AS ELEMENTS OF HUMAN-DOMINATED LANDSCAPES

The network of powerline corridors in the United States is extensive, estimated at 500,000 miles of powerlines with 7,000,000 acres of managed powerline corridors (28). Thus in many regions, powerline corridors are major components of human-dominated landscapes with the potential to affect ecological patterns and processes at various scales. This necessitates a hierarchical approach to assessing the impacts of powerline corridors on forest ecosystems at scales ranging from forested landscapes to forest stands (48).

At a landscape level, corridors are considered to be linear patches, composed of discrete communities or species assemblages, embedded in a matrix, an expanse of dissimilar habitat that surrounds the patch (16, 17). Forman and Godron (16, 17) classified landscape corridors based on: 1) height above the surrounding matrix: convex corridors have a height greater than the adjacent matrix; concave corridors are lower in height than the surrounding matrix; and 2) width and habitat: line corridors are narrow and consist primarily of edge habitat supporting edge species [species whose daily activities are concentrated in edges (54)]; strip corridors contain edge habitat but are wide enough to contain an interior habitat supporting interior species [species whose daily activities are concentrated away from edges (54)]. Thus, powerline corridors are examples of concave line or strip corridors depending on their width and the habitat they contain.

Fluxes of energy, nutrients and species link corridors with the surrounding landscape matrix (14, 15). Theoretically, powerline corridors could have four main functions in a forested landscape (15): 1) as specialized habitat for certain species; 2) as travel lanes that enhance species movement; 3) as barriers to the movement of species, energy or nutrients; and 4) as a source of biotic and abiotic effects on the matrix. The degree to which a powerline corridor provides specialized habitat, enhances or restricts movement of species or materials, or influences the biotic or abiotic environment of the matrix depends on the relative sizes of the corridor and the matrix and the habitat contrast (structural or physiognomic contrast) between them (16, 19).

IMPACTS OF POWERLINE CORRIDORS ON FOREST ECOSYSTEMS

Powerline corridor impacts on forest ecosystems can be segregated into two broad classes: 1) impacts that occur during the construction of corridors; and 2) impacts associated with the maintenance of established corridors.

IMPACTS OF POWERLINE CONSTRUCTION

Construction of powerline corridors in forested regions usually occurs in three stages (18): 1) all woody vegetation with the corridor is shearcleared to ground level; 2) transmission towers are erected and powerlines strung; and 3) a seedbed is prepared within the corridor and planted with desirable vegetation (usually perennial grasses) to control erosion and speed site recovery. Heavy machinery is used in virtually all stages of powerline corridor construction, especially in placement of transmission towers (18).

Initial powerline corridor clearing during construction can induce stress, mortality and ultimately, structural and compositional changes in forest vegetation adjacent to the cleared corridor. Stress in forest plants may be due to single or multiple alterations in the abiotic environment at the forest-powerline corridor edge induced by clearing, particularly alterations in light, wind, soil and air temperature, and soil moisture (34, 41, 55, Table 1) or by direct mechanical damage caused by heavy machinery. The general biotic response to powerline corridor clearing by vegetation at the forest edge would be a shift in species composition from shade-tolerant, mesophytic species to shade-intolerant, xerophytic species (44, Table 1). The magnitude of this shift should be a function of powerline corridor size: the larger the corridor, the greater its impact on the abiotic environment at the forest edge and the greater the compositional change in the vegetation. Thus, wider strip powerline corridors should have greater impact on the forest matrix than narrow line corridors depending on the size of the forest habitat that is traversed by the corridor (14, 15, 19).

Concentration of heavy equipment activity in the vicinity of transmission towers during powerline construction has a major impact on rates of vegetation recovery and soil erosion. Studies have generally noted a sparse ground cover and highly variable rates of vegetation recovery near transmission towers, presumably the result of soil compaction and (or) removal of the upper soil horizons, soil seed banks and vegetative propagules by heavy equipment (18, 31). Erosion rates at these sparsely vegetated sites can be high (18) and the resultant rill or gully erosion can cause local loss of soil in adjacent forest, decreasing soil fertility and habitat quality, particularly for mesophytic forest plants. Alternately, soil lost from the powerline corridor can accumulate in the adjacent forest and stimulate productivity, bury existing vegetation, or provide open substrate for the establishment of new or existing plant species. Finally, sparsely vegetated sites in newly constructed powerline corridors can also become foci for alien invasive plants (e.g., the tree-of-heaven, *Ailanthus altissima*) which generally need open, scarified sites for seed germination and seedling establishment (50, 51).

IMPACTS OF ESTABLISHED POWERLINES

Management Activities and Impacts

In forested regions, vegetation within powerline corridors must be managed periodically to maintain a composition and stature that does not interfere with overhead transmission lines or obstruct servicing by line maintenance crews. Development of a stable, treeless plant community is the management goal in powerline corridors; thus control of woody vegetation, particularly trees, is the main management activity. Control of woody vegetation within powerline corridors has been attempted in many ways including: 1) by blanket spraying of herbicides (35); 2) by frequent cutting (36); 3) by use of selective herbicides to develop an invasion-resistant community of herbs or shrubs (8, 9, 13, 25, 39); and 4) by prescribed burning (40). Each of these control methods may have different impacts on the integrity of adjacent forest ecosystems.

TABLE 1
Some potential effects of powerline clearing on adjacent forest vegetation by alteration of the abiotic environment

Environmental alteration	Biotic response	Effects on integrity
1. increased light	photoinhibition; damage to photosynthetic apparatus in forest plants	shift to shade-intolerant species; flush of growth of vegetation at the forest edge
2. decreased soil moisture 3. increased wind 4. increased temperatures	stomatal closure and shut down of photosynthesis; dehydration	increased mortality of mesic species; shift to xeric "weedy" species
5. increased wind exposure	increased treefalls on windward exposure; dehydration; stomal closure	loss of shallow-rooted tree species

Blanket spraying of herbicides is still used as a management technique in some areas of the United States (35). This vegetation management method may have the greatest impact on adjacent forest vegetation than any other technique, particularly if broad spectrum herbicides are used and herbicide drift is not properly controlled (10, 35). Forest edge and interior plant species can be damaged or destroyed by drifting or misapplied herbicide (10). Also, blanket spraying is not a particularly effective method for controlling woody vegetation in powerline corridors, and in many instances post-treatment woody stem densities exceed pretreatment values, perhaps due in part to herbicide resistance (24, 35).

Frequent cutting of woody vegetation in powerline corridors produces a community composed of tree sprouts dominated by a few prolific sprouting species (35, 36, 37). Frequent cutting is a preferred management technique in ecologically sensitive areas where herbicides cannot be used due to close proximity to special concern species or water supplies (36, 37). However, soil scarification, debris accumulations, and other disturbances caused by cutting machinery can provide microsites for establishment of alien invasive plant species which may become management problems in the powerline corridor and (or) the surrounding forest matrix.

Development of self-maintaining, invasion-resistant plant communities in powerline corridors may be the vegetation management technique most compatible with sustaining integrity in adjacent forest ecosystems. Trees in the powerline corridor are initially controlled by basal spraying of selective herbicides. The shrub community is left intact to spread through the corridor with occasional follow-up basal spraying of trees (8). Dense, clonal growths of shrubs such as blueberries (*Vaccinium* spp.) effectively restrict tree invasion by limiting the availability of tree seedling microsites, by intense competition for light, nutrients and moisture, or by allelopathy (8, 13, 39, 47). Selective basal spraying of herbicides eliminates the drift problems in adjacent forest often caused by blanket spraying, and the relative stability of established, invasion-resistant shrub communities minimizes the need for repeated management and potential disturbance by maintenance crews.

Prescribed burning is infrequently used as a powerline management technique probably because most maintenance crews generally have little experience with fire use and management (40). Prescribed burning may be a useful vegetation management tool in areas dominated by fire-dependent vegetation. Due to low fuel loads, prescribed burned powerline corridors can function as fire breaks, facilitating fire management and control in the larger forest matrix (15). Likewise, prescribed burned powerline corridors can be managed to preserve fire-dependent communities or species that are rare or absent in the adjacent forest matrix due to fire suppression.

Powerline Corridors as Movement Barriers, Travel Lanes or Filters

The degree to which a powerline corridor fragments or isolates adjacent forest depends on several factors including the size of the forest matrix or patch traversed by the corridor, corridor width, the habitat contrast between the corridor and the forest matrix (i.e., edge permeability; this increases as the contrast between corridor and matrix decreases; 16), disturbance within the corridor, and the size and vagility of organisms that exist within the forest matrix (15, 19). Small populations isolated by corridors may be susceptible to further declines due to inbreeding depression, random genetic drift and subsequent loss of genetic variability, demographic stochasticity, or catastrophic events.

Most bird species traverse powerline corridors when necessary, and their movements do not appear to be adversely affected by powerline corridors (2), except for collisions with high tension wires (7). Small forest mammals, however, often avoid crossing open habitat (49) and may suffer isolation effects due to powerline corridors. Schreiber and Graves (42) found that white-footed mice (*Peromyscus leucopus*) and short-tailed shrews (*Blarina brevicauda*) easily crossed powerline corridors of more than 100 m in width, indicating that powerline corridors may have no significant effect on movement in these species. Gates (19) observed that composition and structure of vegetation at the corridor edge were important factors barring movement of small- and medium-sized mammals across powerline corridors; in particular, dense grass swards at the edge of one corridor formed a barrier to movement for some species. However, Johnson et al. (23) cautioned that most studies of animal activity in regard to powerline corridors have been conducted at the stand level, and regional studies at the landscape level are needed to better assess the overall impacts of powerline corridors on animal movement and abundance.

Effects of powerline corridors on movements of invertebrates are poorly known. For example, populations of microhabitat-specific species like land snails, which generally require moist habitats, may be isolated by inhospitable, xeric corridors (6). Some forest butterflies, like the West Virginia white (*Pieris virginiensis*), will not cross open habitats and its current rarity may be a function of habitat fragmentation and isolation (11). Although not documented, isolation of insect pollinators and other key invertebrate symbionts (e.g., seed predators, seed dispersers, herbivores) by powerline corridors may also have demographic consequences for some forest plants. Research on potential powerline corridor effects on the distribution and abundance of forest invertebrates and their impact on ecosystem function is sorely needed.

The use of powerline corridors as travel lanes by animals has been suggested but little documented (15). Gates (19) noted that some medium-sized mammals, which could have easily crossed powerline corridors, moved preferentially along the corridor-forest edge. Edges are generally considered to be high productivity habitats and it is possible that some animals, particularly medium-sized mammalian predators, concentrate at edges not for ease of travel but because of greater prey densities (12, 20).

Forman (15) suggested that landscape corridors could function as semipermeable barriers or filters, restricting the movement or flow of energy, nutrients or species from one area to another. Gates (19) observed that some animals responded to powerline corridors as an "activity filter" which acted to decrease activity with distance to the corridor. Animals were generally most active in adjacent habitat and least active in the corridor.

Powerline Corridors as Specialized Habitat or Species Sources

The extent to which powerline corridors provide specialized habitat for certain species is a function of corridor width, internal habitat heterogeneity, and (or) the uniqueness of relatively open corridor habitat within a forested landscape (15).

Habitat relationships of bird communities in powerline corridors and adjacent forest have been extensively studied. These studies have generally shown that: 1) bird species diversity and (or) density in corridors increases with increased corridor width and habitat heterogeneity (2, 29, 30); 2) diversity of bird communities in forests is higher than that in adjacent powerline corridors of any width (2); and 3) the degree of compositional similarity between bird communities in corridors and forest is low (2). Many bird species living in powerline corridors are primarily associated with early successional or edge habitats. Thus, powerline corridors can provide specialized, early successional habitats required by some species.

The role of powerline corridors as species sources in forested landscapes has been little studied. Powerline corridors could serve as a source of early successional plant species for colonization of disturbed habitats in the forest matrix. Likewise, powerline corridors could also serve as "invasion foci" for alien plants. Anderson et al. (2) found that grassy strip powerline corridors $>$ 30 m in width supported grassland bird species that were absent from adjacent line corridors and forest. Theoretically, if grassland was an uncommon habitat in a forested region, then a grassy powerline corridor could serve as a source of birds for colonization of new grasslands that arise due to disturbance or habitat conversion. However, this new source of grassland birds could come at a cost. Loss and fragmentation of forested habitat due to powerline construction could cause an overall decrease in forest bird species in a region, particularly among neotropical migrants (1, 46). In general, if powerline corridors can function as species sources, they will primarily supply early successional species for disturbed patches within the landscape matrix.

CONCLUSIONS

Despite the voluminous literature on powerline corridors, we know comparatively little about their impacts on forest ecosystem integrity. The bulk of studies has focused on biotic responses to powerline construction and maintenance, and most have centered solely on birds and (or) mammals from a species richness or diversity standpoint. While birds and mammals are among the most conspicuous components of the forest biota, their contribution to ecosystem function may be relatively small compared to other organisms (52). What are the potential effects of powerline corridors on the forest invertebrates, fungi and bacteria that play vital roles in energy transformations and nutrient cycles in the ecosystem? Can powerline corridors impact nutrient cycling or hydrology of adjacent forests? In the absence of detailed studies that address powerline impacts to ecosystem structure *and* function, I can say "we don't know".

Powerline corridors, like highways and agricultural fields, can contribute to the large-scale problem of forest fragmentation and subsequent ecosystem decay (3). Thus, it is imperative that siting of powerline corridors considers not only potential impacts to special concern species (e.g., 45) but also impacts to ecosystem integrity. Detailed ecosystem monitoring programs, including pre- and post-construction assessment studies in selected forest types (38), must be implemented to ascertain the impacts of powerline corridors on forest ecosystem integrity.

LITERATURE CITED

1. Anderson, S.H. 1979. Changes in forest bird species composition by transmission-line corridor cuts. *American Birds* 33:3-6.
2. Anderson, S.H., Mann, K. and H.H. Shugart, Jr. 1977. The effect of transmission line corridors on bird populations. *Am. Midl. Nat.* 97:216-221.
3. Andrews, A. 1990. Fragmentation of habitat by roads and utility corridors: a review. *Australian Zoologist* 26(3/4):130-141.
4. Angermeier, P.L. 1994. Does biodiversity include artificial diversity? *Conservation Biology* 8:600-602.
5. Anonymous. 1994. Electromagnetic fields. *Consumer Reports* 59(5):354-359.
6. Baur, A. and B. Baur. 1990. Are roads barriers to dispersal in the land snail *Arianta arbustorum*? *Can. J. Zool.* 68:613-617.
7. Bevanger, K. 1990. Topographic aspects of transmission wire collision hazards to game birds in the central Norwegian coniferous forest. *Fauna Norv. Ser. C., Cinclus* 13:11-18.
8. Bramble, W.C. and W.R. Byrnes. 1983. Thirty years of research on development of plant cover on an electric transmission right-of-way. *J. Arboric.* 9:67-71.
9. Bramble, W.C., Byrnes, W.R. and R.J. Hutnik. 1990. Resistance of plant cover types to tree seedling invasion on an electric transmission right-of-way. *J. Arboric.* 16:130-135.
10. Buffington, J.D. 1974. Assessment of the ecological consequences of herbicide use along transmission line rights-of-way and recommendation for such use. Argonne National Laboratory, Environmental and Earth Sciences Division. Report ANL/ES-34. Argonne, IL, pp. 44.
11. Cappuccino, N. and P. Kareiva. 1985. Coping with a capricious environment: a population study of rare pierid butterfly. *Ecology* 66:152-161.
12. Chasko, G.G. and J.E. Gates. 1982. Avian habitat suitability along a

transmission-line corridor in an oak-hickory forest region. *Wildl. Monographs* 82:1-41.
13. Dreyer, G.D. and W.A. Niering. 1986. Evaluation of two herbicide techniques on electric transmission rights-of-way: development of stable shrublands. *Environ. Manage.* 10:113-118.
14. Forman, R.T.T. 1981. Interactions among landscape elements: a core of landscape ecology, pp. 35-48. In: *Perspectives in landscape ecology*. Proc. Intl. Congr. Nethl. Soc. Landscape Ecol. Pudoc, Wageningen, The Netherlands.
15. Forman, R.T.T. 1983. Corridors in a landscape: their ecological structure and function. *Ekologia* (CSSR) 2:375-387.
16. Forman, R.T.T. and M. Godron. 1981. Patches and structural components for a landscape ecology. *BioScience* 31:733-740.
17. Forman, R.T.T. and M. Godron. 1986. *Landscape ecology.* John Wiley and Sons. New York, pp. 619.
18. Fowler, D.K., Turner, L.J. and L.F. Adkisson. 1979. Utilization of volunteer vegetation in ground cover restoration operations on new transmission line rights-of-way, pp. 27-1 - 27-22. In: *Proc. second symposium on environmental concerns in rights-of-way management.* University of Michigan. Ann Arbor, MI.
19. Gates, J.E. 1991. Powerline corridors, edge effects and wildlife in forested landscapes of the central Appalachians, pp. 13-32. In: J.E. Rodiek and E.G. Bolen (Ed.), *Wildlife and habitats in managed landscapes.* Island Press. Washington, DC, pp. 219.
20. Gates, J.E. and L.W. Gysel. 1978. Avian nest dispersion and fledgling success in field-forest ecotones. *Ecology* 59:871-883.
21. Herrington, L.P. and G.M. Heisler. 1973. Microclimate modification due to power transmission rights of way, pp. 36-57. In: R. Goodland (Ed.), *Powerlines and the environment.* Cary Arboretum, New York Botanical Garden. Millbrook, NY, pp. 170.
22. Hester, G.L. 1992. Electric and magnetic fields: managing an uncertain risk. *Environment* 34(1):6-11, 25-32.
23. Johnson, W.C., Schreiber, R.K. and R.L. Burgess. 1979. Diversity of small mammals in a powerline right-of-way and adjacent forest in east Tennessee. *Am. Midl. Nat.* 101:231-235.
24. Johnston, P.A. 1973. Ecological effects of herbicide sprayings in shaping plant communities on transmission line rights-of-way. Thesis. West Virginia University. Morgantown, WV, pp. 196.
25. Johnstone, R.A. 1990. Vegetation management: mowing to spraying. *J. Arboric.* 16:186-189.
26. Karr, J.R. 1990. Biological integrity and the goal of environmental legislation: lessons for conservation biology. *Conservation Biology* 4:244-250.
27. Karr, J.R. and D.R. Dudley. 1981. Ecological perspective on water quality goals. *Environ. Manage.* 5:55-68.

28. Kitchings, J.T., Shugart, H.H. and J.D. Story. 1974. Environmental impacts associated with electric transmission lines. Oak Ridge National Laboratory, Environmental Sciences Division. Contract Report No. W-7405-ENG-26, pp. 100.
29. Kroodsma, R.L. 1982. Bird community ecology on power-line corridors in east Tennessee. *Biol. Conserv.* 23:79-94.
30. Kroodsma, R.L. 1984. Ecological factors associated with degree of edge effect in breeding birds. *J. Wildl. Manage.* 48:418-425.
31. Lathrop, E.W. and E.F. Archibold. 1980. Plant response to utility right of way construction in the Mojave Desert. *Environ. Manage.* 4:215-226.
32. Lee, J.M., Jr., Burns, A., Lee, G.E., Reiner, G.L., Shon, F.L. and R. Stearns. 1985. Electrical and biological effects of transmission lines: a review. U.S. Dept. of Energy, Bonneville Power Administration. Portland, OR, pp. 84.
33. Lee, J.M., Jr. and D.G. Griffith. 1978. Transmission line audible noise and wildlife, pp. 105-168. In: J.L. Fletcher and R.G. Busnel (Ed.), *Effects of noise on wildlife.* Academic Press. New York, pp. 305.
34. Lovejoy, T.E., Bierregaard, B.O. Jr., Rylands, A.B., Malcolm, J.R., Quintela, C.E., Harper, L.H., Brown, K.S. Jr., Powell, A.H., Powell, G.V.N., Schubart, H.O.R. and M.B. Hays. 1986. Edge and other effects of isolation on Amazon forest fragments, pp. 257-285. In: M.E. Soule (Ed.), *Conservation biology: the science of scarcity and diversity.* Sinauer Inc. Sunderland, MA, pp. 584.
36. Luken, J.O. 1990. *Directing ecological succession.* Chapman and Hall. New York, pp. 251.
37. Luken, J.O., Hinton, A.C. and D.G. Baker. 1991. Assessment of frequent cutting as a plant community management technique in power-line corridors. *Environ. Manage.* 15:381-388.
37. Luken, J.O., Hinton, A.C. and D.G. Baker. 1992. Response of woody plant communities in power-line corridors to frequent anthropogenic disturbance. *Ecol. Appl.* 1:356-362.
38. Niemi, G.J. and J.M. Hanowski. 1984. Effects of a transmission line on bird populations in the Red Lake peatland, northern Minnesota. *The Auk* 101:487-498.
39. Niering, W.A. and R.H. Goodwin. 1974. Creation of relatively stable shrublands with herbicides: arresting "succession" on rights-of-way and pastureland. *Ecology* 55:784-795.
40. Olson, D.P., Alexander, L. and S. Macrigeanis. 1984. Use of prescribed burning for managing rights-of-way in central New England-preliminary results, pp. 428-445. In: A.F. Crabtree (Ed.), *Proceedings of the third international symposium on environmental concerns in rights-of-way management.* Mississippi State University. Starkville, MS, pp. 689.
41. Osmond, C.B., Austin, M.P., Berry, J.A., Billings, W.D., Boyer, J.S., Dacey, J.W.H., Nobel, P.S., Smith, S.D. and W.E. Winner. 1987. Stress physiology and the distribution of plants. *BioScience* 37:38-48.

42. Schreiber, R.K. and J.H. Graves. 1977. Powerline corridors as possible barriers to the movements of small mammals. *Am. Midl. Nat.* 97:504-508.
43. Shrader, V.L., Jr. 1978. The effects of high voltage transmission lines on some aspects of plant growth and development. M.S. thesis. California State Polytechnic University. Pomona, CA, pp. 125.
44. Stalter, R. 1972. The revegetation of a Rhode Island right-of-way, Exeter, Rhode Island. *Castanea* 37:77-83.
45. St. Clair, G.T. and B.S. McFarlane. 1979. A transmission corridor planning approach to threatened and endangered species, pp. 47-1 - 47-7. In: *Proc. second symposium on environmental concerns in rights-of-way management.* University of Michigan. Ann Arbor, MI.
46. Terborgh, J. 1990. *Where have all the birds gone?* Princeton University Press. Princeton, NJ, pp. 202.
47. Tillman, R.E. 1984. Potential role of allelopathy in ROW vegetation management, pp. 416-420. In: A.F. Crabtree (Ed.), *Proceedings of the third international symposium on environmental concerns in rights-of-way management.* Mississippi State University. Starkville, MS, pp. 689.
48. Urban, D.L., O'Neill, R.V. and H.H. Shugart, Jr. 1987. Landscape ecology. *BioScience* 37:119-127.
49. Wegner, J.F. and G. Merriam. 1979. Movements by birds and small mammals between a wood and adjoining farmland habitats. *J. Appl. Ecol.* 16:349-357.
50. Williams, C.E. 1993a. Alien invasion. *Pennsylvania Wildlife* 14(5):16-19.
51. Williams, C.E. 1993b. The exotic empress tree, *Paulownia tomentosa:* an invasive pest of forests? *Nat. Areas J.* 13:221-222.
52. Wilson, E.O. 1992. *The diversity of life.* W.W. Norton and Co. New York, pp. 424.
53. Woodley, S., Kay, J. and G. Francis (Ed.). 1993. *Ecological integrity and the management of ecosystems.* St. Lucie Press. Ottawa, Canada, pp. 224.
54. Yahner, R.H. 1988. Changes in wildlife communities near edges. *Conservation Biology* 2:333-339.
55. Young, H.E. 1984. Management of forest stands on highway rights-of-way, pp. 276-278. In: A.F. Crabtree (Ed.), *Proceedings of the third international symposium on environmental concerns in rights-of-way management.* Mississippi State University. Starkville, MS, pp. 689.
56. Young, L.B. 1973. *Power over people.* Oxford University Press. New York, pp. 216.

Environmental Contaminants, Ecosystems and Human Health. Edited by S.K. Majumdar, E.W. Miller, and Fred J. Brenner. © 1995, The Pennsylvania Academy of Science.

Chapter Seven

IMPACT OF NONPOINT SOURCE CONTAMINANTS ON ECOSYSTEMS AND HUMAN HEALTH

FRED J. BRENNER, STEPHANIE D. YODER and REBECCA J. BLAIR

Biology Department
Grove City College
100 Campus Drive
Grove City, PA 16127

INTRODUCTION

Over the last two decades, society has become increasingly concerned with the release of potentially toxic compounds and other pollutants into aquatic and terrestrial ecosystems. Although many of these contaminants are released from point sources, once they are dispersed into the environment, it often becomes difficult to locate the point of origin and, therefore, take corrective action. Thus, after dispersal, these contaminants are often intermixed with those from nonpoint sources and, either individually or collectively, adversely impact ecosystem function and human health. Although the adverse impacts of PCB'S, DDT, heavy metals and many other contaminants on ecosystem function and human health is well documented in the literature, the effects of other industrial and agricultural chemicals on ecosystems and human health is speculative and based largely on epidemiology rather than confirmed by laboratory studies.

In watersheds with a low potential for pollution influx, organic and inorganic chemical wastes are often degraded by microbes through natural self-purification (LaRiviere 1989). However, when the self-purifying capacity of the watershed is exceeded, large quantities of these contaminants accumulate in freshwater and marine environments, thereby degrading aquatic life (LaRiviere 1989). Likewise, Peirls et al. (1991) reported that on a global scale there is a strong relationship between the size of the human population within a drainage basin and nitrate concentrations in rivers discharging into coastal ecosystems. Excess nutrient and bacteriological discharges into waterways by human and agricultural sources not only represents potential adverse impacts on aquatic ecosystems but may also be a source of water borne diseases. Organic waste is biodegradable but if the organic loading exceeds the biological capacity of the stream or lake to decompose these materials, then water quality becomes degraded resulting in a decrease in biological diversity. In addition, human and animal wastes may contain pathogenic microbes which are the waterborne agents of cholera, typhoid fever, dysentery and other diseases (LaRiviere 1989). In developing countries with growing populations and nonexistent waste treatment programs, water pollution by organic wastes is a major concern affecting millions of people, especially children, who die annually from water-related diseases. Unlike organic wastes, synthetic chemicals and heavy metals from industrial and agricultural sources are not readily degraded by biological processes, and, therefore, may have a long term impact on aquatic and terrestrial ecosystems and human health.

INDUSTRIAL AND URBAN IMPACTS

Toxic chemical discharges and waste dumps have become a major source of groundwater and surface water contamination throughout the world (LaRiviere 1989). Metal drums containing toxic chemicals are essentially "time bombs" and incidents such as Lekkerkerk in the Netherlands and Love Canal in the United States are examples of the worldwide potential for this type of pollution. In addition, the ocean dumping of human and industrial wastes is a major source of contamination of marine ecosystems around the world. Marine pollution emanates from multiple sources and, hence, is often difficult if not impossible to separate the individual contaminants, let alone trace them to there point of origin. In the United States alone, approximately one third of the shellfish-producing areas have been closed to commercial harvest because of actual or potential contamination of coastal ecosystems (Anderson and Leal 1991). In addition, major fish kills have occurred in the Gulf of Mexico and along the southern Atlantic Coast as a result of low oxygen (hypoxia) caused, in part, by wastes from municipal treatment plants (Anderson and Leal 1991). Pollutants in marine environments also pose hazards to human health. In 1982, the state of New York reported 103 different outbreaks of viral gastroenteritis involving over 1000 people, with all of the cases related to the consumption of contaminated shellfish (U.S. Congress, Office of Technology Assessment 1987). Likewise the release of toxic chemicals, radioactive materials and other nondegradable materials is an international concern for the health of marine ecosystems and human populations throughout the world.

In coastal areas adjacent to urban centers, excessive sewage discharges have been shown to be directly related to eutrophication and changes in the composition of plankton communities (LaRiviere 1989). In England, female sex steroids (or their analogs) have been isolated from sewage discharges in concentrations that could potentially alter the reproductive physiology of fish. However, at this time it has not been determined if the reproduction of wild fish populations have been adversely affected. Likewise, eels in the Seine also appear to be adversely impacted by sex steroids in sewage discharges. In the Florida Everglades, synthetic estrogen compounds from nonpoint sources are also suspected to cause sterility in the Florida panther (*Felis concolor*) (David P. Flemming, Personal Communication).

Abyssal plains around the world have become dumping grounds for human waste thereby affecting the function of deep water marine ecosystems. In the United States the dumping of sewage sludge was to cease in 1988; however, the bioaccumulation of contaminants continues along with changes in the composition of fish communities primarily by the increase in scavenger species (Holloway 1992). Rodger et al. (1992) reported the Firth of Cloye has received sewage sludge at a rate of 10^6 metric tons/year since 1974 resulting in changes in the benthic community along with changes in the chemical characteristics of the sediment. In the Izmir Bay of the Aegean Sea, annual red-tide blooms has been shown to be related to the sewage discharges which alter the composition plankton communities by either forming septic conditions or by the addition of more tolerant and noxious competitor species (Koray et al. 1992).

As indicated previously, the discharge of human waste not only affects the structure and function of aquatic ecosystems, but may impact human health as well. Individuals using these environments for recreation may be exposed to a variety of pathogenic organisms. Both swimmers using coastal beaches and canoeists using fresh water rivers are exposed to human sewage but, according to Fewtrell (1992), there was a higher prevalence of illness after canoeing in fresh water than there was among swimmers in marine systems. This may be due in part to higher enterovirus concentrations in fresh water ecosystems as compared to saltwater (Alexander et al. 1992) and to the method of the sewage treatment which produces a greater attenuation of bacteria than viral concentrations (Fewtrell 1992). Another area of concern is that in both point and nonpoint sewage discharges, bacteria are present in high enough concentrations and in close proximity to one another to affect the transfer and retransfer of R plasmids and other extrachromosomal genetic elements including chimeric plasmids. Heavy metals, which exert a selective pressure for antibiotic resistance, tend to concentrate in sewage sludges, thereby providing an environment selective of these bacteria. The ability of bacteria to transfer resistance should be a major public health concern, since the transfer pathogenity from one species or strain to another is well documented (Grabow et al. 1973). In Sweden for example, polioviruses were isolated from sewage sludges five years after the use of killed poliovirus inoculations were terminated (Bottinger and Herstrom 1992).

The influx of sewage discharges into freshwater ecosystems not only degrade these ecosystems but also represent a major public problem, especially in developing countries. Lake Chivero is the major source of drinking water for the city of Harare, Zimbabwe. The discharge of sewage into this ecosystem has resulted in the development of eutrophic conditions due to cyanobacteria blooms which have been linked to outbreaks of gastroenteritis among children. A two month epidemic of NANB-hepatitis in South Delhi was linked to sewage contamination of the drinking water supply. NANB was especially prevalent among the 11-20 year old age group. In Alexandria, Egypt, Hammouda (1992) reported that sewage workers had a higher risk of developing intestinal parasitic infections than the general population. However, sewage contamination of drinking water is not limited to developing countries. For example, 1400 persons developed Giardiasis when a sewage overflow contaminated the water supply at a Swedish ski resort (Ljungstrom and Castor 1992).

In addition to the contamination of fresh water and marine ecosystems, non-point source pollutants are major contributors to the degradation of groundwater. In the United States and Europe alone, LaRiviere (1989) indicated that between 5 and 10 percent of the wells had nitrate levels in excess of 45 mg/l. However, the European water industry is arguing that the current nitrate level of 50 mg/l does not take into account the virtual elimination of nitrate when public water supplies are treated (Dickson 1993). LaRiviere (1989) stated that organic pollutants enter ground water as seepage from waste dumps, leakage from sewage and fuel tanks, runoff from agricultural lands or from paved surfaces in our ever proliferating urban areas.

In the United States alone, there are over 40 million private wells which constitutes approximately 95 percent of the water supplies for rural communities and, according to McAllister (1990), these wells are often contaminated with varying amounts of nitrites, nitrates, and pesticides. These contaminants may originate from a variety of sources including spills around well heads, back-siphoning during tank filling, irrigation systems or the improper storage and disposal of pesticides (McAllister 1990). In general, there is little if any monitoring of these water supplies nor are they subjected to state or federal water standards. However, groundwater contamination is not limited to rural areas, Goyer (1984) estimated that 50 percent of the United States population receive their drinking water from groundwater wells and many of these wells may contain an excess of 700 specific chemicals. These chemicals are known to include pesticides, organic solvents, and numerous hallogenated compounds, many of which are known carcinogens or have other toxicological effects. For example, Trichlorethylene (TCE) has been found in concentrations of 300 ppb in 13 percent of the wells in Nassau County, NY.

In the Great Lake Region of North America, Park (1984) reported that chemical contaminants in water supplies are responsible for insidious neurological damage, particularly in children, and infertility among adults. The presence of PCB's, organochlorine insecticides, such as dieldrin, and DDT have been isolated from both surface and groundwater in the Great Lake Regions (Park 1991). Children exposed to PCB's in drinking water have been shown to have a limited attention span and deficits in short term memory (Park 1991). According to Raloff (1986), between 200 and 1000 individuals may die annually in the United States from ingesting chemicals such as chloroform and TCE in drinking water. Likewise, Barton (1985) reported that miscarriages and birth defects occurred at two or three times the national average in areas where drinking water was contaminated by toxic chemicals. The California Department of Health reported that in the Silicon Valley, miscarriages and birth defects were 2.4 and three times higher than they were in nearby, uncontaminated communities (Barton 1985). Regardless of whether people use surface or groundwater as their water supply, there exists a potential for a variety of biological and chemical contaminants that pose a threat to the health and welfare of human populations in developed and undeveloped countries alike.

AGRICULTURAL IMPACTS

Over the last decade, it has been documented that agriculture is a major contributor to the degradation of surface and ground water systems in the United States, as well as elsewhere in the world (Brenner et al. 1990, 1991; Clark et al. 1985; Diebel et al. 1992; EPA 1983, 1987; Hill 1987; Macharis 1985; OEDA 1985; Reay et al. 1992; Schaller and Bailey 1985 and Tim et al. 1992). Clark et al. (1985) estimated that nonpoint pollutants (NPS), primarily from agriculture, account for 73 percent of the Biological Chemical Oxygen Demand (BOD), 83 percent of the bacterial loads, and 92 percent of the suspended sediments in the waterways of the United States. Moreover, EPA (1987) estimated that 57 percent of the lake area (km^2), 64 percent of the river miles, and 19 percent of the estuarine areas (km^2) are adversely impacted by discharges from agricultural lands. In Wisconsin alone, 40 percent of the lakes are affected by nonpoint source pollution (Odgers 1991). Macharis (1985) indicated that croplands generate a large portion of the nutrient load in the Chesapeake Bay drainage basin and agriculture is by far the major contributor to non-point pollution throughout the watershed. Hill (1987) reported that nitrogen concentrations in the runoff from agricultural lands in North Carolina were four to five times greater than they were from a forested watershed. However, nutrient and bacterial loading is not entirely a result of agriculture activity but, as indicated in the previous discussion, urban runoff and drainage from on lot sewage systems are also major contributor to the degradation of aquatic ecosystems.

Section 208 of the Federal Water Pollution Control Act (PL 92-500), as amended by the United States Congress in the Clear Water Act of 1977, recognizes that nonpoint sources are major contributors to sediment, bacteria and nutrient loads of fresh water ecosystems. In addition, these nonpoint sources from both agricultural and urban areas often contain a variety of other pollutants that adversely impact aquatic and terrestrial ecosystems as well as human health. Regardless of its origin, nonpoint source pollution is a result of human activity, especially improper land use. Griffin (1991) estimated that agriculture affects 50-70 percent of the surface and groundwater degradation and therefore, it requires that greatest attention.

TABLE 1.
Number of Animal Units, production and nutrient value of manure available for land application in the Shenango River Watershed.

Livestock Information					Manure Production			Nutrient Value of Manure						
Type	Ave. Wt. (Kg)	Farms (No.)	Animals (No.)	An. Units (No.)	Rate Kg/A.U.	Daily Tons/Day	Annual Tons/Yr	Kg/Ton			Tons/Year*			%
								N	P	K	N	P	K	N-P-K
Dairy	595	645	11,007	14,430	37.2	537.2	196,080.4	5	2	4	980.4	392.2	784.3	0.5-0.2-0.4
Heifers	318		10,703	7,492	37.2	278.9	101,804.3	5	2	4	509.0	203.6	407.2	0.5-0.2-0.4
Beef	454	190	9,762	9,762	27.2	265.9	97,059.7	6	4	5	533.8	339.7	485.3	0.55-0.35-0.5
Swine	68	31	5,908	886	29.5	26.2	9,545.4	7	6	6	66.8	52.5	52.5	0.7-0.55-0.55
Poultry	2	6	337,320	1,687	27.7	46.7	17,048.7	15	10	5	255.7	170.5	85.2	1.5-1.0-0.5
Sheep	45	26	4,162	416	18.2	7.6	2,758.7	12	4	10	33.1	11.0	2.8	1.2-0.4-1.0
Horses	499	52	1,392	1,531	20.4	31.3	11,418.1	6	3	5	68.5	28.5	51.4	0.6-0.25-0.45
Totals		950	380,254	36,204		1,193.7	435,715.3				2,447.4	1,198.0	1,868.7	

*Based on Pennsylvania Agronomy Guide - 1992-93

Animal Unit = Animal number x Average weight

Annual Manure Production = Animal Units x Rate x 365

A variety of management practices have been recommended to reduce erosion and the overland movement of nonpoint pollutants from agricultural lands, but their effectiveness varies seasonally as well as among practices (Brenner et al. 1990). Sediment and nutrient influx from croplands and overgrazed pastures not only adversely impact aquatic ecosystems but, unless corrected, will reduce productively as well. In order to reduce erosion and nutrient loss, many farmers have resorted to no-till or conservation tillage but unfortunately, these practices often require an increase use of pesticides and herbicides. In conservation tillage, crop residues along with weeds remain on the field after harvesting and the new crop is seeded through the crop residues, thereby providing ground cover which reduces soil loss and nutrient transport to receiving streams. Currently, conservation tillage is being used on approximately 73 million acres and, along with planting high-residue crops such as corn or sorghum (in proper rotation), have been shown to be effective in reducing erosion. But at present, a long term data base does not exist on their effectiveness or on the bioaccumulation of herbicides and pesticides in aquatic and terrestrial ecosystems.

In addition to conservation tillage, other best management practices currently being used to maximum crop yields while at the same time protecting the environment include crop rotation (corn, grass legume, corn) and contour farming. On steep slopes, terracing is an effective method to reduce overland water flows, thereby reducing soil and nutrient loss. The combination of grass water diversions and grassland or riparian buffer strips are effective methods to decrease the influx of nutrients and other pollutants into streams (Brenner et al. 1990, 1991). In addition, the restoration and enhancement of riparian wetlands, as well as constructive wetlands are cost effective methods of reducing the adverse impacts of nonpoint pollutants of aquatic ecosystems (Brenner 1994, Brenner et al. in press).

The excess use of commercial fertilizers is a major source of pollution of both ground and surface water in agricultural regions. An excess of nitrogen beyond that required by plants is readily converted to nitrites and nitrates (Addiscott and Powlson 1989) and, along with excess phosphorus, is either transported off site or leaches into the groundwater (Moseley 1989). In a survey of 62 wells in Mercer County, Pennsylvania, 8.1 percent contained nitrates which have been shown to be harmful to both animals and young children. A practical method to reduce nitrate leaching is the use of rye as a cover crop because of its high utilization of nitrogen. For example, studies in Maryland found that when rye was planted during the fall in a no-till corn field, the rye biomass contained 100 lb of nitrogen/acre the following spring. Another benefit to using rye as a cover crop is that apparently the utilization of soil nitrogen by rye resulted in a reduction of nitrogen concentrations in the groundwater from 15 to 5 mg/l over a five year period. Unfortunately, the use of rye requires a herbicide application during the flag leaf stage (April 1) which is usually the period of intense rainfall and maximum overland flows, thereby, potentially adversely impacting adjacent streams.

The use of natural fertilizers will also reduce the amount of commercial nitrogen applied to croplands which has been estimated at 60,000 lbs/farm/year

(Hillgren 1993). In many agricultural regions of the United States, the use of commercial fertilizers has decreased in recent years. For example, in the Chesapeake Bay watershed, sales of commercial fertilizers has decreased 24 percent and there has been a 16 percent decrease in fertilizer sales nationally (Moseley 1991). Although, the use of natural fertilizers may be beneficial in reducing commercial nitrogen applications, their use also represent potential adverse impacts on aquatic ecosystems as well as human health. Brenner and Mondok (in press) indicated that over 436,000 metric tons of manure was produced annually in 155,947 ha watershed in western Pennsylvania which translates into over 2,447 and 1,989 metric tons of nitrogen and phosphorus, respectively. If both the livestock and the manure are not handled in a manner to safeguard adjacent waterways, then nutrient loading and other adverse impacts to these aquatic ecosystems will result. Brenner and Mondok (in press) also reported a direct correlation between livestock concentrations and the degradation of streams within the watershed.

Numerous studies have indicated that natural and constructed wetlands are an effective means of treating animal wastes from feedlots and manure storage areas (Brenner 1993). Hillgren and Mills (1993) reported a 90 percent reduction in total pollutants and a 76 percent reduction in phosphorus in constructed wetlands. Likewise, Brenner et al. (in press) indicated that fencing cattle out of streams and restoring riparian wetlands were cost effective methods of reducing fecal coliform and nutrient concentrations in adjacent streams. In a period of 13 months after cattle were fenced out of the stream and the adjacent wetland restored, fecal coliform and phosphorus concentrations were reduced 60 and 40 percent, respectively and nitrates were not detected directly downstream from the site. At Norwood farms in Indiana, constructed wetlands reduced pollutants by 40-60 percent and fecal coliforms by 90 percent during the summer months. In addition to their benefits in reducing pollution impacts, constructed and restored wetlands also provide for a diversity of wildlife and fishery habitats. It should be pointed out, however, that constructed wetlands should be properly matched with the specific needs of the site and, whenever possible, integrated into natural systems to ensure environmental standards and the overall success of the project.

Although fertilizer and manure applications may adversely impact natural ecosystems and public water supplies, the greatest danger to both natural systems and human health is the excessive and/or improper use of herbicides and insecticides. Hillgren (1993) reported that the average farm uses approximately 500 pounds of pesticides annually which contributes to the accumulation of toxic materials in aquatic, terrestrial and groundwater ecosystems. Graham (1991) in a report on the groundwater monitoring of high-volume crop production chemicals, indicated that pesticides are delivered to groundwater systems via leaching after application. Graham (1991) also reported that groundwater concentrations of pesticides may increase over time in a manner analogous to the rise in nitrate in groundwater between 1950 and 1980. Moreover, the occurrence of pesticides in spring water is usually associated with storm events resulting in surface water contamination of groundwater systems. Once these chemicals enter the groundwater they will remain for long periods of time since microbial bioremediation of these chemicals occur under aerobic conditions.

The adverse impact of agricultural chemicals is not restricted to North America, but should be a matter of global concern. Muir et al. (1992) reported that the current state of knowledge of the levels, spatial and temporary trends of contaminants in arctic marine ecosystems varies both among pollutants as well as among environmental compartments. Varying concentrations of polychlorinated biphenyls (PCB's), organochlorine pesticides along with mercury and lead have been documented in arctic marine mammals and fish (Muir et al. 1992). In addition, Muir (1992) reported the bioaccumulation of polychlorinated camphenes, DDT, and chlordane-related compounds in phytoplankton, marine fish, mammals and seabirds. It is important to note that the temporal trends in organochlorine pesticides and DDT suggest a decline in these compounds; however, there is a lack of data on the fate of other pesticides in arctic environments.

The adverse impact of pesticides however, is not limited to arctic and temperate environments. Organophosphorous insecticides are extensively being used in tropical environments at a rate that adversely impacts sensitive tropical marine ecosystems (Readman et al. 1992). According to Barton (1985), chemical defoliants are poisoning a reservoir behind the Tucurui hydroelectric dam on a tributary to the Amazon River which may have resulted in perhaps over 100 deaths among those who rely on the reservoir for drinking water. The continued use of defoliants not only represents a threat to the Amazon but to tropical rainforests and other tropical ecosystems around the world.

In addition to the adverse impacts of pesticides and herbicides on natural ecosystems, these agricultural chemicals pose a serious threat to human health. According to a report issued by the Council of Scientific Affairs of the American Medical Association (JAMA 1988), the primary hazard of pesticide exposure is the development of acute toxic reactions as a result of dermal contact or inhalation of relatively large doses of these compounds. The effects are usually manifested within minutes or hours of contact. Mortality rates have increased for leukemia, all lymphopoietic cancers, and cancers of the stomach, rectum, and eyes of farmers born between 1905 and 1958 (JAMA 1988). A large number of pesticide compounds have been shown to cause genotoxicity or carinogenicity in animals and in vitro screening tests, but only arsenic and vinyl chloride compounds have been definitely linked to carcinogenicity in man (JAMA 1988). Therefore, epidemiological studies offer the only conjectural evidence that pesticides may posses carcinogenic properties. On the other hand, according to a two year study by the Natural Resources Defense Council (1989), adults may face a cancer risk from pesticides on the fruits and vegetables they consumed as children. This study concluded that fruit usually constitutes a large portion of the diet of pre-school children and is likely to harbor toxic pesticide residues. Therefore, more than half of the lifetime cancer risk from fruit is typically acquired before the age of six. Based on preliminary data, metabolites of daminozide contributes between 86 and 96 percent of the cancer risk by fruit consumed by pre-school children.

In a study released by EPA in 1986, a definite link exists between exposure to organophosphate pesticides and a variety of neuropsychological problems, including depression, irritability, and difficulties in thinking and memory, and communication. Silberner (1986) reported that exposure to the herbicides 2,4,-D (2-4 dichlorophenoxyacetic acid), commonly referred to as agent orange, significantly increases the risk of non-Hodgkin's lymphoma (NHL). Likewise, Swedish researchers reported associations between phenoxyherbicides and not only NHL but also soft-tissue sarcoma and Hodgkin's disease (JAMA 1988). It is also of interest to note that Lewin (1985) reported that exposure to the herbicide paraquat and other industrial chemicals increased the incidence of Parkinson's disease. Lewin (1985) concluded that Parkinson's is an environmentally induced disease, but it must be remembered that pesticides are just one class of environmental neurotoxins and compounds such as pyridine compounds may be important in the ethiology of Parkinson's disease. It is also of interest to note that Parkinson's disease was unknown prior to the industrial revolution and its incidence increased rapidly in the early decades of this century (Lewin 1985).

In conclusion, a variety of contaminants from industrial, urban and agricultural point and nonpoint sources represent the potential to adversely impact natural ecosystems, as well as posing a threat to human health and welfare. It is imperative, therefore, that society address these problems on a global scale, not only to protect natural ecosystems but to safeguard the health of present and future generations. The degradation of surface and groundwater not only affects these aquatic systems but represents a danger to downstream communities, as well as receiving lakes and coastal areas. The discharge of contaminants into rivers and coastal ecosystems and the dumping of toxic materials into the open oceans must be addressed on an international level in order to protect the health and welfare of the entire global community and ecosystems.

LITERATURE CITED

Addiscott, T. and D. Powson. 1989. Laying down the ground rules for nitrate. New Scientist 122:28-29.

Alexander, L.M., Heaven, A., Tenvant, A. and R. Morris. 1992. Symptomatolgy of children in contact with sea water contaminated with sewage. J. Epidemiology and Community Health. 46:340-344.

Anderson, C. 1993. Twisted fish sex scare. Science 259:1119.

Anderson, T.L. and D.R. Leal. 1991. Free Market Environmentalism. Pacific Research Institute For Public Policy. Westview Press. Boulder, Colorado. 192 pp.

Barton, K. 1985. Health effects of high tech. Environment 27:22.

Bottiger, M. and E. Herrstrom. 1992. Isolation of polio-viruses from sewage and their characteristics: Experience over two decades. Scandinavian J. Infect. Diseases. 24:151-155.

Brenner, F.J. 1994. Increasing biodiversity through restoration of disturbed ecosystems. IN Biological Diversity: Problems and Challenges. S.K. Majumdar, F.J. Brenner, E.W. Miller, J. Lovich and J.F. Schalles (Eds.) Penn. Acad. Science. pp. 281-301.

Brenner, F.J., Steiner, R.P., Mondok, J.J. and R.J. McDonald. 1990. Non-point source pollution: A model watershed approach to improve water quality. IN Water Resources in Pennsylvania: Availability, Quality and Management. S.K. Majumdar, E.W. Miller and R.R. Parizek (Eds.). Penn. Acad. Science. pp. 70-84.

Brenner, F.J., Mondok, J.J. and R.J. McDonald, Jr. 1991. Impact of riparian areas and land use on four non-point source pollution parameters in Pennsylvania. J. Penn. Acad. Sci. 65:65-70.

Brenner, F.J. and J.J. Mondok. Evaluation of non-point source pollution potential in an agricultural watershed in northwestern Pennsylvania (in press).

Brenner, F.J., Steiner, R.P. and J.J. Mondok. Groundwater-surface water interaction in an agricultural watershed (in press).

Council on Scientific Matters. American Medical Association. 1988. Cancer risk of pesticides in agricultural workers. J. Amer. Med. Assoc. 260:959-963.

Clark, E.H., Haverkamp, H.H. and W. Chapman. 1985. Eroding soils: The off farms impacts. The Conservation Foundation. Washington, D.C.

Dickson, D. 1993. Clash over two demanding Euro-water standards. Nature 364:270.

Diebel, P.L., Taylor, D.B., Batie, S.S. and C.D. Heatwole. 1992. Low-input agriculture as groundwater protection strategy. Water Res. Bull. 28:755 762.

Environmental Protection Agency. 1983. Chesapeake Bay: A frame work for action. Chesapeake Bay Program. Environmental Protection Agency Chesapeake Bay Liaison Office. Annapolis, MD.

Environmental Protection Agency. 1987. Nonpoint source guidance. Nonpoint source branch. Office Water Regulations and Standards. U.S. Environmental Protection Agency. Washington, D.C.

Fewtrell, L., Godfree, A.F., Jones, F., Kay, D., Salmon, R.L. and M.D. Wyer. 1992. Health effects of white-water canoeing. Lancet. 339:1587-1589.

Goyer, R.A. 1984. Potential health effects from groundwater pollution. EPA Journal 10:22-28.

Grabow, W.O.K., Middendorff, I.G. and O.W. Prozesky. 1973. Survival in maturation ponds of coliform bacteria with transferable drug resistance. Water. Res. 7:1589-1597.

Graham, J.A. 1991. Monitoring groundwater and well water for crop production chemicals. Anal. Chem. 63:613-621.

Griffin, R. Jr. 1991. Introducing NPS water pollution (non point source pollution). EPA Journal 17:6-7.

Hammouda, N.A., Gebali, W.M. and M.K. Razek. 1992. Intestinal parasitic infection among sewage workers in Alexandria, Egypt. J. Egyptian Soc. of Parasitology. 22:299-303.

Hill, C.L. 1987. Monitoring the effects of land management practices on water quality in Gullford County, North Carolina. IN Monitoring, Modeling and Mediating Water Quality. S.J. Nix and P.E. Black (Eds.). Amer. Water Res. Assoc. pp. 137-148.

Hillgren, S. 1993. Caught in the clean water act. (Congress trolls for "dirty watersheds". Farm Journal Dec. pp. 12-16.

Hillgren, S. and B. Mills. 1993. Manure handling spreads to wetlands (man-made marshes filter livestock waste). Farm Journal. Dec. pp. 1-4.

Holloway, M. 1992. Are the ocean depths a safe place for sewage sludges? Scientific American. 266:30-31.

Koray, T., Buyukisik, B., Parlak, H. and S. Gokpinar. 1992. Unicellular organisms affecting sea water quality in Bay of Izmar: Red-tides and other blooms. Doga. Turk. Bajologi Dergisi. 16:135-157.

LaRiviere, J.W. 1989. Threats to the world's water. Scientific American. 261: 80-94.

Lewin, R. 1985. Parkinson's disease and environmental cause. Science 229: 257-258.

Lewis, J. 1991. Some other options (on the farm). EPA Journal 17:47.

Ljungstrom, I. and B. Castor. 1992. Immune response to *Giardia lamblia* in a water-borne outbreak of giardiasis in Sweden. J. Medical Microbiol. 36: 347-352.

Marcharis, J. 1985. Chesapeake Bay non-point source pollution. IN Perspectives on non-point source pollution. U.S. EPA pp. 44015-85001.

Marshall, B.E. 1991. Toxic cyanobacteria in Lake Chivero, Zimbabwe: A possible health hazard. Trans. Zimbabwe Scientific Assoc. 65:16-19.

McAllister, A.L. 1990. Well water contamination. J. Environ. Health 53-54.

Moseley, J.R. 1991. View from USDA: Farms only five miles apart may need different solutions. EPA Journal 17:25-27.

Muir, D.C., Wageman, G.R., Hargrave, B.T., Thomas, D.J., Peakall, B.D. and R.J. Norstrom. 1992. Arctic marine ecosystem contamination. Science for the Total Environment. 122-75-134.

Natural Resources Defense Council. 1989. Kids at risk. Natural Resources Defense Council. New York.

OECD. 1985. The state of the environment. Organization for Economic Cooperation and Development. Paris, France.

Odgers, E. 1991. Wisconsin's bad actors program: The bad actors lost in the manure wars. EPA Journal 17:51-52.

Offutt, S. 1991. View from OMB: Must farmers prevent pollution to escape regulation? EPA Journal 17:2 8-30.

Park, P. 1991. Great lakes pollution linked to infertility (noting neurological damage to children). New Scientist 131:18.

Peirls, B.L., Caraco, N.F., Pace, M.L. and J.J. Cole. 1991. Human influence on river nitrogen. Nature 350:386-387.

Peskin, H.M. 1986. Cropland sources of water pollution. Environment 28:30-34.

Raloff, J. 1986. Toxic showers and soils. Science News 130:190.

Readman, J.W., Kwong, L.L.W., Mee, L.D., Bartocci, J., Nilve, G., Rodriguez-Solano, J.A. and F. Faris-Gonzalez. 1992. Persistent organophosphorus pesticides in tropic marine environments. Marine Pollution Bull. 24:398-402.

Reay, W.G., Gallagher, P.L. and G.M. Simmons, Jr. 1992. Groundwater discharge and its impact on surface water quality in a Chesapeake Bay Inlet. Water Res. Bull. 28:1121-1134.

Rogers, G.K., Davies, I.M. and J.G. McHenery. 1992. The effects of sewage sludge disposal on the sediments at the Garroch Head dump site, Firth of Clyde, Scotland. The Science of the total environment 119:133-156.

Schaller, F.W. and G.W. Bailey. 1985. Agricultural management and water quality. Iowa State University Press. Ames, Iowa.

Silberner, J. 1986. Common herbicide linked to cancer. Science News 130:86-87.

Smith, D. 1993. Rye to the rescue (cover crops tie up root-zone nitrate, lowering its concentrations in groundwater). Farm Journal Dec. p. D-2.

Tim, U.S., Mostaghim, S. and V.O. Shanholtz. 1992. Identification of critical nonpoint pollution source areas using geographic information systems and water quality modeling. Water Res. Bull. 28:877-887.

U.S. Congress. Office of Technology Assessment. Wastes in marine environments. OTA-0-334. U.S. Government Printing Office. Washington, D.C.

Vrati, S., Giri, D.K., Parida, S.K. and G.P. Talwar. 1992. An epidemic of non A, non B hepatitis in south Delhi: Epidemiological studies and transmission of the disease in rhesus monkeys. Archives of Virology 125:319-326.

Environmental Contaminants, Ecosystems and Human Health. Edited by S.K. Majumdar, E.W. Miller, and Fred J. Brenner. © 1995, The Pennsylvania Academy of Science.

Chapter Eight

POTENTIAL EFFECTS OF MINING CONTAMINANTS ON ECOSYSTEM FUNCTION AND HUMAN HEALTH

FRED J. BRENNER, HEATHER A. GRAY, BARBARA R. PORTER and VICTORIA A. TILMANS

Biology Department
Grove City College
100 Campus Drive
Grove City, PA 16127

INTRODUCTION

The adverse impacts of mining, along with the improper disposal of the by-products of coal and ore processing on natural ecosystems, have been documented for over 100 years. The environmental problems associated with deep and surface coal and mineral mining in Europe, North America, Africa and elsewhere in the world have been the result of both improper mining, reclamation and environmental regulations to control emissions and discharges from coal and ore processing facilities. It is possible, however, that *Homo sapiens* have been impacted by environmental contaminants throughout their evolutionary history. Bartsiokas and Day (1993) suggest that the presence of dental caries in hominid fossils from the Broken Hill region of the present Zambia were the result of chronic lead poisoning. This is of special interest since these fossils were uncovered during lead and zinc mining between 1925 and 1931.

The adverse impacts on aquatic ecosystems by acid and heavy metal discharges (AMD) from coal and metal ore mining are classic examples of the degradation of aquatic environments. Moreover, the bioaccumulation of these metals in aquatic and terrestrial food chains may be lethal to organisms in the upper levels of the food chain, including man. The poisoning of aquatic and terrestrial ecosystems by cyanide from gold mining and processing facilities is yet another example of the adverse impact of past mining practices on natural ecosystems and human society. In addition, the sterile environment at Copper Mountain, Tennessee and Sudbury, Ontario from previous copper smelting activities are classic examples of the long term adverse impact of ore processing on aquatic and terrestrial ecosystems.

It is impossible within one chapter to address all the possible effects of materials associated with mining on natural environments and human health. Other chapters within this text address specific topics associated with the mining, processing, reclamation and other aspects associated with the mining industry in different regions of the world. The objective of this chapter is to present an overview along with some specific examples of the impact of mining on natural and human ecosystems, as well as suggesting mechanisms to minimize impacts and maintain sustainable systems.

IMPACTS OF ECOSYSTEM FUNCTION

Acid mine drainage (AMD), which is produced when sulfur compounds come into contact with air and water, results in sulfuric acid and, along with associated heavy metals, represents a major potential source of contamination of aquatic environments. When these acidic drainages enter streams with sufficient buffering capacities, the pH is lowered and the resulting bioaccumulation of heavy metals is often toxic to aquatic life. As the pH increases in these environments, metal precipitation occurs which covers the stream substrate with yellow-orange material or "yellow boy", resulting in a mortality of aquatic life because of decreased dissolved oxygen concentrations. The species and concentrations of these heavy metals contained within mine tailings depend on the type of material and geologic strata being mined, as well as the mining procedures and degree of reclamation of the site. For example, the concentrations of copper, zinc, lead, and cadmium in the Lean River of China remained in solution at a pH value of 6.5 or less; but at pH's above 6.5 these heavy metals precipitated and accumulated in the bottom sediments.

In addition, the bacterial leaching of arsenopyrite will produce acid leachates containing concentrations of copper (Cu), nickel (Ni), Zinc (Zn), cadmium (Cd), lead (Pb), arsenic (Ar) and sulfate (SO_4) (Uminska 1993). The bioaccumulation of these metals in aquatic and terrestrial ecosystems has a direct and adverse impact of all levels of the food chain, including man.

Aquatic insects have been used as bioindicators to assess the impact of AMD on aquatic environments. Heavy metal bioaccumulation in immature benthic insects was assessed in two river systems impacted by copper and gold mining. All insect taxa collected from a 381 km section of river impacted by copper mining, as well as a 40 km section of another river receiving discharges for a gold mine, had elevated levels of copper, cadmium, lead and zinc. However, within the taxa

Hydropsyche (caddisfly), individual element concentrations were greater than it was among the individual species of the taxa. Based on the assumption that this taxa is representative of the stream fauna, it may be possible to use this taxa for rapid biomonitoring of the impact of heavy metal discharges on stream ecosystems (Cain et al. 1992, Usis and Foote 1991). Likewise, stone fly nymphs (Plecoptera) have been used as bio-indicators of heavy metal toxicity associated with coal mining since they are more tolerant than other insect taxa and fish species (Whipple and Dunson 1993). Diversity indices that incorporate pollution tolerance have become an accepted practice in biomonitoring of the impact of acidic and heavy metal discharges into aquatic ecosystems (Brenner and Helm 1991).

Over the last 100 years, the infiltration of acidic and heavy metal discharges of coal and ore mining have resulted in a decrease in the biodiversity of aquatic species. These drainages have often reduced aquatic flora and fauna to a few species of acid tolerant bacteria, algae and invertebrates. In addition, these discharges often result in siltation, smothering bottom dwelling organisms and increasing the mortality of filter feeding organisms. Freshwater bivalves (*Anodonta grandis*) collected in a stream impacted by zinc and copper mining and smelting were analyzed for cadmium, copper, zinc, while metallothioneins (MT) and intersite variations in MT concentrations were related to free cadmium (Cd_2) concentrations at the sediment-water interface, suggesting that Cd_2 activity is the key environmental factor in the response of *A. grandis* to cadmium discharges in aquatic environments (Couillard et al. 1993). In addition, these discharges also interfere with oxygen exchange on the gill surface, resulting in suffocation. These authors also reported that discharges from copper and zinc mining in the upper Sacramento River basin resulted in mortality of rainbow trout (*Onchorynchus mykiss*) and chinook salmon (*Onchorynchus tshawytscha*) (Couillard et al. 1993).

In addition to the adverse impact of heavy metals on natural ecosystems, the bioaccumulation of these elements in both aquatic and terrestrial food chains is also a concern for human populations in mining regions of the world. For example, the bioaccumulation of heavy metals in fish represents a public health concern since they are a major source of protein for many human populations, especially in many third world countries. Although different species of fish show varying sensitivities to acidic conditions, salmonids including whitefish (*Coregonus clupeaformis*), lake trout (*Salvelinus namaycush*), rainbow trout, and Atlantic salmon (*Salmo salar*), appear to be more sensitive than other species to acidic drainages. At thresholds of acidic drainage with pH's less than 6.5, the related heavy metal concentrations result in a disturbance in the reproductive physiology of adults (Brenner et al. 1976, 1977, Sykora et al. 1972a, b, 1975), as well as in the development and survival of embryos (Brenner and Cooper 1978).

In addition to their impact on aquatic ecosystems, surface or open cast and deep mines generate tons of waste rock annually and the handling of this overburden and tailing not only result in adverse environmental impacts, but add to the expense of mineral and coal extraction. This spoil material is often prone to contamination by acidic material (Miller et al. 1919, Wang et al. 1993 et al. 1993) along with heavy metal deposition in alluvial soils (Pierzynski and Schwab 1993). In terrestrial systems, contaminated water filtering through these acidic spoils may result in damage to flora and fauna alike. The bioaccumulation of heavy metals by the Rhizobium, associated with legumes subjected to long term sewage sludge amendments, indicated that heavy metals may have an adverse quantitative effect on denitrification by rhizobia, rather than a genetic effect as suggested in previous studies (Obbard and Jones 1993). Pierzynski and Schwab (1993) reported total soil concentrations of 1165 mg/kg for zinc phytotoxicity was a limiting factor in soybean growth. In addition, other studies have indicated the existence of manganese and aluminum toxicity in legumes on surface coal mine spoils (Brenner 1984).

Human Health Impacts

As discussed previously, in those regions of the world where fish serve as a major food source, the bioaccumulation of heavy metals in aquatic food chains pose a concern for the health and welfare of society since humans ultimately consume organisms at the upper levels of the food chain. In the Pappas lagoon in Greece, copper, zinc, cadmium, and lead were found in all species examined but at concentrations below the allowable limit for human consumption. The ingestion of heavy metal may result in physiological and toxicological problems in pregnant and newborn infants. In a study of 51 Austrian women and newborn infants tested for lead, mercury and cadmium concentrations of early breast milk, blood, and urine post- partum using atomic absorption spectrophotometry, these elements did not occur in elevated concentrations in the breast milk. However, high concentrations of mercury were found in both the mothers and infant umbilical cord blood (Ploeckinger et al. 1993) and significant correlations occurred in toxic lead levels between maternal and neonatal blood concentrations. In another study involving pregnant women, heavy metals were correlated with coartation of the aorta (CoA). Tikkanen and Heinonen (1993) analyzed 50 patients of which 18 had CoA and the other 32 had at least one cardiac anomaly. They concluded that genetic factors explained only a small fraction of CoA, whereas exposure to heavy metals poses a major risk for the development of CoA. Human populations may also be exposed to heavy metals via cattle and other animals bred for human consumption that may be susceptible to toxic levels of these metals in their water supply.

In the eastern Carpathian Mountains of Romania, certain rock strata contain high concentrations of heavy metals and radioactive elements, but lack sufficient soluble phosphorus and molybdenum to meet the physiological requirements of livestock. Chronic haematuria of cattle occurring in this region is a direct result of exposure to heavy metals. This chronic disease has been termed an ecosystem disease since it is a result of biogeochemical features caused by alternations in the balance of the natural processes occurring in this particular ecosystem (Lacatusu et al. 1993).

Coal and other types of mining activities have been associated with a variety of respiratory ailments such as emphysema and black lung among others (Souter et al. 1993). Harrison et al. (1993) found that miners were more susceptible to gastritis and intestinal metaphases than the general population. In three collieries in Great Britain, Souter et al. (1993) found that gastritis and intestinal metaplasia were correlated with age and the cumulative exposure to coal dust. In addition, carcinogenic polycylic aromatic hydrocarbons are often present in elevated concentrations in coal mining and processing regions. It appears, therefore, that heavy metal discharges from coal and metal mining not only adversely impacts aquatic and terrestrial ecosystems, but represent a potential public health concern as well.

Treatment of Acid Mine Discharges

In the past, the standard method of treating acidic discharges was the use of hydrated lime, soda ash or another type of basic compound. However, in many cases, the discharges from these treatments have adverse impacts on aquatic environments (Brenner et al. 1993). In recent years, passive treatment of acidic drainages has, in some cases, been shown to be an effective alternative to chemical treatment (Brenner et al. 1993, in press). For example, the recovery of biota in the North Branch of the Potomac River was attributed to a decrease in acidic influx from mining activities in conjunction with dilution from the Savage River (Diamond et al. 1993). Studies on a reservoir constructed on the North Branch revealed high concentrations of manganese, iron, zinc, and aluminum and supported few macrophytes, benthic, and plankton communities, suggesting either metal toxicity and/or nutrient limitations (Diamond et al. 1993). The rapid recovery of the lower North Branch of the Potomac Rivers as compared with previous studies suggest that decreased acid influx from mining operations and dilution from two major tributaries contributed to the improvement in water quality.

Both natural and constructed wetlands are effective sinks for heavy metals and therefore, offer potential for concentrating heavy metals from mine discharges. These constructed wetlands are commonly planted with *Typha, Sphagnum, Phragmites,* or *Cyperus*, all of which have been shown to be acid tolerant and concentrate heavy metals in the roots and rhizomes (Dunbabin and Bowmer 1992). However, discharges from these wetlands may also adversely impact aquatic biota. For example, Usis and Foote (1991) reported that discharges from a coal mine two miles upstream from the Stillfork Swamp Nature Preserve in Carrol County, Ohio resulted in a reduction of the caddisfly (*Limnephillus indivisus*) in the southern half of the preserve. This reduction was most likely due to increased larva mortality as a result of the precipitation of heavy metals and elevated total dissolved solids. In addition to being an effective passive treatment system for acid mine drainages, they also have minimum operational and maintenance cost and, hence, are a cost effective means of treating these discharges (Brierley 1990, Brenner et al. 1993). However, at this time, a long term data base does not exist on effectiveness of these beyond 10 years of operation (Brenner et al. in press).

URANIUM MINING

Contamination from uranium mining and processing, along with leakage from storage, not only impact aquatic and terrestrial ecosystems, but represent a public health concern for the surrounding communities. Uranium concentrations tend to be higher in fine silt and clay fractions of sediments especially in backflow billabongs and floodplains. It appears that uranium concentrations are correlated with the organic contents of these fine sediments (Noller and Hunt 1993). For example, profiles from Long Island Sound and the Amazon Shelf sediments (Kalliolar et al. 1993) suggest that both removal and release processes control uranium geochemistry in estuarine sediments. The release of uranium from sediments appear to occur in association with iron reduction, whereas, the removal of uranium occurs under conditions of sulfate reduction. These estuarine sediment sinks appear to be an important aspect of uranium balance in marine ecosystems. Likewise, uranium as a result of mine discharges have also been identified in marine environments. McKee (1993) suggest that the microbial population in the anoxic waters in the Framvaren Fjord in Norway exerts an effective control on the aquatic biogeochemistry or uranium discharges into this environment.

While in natural environments, Kaul (1993) indicated that the mineralization of uranium in the Siwalike region of the Himalayans was attributed to concentrations of intrinsic uranium in the sediments and subsequent remobilization by groundwater. In marine environments & geochemistry of near anoxic conditions that existed in the Arabian Sea during the entire last glacial cycle has recently been identified in surface waters. This anoxia is inferred from the concomitant enrichment of organic carbon and authentic uranium in the glacial sections of a core from the bottom sediments. The anoxic conditions during the last glacial period probably were caused by a change in deep water circulation, evidently enhancing preservation of organic matter and simultaneous removal of uranium from sea water. In addition to natural biogeochemical processes and mine discharges operating in freshwater and marine ecosystems, discharges from mining operations also have been shown to contaminate groundwaters. Discharges from uranium mine dewatering increased uranium concentrations in the groundwater under the Puerco River 65 Km downstream from the point of discharge (Van-Metre and Gray 1992). Liu and Wu (1993) reported that uranium mine waste discharged into freshwater ecosystems will be concentrated in the algal biomass but 82 percent of the uranium contamination can be recovered by elution with salt solution (Liu and We 1993). The optimum uptake of uranium by anacystis biomass has been reported to occur under acidic conditions at a pH range between 3.0 and 5.0. It would appear, therefore, that the neutralization of uranium discharges will reduce these concentrations in streams receiving mine wastes.

In terrestrial environments, uranium in mine waste may result in destruction of plant tissues because of the alpha-radiation of the uranium nuclei. To reduce contamination of terrestrial ecosystems, mine spoil should be covered with uncontaminated material and the site revegetated, preferably with natural vegetation, but *Cynodon dactylon* and *Panicum coloratum* have been recommended for stabilizing lands disturbed by uranium mining (Hossner et al. [1992]).

HUMAN HEALTH IMPACT

Shields et al. (1993); investigating the effects of uranium mining on the ethiology of birth defects in Navajos in the Shipwreck uranium mining region, found a significant correlation in the number of birth defects occurring in women living near mine tailings. Apparently during sedimentation, uranium enriched clay sediments along with the associated shales, are a potential local source of uranium contamination. In an epidemiological study, French underground uranium miners showed a latency period of 10-20 years for lung cancer (Tirmarche et al. 1992). In another study, a statistically significant increase in lung and laryngeal cancer deaths among French uranium miners (Sevic et al. 1992, Vich and Koskova 1992) and childhood leukemia has been associated with uranium mining and processing in Ontario, Canada (McLaughlin et al. 1993). On the other hand, in West Bohemia, there was no evidence among a sample of 4320 miners that the radon-rich atmosphere increased the risk of any cancer except lung cancer (Tomasek et al. 1993).

However, uranium mining has been associated with other health risks besides cancer. For example, silicosis has been associated with dust concentrations and corresponding radon exposure (Mehlhorn et al. 1993). In addition, Mehlhorn (1992) also reported an increased frequency of rare autoimmune diseases generally associated with progressive systemic sclerosis and lung fibrosis. In conclusion, a variety of human risks have been associated with exposure to uranium mine wastes, but under proper management and reclamation, these risks can be reduced but probably not eliminated.

OIL SHALE

Oil shale is lithified rock which may range from soft "punky" rocks to consolidated resistant strata. These strata are characterized by high porosities but with low permeability because of the small interstices among the pore spaces which account for the high petroleum content. In addition, these strata contain abundant aluminum silicate clays, along with sand, limestone and iron oxides (Buckley 1951). These shales developed from the accumulation of organic matter in historic inland seas and lakes. Over time, these organic petroleum compounds became mixed with bottom sediments producing kaolinote which is comprised of 80 percent carbon and 10 percent hydrogen. Kaolinote then combines with sediments forming an insoluble rock.

Although oil shale is an abundant and potential energy source, the mining and processes poses environmental and public health concerns. Oil shale mining not only disturbs the landscape and produces waste material that must be handled in an environmentally safe manner, but also may contaminate ground and surface waters. If proper environmental safeguards are used during mining, the long-term environmental impact is minimal (Welles 1970). Moreover, it is generally easier to mine solids in an environmentally safe manner than is the pumping and transportation of liquid fuels. On the other hand, waste waters produced during oil shale mining often contain ammonium gas and ammonium ions in concentrations inhibitory to biological treatments, but this water generally contains nitrogen compounds that offer a potential recoverable fertilizer resource (Daughton et al. 1986).

Although oil shale may be mined in an environmentally safe manner, once a natural area is contaminated by petroleum products, it will remain in the environment for extended periods of time. In aquatic ecosystems contaminated by petroleum products, the diversity and species richness are linked to the level of hydrocarbons in the sediments and these hydrocarbons enter the food chain, and through bioaccumulation, represent a potential public health risk (Murdock 1971). Other toxic effects of hydrocarbons involve the resiliency of aquatic organisms during the embryonic and larval stages. Paine et al. (1992) found that concentrations of water soluble hibernia crude oil of between 27 and 37 mg/l was necessary to kill embryos while 1.3 to 7.1 mg/l were lethal to larval. In addition, sublethal effects were observed on the growth, pigmentation, development and hatching of aquatic organisms at concentrations of 10-50 percent the lethal exposure (Paine et al. 1992).

Regardless of the safeguards, accidental discharges of hydrocarbon compounds will continue to occur as long as society continues to rely on crude oil and oil shale as a source of energy. For both crude oil spills and the byproducts of oil shale production and processing, hydrocarbon compound degradation can be increased by the use of fertilizers which enhance bacteria growth. The bioremediation of hydrocarbons occurs with the addition of inorganic nutrients which enhance the growth of hydrocarbon-oxidizing bacteria and fungi especially *Pseudomonas, Corynebacteria, Mycobacterium,* and yeasts (Brock and Madigan 1991, Lindstrom et al. 1991, Amund and Akangbou 1993). It appears, therefore, that, although hydrocarbon compounds may represent a potential source of contamination of natural environments, overtime bioremediation of these compounds occurs by natural occurring microorganisms which can be enhanced by the addition of nutrients.

In conclusion, the adverse impacts of mining and processing addressed in the preceeding discussions are due at least in part to a lack of basic research on the effects of these contaminants on ecosystem function and human health. However, the failure to provide safeguards to protect natural biological systems and the health of human populations cannot continue to occur if we are to have sustainable development and uncontaminated natural environments. In this regard, two basic questions must be addressed by the mining industry, governments and the general public. First of all, in all current and future fossil fuel and mineral developments, we must protect critical and unique ecosystems and restore all disturbed lands with the same community diversity that existed prior to mining. Secondly, we must continue our research effects to develop cost effective procedures to restore previously disturbed lands to functioning ecosystems. In other words, we must have both our natural house (ecology) and our economic house in order to have sustainable resource development while at the same time providing uncontaminated ecosystems for future generations.

LITERATURE CITED

Amund, O.O. and T.S. Akangbou. 1993. Microbial degradation of four Nigerian crude oils in an estuarian microcosm. Letters in Applied Microbial. 16:118-249.

Bartsiokas, A. and M.H. Day. 1993. Lead poisoning and dental caves in the Broken Hill hominid. J. Human Evol. 24:243-249.

Barnes, C.E. and J.K. Cochran. 1993. Uranium geochemistry in estuarine sediments: Controls on removal and release processes. Geochemica et Cosmochimica. ACTA 53:555-569.

Brenner, E.K., Brenner, F.J., Bovard, S. and T.E. Schwartz. 1993. Analysis of wetland treatment systems for acid mine drainage. J. Penn. Acad. Sci. 67:85-93.

Brenner, E.K., Brenner, F.J., and S. Bovard. Comparison of bacteriological activity in two acid mine drainage wetland systems, in press.

Brenner, E.K. and F.J. Brenner. Analysis of wetland treatment systems for heavy metal remediation of acid mine drainage, in press.

Brenner, F.J. 1984. Ecosystem development and natural succession in surface coal mine reclamation. Minerals and Environment. 6:10-22.

Brenner, F.J. and J. Helm. 1991. Macroinvertebrate recolonization and water quality characteristics of a reconstructed stream after surface coal mining in northwestern Pennsylvania, USA. Intern. J. of Surface Mining and Reclamation. 5:31-35.

Brenner, F.J. and W.L. Cooper. 1978. Effect of suspended iron hydroxide on the hatchability and embryonic development of the coho salmon. Ohio J. of Science. 78:34-38.

Brenner, F.J., Corbett, S. and R. Shertzer. 1976. Effect of ferric hydroxide suspension on blood chemistry in the common shiner, *Notropus cornutus.* Trans. Amer. Fish. Soc. 105:450-455.

Brenner, F.J., Shertzer, R., Corbett, S. and J. Centofani. 1977. Partial correlation analysis of some chemical and biological parameters of two streams receiving mine drainage. Proceed. Penn. Acad. Sci. 51:125-130.

Brierley, G.C.L. 1990. Bioremediation of metal contaminated surface and groundwaters. Geomicrobiology J. 8:201-204.

Brock, T.D. and M.T. Madigan. 1991. Biology of microorganisms. Prentice Hall. Englewood Cliffs. N.J.

Buckley, E. 1951. Petroleum conservation. American Institute of Mining and Metallurgical Engineers. Dallas, TX.

Cain, D.J., Luoma, S.N., Carter, J.L. and S.V. Fend. 1992. Aquatic insects as bioindicators of trace element contamination in cobble-bottom rivers and streams. Can. J. Fisheries and Aquatic Sciences. 49:2141-2154.

Couillard, Y., Cambell, P.G.C. and A. Tressier. 1993. Response of metallothionein concentrations in a fresh water bivalve (*Anodonta grandis*) along an environmental cadmium gradient. Limmol. and Oceanography. 38:299-313.

Daughton, C.G., Sakaji, R.H. and G.W. Langlois. 1986. Oil shale retort water ammonia determination by titrimetry, phenate colorimetry, enzymatic analysis, and chromatographic fractionation/chemiluminescence. Analytical Chem. 58:1556-1561.

Diamond, J.M., Bower, W. and D. Gruber. 1993. Use of man made impoundment in mitigating acid mine drainage in the North Branch Potomac River. Environ. Manage. 17:225-238.

Dunbabin, J.S. and K.H. Bowmer. 1992. Potential use of constructed wetlands for treatment of industrial waste-waters containing metals. Science for the Total Environment. 111:151-168.

Harrison, J.D., Morris, D.L. and J.D. Hardcastle. 1993. Screening for gastric carcinoma in coal miners. Gut 34:449-498.

Hossner, L.R., Woodwind, H.J. and J. Bush. 1992. Growth and selenium uptake of range plants propagated in uranium mine spoils. J. Plant Nutrition. 15:2743-2761.

Kalliola, R., Linna, A., Puhakka, M., Salo, J. and M. Rasanen. 1993. Mineral nutrients in fluvial sediments from the Peruvian Amazon. Catnap 20:333-349.

Kaul, R., Umamaheswar, K., Chandrasekaram, S., Deshmukh, R.D. and B.M. Swarnkar. 1993. Uranium mineralization in the Siwaliks of northwestern Himalaya, India. J. Geological Soc. of India. 41:243-258.

Lacatusu, R., Rauta, C., Grigore, C., Carstea, S. and I. Ghelase. 1993. Geochemical and biogeochemical features of areas of haematuria in cattle in the eastern Carpathian Mountains of Romania. Environ. Geochem. and Health. 15:3-13.

Larrsson, E., Gunneriusson, E. and E.B. Lindstrom. 1992. Metal removal from bacterial leaching effluents by precipitation and biosorption. Biorecovery 2:107-119.

Lin, Y. and Q. Li. 1992. Influence of acidic mining waste water on water quality of the Lean River. Chinese J. Environ. Science (Beijing). 13:9-14.

Lindstrom, J.F., Prince, R.C. and J.C. Clark. 1991. Microbial populations and hydrocarbon biodegradation potentials in fertilized shoreline sediments affected by the T/V Exxon Valdez oil spill. Appl. Environ. Microbiol. 57:2514-2522.

Liu, H.H. and J.T. Wu. 1993. Uptake and recovery of americium and uranium by *Anacystis* biomass. J. Environ. Sci. and Health. Part A Environ. Sci. and Engineer. 28:491-504.

Llopoulou-Georgudaki, J., Tsonis, S.P., Kostoglou, E., Alexopoulos, A. and P. Kaspiris. 1992. Trace elements in fishes collected from "Pappas" lagoon in

Greece. Revue Internal. P'Oceanographia Medicale. 107-108:59-66.
McRae, B.A. and J.F. Todd. 1993. Uranium behavior in a permanently anoxic Fjord:Microbial control? Limmol. and Oceanography. 38:408-414.
McLaughlin, J.R., Clark, E.A., Nishri, E.D. and T.W. Anderson. 1993. Childhood leukemia in the vicinity of Canadian nuclear facilities. Cancer Causes and Control. 4:51-58.
Mehlhorn, J. 1992. Professional lung diseases in Uranium mines of the former SDAG at Kismet. Atemswegs und Lungenkrankheiten. 18:414-419.
Miller, J.J., Pawlulk, S. and G.J. Beke. 1993. Soil salinization at a side-hill seep and closed basin in southern Alberta. Can. J. Soil Science. 73:209-222.
Murdock, B.N. 1971. Environment: Resources, Pollution and Society. Sinauer Associates Inc. Stamford, Connecticut.
Noller, B.N. and B.T. Hart. 1993. Uranium sediments from the Magela Creek catchment, Northern Territory Australia. Environmental Technology. 14:649-656.
Obbard, J.P. and K.C. Jones. 1993. The effect of heavy metals on denitrogen fixation of Rhizobium white clover in a range of long term sewage sludge amended and metal contaminated soils. Environmental Pollution. 79:105-112.
Paine, M.D., Laggett, W.C., Mcruer, J.K. and K.T. Frank. 1992. Effects of hibernia crude oil on capelin (*Mallotus villosus*) embryos and larvae. Marine Environmental Research. 33:159.
Pierzylnski, G.M. and A.P. Schwab. 1993. Bioavailability of Zinc, cadmium, and lead in a metal contaminated alluvial soil. J. Environmental Quality. 22:247-254.
Ploeckinger, B., Dadak, C. and V. Meisinger. 1993. Lead, mercury and cadmium in mothers and their newborn babies. Zeitschrift Fuer Gebustshilfe und Perinatologie. 197:104-107.
Sarkar, A.S., Bhattacharya, K. and M.M. Sarin. 1993. Geochemical evidence for anoxic deep water in the Arabian Seas during the last glaciation. Geochimica et CosochimicaACTA 57:1009-1016.
Sevc, J., Tomasek, L., Kunz, E., Placek, V., Chmelevsky and A.M. Kellever. 1993. A Survey of the Czechoslovak follow-up of lung cancer mortality in Uranium Miners. Health Physics. 64:355-369.
Soutar, C., Campbell, S., Gurr, D., Lloyd, M., Lover, R., Cowie, H., Cowie, A. and A. Seaton. 1993. Important deficits of lung function in three modern colliery populations: Relations with dust exposure. Amer. Rev. Respiratory Disease. 147:797-803.
Stone, R. 1992. Environmental research: Oil clean-up method questioned. Science. 257:320-321.
Sykora, J.L., Smith, E.J., Shapiro, M.A. and M. Synak. 1972a. Chronic effect of ferric hydroxide on certain species of aquatic animals. Fourth Symposium on Coal Mine Drainage Research. Proceed. pp. 347-369.
Sykora, J.L., Smith, E.J. and M. Synak. 1972b. Effect of lime neutralized iron hydroxide suspensions on juvenile brook trout (*Salvelinus fontinalis* Mitchill). Water Research. 6:935-950.
Sykora, J.L., Smith, E.J., Synak, M. and M.A. Shapiro. 1975. Some observations

on spawning of brook trout (*Salvelinus fontinalis* Mitchill) in lime neutralization iron hydroxide suspensions. Water Research 9:951-958.

Tikkanen, J. and O.P. Heinonen. 1993. Risk factors for co-arctation of the aorta. Teratology 47:565 572.

Tirmarche, M., Raphalen, A. and J. Chanmeaud. 1992. Epidemiological study of French uranium miners. Cancer Detection and Prevention. 16:169-172.

Tirmarche, M., Raphalen, A., Allen, P., Chameaud, J. and R. Bredon. 1993. Mortality of a cohort of French uranium miners exposed to relatively low radon concentrations. British J. of Cancer. 65:1090-1097.

Tomasek, L., Darby, S.C., Sivaerdlow, A.J., Placek, V. and E. Kunz. 1993. Radon exposure and cancers other than lung cancer among uranium miners in West Bohemia. Lancet. 341:919-923.

Uminska, R. 1993. Cadmium contents of cultivated soils exposed to contamination in Poland. Environmental Geochem. and Health. 15:15-19.

Usis, J.D. and B.A. Foote. 1991. Influence of strip mining on the mortality of a wetland caddisfly, *Limnephilus indivisus* (*Trichoptera:Limephilidae*). Great Lakes Entomolgist. 24:133-144.

Van Metre, P.C. and J.R. Gray. 1992. Effects of uranium mining discharges on water quality in the Puerco River Basin, Arizona and New Mexico. Hydrological Sciences J. 37:463-480.

Vich, Z. and D. Koskova. 1992. On the causes of death in former miners of Uranium mines. Pracovni Lekarstvi. 44:225-228.

Wang, J., Evangelou, V.P. and B. Creech. 1993. Characteristic of manganese-calcium exchange behavior on Koalinite and Illite and pH influence. J. Environ. Sci. and Health. Part A. Environmental Science and Engineer. 28:1381-1391.

Welles, C. 1970. The elusive bonanza: The story of America's richest and most neglected natural resource. E.P. Dutton and Co. New York.

Whipple, A.V. and V.A. Dunson. 1993. Amelioration of the toxicity of hydrogen to larva stoneflies by metals found in coal mine effluent. Archives of Environmental Contamination and Toxicology. 24:194-200.

Environmental Contaminants, Ecosystems and Human Health. Edited by S.K. Majumdar, E.W. Miller, and Fred J. Brenner.

Chapter Nine

DISTRIBUTION OF ESSENTIAL ELEMENTS IN SEAWEEDS ALONG THE SOUTH-WEST COAST OF INDIA

TRESA V. FERNANDEZ and GEORGE THOMAS

Department of Aquatic Biology and Fisheries
University of Kerala
Trivandrum-7, India

INTRODUCTION

Seaweeds or marine algae are primitive plants and they constitute one of the commercially important marine living resources. They grow in the littoral and sublittoral regions up to 20 or 25 m deep in the sea, and also grow in the estuaries and backwater areas. They belong to four groups namely green, brown, red and blue green algae. The groups are based on the kind of pigments present in them, and their morphological and anatomical characteristics.

DISTRIBUTION OF SEAWEEDS

Around 700 species of marine algae have been reported along the east and west coast of India. As per CMFRI (Central Marine Fisheries Research Institute), the total standing crop of all seaweeds in Indian waters is more than one lakh tons (wet weight) consisting of 6,000 tons (wet weight) of agar yielding seaweeds and 16,000 tons (wet weight) of algin yielding seaweeds. The quantities of different groups of seaweeds comprising 35 species were green algae 245.94 tons, red algae 91.63 tons and brown algae 18.11 tons, in order of abundance. Chennubhotla et al. (1988) stated that the economically important seaweeds contributed about 15.31% and the other seaweeds to 84.69% of the total seaweed population. The individual values for the commercially important seaweeds were agarophytes of 27 tons, followed by alginophytes of 18.12 tons and agaroidophytes of 9.28 tons. The other seaweeds had a value of 300.83 tons (including edible species) making a total of 355.23 tons for an area of 315,750 sqm. According to Chennubhotla et al., the annual production of seaweeds for the entire coast is estimated around 1,000 tons depending upon the number of harvests the seaweeds usually sustain.

The south west coast of India, especially the Kerala coast, is unexplored with regard to its algal distribution, chemistry, and ecology. A preliminary report on the incidence of marine algae of the Kerala coast has been given by Nair et al. (1982). A detailed survey has been made to study the seaweed resources along the two extremes of Kerala coast by Chennubhotla et al. (1988).

The edge of the sea, a comparatively narrow zone, is one of the best habitats of plant growth (Mann, 1973). The marine macroscopic algae form the dominant lithophytic communities and have a high rate of productivity. Dawes (1981) has described the factors affecting the zonation and distribution of algal communities in inter-tidal and sub-tidal regions. About 20,000 marine algal species were distributed throughout the world and they form one of the major renewable resources of the ocean. The algae that occur along the Indian coasts have been studied. Information is available on certain aspects of their systematic distribution (Biswas, 1945, Biswas and Sharma, 1950; Dixit, 1964 & 1968 and Srinivasan, 1965). Along the Indian coastline, certain areas are well known for their luxurious growth of seaweeds. (Agadi et al. 1978, Untawale et al. 1983 have listed 623 species of marine algae from India.)

An academic survey revealed that Kerala has a rich algal flora along its coasts comprising of 30 species of Chlorophyta, 13 species of Phaeophyta and 58 species of Rhodophyta. A checklist of Indian marine algae including 520 species was prepared by Krishnamurthy and Joshi (1970). An elaborate list of 623 marine algae from India was published by Untawala et al. (1983).

SEAWEEDS AND THEIR USES

Marine algae are used for various purposes all over the world for food, medicine, fodder . . . etc. Recently, many algae are used in industry as well as for the production of fuel. Macroscopic algae are also used as a source of salts and phycocolloids (Bonotto, 1976). The algal genera such as *Porphyra, Eucheuma, Laminaria* and *Undaria* are exclusively grown under agronomic conditions. The other cultivated algae are *Enteromorpha, Gelidium* and *Gracilaria* (Dawes, 1981).

The phytochemicals such as agar, carrageenan and sodium alginate are extracted from the seaweeds. They are used as gelling, stabilizing and thickening agents in food, confectionery and pharmaceuticals, as well as in the dairy, textile, paper, paint and varnish industries. Many protein rich seaweeds such as Ulva,

Enteromorpha, Caulerpa, Codium, Monostroma, Sargassum, Hydroclathrus, Laminaria, Undaria, Macrocystis, Porphyra, Gracilaria, Eucheuma, Laurencia, and Acanthophora are used for human consumption in the form of soup, salad, curry, etc. Jelly, jam, pickle, chocolate and wafer can be prepared from certain seaweeds. Marine algae are also utilized in different parts of the world as animal feed and fertilizers for land crops. They contain more than 60 trace elements, carbohydrates, iodine, bromine, vitamins, and certain antibiotic substances.

SEAWEEDS AS BIO-INDICATORS

Industrial, domestic and other types of pollutants find their way into the water bodies which have created a cause for concern in recent years. Aquatic plants, especially marine algal species, have been suggested to be the indicators of pollution (Bryan and Hummerstone, 1973; Palmer, 1980; Preston et al. 1972; Nickless et al. 1972; Melhar et al. 1978 and Morris and Bale, 1975). It was also reported that the accumulation of trace metals in the marine waters in seaweeds will check the pollution in seawater. The accumulation of metals in seaweeds is probably by non-metabolic absorption or an ion exchange process (Gutknecht, 1955 and Baroda et al. 1964) and are retained by biota by strong chemical bonds (Goldberg, 1957).

It is also reported that some marine algae are useful as primary producers in coastal waters (Mann, 1973). Their reduction, due to pollution, would pose a threat to the next link in the food chain. There are reports from different parts of the country that the algal flora which flourished are now drastically reduced to few species, mainly as a result of increased pollution or human interference. Whatever the origin of pollution, control has to be exercised at the source. It is also important that there must be a program of environmental monitoring, both chemical and biological. This is to ensure that water-based life forms harvested for human consumption should undergo tissue analysis if the reality predicted earlier proves correct.

ESSENTIAL ELEMENTS AND SEAWEEDS

Plants as a whole (both terrestrial and aquatic) are known to contain not less than 60 elements. However, around 16 of them are needed by the plants for a normal and healthy life. Out of the 16, 9 elements are required by plants in relatively large quantities. These are called macronutrients or major elements. The major elements are C, H, O, N, S, P, Ca, K and Mg. The remaining 7 elements are required in relatively small quantities. These are called trace elements, micronutrients, or minor elements. The micronutrients are Fe, Mn, B, Cu, Zn, Mo and Cl. In addition to the 16 elements required by all plants, certain species of plants have some special requirements. Such plants require some other elements. Silica, selenium and cobalt are examples of such elements.

It must be emphasized that each plant absorbs very many elements than those mentioned above. Most of the elements that occur in soil are present in the plants. It seems probable that most of the elements like iron, zinc, etc., even in small quantities, are indispensable to normal plant growth or they influence the development and maturity of plants.

DETAILED DESCRIPTION OF THE STUDY AREA

The coastline of Kerala is rich with sandy stretches, natural rocks, and also artificial granite sea walls on 'groynes' and spurs which protrude into the sea at different places (Pl. I). The natural rocks are of laterite, granite or calcareous origin of different kinds. The sandy beaches are interrupted by rocky cliffs, peaks and creeks, and are observed to have green, yellow, red, black and grey colored stones, calcareous and mineral matter which are exposed to the sea.

While on a visit for seaweed collection, chlorine, sulphur, and acid fumes were noticed at certain locations. The tidal amplitude varied from 1-1.5 m at most of the sampling sites. The wind velocity ranged from 0.10m/s to 3.90m/s at different times of sampling at different sites. At most of the places the wind direction is in general from west to east.

The color of the soil is pink, black, grey, whitish black and so on ranging from

Pl.1. Coastline showing artificial sea walls built by stacking granite rocks which in many places protrude into the sea.

blackish white to pink. Most of the sediment substrates contain shell bits or detritus. The sediment texture varied in grain size from medium coarse to medium fine sand.

Climatic conditions along the Kerala coast are characterized by pre-monsoon (Feb-May), monsoon (June-September) and post-monsoon (October-Jan) respectively. The annual rainfall over the coast is high, ranging from 1000 to 2000 mm.

Along the entire coastline of nearly 580 km, the rocks of the different kinds and granite stones are found in patches in the inter tidal and sub-tidal regions and many species of green, brown and red algae were found to grow luxuriously on them (Pl. II).

Pl. 2. Marine algae *Ulva* and *Rhizoclonium* growing on surfaces of rocks which are partially covered by sand due to wave action.

COMPOSITION OF ESSENTIAL ELEMENTS IN SEAWEEDS OF KERALA COAST

The metal composition of marine algae has been studied by Zingde et al. (1976) and Agadi et al. (1978) along the Goa coast. A lack of knowledge about the occurrence of essential elements and trace metals in the marine algae of the south-west coast of India was apparent. The present chapter is intended to give a report on the incidence of essential elements and trace metals in the major algal species occurring in 15 stations along the south-west coast of India (Figure 1).

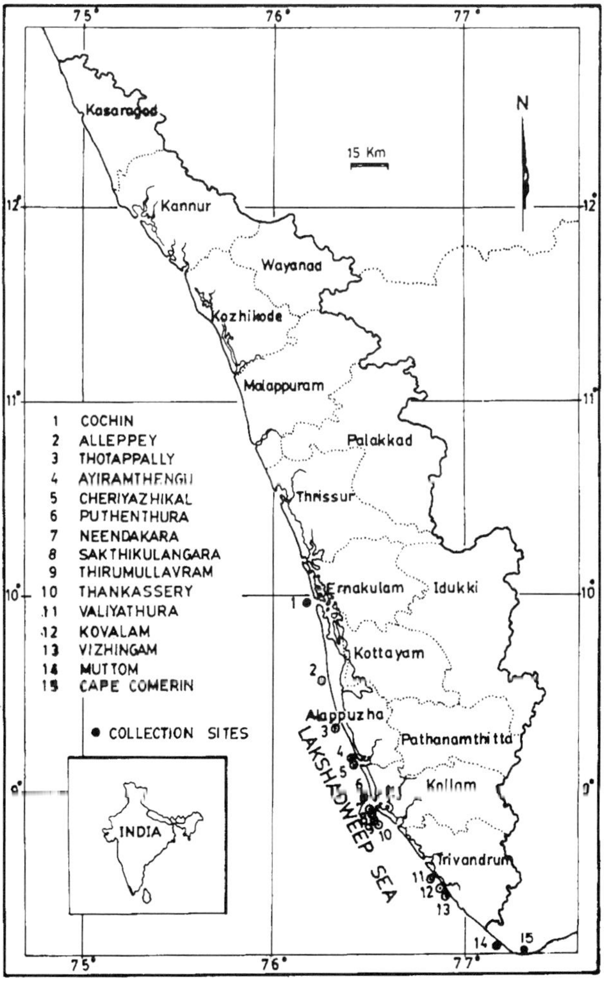

FIGURE 1. Map of South-West coast of India showing the collection sites.

METHODOLOGY ADOPTED

Algal samples were collected from the inter-tidal regions of the 15 stations along the south-west coast of India from Cochin to the farthermost tip of the Indian Peninsula, Cape Comerin (Figure 1). The sampling was done during the month of October and November. Algae were hand-picked from the substratum, washed in fresh water and sorted out into different species. Species identification was done with the help of published literature. Samples were dried in an oven at 80±1°C to constant weight. Then they were powdered and weighed to sub samples of approximately 0.5 gms each. These samples were then digested with a combination of analar grade nitric acid, hydrochloric acid, and perchloric acid (EPA, 1983). Digested material was filtered and made up to a known volume using glass distilled water. The analysis for Na, K, Ca, Mg, Sr, Mo, Fe, Mn, Cu and Zn was carried out using an inductively coupled plasma emission spectrophotometer (Model No. A.R.L 3410 I.C.P.). The standards and blanks were also treated in the same way as test samples except the addition of algal material.

RESULTS AND DISCUSSION

The species distribution of marine algae along the 15 stations are shown in Table 1. A total of 18 genera of algae were present in these coastal areas studied. Among the marine algae collected Chlorophyta was represented by seven genera,

TABLE 1

Species distribution of marine algae in different stations along the South West Coast of India

Name of Species		Stations														
Family	Species	1	2	3	4	5	6	7	8	9	10	11	12	13	14	15
Cladophoraceae	*Rhizoclonium kerneri*	-	+	+	-	+	-	+	+	+	+	+	+	-	-	-
	Chaetomorpha antennina	-	+	-	+	+	+	+	-	-	+	+	+	+	+	+
	Cladophora fascicularis	-	-	-	-	-	-	-	-	-	-	+	-	-	-	-
Ulvaceae	*Ulva fasciata*	-	-	-	+	+	+	+	+	+	+	+	+	+	+	+
	Enteromorpha linza	+	-	-	-	-	-	-	-	-	-	-	-	-	-	-
Valoniaceae	*Valaniopsis pachynema*	-	-	-	-	-	-	-	-	+	-	-	-	-	-	+
Caulerpaceae	*Caulerpa chemnitzia*	-	-	-	-	-	-	-	-	-	-	-	-	-	-	+
Sargassaceae	*Sargassum wightii*	-	-	-	-	-	-	+	-	+	-	-	-	-	-	+
Scytosithonaceae	*Chnoospora minima*	-	-	-	-	-	-	-	-	-	-	+	+	-	-	-
Hypneaceae	*Hypnea musciformis*	-	-	-	-	-	-	-	-	-	-	-	+	-	-	+
Rhodomelaceae	*Laurencia obtusa*	-	-	-	-	-	-	-	-	-	-	-	-	-	+	-
Ceramiaceae	*Ceramium fimbriata*	-	-	-	-	-	-	-	-	+	+	-	+	+	-	-
	Spyridia insignis	-	-	-	-	-	-	-	-	-	-	-	-	-	-	+
Gracilariaceae	*Gracilaria corticata*	-	-	-	-	-	-	-	-	+	-	-	+	-	+	+
Gelidiaceae	*Gelidiopsis repens*	-	-	-	-	-	-	-	-	-	-	-	-	-	-	+
	Gelidium pusillum	+	+	-	-	-	+	-	-	-	-	-	-	-	-	-
Delesseiraceae	*Caloglossa pusillum*	+	-	-	+	-	-	-	-	-	-	-	-	-	-	-
Cryptonemiaceae	*Grateloupia comoroni*	-	-	-	+	+	+	-	-	-	-	-	-	-	-	-

+ Present
- Absent

Phaeophyta by two genera, and Rhodophyta by nine genera. The most widely distributed algae were *Ulva fasciata, Chaetomorpha antennina* and *Rhizoclonium kerneri* which were present in twelve, eleven and nine stations respectively. The algal diversity was found to be maximum in station 1, that is, Cape Comerin, where a total of nine algal species were present. Kovalam also showed a high degree of species diversity where 7 species of algae were reported. Only a single species of algae, *Rhizoclonium kerneri* was present in Thottapally coastal area.

The elemental composition in these marine algae throws some light on their adaptability in concentrating the essential elements and also shows a reflection of the trace elemental availability in the different coastal zones of Kerala. Figure 2, Figures 3.1 to 3.6 and Table 2 depict the relative concentrations of 10 elements in the algal species and the nature of accumulation of these elements in different algal species present in these areas. From Figure 2, it can be noted that Station 9

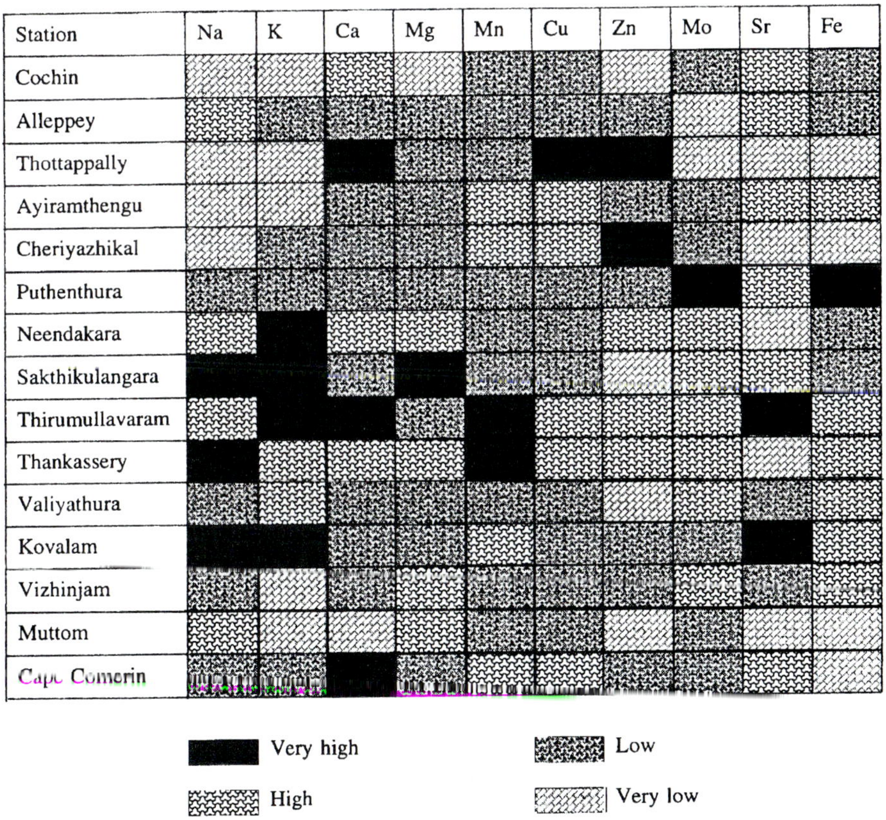

FIGURE 2. Relative concentrations of the elements in the algae along the 15 sampling stations. The concentrations of the 10 elements in the different stations expressed as mean value for all the algal species present in that station. The concentration levels are classified as very high, high, low and very low by taking considerations of the overall values obtained in the present study.

(Thirumullavaram) is characterized by the presence of all the 10 elements in relatively high quantities in the algal species collected. Station 9 (Thirumullavaram) is characterized by a substratum formed of laterite rocks. In all the other stations the substratum is formed of granite rocks. The nature of substratum determines the occurrence and distribution of the algal species and also their ability to concentrate the elements from the surroundings. The red and brown algae were found to effectively concentrate some elements such as Mn and Sr. These elements are found in higher concentrations in algae from Thirumullavaram. The species composition of this station shows that the majority of algal species come under the division Rhodophyta and Phaeophyta. At station 2 (Muttom) all the 10 elements were present in lesser concentrations. No element occurred in 'Very High' concentration in the stations Muttom, Vizhinjam, Valiathura, Ayiramthengu, Alleppey and Cochin (Figure 2). Fe occurred in very high concentrations at Puthenthura, whereas at Thottapally, Cu and Zn are in very high concentration.

Fig. 3.1

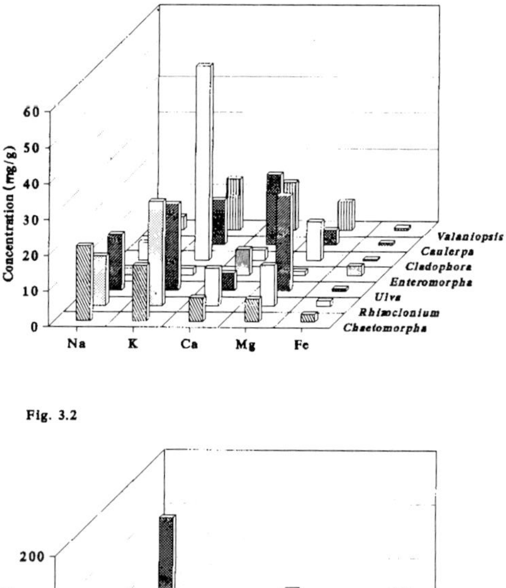

Fig. 3.2

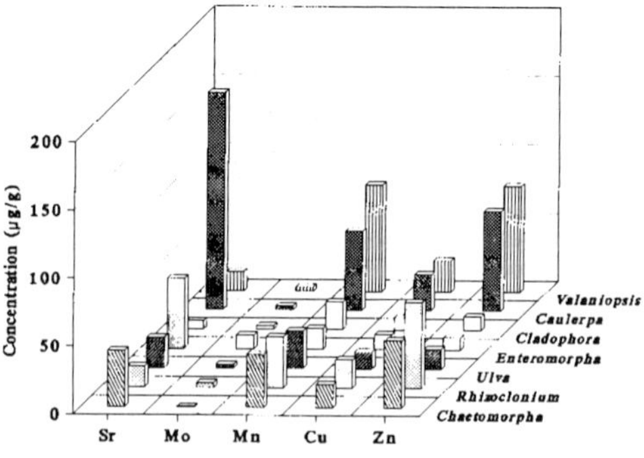

FIGURES 3.1 TO 3.6. Concentration of the elements in various algal species along the collection sites. The concentrations are expressed as mean value for each algal species collected from all the stations where it was present.

Sakthikulangara is characterized by very high contents of Mg whereas high contents of Mn are observed at Thankassery and Thirumullavaram. Sr is present in very high concentration at Kovalam and Thirumullavaram.

The concentrations of 10 elements in different algal species are depicted in Figures 3.1-3.6. The concentrations have been expressed as mean value for each algal species collected from all the stations where it was present. Among the algal species analyzed the red and brown algae are found to contain more elements in higher concentrations. *Ceramium fimbriata* was found to be efficient in concentrating Ca, K, Fe, Sr, and Mo. Very high concentrations of Fe are recorded in this species (10172.37 ppm). Very high concentrations of Fe are also recorded in *Cladophora fascicularis* at Thankassery and Puthenthura. Agadi et al. (1978) reported high values of Fe in the algal species collected from Chapora Bay in Goa. *Ulva fasciata* is found to be exceptionally richer in Mg content. *Ceramium fimbriata* and *Spyridia insignis* are richer in Ca content. *Gracilaria corticata, Grateloupia comoronii, Caulerpa cheminitzia* and *Valaniopsis pachynema* are found to contain high concentrations of Mn, Cu and Zn. The high level of Cu

Fig. 3.3

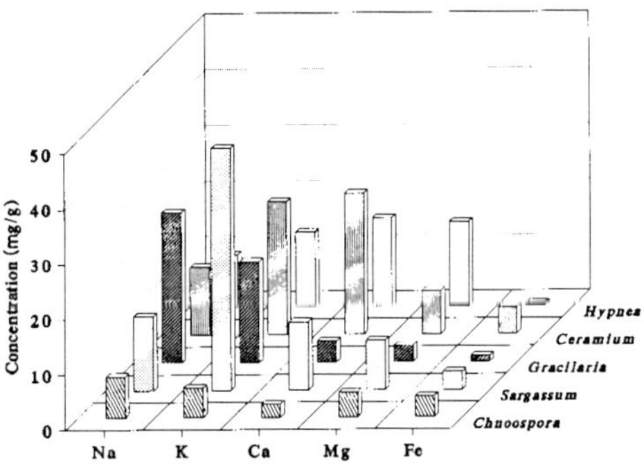

Fig. 3.4

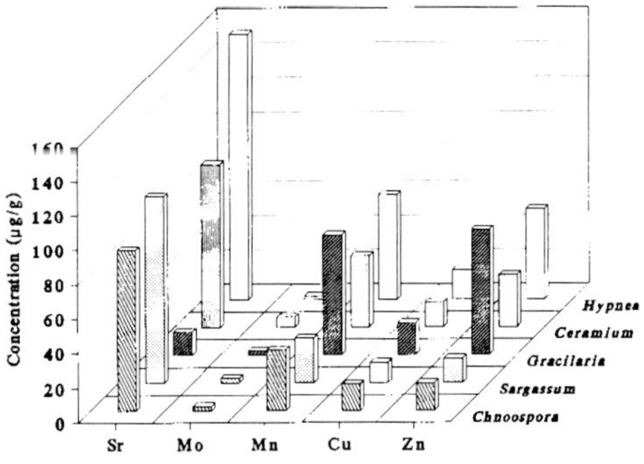

content is in accordance with the reported value of Agadi et al. (1978) for the Goa region. Bryan (1969) had reported values of Zn greater than 6(µg/g in the brown algae, *Laminaria digitata,* in the British coastal area. Sr is present in relatively higher concentrations in the red and brown algae such as, *Ceramium fimbriata, Hypnea musciformis, Chnoospora minima,* and *Sargassum wightii.* This metal was not determined in the green algae except in *Caulerpa cheminitzia.* Na and K are present in relatively high concentrations in *Gracilaria corticata* and *Ulva fasciata.* Na is in extremely low concentration in *Enteromorpha linza, Cladophora fascicularis, Caulerpa cheminitzia, Valaniopsis pachynema, Chnoospora minima, Gelidium pusillum, Gelidiopsis repens, Spyridia insignis,* and *Laurencia obtusa*.

This study revealed that many algae contain fair concentrations of essential elements. This study also indicated that the essential elements in the marine algae of the Kerala coast are on the higher side as compared to the earlier reports. The reason for higher concentrations may be due to the difference in pH, higher temperature, salinity and other metal binding nutrients in sea water (Bryan, 1976 and

Fig. 3.5

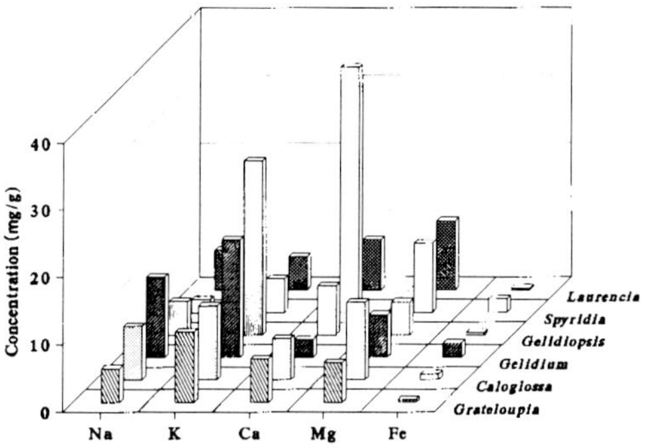

Fig. 3.6

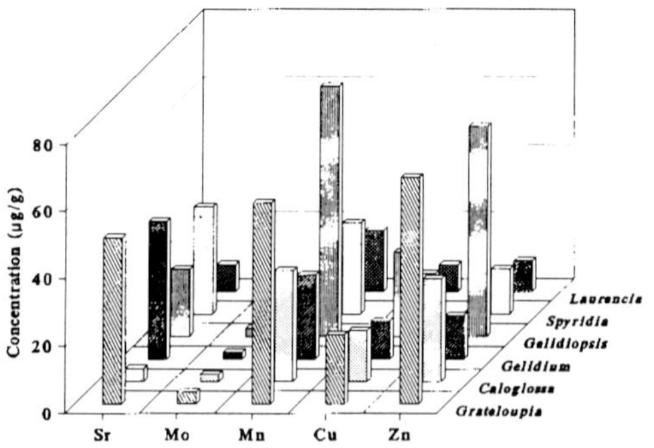

TABLE 2

Concentration of the elements in various algal species along the collection sites

Species of algae	Na	K (mg/g)	Ca	Mg	Fe	Sr	Mo (μg/g)	Mn	Cu	Zn
1 *Chaetomorpha antennina*	21.10	15.59	6.49	6.25	2.05	41.09	0.39	38.72	16.93	49.43
2 *Rhizoclonium kerneri*	13.83	29.17	10.54	11.52	1.57	15.55	4.03	37.70	20.80	63.27
3 *Ulva fasciata*	15.71	24.21	4.91	25.56	0.70	22.33	2.73	27.53	12.48	14.53
4 *Enteromorpha linza*	2.00	2.00	7.30	1.20	2.80	52.00	9.90	15.20	11.00	9.00
5 *Cladophora fascicularis*	5.00	54.00	2.80	10.80	0.40	6.00	2.70	20.70	9.00	9.90
6 *Caulerpa cheminitzia*	2.00	12.50	19.60	4.30	0.68	160.00	2.85	58.55	26.80	73.90
7 *Valaniopsis pachinema*	3.50	14.00	13.15	7.90	0.70	14.00	3.08	78.45	23.20	77.90
8 *Chnoospora minima*	7.50	5.50	2.45	4.60	3.84	93.00	2.55	34.90	15.30	15.90
9 *Sargassum wightii*	13.33	43.67	12.27	8.87	3.16	108.66	2.97	25.80	11.50	13.83
10 *Gracilaria corticata*	27.25	18.25	3.78	2.98	1.04	13.50	2.33	69.73	18.53	72.60
11 *Ceramium fimbriata*	12.25	24.13	25.55	7.78	4.68	95.00	5.85	41.30	14.53	30.08
12 *Hypnea musciformis*	9.50	13.50	16.00	15.20	0.48	154.00	2.18	60.95	17.00	52.65
13 *Grateloupia comoroni*	5.00	10.50	6.50	5.93	0.41	49.33	3.55	59.67	20.53	67.47
14 *Caloglossa pusillum*	8.00	11.00	6.20	11.50	0.88	4.00	2.25	32.85	15.10	30.45
15 *Gelidium pusillum*	12.00	17.50	2.70	6.35	2.12	41.00	2.03	24.85	11.25	12.90
16 *Gelidiopsis repens*	5.00	26.00	7.30	4.90	0.36	20.00	2.40	74.50	25.00	62.70
17 *Spyridia insignis*	2.00	5.00	36.40	10.30	2.00	32.00	3.00	27.20	11.90	13.20
18 *Laurencia obtusa*	6.00	5.00	7.50	10.30	0.28	8.00	2.40	18.30	8.00	9.20

Sivalingam, 1978). This may also be due to high water movements and recycling of the mineral elements after the decomposition of organisms in the sea.

It is a well known fact that in minute quantities many trace elements are nutritive to the growth of marine algae (North et al., 1972), and the present values are far exceeding as compared to the values for other marine plants (Browen, 1966).

Relatively high concentrations of all 10 elements were determined in most of these species at station 9. Except for some fish processing units at this region, there was no other prominent means of causing water contamination and hence the high metal concentration could not be explained.

CONCLUSION

Considering the ever growing demand of food for human consumption and for other purposes, it is essential to locate the non-conventional sources. In this context, the incidence of essential elements in the marine seaweeds occurring along the Kerala coast indicates that at least some species could be utilized as additional sources of essential elements when the need arises. Based on the present investigation, in general, we can conclude that red and brown algae showed ability to concentrate a greater amount of trace elements than those of the green algae.

REFERENCES

Agadi, V.V., Bhosle, N.B. and A.G. Untawale. 1978. Metal Concentrations in Some Seaweeds of Goa (India). *Bot. Mar.* Vol. XXI, pp. 247-250.

Biswas, K. 1945. A General Review of the Marine Algae of the Western Coast of India. *J. Bombay Nat. Hist. Soc.* 45:515-530.

Biswas, K. and A.K. Sharma. 1950. *Sargassa* of Indian Sea. *J. Roy Asiat. Soc. Bengal (Science).* 16(1):79-97.

Bonotto, S. 1976. Cultivation of Plants. Multicellular Plants. In: Kinne, O. (Ed.) *Marine Ecology,* Vol. 3. *Cultivation* Part I, Wiley, New York.

Bowen, H.J.M. 1966. *Trace Metals in Biochemistry.* Academic Press Inc., London. P. 241.

Broada, E., Dresser, H. And G. Findenegg. 1964. Wirkung von Dinitropenol, Azid und Anaerobiose suf die Zinkaufnahme durch Algan. *Naturwissen-schaffen.* 51:361-362.

Bryan, G.W. 1969. Absorption of Zn and other Metals by the Brown Seaweed. *Laminaria digitata. J. Mar. Biol. Ass. U.K.* 49:225-243.

Bryan, G.W. 1976. Heavy Metal Concentration in the Sea. In: R. Johnston (Ed.) *Marine Pollution.* Academic Press Inc., London. Pp. 185-302.

Bryan G.W. and Hummerstone, L.G. 1973. Brown Seaweeds as Indicators of heavy metals in Estuaries in South West England. *J. Mar. Biol. Ass. U.K.* 53:705-720.

Chennubhotla, V.S.K., Ramachandrudu, B.S., Kaladharan, P. and S.K. Dharmaraja. 1988. Seaweed Resources of Kerala Coast. *Aqua. Biol.* VII:69-74.

Dawes, C.J. 1981. *Marine Botany.* John Wiley and Sons, Holland, p. 628.

Dixit, S.C. 1964. Species List of Indian Marine Algae determined by Boergessen. *J. Univ. Bombay.* 32(2-5):1-23.

Dixit, S.C. 1968. Species List of Indian Marine Algae II. *J. Univ. Bombay.* 32 (3-5):9-24.

EPA. 1983. Methods for Chemical Analysis of Water and Wastes. United States Environmental Protection Agency. EPA-6004-79-020. USA. *Methods.* 110.1-430.2.

Goldberg, E.D. 1957. Biochemistry of Trace Metals. *Geol. Soc. Amer. Memoir.* 67(1):345-356.

Gutknecht, J. 1965. Uptake and Retention of Cesium-137 and Zinc-65 by Seaweeds. *Limnol. Oceanogr.* 10:58-66.

Krishnamurthy, V. and M.V. Joshi. 1970. A Checklist of Indian Marine Algae. Central Salt and Marine Chemicals Research Institute, Bhavnagar, India. P. 37.

Mann, K.H. 1973. Seaweeds: Their Productivity and Strategy for Growth. *Science.* 182:975-981.

Melhuus, A., Seip, K.L, Seip, H.M. and S. Myklestad. 1978. A Preliminary Study of the use of Benthic Algae as Biological Indicators of Heavy Metal Pollution in Sorfjorden, Norway. *Environ. Pollut.* 15:101-107.

Morris, A.W. and A.J. Bale. 1975. The accumulation of Cadmium, Copper,

Manganese and Zinc by *Fucus vesiculosus* in the Bristol Channel. *Estuar. Coastal Mar. Sci.* 3:153-163.

Nair, N.B., Sobha, V. and M. Arunachalam. 1982. Algae from the Southern Kerala Coast. *Indian J. Mar. Sci.* 11:266-269.

Nickless, G., Stenner, G. and N. Terrille. 1972. Distribution of Cadmium, Lead and Zinc in the British Channel. *Mar. Poll. Bull.* 3:188-190.

North, W.J., Stephens, G.C. and B.B. North. 1972. Marine Algae and their Relation to Pollution Problems. In: M. Ruivo (Ed.) *Marine Pollution and* Sea *Life*. F.A.O. Publication, London.

Palmer, C.M. 1980. *Algae and Water Pollution,* Castle House Publications Ltd. London, p. 123.

Preston, A., Jeffries, D.F., Harvey, B.R. and A.K. Steele. 1972. British Isles Coastal Waters: the Elemental Concentration of the Selected Heavy Metals in Sea Waters, Suspended Matter and Biological Indicators-A Pilot Survey. *Environ. Poll.* 3:69-82.

Sivalingam, P.M. 1978. Bio-deposited Trace Metals and Mineral Content of some Tropical Marine Algae. *Bot. Mar.* 21:327-330.

Srinivasan, K.S. 1965. Algarum Species ex India-Oriundae. *Bull. Bot. Surv.* India. 67(1-4):188-266.

Untawale, A.G., Dhargalkar, V.K. and V.V. Agadi. 1983. List of Marine Algae from India. *National Institute of Oceanography Report.* Dauna Paula, Goa, India, p. 45.

Zingde, M.D., Singbal, S.Y.S., Moraes, C.F. and C.V.G. Reddy. 1976. Arsenic, Copper, Zinc and Manganese in the Marine Flora and Fauna of Coastal and Estuarine Waters around Goa. *Indian J. Mar. Sci.* 5:212-217.

Environmental Contaminants, Ecosystems and Human Health. Edited by S.K. Majumdar, E.W. Miller, and Fred J. Brenner.

Chapter Ten

IMPACTS OF CONTAMINANTS ON MARINE COASTAL ECOSYSTEMS*

HERBERT M. AUSTIN

School of Marine Science/Virginia Institute of Marine Science
College of William and Mary
Gloucester Pt., VA 23062-1346

INTRODUCTION

Magnitude of the Problem

Marine resource distribution and abundance depend upon fluctuations in the natural environment (weather or climate), anthropogenic input (pollution), and harvest (fishing) (1). Partitioning the effects of each is nearly impossible and creates a frustrating situation for managers and politicians alike. In the open ocean, where most American fisheries management was founded, the effects of pollutants are limited and fluctuations in weather are damped by the slow thermal response time of the ocean. Here the effects of harvest can be demonstrated and controlled. Management principles developed for the open ocean have often been applied to estuarine resource problems. Yet it seems unfair to restrict harvest by commercial fishermen and recreational anglers in order to reverse a decline in stock abundance if the cause of the decline may be in fact, reduced fecundity due to contaminants.

The physical geography of an estuary, coupled with its proximity to anthropogenic sources, makes it extremely susceptible to collection and retention of contaminants. By its very definition an estuary is a region where the river meets the sea, and the sea water measurably diluted. Estuarine flow creates what is known as a salt wedge (Figure 1) (2) which provides a constant inflow of clean oceanic waters, but which also provides for a retention mechanism for sediments, plankton, and contaminants. Estuaries, often with extensive river systems (e.g.

*VIMS Contribution Number 1878.

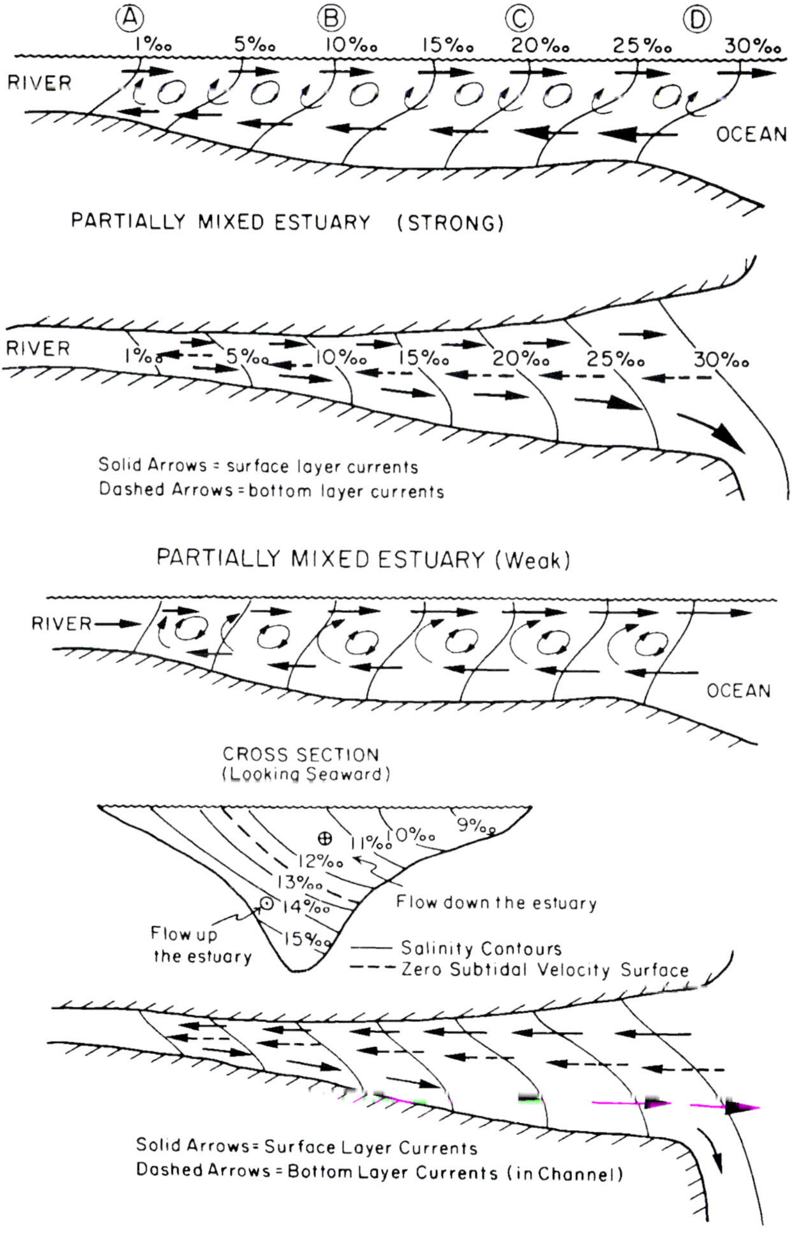

FIGURE 1. Vertical section of a Partially Mixed (Strong) and a Partially Mixed (Weak) estuary showing surface and bottom layer circulation differences, midwater mixing zone, and distribution of salinity (‰). Reprinted from J.R. Schubel and D.W. Pritchard, *Contaminant Problems and Management of Living Chesapeake Bay Resources,* 1987, The Pennsylvania Academy of Science.

Chesapeake Bay) drain large areas of agricultural land and provide a meeting place for commerce transportation routes (e.g. Valdez, Alaska) and urban and industrial development (New York Bight, Zuider Zee, Minimata). Consequently, terrestrial pollution/contaminant inputs are large, and coupled through binding to sediments and biota, with the circulation system, are retained, in enormous local concentrations (3) within the estuary.

Contaminants from point sources are generally restricted to within a few kilometers of the source as dilution, sedimentation and chemical processes act to impede their spread. Important contaminants from point sources include chlorine, halogenated hydrocarbons (pesticides and PCB's) and trace metals from industry (3, 4). Nutrients are also significant point source pollutants when they enter the system from sewage discharges. Nutrient inputs can unbalance communities of primary producers, often leading to dramatic dissolved oxygen fluctuations.

Non-point sources of contamination (see Brenner, et al) include land run-off and atmospheric deposition. Important contaminants from these sources include herbicides and pesticides from agriculture, and metals from industry and land fill leaching. Globally distributed compounds include polychlorinated biphenyls (PCB's), and Polycyclic Aromatic Hydrocarbons (PAH's) from air borne sources (5). Localized high concentrations of these pollutants are normally attributed to point sources. Air borne sources are a greater problem along east coast estuaries in the industrially developed northwestern hemisphere where zonal westerly winds transport air masses, and their pollutant loads, across the continent. Northern hemisphere west coasts are spared as the zonal westerlies arrive from over the open ocean.

Contaminants affect organisms several ways. Aside from direct lethality, dissolved fractions in the water column may interfere with an organism's ability to "smell" thereby reducing its opportunities to reproduce, migrate, detect food or avoid predators. Genetic effects can result in abnormalities or death. Exposure to dissolved contaminants leading to high body burdens can reduce fecundity, growth and primary production rates (6). Direct contact with contaminated sediments or surface slicks, can cause body lesions, fin rot, liver and kidney alterations, and cataracts (3, 7, 8, 9); lymphocystis (10); reduced phagocytic activity (11); and reduction in intestinal activity (12). Tumorilytic activity can be suppressed resulting in increased cancer rates (13), and when concentrations are high enough, neoplasms and carcinomas can form in as little as a single season (14, 15). Recently, it has become apparent that immune responses to disease and parasites can also be supressed by the presence of contaminants (15, 16, 17).

Many contaminants move through an ecosystem via the food chain. At the base of the food chain are the primary producers. Primary producers in a temperature or subtropical riverine estuary include: The emergent wetlands (e.g., *Spartina alterniflora*), Submerged Aquatic Vegetation, SAV's (e.g. *Zostera marina*), and pelagic phytoplankton (e.g. diatoms and dinoflagellates) (Figure 2) (6). Estuarine macroalgae play a more significant role in rocky estuaries where the hard substrate provides a site for attachment, but little littoral soil for roots or rhizomes. In the open ocean or over the continental shelf phytoplankton provide the major source of fixed carbon to the food chain. In the estuary however, there is a seasonal progression as the detritus from the emergent wetlands (*Spartina*) assumes a dominant role in early spring (18), followed by phytoplankton (diatoms) in late-spring to early-summer, then a microbial food web production (picophytoplankton) (19) after summer stratification. Tropical systems are supported by benthic algae, mangrove prop root epiphytic algae, extensive symbiotic green algae living within coral heads, and reef-building coralline algae.

The estuary is one of the most productive ecosystems in the world. Phytoplankton, which normally provide production in three dimensions in the aquatic/marine environment, often provide only near-surface two dimensional photosynthesis in the nutrient rich, but light-limiting estuary. Nevertheless, this producing is high, and with coastal wetlands out-producing most heavily cultured agricultural lands.

DISCUSSION

The author's experience, familiarity, and contacts with the problem of contaminants in the estuary are somewhat parochial in that they are limited to east coast United States estuaries, as evidenced by the citations. Nevertheless, a contaminant produces similar impacts on the flora and fauna regardless of whether it is in the Chesapeake Bay, North Sea, or Indian Ocean. Therefore, the examples given in the following discussion apply, regardless of geography.

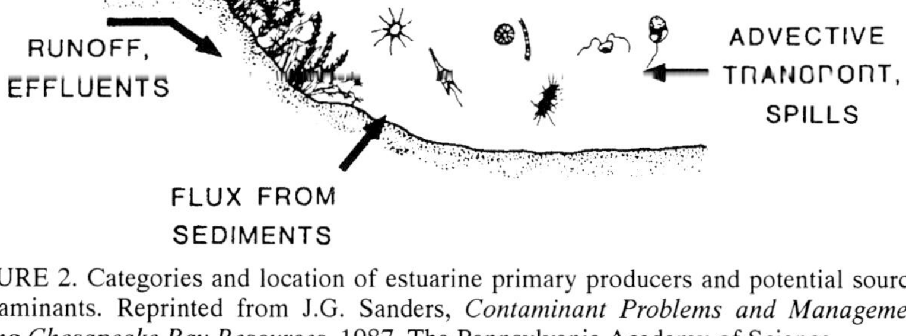

FIGURE 2. Categories and location of estuarine primary producers and potential sources of contaminants. Reprinted from J.G. Sanders, *Contaminant Problems and Management of Living Chesapeake Bay Resources,* 1987, The Pennsylvania Academy of Science.

PLANTS/PRIMARY PRODUCTIVITY

Impacts on the Emergent Vegetation

Marsh grasses live in an environment subject to large daily and seasonal fluctuations of temperature, salinity and periods of inundation. Typical of ecosystems that span environmental extremes, they are extremely hardy and tolerant of the wide fluctuations, including chemical stress (6). Notable is the ability of *Spartina alterniflora,* the salt marsh cord grass, to accumulate heavy metals (20) and to degrade contaminants such as the herbicide atrazine into less toxic compounds (21). It has been suggested that the "... large degradation capacity of estuarine communities may act to prevent transport of labile organics from land to the oceans (22)." It is for this reason that salt marshes, in addition to their flood buffering ability, serve as valuable chemical filters.

Studies of the actual impacts of contaminants on marsh grasses is limited, although simulated oil spills have been shown to cause reductions in growth, and when introduced to the rhizome system resulted in death (23).

Wetlands provide the bulk of the annual detritus load to estuarine systems which is consumed by detritivor zooplankters (e.g. *Eurytemora*) (18). This establishes an efficient short food chain for rapid energy transfer from the primary producer to higher trophic levels, and is unlike the more traditional open water pelagic primary production provided by phytoplankton. It also provides, however, for a rapid transfer of contaminants.

Impacts on Submerged Aquatic Vegetation (SAV)

Best known of the SAV's are the temperate eel grass, *Zostrea*, and tropical turtle grass, *Thallassia*. As with the emergent grasses, SAV's concentrate contaminants, particularly heavy metals, which are then passed up the food chain. Chlorine discharges have been implicated in SAV decline in the Chesapeake Bay, U.S.A., but there is no direct proof of this assertion (6).

Impacts on the Phytoplankton

Little progress has been made toward generalizing the response of phytoplankton communities to toxic substances (6). The coastal region is however, an area of water mass mixing, where estuarine "fronts" are common (23, 24, 26). These fronts, or areas of convergence, are visually apparent as slicks and foam lines, both of which include a complex mixture of organic and metallic compounds. Contaminants may be concentrated 10 to several thousand fold over background levels (27). Both phytoplankton, and their predators, the zooplankton, also concentrate in slicks associated with fronts. As a result, phytoplankton may accumulate body burdens that are several times the concentration in the surrounding water (27).

As with the vascularized plants the primary impact of contaminants is decreased growth and photosynthesis. Not only does the response result in a seasonal alteration of the succession or domination of species, resulting in a phytoplankton assemblage that may affect the predator-prey interaction with zooplankton or bivalves, but the resistance of phytoplankton to many contaminants, and their ability to concentrate toxic materials, may result in magnified concentrations in higher trophic levels.

INVERTEBRATES

Impacts on the Zooplankton

In an excellent review of the effects of contaminants on estuarine zooplankton, Bradley and Roberts (28) report that most studies have been conducted on larvae of clams, oysters and crabs or on copepods and mysids. Further, that most studies have focused on the effects of heavy metals, pesticides, oxidants, and the search for an effective integrated laboratory/field assay. Zooplankton, like phytoplankton, may not only be killed by contaminants found near discharges from industry (29), but also exhibit reduced rates of growth (30), and in some cases develop abnormal growth patterns (31). They may also change their behavior making them more susceptible to predation by jellyfish and fish larvae (32). Further, reduction in the rate of larval growth results in a longer planktonic existence, and therefore greater exposure to predation (31). Many impacts are most noticeable shortly after fertilization, and become less so as time progresses (33).

Environmental factors also play a role in determining the degree of sensitivity of plankton to contaminants. At low temperatures many species are less sensitive because physiological responses are slowed resulting in less uptake of contaminants. Sometimes reduced salinity increases sensitivity to contaminants. Tolerance to some heavy metals (noticeably Cd) is most pronounced in freshwater, less so in the saline estuary. The synergistic effects of temperature, salinity and contaminants, particularly heavy metals, on metabolism is sometimes hard to partition because the contaminant acts in different ways in different environments (34).

Impacts on Benthic Invertebrates

Heavy metals, particularly methyl mercury, can cause outright mortalities in crustaceans (*Uca*) (34); impair the avoidance response of grass shrimp, *Palaemonetes* (32); and reduce larval growth of oysters exposing them to predators for a longer period (31). Chlorine, associated with sewage discharge, has been shown to cause a loss of osmotic integrity in blue crabs (*Callinectes*) (35) and produces a reduced enzymatic response in oysters (35). Oyster larvae experimentally exposed to chlorine from sewage discharge exhibited no growth (36).

Water soluble polyaromatic hydrocarbons (PAH's) have been documented as capable of decreasing the number of hemocytes in the American oyster, *Crassostrea virginica* (37), and significantly suppressing their overall immune response to protozoan infection by *Perkinsus marinus* (17).

Organotins (TBT or tributyltin) was the active ingredient in many antifouling paints until banned. TBT based paints were proposed for wide-spread use by the U.S. Navy prior to the ban. In carrying out its role as an antifouling compound, TBT also affected other, non-target, organisms (38, 39). The primary observed affects of TBT were decreased spatfall and growth in oysters and other commercially important shellfish, and some degree of male sex organ development on females producing various degrees of sterility (39).

FINFISH

Impacts on the Finfish

Pollution has been increasingly the alleged cause of the decline of estuarine fish stocks in recent years. Often this is allegation is not accompanied by supporting data, or the concentrations required to produce the effects in a laboratory experiment are so far above ambient that no realistic ecosystem impact is justified. Nevertheless, this attribution of stock decline to pollution continues (40, 41).

Data on body burdens of pollutants, in animals exposed to less than the massive levels associated with accidental spills, became available during the mid-1960's with the development of more sensitive instrumentation. When concern was raised over the apparent build up of heavy metals, particularly mercury, and PCBs (Polychlorinated Biphenyls) during the late 1960's and early 1970's in seafood, it was possible to trace the concentrations both within species and individual organs. Generally concentrations in estuarine organisms far exceeded the background environmental levels. Since compounds may accumulate in higher concentrations as one moves up the food chain, U.S. federal agencies, charged with regulating and protecting the consumer from contaminated seafood, required warning about per dium consumption on some seafoods (e.g. swordfish) and an outright ban on others taken from particular water bodies (e.g. bluefish, *Pomatomus saltatrix* from the James River, VA, U.S.A. and striped bass, *Morone saxatilis* from the Hudson River, NY, U.S.A.).

Results of experimental exposure of finfish to waters from PAH contaminated rivers have prompted some scientists to speculate that there is a reduced phagocytic efficiency, thereby predisposing the organism to disease (42). Further studies have documented that they cause a release of Hemoglobin into the mucus, and destroy the mucus forming goblet cells in the skin (43). PAH contamination also has been shown to result in the formation of cataracts (44), and to interfere with synthesis of DNA and RNA as well as proteins in general in species of the family Sciaenidae (45) (drums and croakers). Striped bass larvae exposed to chlorine have shown reduced vigor (46) and therefore an increased susceptibility to predation. Exposure of juvenile fish (spot, *Leiostomous xanthurus,* Fa. Sciaenidae) to Kepone (pesticide/rodenticide) resulted in skeletal abnormalities, and often autofracturing of the vertebrae (47). Most direct however, is damage at the genetic level. Chromosome mutation has resulted in abnormalities in fishes exposed to crude oil (48, 49). Less apparent, but equally important can be the reduction in growth of larval fish exposed to crude oil during a spill (50). Some juvenile flatfish, in fact, prefer "oiled" sediments to natural sediment in a laboratory experiment (51).

Ecosystem Interactions

While contaminants have been documented as capable of impacting particular organisms in laboratory experiments, and have been directly and circumstantially linked to *in situ* degradation, the measured levels of contaminants in the environment are often well below the measurable impact level. At these low but persistent levels fecundity or growth are reduced, feeding habits change, and many contaminants migrate up the food chain, increasing in concentration at each level, and potentially causing impacts several trophic levels away. One of the best examples of this is the thinning of raptor and heron egg shells due to consumption of fish containing pesticides (52). Some contaminants are even transported globally. The carnivorous ice fish (Notothinidae), only found in the Antarctic Ocean, has been found to have measurable levels of pesticides. Even sea turtles, feeding on jellyfish have been shown to exhibit elevated levels of pesticides (53). The entrance and route of contaminants through the ecosystem is varied, but one sure route is the food chain. Many contaminants, chlorinated hydrocarbon pesticides for example, remain in the body and as contaminated food are ingested the body burden remains, passed on from one level to another, becoming more concentrated with each level. In a classical study on Long Island, NY, U.S.A. it was shown that predator pesticide concentrations were from five to ten times that of the prey (54).

Many recent studies have pointed to the developing concern that contaminants reduce the ability of an organism to battle disease or parasitism (7, 8, 11, 17). This results in mortality and reduced vigor in spite of an otherwise healthy appearance (55). Oysters and other bivalves can close and initiate an anaerobic metabolism when challenged with high levels of contaminants. For this reason they are often affected at low concentrations, but seem to survive high levels of contaminants when exposure periods are short.

Fishery Impacts

The impacts of contaminants on coastal marine organisms is two fold. Not only are there direct biological consequences, but also economic losses resulting from closures of shellfish beds or removal of a species from the market place because of contaminant levels in edible meats deemed harmful to humans (39). For example, Kepone, a pesticide, identified in the James River, Virginia, USA in 1975 resulted in a closure of the entire river to the taking of seafood. Levels of Kepone harmful to blue crabs (56) were several times that found in the environment, yet James River blue crabs were considered contaminated, and so taken off the market (39). Harvest of both recreational and commercial James River finfish, including the valuable striped bass, *Morone saxatilis*, was prohibited. Ironically, the striped bass is anadromous, spawning in the river before migrating out to sea. The moratorium therefore had the effect of creating a sanctuary that lead to a population resurgence.

Even when there is no observable impact on the organism, nor a documented danger to humans, the perception by the public that the seafood is "tainted" or contaminated leads to reduced demand (57, 58). This leads to the "seafood paradox". Smart consumers know that seafood is inherently healthy, but they have a distrust for products that come from an uncontrolled environment (58).

REFERENCES

1. Austin, H.M. 1983. Ecosystem management: An alternative to fisheries management, Va. Marine Res. Rept. 83-10, SMS/VIMS, Coll. William and Mary, 11 pages.
2. Schubel, J.R. and D.W. Pritchard. 1987. A brief physical description of Chesapeake Bay, pp. 1-32, In: S.K. Majumdar, L.W. Hall and H.M. Austin (Ed.) *Contaminant Problems and Management of Chesapeake Bay Resources,* Pennsylvania Academy of Sciences Publ., Easton, PA, 573 pages.
3. Huggett, R.J., Van Veld, P.A., Smith, C.L., Hargis, W.J. Jr., Vogelbein, W.K. and B.A. Weeks. 1992. The effects of contaminated sediments in the Elizabeth River., Chapter 17, pp. 403-430, In: G.A. Burton, Jr. (Ed.) *Sediment Toxicity Assessment,* Lewis Publ., Boca Raton, FL.
4. Helz, G.R. and R.J. Huggett. 1987. Contaminants in the Chesapeake Bay: A Regional Perspective, pp. 270-297, In: S.K. Majumdar, L.W. Hall and H.M. Austin (Ed.) *Contaminant Problems and Management of Living Chesapeake Bay Resources,* Pennsylvania Academy of Sciences Publ., Easton, PA, 573 pages.
5. Dickhut, R.M. and K.E. Gustafson. (In press). Atmospheric inputs of selected polycyclic aromatic hydrocarbons and polychlorinated biphenyls to southern Chesapeake Bay. Mar. Poll. Bull.
6. Sanders, J.G. 1987. Contaminant effects on primary producers in Chesapeake Bay, pp. 394-416, In: S.K. Majumdar, L.W. Hall and H.M. Austin (Ed.) *Contaminant Problems and Management of Living ChesapeakeBay Resources,* Pennsylvania Academy of Sciences Publ., Easton, PA, 573 pages.

7. Hargis, W.J. Jr., Roberts, M.H. Jr. and D.E. Zwerner. 1984. Effects of contaminated sediments and sediment-exposed effluent water on an estuarine fish: acute toxicity. Mar. Environ. Res., 14:337-354.
8. Weeks, B.A., Warinner, J.E., Mason, P.L. and D.S. McGinnis. 1986. Influence of toxic chemicals on the chemotactic response of fish macrophages. J. Fish. Biol. 28:653-658.
9. Van Veld, P.A., Vegelbein, W.K., Smolowitz, R., Woodin, B.R. and J.J. Stegeman. 1992. Cytochrome P450IA1 in hepatic lesions of a teleost fish (*Fundulus heteroclitus*) collected from a polycyclic aromatic hydrocarbon-contaminated site. Carcinogenesis 13:505-507.
10. Faisal, M., Marzouk, M.S.M., Smith, C.L. and R.J. Huggett. 1991. Mitogen induce proliferative responses of lymphocytes from spot (*Leiostomus xanthurus*) exposed to polycyclic aromatic hydrocarbon contaminated environments. Immunopharm. and Immunotox. 13:311-327.
11. Weeks, B.A. and J.E. Warinner. 1986. Functional evaluation of macrophages in fish from a polluted estuary, Veterin. Immuno. and Immunopath. 12:313-320.
12. Van Veld, P.A., Ko, U., Vogelbein, W.K. and D.J. Westbrook. 1991. Glutathione S-transferase in intestine, liver and hepatic lesions of mummichog (*Fundulus heteroclitus*) from a creosote-contaminated environment, Fish Physiol. and Biochem. 9:369-376.
13. Faisal, M., Weeks, B.A., Vogelbein, W.K. and R.J. Huggett. 1991. Evidence of aberration of the natural cytotoxic cell activity in *Fundulus heteroclitus* (Pisces: Cyprinodontidae) from the Elizabeth River, Virginia, Veterin. Immunol. and Immiunopath. 29:339-351.
14. Volgelbein, W.K., Fournie, J.W., Van Veld, P.A. and R.J. Huggett. 1990. Hepatic neoplasms in the mummichog, *Fundulus heteroclitus,* from a creosote-contaminated site, Cancer Res. 50:5978-5986.
15. Hargis, W.G., Zerner, D.E., Thoney, D.A., Kelly, K.L. and J.E. Warinner III. 1989. Neoplasms in mummichogs from the Elizabeth River, Virginia, J. of Aquat. Animal Health. 1:165-172.
16. Weeks, B.A., Huggett, R.J., Warinner, J.E. and E.S. Mathews. 1990. Macrophage responses of estuarine fish as bioindicators of toxic contamination, In: J.F. McCarthy and L.R. Shugart (Ed.) *Biomarkers of environmental contamination.* Lewis Publ., Boca Raton, FL, pages.
17. Chu, F.-L. and R.C. Hale. 1994. Relationship between pollution and susceptibility to infectious disease in the eastern oyster, Mar. Environ. Res., *In Press.*
18. Heinle, D.R., Flemer, D.A. and J. Ustach. 1976. Contribution of tidal marshlands to mid-Atlantic estuarine food chains. pp. 309-320. In: M. Wiley (Ed.) *Estuarine Interactions*, Vol. 2, Academic Press, NY.
19. Malone, T., Kemp, W.M., Ducklow, H.W., Boynton, W.R., Tuttle, J.H. and R.B. Jonas. 1986. Lateral variation in the production and fate of phytoplankton in a partially stratified estuary. Mar. Ecol. Prog. Ser. 32:149-160.

20. Kraus, M.L., Weis, P. and J.H. Crow. 1986. The excretion of heavy metals by the salt marsh cord grass, *Spartina alterniflora* and *Spartina's* role in mercury cycling. Mar. Environ. Res. 20:307-316.
21. Davis, D.E., Weete, J.D., Pillai, C.G.P., Plumley, F.G., McEnerney, J.T., Everest, J.W., Truelove, B. and A.M. Diner. 1979. Atrazine fate and effects in a salt marsh. EPA-600/3-79, 84 pp.
22. Means, J.C., Wijayaratne, R.D. and W.R. Boynton. 1983. Fate and transport of selected pesticides in estuarine environments. Can. J. Fish. Aquat. Sci. 40(Supple. 2):337-345.
23. Hershner, C. and J. Lake. 1980. Effects of chronic oil pollution on a salt-marsh grass community. Mar. Biol. 56:163-173.
24. Bowman, M.J. and R.L. Iverson. 1978. Estuarine and Plum Fronts. pp. 87-104, (In): Bowman, M.J. and W.E. Esaias (Ed.) Oceanic Fronts in Coastal Processes, Proc. Workshop at Mar. Sci. Res. Cntr, Stony Brook, L.I., NY, 113 pp.
25. Simpson, J.H. and R.D. Pingree. 1978. Shallow sea fronts produced by tidal stirring, pp. 29-42. (In): Bowman, M.J. and W.E. Esaias, (Ed.) Oceanic Fronts in Coastal Processes, Proc. Workshop at Mar. Sci. Res. Cntr, Stony Brook, L.I., N.Y., 113 pp.
26. Faller, A.J. and A.H. Woodcock. 1978. The spacing of windrows of *Sargassum* in the ocean. J. Mar. Res., 22-29.
27. Klemas, V. and D.F. Polis. 1977. A study of density fronts and their effects on coastal pollutants. Remote Sens. Envir. 6:95-126.
28. Bradley, B.P. and M.H. Roberts. 1987. Effects of contaminants on estuarine zooplankton, pp. 417-441. (In): S.K. Majumdar, L.W. Hall, and H.M. Austin (Ed.) *Contaminant Problems and Management of Chesapeake Bay Resources,* Pennsylvania Academy of Science Publ., Easton, PA, 573 pp.
29. Heinle, D.R. and M.S. Beaven. 1977. Effects of chlorine on the copepod *Acartia tonsa.* Ches. Sci. 18:140-145.
30. Sullivan, B.K., Buskey, E., Miller, D.C. and P.J. Ritacco. 1983. Effects of copper and cadmium on growth, swimming, and predator avoidance in *Eurytemora affinis* (Copepoda). Mar. Biol. 77:299-306.
31. Calabresse, A., Collier, R.S., Nelson, D.A. and J.R. MacInnes. 1973. The toxicity of heavy metals to embryos of the American oyster, *Crassostrea virginica.* Mar. Biol. 18:162-166.
32. Barthalmus, G.T. 1977. Behavioral effects of Hg on grass shrimp. Mar. Poll. Bull., 8:87-90.
33. Glickstein, N. 1978. Acute toxicity of mercury and selenium of *Crassostrea gigas* embryos and *Cancer magister* larvae. Mar. Biol. 49:113-117.
34. Vernberg, W.B. and J. Verberg. 1972. The synergistic effect of temperature, salinity and Hg on survival and metabolism of adult fiddler crab, *Uca pugilator.* Fish. Bull. 70:415-420.
35. Block, R.M. 1977. Physiological response of estuarine organisms to chlorine. Ches. Sci. 18:158-160.

36. Roberts, M.H. and B.B. Casey. 1984. Depression of larval growth and metamorphosis of oysters exposed to chlorinated sewage, pp. 509-520. (In): R.L. Jolley, R.J. Bull, W.P. Davis, S. Katz, M.H. Roberts, Jr. and Jacobs, V.A. (Eds.), Water Chlorination, Chemistry, Environmental Impacts and Health Effects, Vol. 5, Proc. Fifth Conf. Water Chlorin., Lewis Publ., Chelsea, Mich.
37. Sami, S., Faisal, M. and R.J. Huggett. 1993. Effects of laboratory exposure to sediments contaminated with polycyclic aromatic hydrocarbons on the hemocytes of the American oyster, *Crassostrea virginica,* Mar. Environ. Res. 35:131-135.
38. Stebbing, A.R.D. 1985. Organotins and water quality - some lessons to be learned. Mar. Poll. Bull. 16:383-390.
39. Bender, M. and R.J. Huggett. 1987. Contaminant Effects on Chesapeake Bay shellfish. pp. 373-392. In: S.K. Majumdar, L.W. Hall, and H.M. Austin (Ed.) *Contaminant Problems and Management of Chesapeake Bay Resources,* Pennsylvania Academy of Science Publ., Easton, PA, 573 pp.
40. Cronin, L.E. 1967. The condition of the Chesapeake Bay. Trans. N. Amer. Wildl. Nat. Res. Conf. 23:137-150.
41. Environmental Protection Agency. 1983. Chesapeake Bay: A profile of environmental change. EPA, Phila., PA, 186 pp.
42. Weeks, B.A. and J.E. Warinner. 1984. Effects of toxic chemicals on macrophage phagocytosis in two estuarine fishes. Mar. Environ. Res. 14:327-335.
43. Mezin, L. 1994. Effects of contaminated sediments on the epidermal goblet cells of the mummichog, *Fundulus heteroclitus.* MA Thesis, SMS/VIMS, College of William and Mary, 107 pp.
44. Hargis, W.J., Jr. 1991. Disorders of the eye in finfish, Annual Rev. of Fish Diseases, 1:95-117.
45. Williams, C.D., Faisal, M. and R.J. Huggett. 1992. Polynuclear aromatic hydrocarbons and fish lens cataract: Effects of Benzo[a]pyrene-7, 8-Dihydrodiol on the macromolecular synthesis of cultured eye cells. Mar. Environ. Res. 34:333-337.
46. Middaugh, D.P., Couch, J.A. and A.M. Crane. 1977. Response of early life history stages of striped bass, *Morone saxatilis,* to chlorine. Ches. Sci. 18:141-153.
47. Stehlik, L.L. 1980. Bioaccumulation and effects of Kepone on spot, *Leiostomus xanthurus.* MS thesis, VIMS, College of William and Mary, 71 pp.
48. Longwell, A.C. 1977. A genetic look at fish eggs and oil. OCEANUS. 20:46-58.
49. Longwell, A.C., Chang, S., Herbert, A., Hughes, J.B. and D. Perry. 1992. Pollution and developmental abnormalities of Atlantic fishes. Environ. Biol. Fishes. 35:1.21.
50. Norcross, B.L. and M. Frandsen. (In press). Distribution and abundance of larval fishes in Prince William Sound, Alaska during 1989 after the

EXXON VALDEZ oil spill. EXXON VALDEZ Oil Spill Symposium. Proc., Amer. Fish. Soc. Symp. No. XXX. 49 pp.

51. Moles, A., St. Rice and B.L. Norcross. (ms). Non-avoidance of hydrocarbon laden sediments of juvenile flatfishes. 32 pp.
52. Fitzner, R.E., Blus, L.J., Henny, C.J. and D.W. Carlile. 1988. Organochlorine residues in great blue herons from the northeastern United States. Col. Waterbirds. 11:293-300.
53. Rybitski, M.J. 1993. Distribution of organochlorine pollution in sea turtles. PhD Diss., VIMS/SMS, College of William and Mary.
54. Woodwell, G.M., Wuster, C.F., Jr. and P.A. Isaacson. 1967. DDT residues in an east coast estuary: A case of biological magnification. Science. 156:821-824.
55. Nelson, K. and H. Austin. 1992. Comparison of Condition Index (K) of spot (*Leiostomus xanthurus*) from the Elizabeth and York Rivers, Virginia, Va. J. Sci. 43(4):381-388.
56. Fisher, D.J., Bender, M.E. and M.H. Roberts, Jr. 1983. Effects of ingestion of Kepone-contaminated food by juvenile blue crabs (*Callincectes sapidus Rathbun*), Aquatic Toxicology. 4:219-234.
57. Lin, C.T., Milon, J.W. and E. Babb. 1991. Determinants of subjective food safety perceptions: A case study of oysters in the southeast. J. Agribusiness. 9:71-84.
58. Chaudhry, R. 1992. Fishing for answers, Price sensitivity and fears of contamination stunt seafood consumption. Rest. and Inst. 102(5):80, 86, 88.

Environmental Contaminants, Ecosystems and Human Health. Edited by S.K. Majumdar, E.W. Miller, and Fred J. Brenner.

Chapter Eleven

POTENTIAL ENVIRONMENTAL IMPACTS OF NUCLEAR POWER

RICHARD WILSON
Mallinckrodt Professor of Physics
Lyman Laboratory
Harvard University
Cambridge, MA 02138

INTRODUCTION

Environmentalists raise at various times a large number of issues about fuel consumption and energy use, of which I discuss thirteen in this chapter. Some of these issues might not qualify to a professional as an environmental issue. Therefore some of them are often omitted in discussions such as this because they are not strictly environmental issues. But each has been called an environmental concern at various times and all are worth considering. They might be called "environmentalist" issues, and thereby qualify for consideration. But they are not always consistent, and it is well to consider them separately in the initial discussions.

In addition, environmentalists have often discussed three strictly economic and resource issues: availability of fuel, availability of resources to exploit the fuel; and cost. Although these are interconnected with the environmental issues I will not discuss them further here.

I address the way in which nuclear power impacts, or might impact, upon each of these issues. I stress the necessity of making comparisons with other alternate fuel sources in order to maintain perspective. The comparisons will vary; between nuclear and fossil fuels on one occasion; one between electricity availability and non-availability on other occasions. This is the procedure followed by many comprehensive studies compiled in (1) and (2). It will be seen that when I make numerical estimates of the total effect upon health, they are dominated by two items that are presently controversial:

(a) What is the probability and effect of a severe nuclear power accident? and
(b) What is the effect upon health of particulate air pollution at today's levels?
In addition it can be seen that the calculations do not justify the public concern about nuclear waste.

IMPACT UPON THE VISUAL ENVIRONMENT

It is said that "Beauty lies in the eye of the beholder". It follows that people will differ in their views on the impact of anything on the visual environment. The principal impact of a power plant are the buildings, of which the turbine building, the cooling structure and the office buildings are the largest part. Although a nuclear power plant does have a small stack to dispose of small amounts of radionuclides, modern fossil fuel power plants have much taller ones that most people consider unsightly. Some consider a hydroelectric dam, and the lake it restrains, a thing of beauty. Some do not; Windmills can obstruct a beautiful view, or they can remind one of the sustainable society they support. However, most people regard coal storage and ash dumps to be very unsightly. In many articles, a large cooling tower has been taken as a symbol of nuclear power; it is more a symbol of any large electricity generating power plant or industrial facility. However, as noted below, environmentalists are more likely to force the use of cooling towers on nuclear plants than on fossil fuel plants. To this extent, therefore, the cooling tower is a partially legitimate symbol of nuclear power.

IMPACT UPON ECOLOGICAL SYSTEMS

Dramatic effects on ecology arise often in the use of energy (common to all fuels) or in mining of fuels. The disturbance of the land that a large mine represents can have considerable ecological impact. The transport of fuel disturbs the ecology considerably -an oil or gas pipeline can interrupt the migration of species. In all of these nuclear fuel comes out favorably. Because of the high energy density of the fission energy in the uranium (3,000,000 times as much for uranium 235 as for coal) much less (100 times less) mining is needed and much less transportation.

The ecological effects of man's activities are receiving increasing political attention. In early April 1993, the President and Vice-President of the USA, 6 Cabinet Secretaries, the administrator of the Environmental Protection Agency, state Governors, and a cast of thousands met for a full day to resolve an ecological issue that had created gridlock in the Pacific Northwest. Should one continue to cut trees extensively, or was it destroying the very existence of endangered species such as the spotted owl?

This dramatic and important development followed a report by a committee of scientific experts assembled by the Environmental Protection Agency to evaluate the risks facing ecological resources. The top four were:

- habitat alteration
- global climate change
- depletion of stratospheric ozone
- decrease in biological diversity

The experts ranked lowest several that are high in the popular imagination:

- herbicides
- pesticides
- acid rain deposition
- airborne toxics
- oil spills
- groundwater pollution

This raises at once a very important distinction-between the views of the general public, who through their elected representatives often decide an issue, and professional experts who are often more powerful in a dictatorship. This distinction appears in many ways.

If we restrict ourselves to effects upon a world ecological situation, such as the number of species, the effects of nuclear energy even in an accident situation are minuscule. A radiation dose high enough to destroy plants or animals directly would also kill humans directly. At Chernobyl, trees downwind of the power plant died after a few months and no doubt many animals died. But the total number of animals later increased, because their major predator (man) had left. The main human concern about Chernobyl was subsequent cancers and few people care whether a few rats or squirrels get cancer.

Ecological effects do arise elsewhere in the fuel cycle-particularly in mining. These, however, are smaller than for fossil fuels because much less material needs to be mined. There is no parallel in nuclear energy to the disruption of migration routes by a pipeline, or flooding by a hydroelectric reservoir. Smaller effects than these do occur at a power station in common with power stations using other fuels. Waste water discharges can affect fish by either their chemical or thermal effects. While thermal effects have often been solely attributed to nuclear energy, and have been used by opponents of nuclear energy as a club with which to attack any nuclear power plant project, it applies to any power plant or industrial activity that uses heat-whether by burning of fossil fuel or operation of a nuclear reactor.

Often a major new power plant will be on the coast and use ocean water or cooling. It is important that this be done in such a way as to protect flora and fauna. This usually entails discharge of the condenser cooling water far enough out in the ocean to ensure rapid mixing. This is not a problem unique to nuclear power plants, it is present for other fossil fuel power plants and for other industries generally. However, the pressure on nuclear power plants is greater than on other power plants. No other industry has been forced to install cooling water pipes four miles long as was the power plant at Seabrook, New Hampshire.

EFFECT UPON WORLD CLIMATE

The first and most obvious climatic effect of energy use comes from the direct heating of the world. However, this, at present energy use is small. More important is the INDIRECT effect of the gaseous emissions on the world's atmosphere. The earth is a greenhouse; heat is absorbed from the sun, but the infrared radiation re-emitted by the earth is trapped by a number of "greenhouse gases" of which carbon dioxide (CO_2) is one. This probably makes the world warmer than otherwise, and therefore sustains life on the planet. As carbon dioxide is emitted from fossil fuel burning, the nature of the greenhouse changes. An increase in the greenhouse effect can result in a global warming which can change life on the planet considerably.

The main greenhouse gases, which reflect infrared rays back to the earth, are water vapor, carbon dioxide, methane, nitrogen oxide, and fluorocarbons. It was Fourier in 1835 who first noticed that carbon dioxide had this property, and Arrhenius (3) who made the first detailed calculations in 1896. Of these gases, carbon dioxide and methane are those which are principally energy use. Every time a carbon atom is burnt, a carbon dioxide molecule is produced. It therefore behooves us to burn fewer carbon atoms-either by changing what we burn, or increasing the efficiency. On both counts the burning of natural gas is touted as only half as bad as coal. But that may be only a half truth from considering only half of the facts. It is important to realize that the effect of excess carbon dioxide emissions is a global one and the citizens living in any country must concern themselves with the whole fuel cycle even though a part of it takes place in other countries.

In one gas well in Indonesia for example, 4 molecules of CO_2 are emitted for every CH_4 molecule released. Also CH_4 is 30 times as bad a greenhouse gas as CO_2 so that even 3% leakage anywhere from the well to the burner can double the greenhouse effect, nullifying the advantage over coal. On this matter there have been few studies. But claims vary from a statement that a bad British plant lost 20% of its gas (4) or the Trans-siberian pipeline loses 12% of its gas along its length, to an engineering use of 3% as a standard leakage factor (5), and a very recent World Bank report (6) that less than 0.1% leakage can be expected with new pipelines. I note that this is parallel to optimistic statements that the severe accident rate (meaning one with measurable public health impact) in the new generation of nuclear reactors should be less than 1 in many millions per year-both possibly true and as yet unprovable.

In contrast to the burning of any carboniferous fuel, burning of nuclear fuel in a reactor does not emit any CO_2. When the full energy chain is considered, there is a small contribution. But nuclear power is not an important cause of global warming. Of course, neither hydroelectric power nor the renewable sources of energy are either, although the intensive material flows of photovoltaic conversion make this source begin to be appreciable. Nonetheless, some public opinion surveys have found that a significant number of people attribute global warming to nuclear energy the inversion of the true situation.

IMPACT UPON WASTE PROBLEMS

The problems of waste in society are very badly characterized and discussed. Although waste problems are often thought to be ecological or environmental issues, no specific ecological or environmental hazard is usually adduced. Nonetheless, waste problems of all industries have attracted public attention, and the "solution" of the problem of nuclear waste is often regarded as essential to the future of nuclear power. A crucial misconception, however, is a common feeling that nuclear waste is uniquely bad. In what follows, I will discuss the limited extent to which this might be true.

Since the amount of energy from fission of uranium is 3 million times the amount from burning of coal, the burning of coal produces enormous quantities of waste, which technically dwarfs that of nuclear wastes. Nonetheless, coal wastes are not perceived by the public as a major problem. It is correct, but perhaps an oversimplification, to state that the problem of nuclear waste is political rather than technical. The fundamental question most raised by activists is: "Is it morally acceptable to leave the burden of waste disposal to future generations?" I keep this question as one to be continually addressed.

One problem of disposal of nuclear waste is a safety problem. Many people are willing to concede that if they are disposed of successfully, there is and will be no problem as shown in the "professional" estimates of effects upon public health presented below. If there is an accident of some sort at a waste disposal facility, there is a fear that radioactive material will enter the environment where it will have a large effect on public health. What would be the *likely* effect of a failure of a given waste disposal site, and how do these effects compare to effects, similarly calculated, of coal and other wastes?

There is very little *ecological* effect of *nuclear* waste disposal in contrast to the possible large regional ecological effects of disposal of chemicals such as arsenic. It is important to realize that the effect on public health is the *only* important effect of any failure to contain the waste. Nonetheless, the criteria that the world is suggesting for the disposal of radioactive waste are greater than ever before imposed upon any segment of society. The world is demanding that it be proven that over the next 50,000 years a waste disposal site will not leak enough to cause even a hundred or so cancers in the population around. The difficulties arise in the nature of such a proof for some catastrophe in the world's future, and in the smallness of the effects being prevented. The disposal of coal waste, which is radioactive with half lives comparable to the long lived nuclear waste, has to meet no such criteria. Moreover, the sheer quantity of coal waste makes it impossible to take such a stringent approach as we will probably successfully meet for nuclear waste. Although in the perception of many members of the public the waste issue is a problem unique to nuclear power that may prevent its continuation, this is NOT a technical problem. Indeed many scientist have said that nuclear waste is the ONLY waste problem in society where society has any idea of a sensible long term solution.

Although NO ONE has come up with a failure scenario for a projected DOE nuclear waste site which comes close to having an effect equal to the Chernobyl accident, accidents did occur in Russia at Kyshtym. In the first accident, 2 million Curies of radioactive material, including 500,000 Curies of strontium 90, were dumped into the Techa river (7). The peasants downstream were not told for reasons of military security, and many got radiation doses of 200 Rems. In another accident, a facility suffered a chemical explosion (after coolant failed), which blew 5 million Curies of radioactivity into the countryside to the Northeast. These were accidents in the military complex, not in the civilian uses of nuclear energy. Moreover, although this caused a major effect upon public health, and considerable dislocation, it is noteworthy that the effects were regional and the details were successfully kept secret from the rest of the world for 35 years. Other regional effects of comparable magnitude came from other, non-nuclear, pollution. A distinction can be made between accidents in the USSR military complex, built in a hurry and in secrecy to catch up with the "wicked" USA, and a civilian complex in the USA. If attention is paid to answering such questions, the direction of the public debate might be changed.

Just as with the effects on ecological systems, the calculations and measurements of experts that nuclear wastes are the ONLY waste problems of society for which there exist good technical solutions for disposal (because of the ability to keep the volume small) are not so perceived. Possible reasons may include:

- nuclear fission is perceived as particularly complicated,
- the understanding of the issues among administrators, politicians and the public is very small,
- new technologies must not merely be better than old technologies, but a lot better—perhaps perfect.

Public policy must address the perception. They are, in the sense of the introduction, *environmentalist* problems rather than *environmental* problems. They are nonetheless real and might be circumvented by logical discussion.

EFFECT UPON OCCUPATIONAL HEALTH

The use of fuels cause as much risk to the workers in the energy industry as the risk sustained by workers in other industries of comparable size. In older coal fired electricity generating plants, personnel are exposed to noxious gases of all sorts. In the steel industry, coke ovens are used, and it is well known that among coke oven workers there was a considerable, statistically significant increase of lung cancer. In nuclear power plants, workers are exposed to radiation from various plant operations. However, their exposure is measured and can be controlled. Adverse health effects attributable to radiation exposure have only been identified in the mining sector, but some have been postulated, based upon a linear dose response relationship, in the power plants themselves.

Health effects from fossil fuel plants are not recorded as well as those from nuclear power. But Coal Workers Pneumoconiosis (CWP) in miners is, by itself, estimated to cause about 1 death per GWe-y, which is greater than any of the entries in Table 1.

EFFECT UPON OCCUPATIONAL SAFETY

The expression occupational safety is normally applied to accidents. These can occur at all parts of the fuel cycle.

The recent fatality rates in coal mines alone are about 1.3 per GWe-y; for natural gas 0.4 per GWe-y, and 0.1 to 0.7 per GWe-y for oil. The most complete listing of accidents is that prepared by the Paul Scherrer Institute in Switzerland (9, 10). It is probable that the lower figures in the estimates of fossil fuel use come from incomplete listings, whereas the listing for nuclear is complete except for the USSR.

The well known accident at Chernobyl does NOT change this overall conclusion. Thirty one firemen and plant workers died as a direct result of the accident, and 150 suffered acute radiation sickness from which most ultimately recovered. 300 workers suffered enough exposure to have, according to present understanding of dose response relationships, an increased probability of developing cancer of 3%.

TABLE 1
Modified from refs 2 and 8

Occupational health risks	Disease	Radiological (deaths/GWe-y)
Mining	0.06	0.15
Processing	-	0.05
Transport	-	0.0005
Power generation		0.13
Decommissioning	0.008	0.0005
Waste management		0.0005
TOTAL	0.06	0.33

TABLE 2
Modified from refs 2 and 8

Occupational accidents; in the nuclear fuel cycle	Deaths/GWe-y
Mining (underground)	0.4
Mining (surface)	0.1
Processing	0.006
Transport	0.001
Power generation	0.01 to 0.1
TOTAL (half underground, half surface mining)	0.26 to 0.35

EFFECT UPON PUBLIC HEALTH

There is no doubt that high doses of radiation have led to acute radiation sickness and death; doses just less than these have led to cancer. But there is much more doubt whether the low doses that arise from normal operation of nuclear power plants, and even doses to most of the people exposed in accident conditions, lead to any health problems. However, it is conventional, for prudent public policy, to assume that there is a linear relationship between dose and response so that even small doses, if widely spread over a population, can produce an appreciable response. This assumption, was originally suggested by Crowther in 1925 (11), and for many years was only made by those concerned about radiation exposure. This lead to an incorrect feeling that anything involving radiation is uniquely dangerous. It is now realized that this assumption is probably equally valid (or invalid) for other carcinogens, and recently it has been realized that it may apply to many other situations such as particulate air pollution also.

In ordinary operation these effects have again been estimated (they are too small to be measure), assuming (pessimistically) a linear dose response relationship.

The exposures from nuclear power plants are kept small by a rigorous attention to detail, incorporated in a system of regulations for radiation protection suggested in the first instance by the International Commission on Radiological Protection (ICRP) and adopted by all important government and international bodies. The failure of this system in an accident is discussed separately in the next section.

The effects of fossil fuel use on public health are primarily those of air pollution: the liberation of gasses from the power plant as a result of fossil fuel burning or of radionuclides from a nuclear plant. Burning of fossil fuels results in emission of gases from incomplete combustion as well as gases from impurities.

There is no doubt that air pollution and in particular the burning of coal HAS killed members of the public outside the plant. The air pollution incidents in the Meuse valley in Belgium in 1930 (60 deaths), in Donora, Pennsylvania in 1948, (20 deaths) and in London in December 1953 (5,000 deaths) are only examples of *recorded* incidents. Air pollution levels are now 100 times smaller than in

TABLE 3
Modified from refs 2 and 8

Calculated Routine radiological mortality	Deaths per GWe-yr
Mining	0.05
Processing	0.013
Transport	0.0005
Power generation	0.02
Waste management	0.000008
TOTAL	0.08

these recorded incidents, but they last for longer and are more geographically widespread. Air pollution in Lanchow or Wuhan in mainland China exceeds that of the western cities mentioned, and in Taipei it is worse than present US EPA standards. There is controversy about how to extrapolate these known hazards to the lower levels of today.

About 1970, a view that *present day* air pollution in eastern US or northern Europe affects 1% of the people exposed was presented by a number of people (12), but was widely attacked as speculative and unproven. The view did not affect any government action. However, this view is receiving much stronger scientific justification (13). If the view is correct, the air pollution from fossil fuels is responsible for several percent of the US death rate, and therefore over 100 deaths per GWe-yr which dwarfs all other health problems of fuel use. The magnitude can be summarized by saying that air pollution, mostly from coal burning, causes more effects on public health than would a Chernobyl-size nuclear accident every year. General models can be developed (14) which suggest that the dose response at low doses could be linear with at least as much justification as the low dose response assumed for radiation exposures. Therefore small palliatives cannot bring the 100 deaths per GWe-yr down to 1 death to 1 death per GWe-yr, which is an upper limit of what one could reasonably calculate for nuclear power; the only hope for a large improvement is to change the technology completely.

EFFECT ON PUBLIC SAFETY

In spite of the greatest possible care, accidents will happen. It is one of the greatest triumphs of modern civilization that we have been able to reduce industrial accidents steadily over the last 150 years. An exception was in the USSR, whose political system seemed to prevent the learning from mistakes that is necessary to have a safe society. For large scale uses of fuels, we store a lot of energy in one place, and often that place is where there are a lot of people. It is this that gives the potential for disaster.

For nuclear power plants the problem is not that of an explosion *per se*. A nuclear power plant cannot blow up like a bomb; it is not easy to configure uranium to make it explode and in a power plant uranium is configured to prevent it from exploding. The concern about safety is that a power plant might be destroyed, and radioactive material enter the environment where it might be destroyed, and radioactive material enter the environment where it might impact people and cause adverse effects upon public health. A large part of the radiation protection of regulations discussed in the previous section are to cope with radioactive releases in case of an accident.

The safety of nuclear power plants has been accomplished in the US by what the Atomic Energy Commission called "Defense in Depth". One tries to imagine the worst possible accident; then one devises a safety device to prevent it from happening. Then one devises mitigation steps (a containment, or strategic location) to mitigate the effects if the safety device fails also, and the accident occurs. This procedure allows some checking; if the "precursors" to an accident happen only rarely, so that the safety device is rarely called upon, then we can be sure that the accident rate is correspondingly diminished. In the nuclear field so far, the record has been outstanding with no loss of life to anyone in the public from a licensed facility outside the USSR.

Even in the Chernobyl accident there was no immediate loss of life to the public by Acute Radiation Sickness (death within a month). However, delayed effects of cancer are expected-20,000 cases worldwide on an estimate prepared for US DOE (15, 16). The (fortunate) lack of a large historical data base poses a dilemma. We can take the accident frequency worldwide (including Chernobyl) and apply it to all reactors, or we can calculate the frequency. The first calculation, together with an (unrealistic) estimate of cancers due to Chernobyl 15 times higher than the DOE estimates, has led to high estimates of 150 deaths per GWe-yr (17). But western reactors are very different from the RBMK reactor at Chernobyl. The science of calculating accident frequency has been greatly developed over the last 20 years, and leads to much smaller estimates of overall death rate (of 0.001 to 0.01 per GWe-yr including calculated delayed deaths) (9).

Accidents can happen in oil or gas use. Accidents of liquefied energy gases and oil happen at a frequency of once a year worldwide. In contrast to accidents of nuclear power plants some of these can be accidents where an explosion occurs and kills people by the explosion. Although accidents with natural gas, (LNG - methane) as distinct from Liquefied Petroleum Gas (LPG - Propane) are rare (since methane rises rapidly and does not collect to form an explosive mixture), a large accident with LNG, is as likely as a *large* accident with LPG and could be very bad. The fuel IF MIXED WITH AIR, would be in its most combustible state and very liable to explosion. Nuclear fuel, in contrast is not so easy to explode. A large modern liquefied natural gas tank contains enough energy to set off an explosion 30 times the size of that at Hiroshima. Fortunately there seems to be no scenario where it could all mix and detonate accidentally. Smaller explosions or a large conflagration are possible. Nonetheless, it is wise to remove all LNG storage facilities, and even more shipment facilities, far from population centers.

EFFECT UPON SABOTAGE

Where there is a potential for accidents, there is also a potential for sabotage and terrorism. I distinguish the two here. Sabotage is merely destroying the facility so that it cannot operate and produce electricity. Most industrial facilities are vulnerable to sabotage, and nuclear power plants are no exception. The unattended substation at the street corner is a particularly vulnerable target. Dispersal of facilities can reduce the effect of a sabotage attempt, but probably increases the frequency by making an attempt easier.

It is important to realize that for nuclear power it is hard for a saboteur to do more than what would happen in an accident. Studies of the potential for sabotage therefore focus on the possibility of increasing the probability of accident, and studies to mitigate the effects of sabotage follow the same paths as mitigation of a natural accident. Analysts of the accident at Three Mile Island were surprised that the safety devices allowed the operators to do what they did; could, or would, a saboteur do as much?

But for natural gas, a terrorist can do much more than would normally considered probable accidentally. This is because as noted in the preceding paragraphs, the fuel only has to be mixed with air to put it in its most combustible, and explosive state. A saboteur might well set off an LNG explosion which is considerably larger than is possible accidentally-and a large LNG tank contains enough material that if mixed with air and exploded, would give an explosion equal to 30 Hiroshima bombs.

A concentration of facilities in large power plants makes a power plant an attractive sabotage target. More damage can be done for a given amount of effort than for small facilities. Again nuclear power plants are not unique, and may be more immune to sabotage than other facilities (the containment vessel, by isolating the inside world from the outside can be particularly helpful here.)

EFFECT UPON TERRORISM

I define terrorism as an action or set of actions by a person or group threatening to cause an environmental or public health mess by destroying a facility. It is usually a threat of sabotage. A terrorist threat has to be believable to be effective.

Terrorism can particularly occur in time of partial war. The possibility of terrorism by their Azeri neighbors has led some Armenians to oppose the reopening of the Medzmor nuclear power plant (which would help to solve their energy crisis). But hydroelectric plants are also particularly liable both to sabotage and to terrorism. Dams can fail by accident. But they have also been blown up in time of war. If one can blow up a dam by dropping a bomb from the air, under the unfavorable circumstance of being shot at by accurate anti-aircraft fire, how much easier is it to blow up an *unattended* dam with a more modern, lightweight bomb that can be carried on one's back? It is hard to quantify these issues, but analysts tend to agree that blowing up a nuclear power plant is far from the easiest way of making a mess, although it might be psychologically very attractive to a terrorist.

At a time of a natural disaster, the potential for sabotage increases. For example in July 1993, the mid-west of the USA is experiencing massive floods; the potential for sabotage by destroying a levee is large, and might be undetected. Likewise this would be the time a terrorist can destroy an upstream hydroelectric dam with maximum effect.

EFFECT UPON LIFESTYLE

One of the issues that is raised in the industrialized world is that we are living in an ecologically non-sustainable lifestyle, and consuming more than our share of the world's resources. While not arguing this point in an absolute sense, it is worth comparing the different fuel sources in this respect. Clearly, the "renewables"-solar energy, hydropower, windpower are preferable. But we all use non-renewables to some extent. With nuclear energy, including using sometime in the future the more abundant uranium 238 in a fast neutron reactor, provides fuel for over 100,000 years-1,000 times more than any other proposed non-renewable fuel source.

EFFECT UPON NATIONAL INDEPENDENCE

The most important issues are left until the last. Will the fuel source chosen make a country overly dependent on another country for its essential supplies? What is the effect of energy dependence or independence upon the probability of war? War is so devastating and so destructive of human life, human spirit, and of property, that any increase in the probability of war must be considered unacceptable.

Nuclear energy has a very positive effect on energy independence of a country in the medium term. This arises because of the 3 million times higher energy density of uranium compared to a fossil fuel. It is hard to find space, and it is very unsightly (damaging to the visual environment) to store even a year's supply of coal, gas or oil. On the other hand it is easy, and common, for a power plant to have several years' supply of uranium on hand, already made into fuel rods.

EFFECT UPON PROBABILITY OF WAR

If the use of an energy source inevitably led to war, the effect on health would be enormous. Many people have argued that the very existence of nuclear electric power encourages proliferation of nuclear weapons throughout the world, and increases the chance of nuclear war. The connection between the civilian use of nuclear fission and the military use is real, and very important. Cottrell (18) argued that it was the only legitimate issue about nuclear energy. It is an issue that seems to dominate public thinking in a way that few people know how to explain or even address. They then transpose their opposition to nuclear power on this ground to making incorrect or illogical statements about one of the other 14 items noted above in their opposition to nuclear energy. It is therefore essential in any discussion of the potential environmental impacts of nuclear power to discuss the connection between electricity and bombs.

The USA built an atomic bomb in four years starting from scratch, and not being absolutely sure how to proceed. Now that most of the information is public knowledge, a national determination and some industrial infrastructure is all that is needed. A moderately industrial country can obviously make a weapon unaided within 10 years. If other countries of the world wish to do so, they can act to prevent that; the first step is to acquire information about what a country is doing, and secondly the world can, by agreements and sanctions endeavor to prevent it. The worldwide nature of the nuclear industry can be used, if the parties so wish in both these steps. Now that the "cold war" is over, professional societies and others can make openness and peacefulness a condition for cooperation. I believe that with care, this argument that the use of nuclear energy leads to war, or even to proliferation of nuclear weapons, can be inverted, and that the existence of a worldwide nuclear power community can inhibit the development of nuclear weapons (19).

Nuclear energy, by helping energy independence, can have a very important positive effect on the probability of war. This can particularly be the case for an island such as Great Britain, Japan, or Taiwan. On the one hand, the island country need not go to war to get immediate sources of fuel, and on the other another country or regions will be less tempted to exercise its will with an oil blockade.

Although the prospect of nuclear war is qualitatively different from the prospect of a war using only dynamite and cordite for explosives, shortages of other fuels have led to warlike acts in the past, and might lead to nuclear war in the future. Fuel supplies and energy use generally are so important to man's activities that it is not surprising that wars have been fought over their use. In 1910, the British Government bought a controlling interest in the Anglo-Iranian oil company to guarantee access to oil for the British fleet. In 1914, the Japanese joined in the first world war-against Germany and on the side of Great Britain, to safeguard supplies of coal from Manchuria. In 1941, Holland and the United States put an oil embargo on Japan, in an endeavor to persuade them to leave China, but when oil was running low, Japan bombed Pearl Harbor, and went south to take the oil they needed from Indonesia. In more recent times, the Republic of Azerbaijan has put a fuel embargo on the Republic of Armenia, and shut off the gas pipeline running through its territory from Russia, in an attempt to prevent Armenia from helping their ethnic brothers in Nagorno-Karabagh, and the whole world (the UN) fought a war to overturn the Iraqi takeover of Kuwait, a war which was largely a war about supplies of oil.

This is of particular concern to an island economy with few indigenous fuel supplies. The degree of short- and even medium-term energy independence that a wise use of nuclear energy can provide, can remove fuel supplies as one of the complicated issues to be considered in international diplomacy. An interruption of shipping from overseas need not interrupt the working of industry provided that the country concerned can obtain an appreciable fraction of its electricity from resources, such as nuclear energy, hydropower or renewables, that cannot be interrupted.

The above paragraphs stress the usefulness of energy independence for a country if a neighbor threatens an embargo, or other act of war. It is important to realize that energy independence can prevent war. It is important to realize that energy independence can *prevent* war. This can occur because an unfriendly neighbor may hesitate to start a war when the easy weapon of an embargo is not effective.

CONCLUSION

In conclusion, it should be apparent that in all the issues of concern to the general class of problems collected together under environmental issues, Nuclear Energy is as favorable as any of the fossil fuels, and should be encouraged provided that proper regard to safety and health continues.

REFERENCES

1. Colloque sur les Risques Sanitaires des Differentes Energies. 1980. Paris, France Jan 24-26. 1980. Gedim, Saint-Etienne, France (in English).
2. Electricity and the Environment. 1991. Proceedings of the Senior Expert Symposium, Helsinki, Finland, 13th-17th May 1991. IAEA, Vienna.
3. Arhenius, J. 1986. On the Influence of Carbonic Acid Upon the Temperature of the Ground. *Philosophical Magazine* 41:237.
4. Chubb, W. 1989. Energy Policy.
5. Private Communication from Stone and Webster Engineering Corporation.
6. World Bank Report.
7. Kossenko, Mira M., Degteva, Maria O. and Nelly A. Petruskova. 1992. Estimate of the risk of leukemia to residents exposed to nuclear radiation as a result of a nuclear accident in the southern Urals. *The PSF Quarterly* 2 (no. 4): 187-197.
8. Hamilton, Leonard D. 1984. Health and Environmental Risks of Energy Systems, pp. 21-57. In Proceedings of a Symposium at Julich, Germany, IAEA, Vienna.
9. Hirschberg, S. and C. Parlavantzas. 1994. Severe Accidents in Comparative Assessment of Energy Sources: Current Issues and Actual Experience Data", PSAM-II, Conference Devoted to the Advancement of System Based Methods for the Design and Operation of Technological Systems and Processes. March 20-25, 1994, San Diego, CA.
10. Hirschberg, S. and E. Cazzoli. Contribution of Severe Accidents to External Costs of Nuclear Power, 1994. In 1994 European Nuclear Society Conference on PSA/PRA and Severe Accidents.
11. Crowther, J. 1924. Some Considerations Relative to the Action of X-rays on Tissue Cells. *Proc. Roy. Soc. Lond. B.* (Biol. Sci.) 96:207-211.

12. Wilson, Richard, Colome, Steven, Spengler, John and David G. Wilson. 1982. *Health Effects of Fossil Fuel Burning.* Ballinger Press, Cambridge, MA.
13. Dockery, Douglas et al., N. Engl. J. Med. 1993.
14. Crawford, Martha, Evans, John and Richard Wilson. Low dose linearity - the rule or the exception? submitted to *Human and Ecological Risk Assessment,* Sept. 1994.
15. Goldman, Marvin, Catalino, R.J. and Lynn Anspaugh. 1987. Health and Environmental Consequences of the Chernobyl Nuclear Power Plant Accident. U.S. DOE Report/ER-0332.
16. Goldman, Marvin. 1987. Chernobyl: A Radiological Perspective. *Science* 238:622.
17. Hohmeyer, O. 1988. *Social Costs of Energy Consumption.* Springer-Verlag, New York.
18. Cottrell, Sir Alan. 1983. How Safe is Nuclear Energy? *Heinemann.*
19. Sokolow, Robert. 1994. Talk at Dedication of Princeton Plasma Laboratory, NJ.

Environmental Contaminants, Ecosystems and Human Health. Edited by S.K. Majumdar, E.W. Miller, and Fred J. Brenner. © 1995, The Pennsylvania Academy of Science.

Chapter Twelve

LAND USE AND HEALTH

JOHN E. BENHART
Geography - Earth Science Department
Shippensburg University of Pennsylvania
Shippensburg, PA 17257

INTRODUCTION

Land use has a major impact on the environmental health in the region where it is located. Geographers look at where a particular land use type is located, why that location was chosen, and what are the consequences of the final locational decision. To fully understand any locational land use decision, the spatial distribution of the land use type and the human and physical processes undertaken at the site must be studied.

For example, in locating a new well site in a Pennsylvania municipality, the geographer or earth scientist would locate and map all existing wells in the political subdivision. Maps of the hydrology, soil, geology, roadways, land use, and waterlines would be reviewed to interpret the best location for an environmentally sound well. The new land use type (a well site) must be close to waterlines, highway access, geologic structure that provides adequate water, and land that can be purchased. The site must avoid nitrate pollution from agricultural land and potential spills from truck stops and manufacturing plants. Essentially, a new well location must have wellhead protection where the surrounding land use is protected from potentially damaging land use types that might contaminate the municipal water supply.

The continued supply of safe drinking water is becoming an important land use issue as growth pressures are increasingly felt by world communities. Without proper wellhead protection, waterborne diseases such as acute gastroenteritis may occur. Even more serious for a community would be complete loss of a water supply system. The cost of providing filtration, biological indicators and chlorination that reduce health problems, can be more expensive than some municipalities or nations can endure. Other land use issues related to water and health would be its importance for crops, livestock, fish, aquatic plants, transport, power generation, sewage disposal, cooling and recreational activities. The emotional and physical health of many of the world's inhabitants is strongly associated with a land use arrangement that facilitates an adequate supply of quality water.

Land use decisions come strongly into play when deciding where to locate locally unwanted land uses (LULU's) such as a Low Level Radioactive Waste Disposal Facility (LLRW). Chem-Nuclear Systems, Inc. of South Carolina has designed a three stage land use process to review potential sites for locating such a facility in Pennsylvania. LLRW includes items such as rags, papers, protective clothing, filters, gloves, and other materials involved in the production or use of items such as smoke alarms, emergency exit signs, medical equipment, and materials from nuclear power plants. The following nine disqualifying land use criteria were used in screening potential sites for location of a LLRW: Active Faults, Carbonate Lithology, Coastal Floodplains, Wildlife Area Boundaries, State Forests and Gamelands, Watersheds, Oil and Gas Areas, Mines and Protected Area Boundaries designated by the national, state and county parks (Chem-Nuclear Systems, Inc., 1994) (1). Most past land use decisions about where to locate a waste disposal dump were made without assessing land suitably for such a facility. Today a more detailed environmental impact analysis is performed on the potential location to reduce hazardous waste dispersion from waste sites through air, soil, or water.

WATER

Water covers approximately 70 percent of the earth and is the fundamental chemical compound required for life. Throughout history humans have based their culture around water by forming cities near water sources such as a river or lake. The relationship of proper land use, water resources and human health have greatly influenced the success or failure of world civilizations. Residential, manufacturing, and agricultural land use patterns throughout the world have been strongly connected to water resources. Water spelled success for agricultural productivity in early civilizations, but waterborne illnesses such as cholera, typhoid, amoebic dysentery and infectious hepatitis caused major epidemics throughout human history (Greenberg, 1987) (2).

Water pollution is largely the result of several serious land dilemmas of our time: urbanization, poor infrastructure, industrialization and overpopulation. The implications of these causes of water pollution are quite staggering. The problems are difficult to correct because they overlap and are universal. Many countries with poor infrastructure problems have poor quality water, and the issue becomes exacerbated by overpopulation and urbanization. The more developed nations of the world are faced with water problems because of industrialization that is often compounded by urbanization (Benhart, 1993) (3). With population growth in world countries, rivers, lakes, estuaries and oceans have become disposal sites for municipal sewage and industrial wastes. Well-designed sewage treatment plants help to eliminate human and industrial waste from entering our water resources. Proper treatment of water prior to distribution among the population is the best method of protecting public health. Many health authorities estimate that approximately 80 percent of all illnesses in the world today could be prevented if all people had access to safe water supplies (Nadakavukaren, 1990) (4).

Water resources fall into one of two categories. Surface water including rivers, streams, lakes, and wetlands; and groundwater sources such as aquifers, layers of rock and earth that could contain sources of drinking water. About half of all drinking water comes from groundwater (Switzer, 1994) (5).

Even though water pollution from human wastes is dangerous, usually only the surface water is contaminated. The groundwater stays markedly more pure because the bacteria from human wastes are filtered out by the minerals in the Earth. Limestone areas with rapid percolation of water however can encounter more serious groundwater contamination. A problem that is much more difficult to deal with and remedy is the wide expanse of chemicals that humans spray over land in the form of pesticides, herbicides, and fertilizers. Some of these chemicals not only contaminate the surface water but also seep into the groundwater quite easily and do not get broken down or filtered out by natural minerals in the Earth. Chemicals go through transformation in the environment, but metabolites may still be toxic, or even more so than the original compound. Some chemicals can get attached to soil particles and remain there for extended periods of time. This poses a threat to surface and groundwaters, especially when the filtering capacity of the soil is overwhelmed. Once in the groundwater, chemicals become dispersed throughout the water table.

A major reason for such water contamination is nonpoint source pollution where pollutants originate from multiple locations such as chemically treated agricultural areas. Land use regulations like nutrient management plans are needed to control nitrate pollution from fertilization of cropland and large chicken, hog, and turkey farms. Problem causing concentrations of nitrate are possible in water from either surface or groundwater sources, but most nitrate related problems in the United States have involved shallow wells located near barnyards, septic tanks, or fertilized and irrigated fields (Delfino, 1977) (6).

Waterborne disease is a potentially catastrophic problem in less developed countries. Because the mode of transmission of diseases is water and many lack proper sanitation facilities, the threat of epidemic becomes quite real. Cholera, typhoid and paratyphoid fevers, bacillary dysentery, amoebic dysentery, and infectious hepatitis are caused by the digestive tracts of humans (Samuel and Bennett, 1983) (7). Human waste carries the organisms and in turn releases them into the water supply. Without proper technology to detect or cure the problem, many less developed nations have large numbers of their population infected by these diseases.

Common organisms found in water are biological contaminates such as bacteria, parasites and viruses. Although the contaminates are more often found in surface water than groundwater, they are a threat to human health. Bacteria is often easily detected, but many people are usually affected before the detection is made. Parasites like *Giardia lamblia*, are a threat because they pass through sophisticated chlorination processes. Viruses are expensive to detect and less contaminates are required to cause potential health problems.

Suspended and total dissolved solids create cloudy or discolored water that may carry bacteria and other substances. The bacteria may pollute the water and cause health problems to humans. Sedimentation, erosion, and stormwater management controls are important land use processes that can reduce potential health problems. Soil erosion can involve the loss of topsoil and impact sedimentation in drainage basins, wetlands, reservoirs, storm sewers and vegetation that can destroy valuable natural and human resources. The Maryland Critical Area Commission designed a land use management program to prevent pollution in the Chesapeake Bay (Critical Area and You, 1993) (8). Limited Development Areas (LDA's) are those areas which are currently developed in low or moderate intensity. LDA's encourage reducing erosion and sedimentation by planting of forests, limiting impervious areas, reducing slope development, conserving natural habitats, and encouraging shoreline vegetative buffers (Figure 1). A viable land use planning program that incorporates proper management of erosion and sedimentation control should include the following procedures at the local and state levels of government: 1) Development of sites with a slope greater than 15 percent should be restricted; 2) Whenever possible vegetative cover should be retained on development sites; 3) During construction, water features and existing drainage facilities should be protected; 4) To preserve surrounding land while the area is under construction, berms should be placed adjacent to the disturbed soil, fiber nets should be used on steep slopes, and filtration ponds, vegetative buffers and sediment basins should be constructed (Marsh 1991) (9).

With the rapid urbanization and development in the world, increased flooding has continued to impact water quality and health standards. More impervious surfaces, deforestation, agricultural activity, and floodplain development have encouraged more stormwater discharge. The substantial increase in the magnitude and frequency of peak discharge in streams encouraged communities in the United States to develop strong subdivision ordinances. These ordinances required that developers use performance standards where no net increase in stormwater can be created on a site after any new land use development has occurred. The land use concepts implemented in many municipalities required storing excess water on the developed site and slowly releasing the stormwater to surrounding land.

Techniques used to implement a stormwater management program include construction of detention basins, rooftop detention, grass swales instead of impervious surfaces and the institution of planned unit development (PUD's) where the cluster concept is encouraged. The design and construction of wetlands is encouraged to reduce the runoff coefficient and to improve water quality (Figure 2).

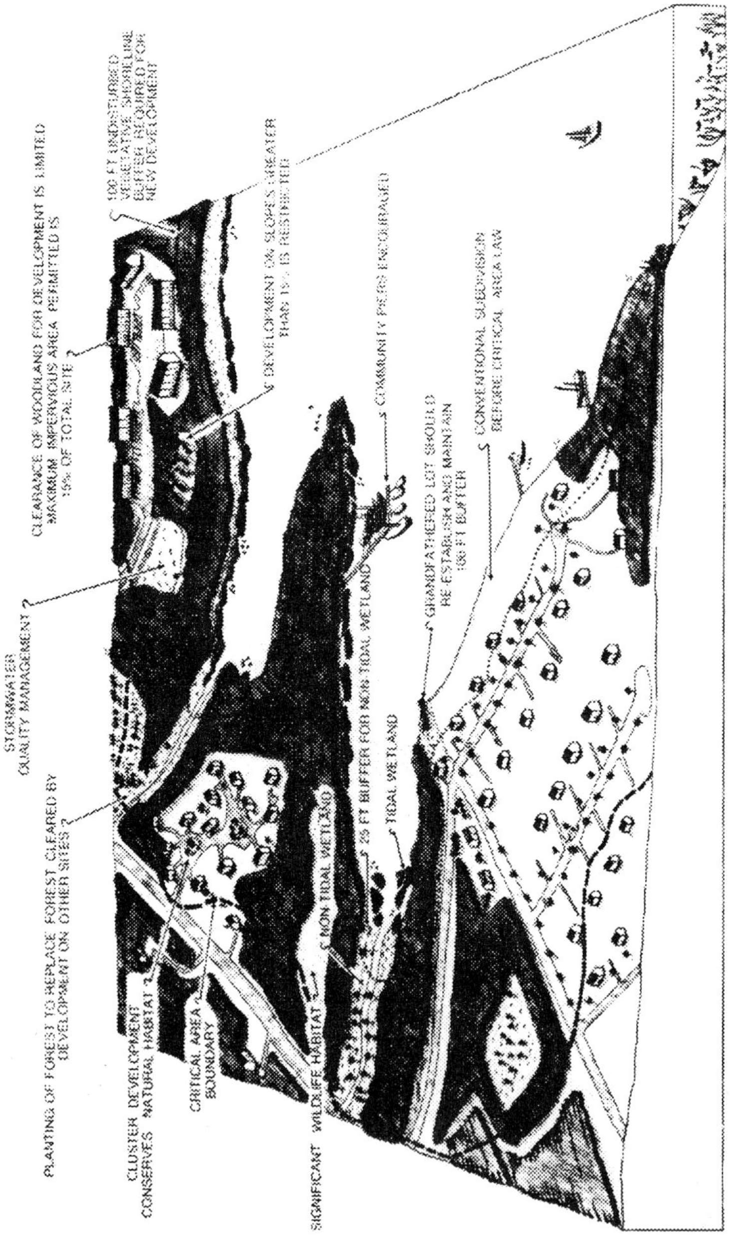

FIGURE 1. Limited Development Area (LDA) Conserving Natural & Cultural Landscape.

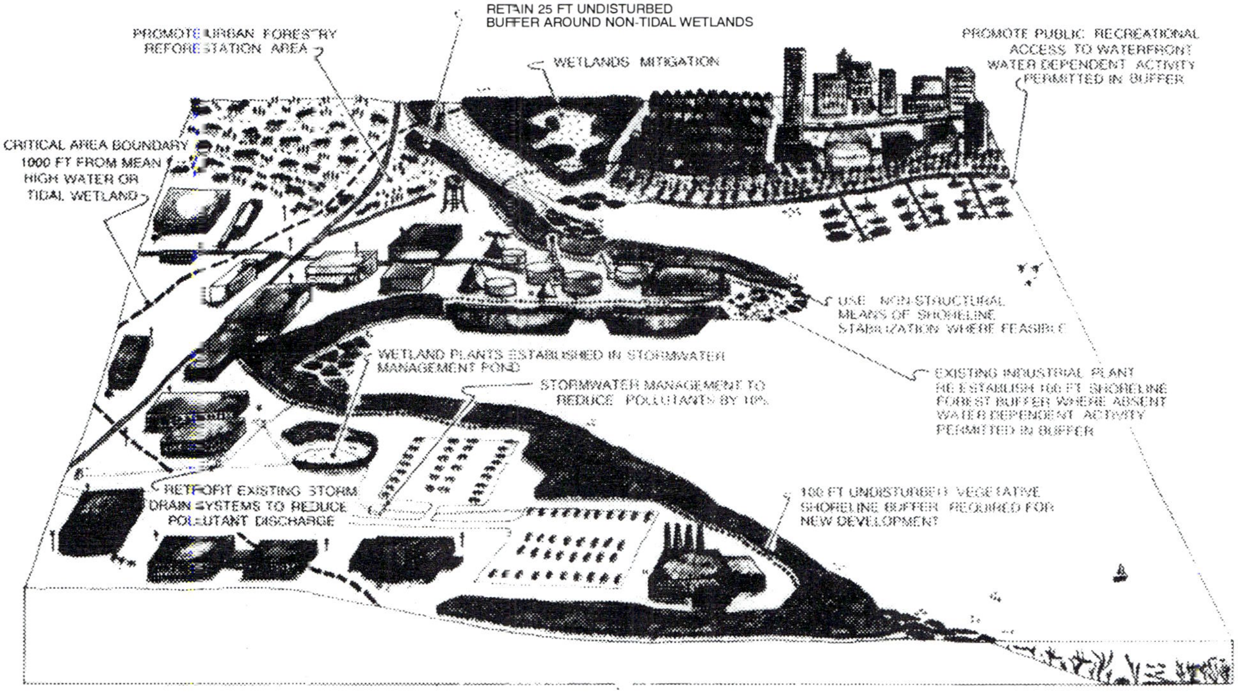

FIGURE 2. Intensely Developed Area (IDA) Residential, Commercial, Industrial & Institutional Uses Dominate.

Intensely Developed Areas (IDA's) are part of the Maryland Critical Areas Commission land use program that includes any area of 20 or more contiguous acres, or the entire upland portion of a municipality within the Critical Area where residential, commercial, institutional or industrial land uses predominate, and where relatively little natural habitat occurs. In Maryland, land within 1,000 feet of the tidal waters' edge or from the landward edge of adjacent tidal wetlands is considered a Critical Area. Stormwater management, wetland mitigation, vegetative shoreline buffers, shoreline stabilization, and reduced pollutant discharge are important land use planning components of IDA's.

Wetlands have become a more important land use type throughout human history. Originally wetland areas were viewed as regions associated with transmitted diseases including malaria and yellow fever. During the 1980's and 1990's, the world community began to realize that wetlands were uniquely important sedimentation control, water filtration and purification, groundwater recharge, filtration of mine drainage and sewage, and recreation. Preservation of upland forest cover, control of on-site sewage disposal systems, proper planning of developed lot size, floodplain and conservation zoning, cluster development and wellhead protection are all land use planning techniques that can assist in protecting wetlands, and improving the water quality and health of various regions of the world.

Because of the increased number of golf courses around the world, the environmental degradation has been under attack by environmentalists. Golf courses as a land use type are increasingly built where the demand for them exists, not necessarily where is the best suited location. In urban areas, many people believe that a golf course represents open space and it does to some extent, but the ramifications of the extensively manicured courses is a potential threat to water supplies.

Golf courses tend to spawn other land use activities that all consume large amounts of water. The artificially created golf course can drain a water supply and potentially pollute it. Extensive use of fertilizers and other lawn chemicals are important to create the atmosphere necessary to attract golfers. More frequently, housing developments are being incorporated into the land use scheme of the golf course and place more strain on the community water supply and enhance the probability of groundwater contamination. Although golf courses can be valuable recreation areas and stormwater catchment basins, extreme care must be taken to protect the quality of water resources in regions where they are located.

SOLID WASTE MANAGEMENT

The changing locational pattern of sanitary landfills in Pennsylvania between 1965 and 1994 has been one of the most important land use changes in the state (Benhart, 1984) (10). In 1965, approximately 1,100 landfill sites were in operation, but the number was reduced to 48 sites by 1994 (Figure 3 & 4). Prior to

1968 in Pennsylvania, there was no true regulatory power over how landfills should operate or any land use controls to reduce environmental threats to the region, where landfills were located. Illegal dumps and poorly operated landfills causing surface and groundwater pollution were a common occurrence. The standard practice for

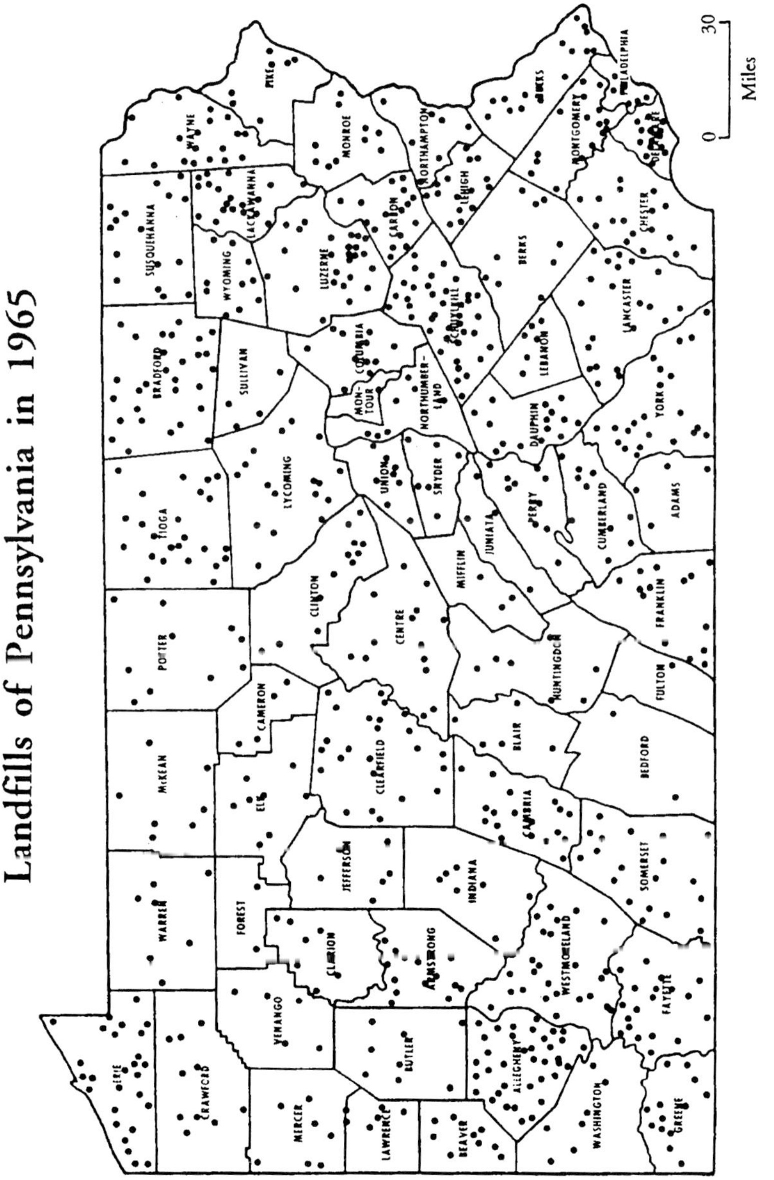

FIGURE 3. Landfills of Pennsylvania in 1965. Source: DER, Bureau of Waste Management.

disposing of municipal solid waste in the United States was to dump it in an open pit and burn it. Organic garbage is a fertile breeding ground for bacteria and viruses that cause disease in humans as well as for flies, mosquitoes, and rodents which transmit the germs to people (Rajagopalan & Schiffman, 1974) (11) (Table 1).

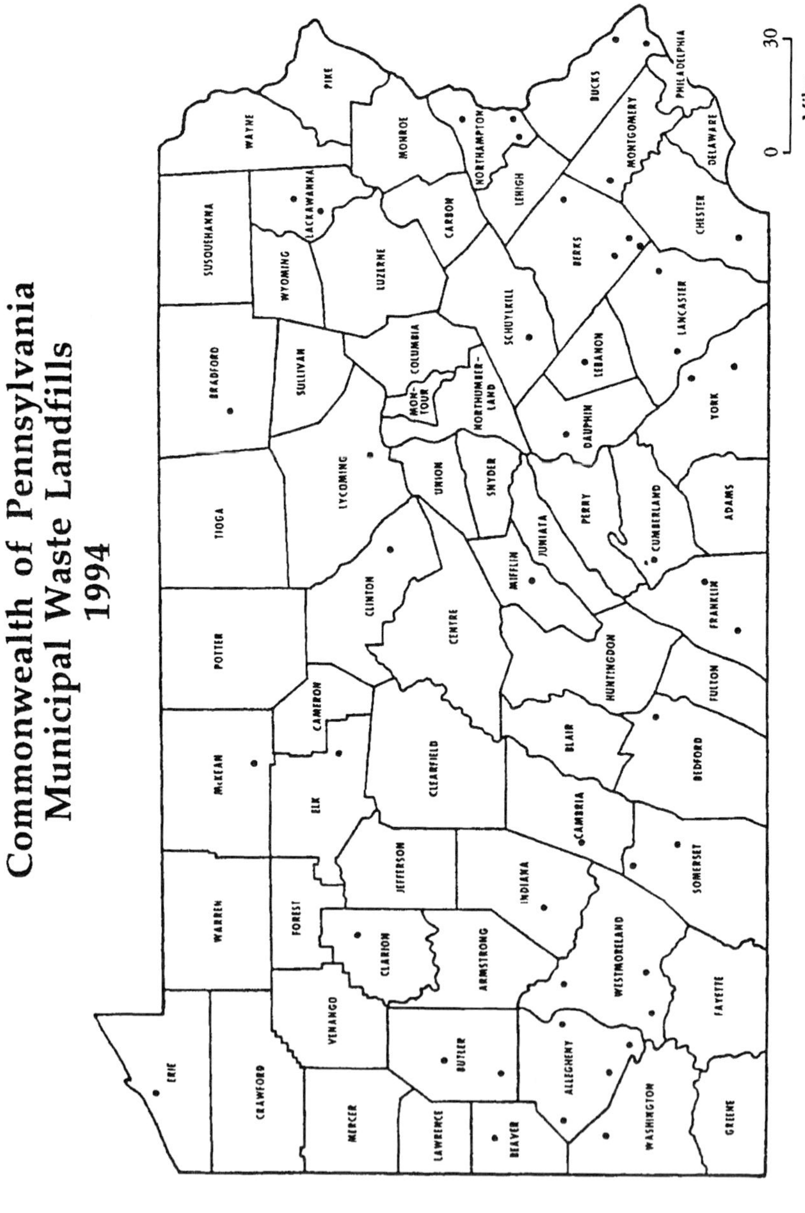

FIGURE 4. Commonwealth of Pennsylvania Municipal Waste Landfills 1994. Source: DER, Bureau of Waste Management.

As a result, on July 31, 1968, the Pennsylvania Legislature enacted the Solid Waste Management Act (Act 241) for the purpose of planning and regulating solid waste storage, collection, transportation, and disposal systems. From this, the Pennsylvania Department of Environmental Resources (DER) issued Chapter 75 Rules and Regulations on Solid Waste Management on August 1, 1991. Among the requirements contained in this document, all waste disposal facilities must have a permit to operate. In order for a permit to be issued, plans and specific design specifications for a landfill must be met. A sample of the information submitted when applying for a permit is: the geology, hydrology, and soils of the site area along with the location of nearby wells and a plan to monitor and

TABLE 1
Solid Waste*, Land Use & Health

Landfill Concern	Neighborhood Concern	Land Use & Environmental Controls
dust or particulate matter	respiratory problems, eyes, nose and throat and aesthetic problems	application of water spray to landfill and use of rip rap
methane gas	potential explosions	proper venting and design of landfill
leachate	potential groundwater contamination	constructing double liners, leachate treatment, and monitoring wells
surface water pollution via runoff	contamination of surrounding wells and destruction of agricultural crops	implement erosion and sedimentation control plan
fires	air pollution	daily soil cover
odors from garbage	reduced air quality	daily soil cover
traffic flow	traffic noise, safety, and property values	implement transportation plan reducing traffic speed and cross-town and neighborhood traffic
siting of landfill	declining property values and the availability of utilities	implement planning program for sound barriers, vegetative buffers, and landscaping plans
vector causing digestive tract diseases	flies, mosquitos, rodents, birds, bacteria, and viruses	daily soil cover

*Solid Waste - Any garbage, rubbish, refuge, sludge or other discarded material, including solid, liquid, semi-solid or contained gaseous material resulting from domestic, industrial, commercial, mining, agricultural or governmental operation.

control all pollutants produced, especially leachate. Leachate is a liquid high in dissolved chemicals as a result of decaying waste. If untreated, this substance could migrate into the groundwater and streams and could contaminate the public water supply. Many of the original landfills in Pennsylvania indicated in Figure 1 lacked the proper facilities to control leachate. Because of the Chapter 75 Regulations, many of the sites noted in the 1965 map have been forced to close and the siting of new landfills has become more difficult.

On July 28, 1988, Governor Robert P. Casey signed Senate Bill 528 making it Act 101. Act 101 is the Pennsylvania Municipal Waste Planning, Recycling, and Waste Reduction Act. Act 101 mandates recycling in Pennsylvania's larger municipalities, requires counties to develop municipal waste management plans, and provides grants to offset expenses. The goal of Act 101 in Pennsylvania is to reduce municipal waste flow, recycle at least 25 percent of waste generated, produce and use recycled and recyclable materials, and educate the public concerning recycling and waste reduction. The land use planning and public health benefits of recycling and waste reduction include less pollution risks, conservation of natural resources, energy and landfill space, and lower disposal costs. Act 101 encourages leaf and waste composting, household hazardous waste collection, recycling lead batteries, county planning, and environmental protection (Pennsylvania Code, 1992) (12).

The new Act 101 regulation encouraged sound land use planning and pollution control of groundwater by prohibiting landfills in floodplains, in coal or limestone areas, and within a specific distance from streams and property lines. To prevent leachate from contaminating groundwater new state regulations require new landfills to have:

1. a double protective liner
2. a leachate collection and treatment system
3. a system for collecting methane gas
4. a terrace system to direct upslope storm or surface water and to reduce erosion and sedimentation.

A detailed operation plan also must be submitted for a landfill and must include:

1. How will traffic enter and exit the facility?
2. How will litter be controlled?
3. How and when will waste be covered?
4. How will soil erosion and sedimentation be controlled?
5. How will leachate and methane gas be collected and treated?
6. How will water and air quality be monitored?
7. How will security be maintained?
8. How will the health and safety conditions be implemented?

The strong environmental controls imposed by Act 241 and Act 101 were responsible for many municipal waste disposal facilities closing in Pennsylvania between 1965 and 1994. By 1994 the 48 municipal landfills implemented sound environmental planning practices including daily soil cover, no open burning and no water pollution problems. The existing landfills have monitoring wells to access groundwater quality adjacent to each site and are regularly inspected by DER officials. With the reduced number of municipal sanitary landfills in Pennsylvania, a new distributional pattern has emerged in 1994 with fewer landfills being located near urban areas. Most are found in rural and non-metropolitan areas with good access to large population centers.

Site selection for a municipal sanitary landfill is an extremely important land use decision. The locational process must include a detailed study of geology, soils, hydrology, mineral deposits, transportation, land use, and solid waste generation. Environmental concerns about contamination of water supplies, litter, vermin control, methane migration and general health and safety designs must be studied. A major site consideration is the geological background in the region. A site that has thick deposits of relatively impermeable material offers more natural protection to groundwater pollution. Deposits such as glacial till, lake silts, and clays are much better than porous material like sand and gravel. Major aquifers below the landfill site can pose a problem unless the overburden soils are particularly thick and impermeable.

Two key health concerns associated with municipal sanitary landfills are methane gas and leachate (Lu, 1985) (13). Methane gas is colorless and odorless and highly explosive in concentrations of 5-15 percent in the presence of oxygen. The primary gas movement mechanism is convection. There are cases where there was a transmission of gas and an accumulation in nearby sewerlines and buildings. These gases kill nearby vegetation by excluding oxygen from the roots. Carbon dioxide can increase the hardness of the water in groundwater and affect nearby wells. Generally the gas can be dispensed into the atmosphere without any problems occurring. Methane gas movement can be controlled by using material that would be more permeable than surrounding soil such as gravel vents or gravel filled trenches. Another control method is using impermeable materials that can contain the gas and vent it through the top cover or simply block the gas flow. The formation of leachate is a major environmental concern associated with landfills. Water percolates down through the soil picking up chemicals and heavy metals from the decomposing waste which eventually encounters the water table. In response to this problem, landfill designers have used sophisticated leachate collection and treatment systems and clay and synthetic liners to protect groundwater quality.

The sanitary landfill is usually the most economical method of acceptable waste disposal. A sanitary landfill is a complete or final disposal method compared to incineration and composting where items such as residue and unusable materials require further disposal. A sanitary landfill is flexible; increased quantities of solid waste can be disposed of with little additional work force and equipment. However, finding suitable site locations for landfills can be difficult, methane gas and leachate control can be expensive, and landfills will settle and cause environmental problems to occur after the landfill is closed.

Because municipal waste landfills have long term environmental and health implications, a closure plan must be submitted in Pennsylvania to show how an operator plans to protect public health, safety and the environment. A closure plan in Pennsylvania must include the following requirements: final cover and grading, sedimentation and erosion control, vegetation, water quality monitoring, leachate management, gas venting, and bonding and insurance.

A strong municipal land use planning program facilitates the successful location of a municipal sanitary landfill. A zoning ordinance assists in the proper arrangement of land use types adjacent to the landfill. Sound barriers, vegetation buffers and landscaping plans can help alleviate land use conflicts. Transportation planning, including safety in residential neighborhoods, road construction, speed limits, and eliminating cross town traffic all contribute to the health and safety of the local population.

SUMMARY AND CONCLUSION

We in the United States have obviously not learned from lessons of poor land use planning in the past. The Dust Bowl in the midwestern United States occurred as a result of poor farming practices and poor planning. Since the Dust Bowl, sweeping changes have occurred in the way we use the land. Development is occurring at an astounding rate. The world has experienced rapid and extensive clearing of rainforest and sudden irrigation in areas where arid land once existed. Rapid clearing of land has the potential to cause water contamination and flooding because of an increase in runoff from lost vegetation. Irrigation of areas, coupled with pesticide use and the presence of farm animals increases the chance of water contamination.

Water is a resource that is essential to life. With the increasing growth of the world population to exceed over 5.6 billion people, many areas of the world are dealing with more health related water problems. The need for drinking water, sewage disposal, recreation, irrigation, and agriculture indicate the importance of having land use regulations to protect sensitive and fragile water environments (14). Diseases such as cholera, typhoid, paratyphoid fevers, amoebic dysentery and infectious hepatitis are all waterborne illnesses. Land use programs such as the Maryland Critical Area Commission are necessary to protect future surface and groundwater supplies.

Landfills are one of the least expensive, most efficient and practical forms of solid waste management. Existing regulations require stringent guidelines for sanitary landfills. Of a larger threat, however, are landfill sites that were in operation prior to regulation and are no longer in use. These sites still produce leachate and methane gas and affect the health of surrounding communities. Locally unwanted land uses are becoming more difficult to locate because of environmental regulations and community health concerns. With increased societal demands on resources, the challenge for modern society will be to find environmentally suitable sites for locally unwanted land uses that fit into the spatial land use scheme of the region.

ACKNOWLEDGMENT

Graduate assistants Patrick Closs and Shannon Miller provided research and cartographic assistance in preparing the chapter. The author would like to thank Dr. Sonia Arbona, Dr. William Shirk, and James and Linda Gemmell for reviewing the chapter and Rebekah Ocker for typing and graphic assistance.

BIBLIOGRAPHY

1. Pennsylvania Low Level Radioactive Waste Disposal Facility Site Screening Interim Report. 1994. Chem-Nuclear Systems, Inc. Harrisburg, PA, pp. 4-5.
2. Greenberg, Michael (Ed.). 1987. *Public Health and The Environment - The United States Experience.* The Guilford Press, New York, pp. 107.
3. Benhart, John E. and C. Robert Scull. 1993. *Regions of the World Today.* Kendall/Hunt Publishing Company, Dubuque, Iowa, pp. 52.
4. Nadakavukaren, Anne. 1990. *Man and Environment.* Waveland Press, Inc. Prospect Heights, Illinois, pp. 440.
5. Switzer, Jacqueline Vaugh. 1994. *Environmental Politics: Domestic and Global Dimensions.* St. Martins Press, New York, pp. 176.
6. Delfino, J. 1977. Contamination of potable groundwater supplies in rural areas, pp. 34. In: R.B. Pojasek (Ed.) *Drinking Water Quality Enhancement Through Source Protection.* Ann Arbor Science, Ann Arbor Michigan.
7. Samuels, Mike and Hal Zina Bennett. 1983. *Well Body, Well Earth.* Sierra Club Books, San Francisco, pp. 162.
8. Critical Area and You - The Chesapeake's First Line of Defense. 1993. Chesapeake Bay Critical Area Commission, Annapolis, Maryland, pp. 2.
9. Marsh, William M. 1991. *Landscape Planning Environmental Applications.* John Wiley & Sons, Inc., NY, pp. 123-125.
10. Benhart, John E. and Donna McCartney. 1984. Solid Waste Management in Southcentral Pennsylvania. *Pennsylvania Geographer*, pp. 1-6.
11. Rajagopalan, S. and M.A. Schiffman. 1974. *Guide to Simple Sanitary Measures for the Control of Enteric Diseases.* World Health Organization, Geneva, pp. 44.
12. Commonwealth of Pennsylvania - Pennsylvania Code. 1992. *Title 25 Environmental Resources.* Department of Environmental Resources, Harrisburg, pp. 273-1-68.
13. Lu, James C.S. and Bent Eichenberger and Robert Stearns. 1985. *Leachate From Municipal Landfills.* Pollution Technology Review, no. 119, Noyes Publications, Park Ridge, NJ, pp. 3-14.
14. Benhart, John F. and Alex Marcin. 1994. *Wetlands: Science, Politics and Geographical Relationships.* The National Council for Geographic Education - Pathways Publication. Indiana, PA

Environmental Contaminants, Ecosystems and Human Health. Edited by S.K. Majumdar, E.W. Miller, and Fred J. Brenner. © 1995, The Pennsylvania Academy of Science.

Chapter Thirteen

A REVIEW OF EPIDEMIOLOGIC EVIDENCE OF HUMAN HEALTH EFFECTS RELATED TO POWER PLANT-GENERATED AIR POLLUTION

DAVID M. MANNINO and RUTH A. ETZEL

National Center for Environmental Health
Centers for Disease Control and Prevention (CDC)
4770 Buford Highway
Mail Stop F-39
Atlanta, GA 30341-3724

INTRODUCTION

Ambient air pollution has been linked to respiratory symptoms both in the United States and abroad (1, 2). Sulfur oxides (SO_x), which are generated by the burning of fossil fuels, have been important pollutants in air pollution episodes in London; Tokyo; the Meuse valley of Belgium; and Donora, Pennsylvania. In the United States electric utilities generate 70% of all SO_x and coal-fired power plants generate 92% of these emissions (3). The U.S. Environmental Protection Agency (EPA) has set primary National Ambient Air Quality Standards (NAAQS) for SO_2 at 80 $\mu g/m^3$ (annual arithmetic mean), and 365 $\mu g/m^3$ (24-hour standard). From 1980 through 1990, the annual mean ambient SO_2 concentration in the United States decreased about 24%. Nevertheless, in 1990, 50 areas in the United States had not attained EPA air quality standards for SO_2.

In this chapter we will review epidemiologic studies that examined the relationship between power plant generated air pollution (SOx) including sulfur dioxide [SO2], sulfite [SO3], and sulfate [SO4], and particulates) and human health effects.

METHODS

We searched the National Library of Medicine database, MEDLINE, for the years 1966 through 1993 for all articles that mentioned either sulfur dioxide, particulate, sulfate, or power plant in the title or the abstract. We then searched these articles for the MEDLINE term "air pollutants". We limited our search to English-language articles that dealt with human subjects, and we found 179 such articles. We then reviewed the abstracts of these articles in order to select epidemiologic studies that identified power plants that burned fossil fuels as the primary polluter in the area studied. We excluded 35 exposure-chamber studies, 75 epidemiologic studies that did not mention power plants, and 51 other papers that were neither epidemiologic studies nor exposure studies. Using this process, we identified 18 articles for review. In addition, we reviewed the bibliographies of the selected articles to identify other articles that dealt with this problem, and found two additional articles.

We used the following criteria for inclusion of a study in our review:

1. The study must have identified power plants that burned fossil fuels as potentially affecting the population studied.
2. The study was reported in the peer-reviewed literature.
3. The study reported concentrations of the air pollutant(s) under investgation.
4. Major confounding factors, particularly temperature, season, and day of week, were accounted for in studies of acute exposure; and smoking, race, occupational exposure, and socioeconomic status were accounted for in studies of chronic exposure (4).

Of the 20 studies identified in the literature review 2 did not meet the third criterion and were excluded. This review is divided into two parts: the first reviews studies that evaluated health effects in the United States, and the second reviews studies that evaluated health effects in other countries.

UNITED STATES STUDIES

One of the first studies of the effects of air pollution on health in the United States was conducted in the Chestnut Ridge area of southwestern Pennsylvania. This is an approximately 700 km^2 rural region that includes four coal-fired power plants. This region was part of an early study of air pollution-related mortality (5). Researchers determined mortality experience for samples of people in 2 communities within this region over a 13-year period. These people were nonrandomly selected volunteers. One community was located close to a coal-fired power plant and had higher mean levels of pollutants compared with the other community, whose location was farther away from the plant. Sulfation rates in the closer community were 3.7 mg SO_3/100 cm^2/day, compared with 0.6 mg SO_3/100 cm2/day for the more distant community (direct correlations to SO_2 are not available, but later studies demonstrated a range of SO_2 measurements from 18 $\mu g/m^3$ to 176 $\mu g/m^3$), and suspended particulates were 151 $\mu g/m^3$ compared with 109 μg/m3. In 1960 the populations studied completed a survey that included questions on smoking status. For a follow-up study conducted in 1972 information was available for 469 of the 533 original subjects from one community and for 366 of the 401 subjects from the other community. This follow-up study demonstrated a clear relationship between smoking and mortality (excess risk 1.9, 80% confidence interval 1.3 to 2.6), and it also demonstrated a slight excess risk for mortality in nonsmokers in the more-polluted community compared with the less-polluted community. This excess risk, although not statistically significant (0.2, 80% confidence interval, -0.1 to 0.6), was approximately one tenth the excess mortality attributed to heavy smoking alone. Researchers also found evidence of an additive effect of both air pollution and smoking.

This region of Pennsylvania was restudied from 1979 through 1981, when four power plants were active. The power plants and their 1980 SO_2 emissions are Conemaugh, 205,200 tons of SO_2; Homer City, 169,100 tons of SO_2; Keystone, 142,300 tons of SO_2; and Seward, 24,400 tons of SO_2 (6). Rcsearchers published two studies regarding children in this region. One study was a cross-sectional study (7) that evaluated symptom prevalence and pulmonary function, and the other was an 8-month long panel study (8) that evaluated pulmonary symptoms and serial peak flows.

In the Chestnut Ridge cross-sectional study, 4,071 children, aged 6 to 11, completed a questionnaire and pulmonary function testing (7). Researchers divided the study population into low-, medium-, and high-exposure categories on the basis of annual mean SO_2 ambient air measurements. Annual SO_2 levels were 65 $\mu g/m^3$ in the low-exposure group, 70 $\mu g/m^3$ in the medium-exposure group, and 91 $\mu g/m^3$ in the high-exposure group. Pulmonary symptom prevalence, pulmonary illness prevalence, and pulmonary function were adjusted for significant predictors, including age, sex, socioeconomic status, gas-stove use, and parental smoking. Multiple logistic regression determined the relative odds of respiratory symptoms or pulmonary illnesses that were due to residing in a high- or medium-exposure area, compared with living in the low-exposure area. This regression was done both for the overall group and for a subgroup of 1,191 subjects who were lifetime residents of the township in which they were residing at the time of the survey. The relative odds for serious chest illness before the age of two years was 1.52 ($p < 0.05$) in the high-exposure group and 1.34 ($p < 0.05$) in the medium-exposure group. Relative odds for chronic cough or phlegm, or persistent wheeze, or chest illness were not significantly elevated. After adjusting for the confounders noted previously, researchers found that pulmonary function was similar in the low-, medium-, and high-exposure areas.

Over an 8-month period in 1980 and 1981, researchers evaluated a subgroup of these children as part of a study that included peak flow measurements and symptom diaries (8). This group had 351 participants, one third of whom had persistent wheeze, another third had cough and phlegm production for most days of the week for at least 3 months, and one third had neither persistent wheeze nor chronic cough. These subjects all lived in the most highly polluted part of the Chestnut Ridge study area. Of the 351 subjects selected for the study, only 128 completed the diaries; 144 completed the serial peak flow measurements. During this study, the mean daily SO_2 level was 51.2 $\mu g/m^3$, with a range of 18 $\mu g/m^3$ to 176 $\mu g/m^3$. Analyses included determining incidence rates of lower respiratory illnesses (LRI), upper respiratory illnesses (URI), and wheeze episodes for the entire sample and for each of the subgroups. The rates for each symptom were divided at the mean into high- and low-incidence groups, and mean daily pollutant concentrations and temperature for each group were compared. A high incidence of LRI and wheeze was associated with a lower minimum temperature, but was not associated with any of the pollutants. A separate multiple logistic regression included maximum daily pollutant values, the minimum daily temperature, and the symptom outcome on the previous day as independent variables. Regression models could only be fitted for subjects who had symptoms; thus, 111 subjects were included in the regression analysis of URI, 55 were included in the regression analysis of LRI, and only 26 were included in the analysis of wheeze. Using multiple logistic regression, a 20°C decrease in minimum temperature was associated with an odds ratio of 1.2 (95% confidence interval, 1.05 - 1.37) for URI and 1.3 (95% confidence interval, 1.04 - 1.68) for LRI. Changes in pollutants were not associated with significant changes in the rate of symptoms. None of the pollution variables was associated with a change in the peak flow rate in this study.

Researchers also administered respiratory questionnaires to 5,557 adult women, aged 17 to 74, as part of the Chestnut Ridge study (9). Respiratory symptoms tested in the multiple logistic regression models included chronic cough, chronic phlegm, dyspnea, and wheeze most days or nights. Analyses were performed separately for nonsmokers and current smokers, and confounders, including age, socioeconomic status, spouse smoking, former cigarette smoking, and cigarettes smoked daily were included in the models. The annual mean SO_2 was 62 $\mu g/m^3$ in the low-exposure region, 66 $\mu g/m^3$ in the medium-exposure region, and 99 $\mu g/m^3$ in the high-exposure region. The risk for "wheeze most days" in nonsmoking women residing in the high- and medium-exposure areas was 1.58 and 1.26 ($p < 0.02$), relative to women living in the low-exposure area. Cigarette smoking characteristics (cigarettes smoked daily) were the major determinant of symptoms in smoking women.

In the Chestnut Ridge studies, researchers based exposure estimates on known community exposures, and not on individual measurements. This could possibly invalidate the findings, as individual exposures could differ markedly from community exposures. The differences between the most polluted and the least polluted regions in the studies were rather small. Such small differences might cause researchers to underestimate the effect of pollution on respiratory symptoms. The panel study had a low participation rate (41%), and the number of subjects analyzed was very small in several analyses.

A second large study of air pollution in the United States was the Six Cities Study of air pollution and health. This was begun in 1974 to provide a contemporary source of epidemiologic data regarding the health effects of the principal constituents of air pollution caused by fossil fuel (10, 11). The cities studied were Watertown, Massachusetts; Kingston and Harriman, Tennessee; Steubenville and Mingo Junction, Ohio; St. Louis, Missouri; Portage, Wisconsin; and Topeka, Kansas. Pollution sources in the most polluted cities (St. Louis and Steubenville) were industrial sites and power plants.

The Six Cities cross-sectional study of children included 10,106 children aged 6 to 11 years, of whom 8380 were seen for follow-up at one year (10). The subjects completed an initial questionnaire and, one year later, a follow-up questionnaire, at which time they reported the incidence of cough, bronchitis, chest illness, LRI, and wheeze during the previous year. Researchers obtained pulmonary function tests at both the initial and follow-up evaluations. Air pollutants monitored at nine different sites within the six cities were total suspended particulates (TSP), the sulfate fraction of TSP (TSO_4), and SO_2.

Annual mean concentrations for the pollutants ranged from 39.3 $\mu g/m^3$ to 114 $\mu g/m^3$ for TSP, from 5.4 $\mu g/m^3$ to 18.8 $\mu g/m^3$ for TSO_4, and from 3.4 $\mu g/m^3$ to 68.3 $\mu g/m^3$ for SO_2. Symptom rates and pulmonary function were adjusted for age, sex, maternal smoking, and parent education. TSP and TSO_4 were positively associated with increased rates of cough, bronchitis, and LRI when between-city differences were evaluated. For a 10 $\mu g/m^3$ increase in TSP level the odds ratio for cough was 1.11 ($p < 0.01$); for bronchitis it was 1.11 ($p < 0.05$); and for LRI it was 1.08 ($p < 0.05$). The within-city analyses of symptom rates in the two most polluted cities were not significant except for a negative association between TSP and wheeze. Pulmonary function, adjusted for age, sex, height, maternal smoking, and parental education, was not associated with any of the pollution measurements.

Researchers followed a group of 5,422 children, aged 10 to 12 years, for 12 months during the 1980-1981 school year (11). Subjects again completed a respiratory questionnaire. Pollution measurements at this evaluation included TSP, SO_2, fine fraction aerosol sulfate (FSO_4), particulate matter less than 15 μm (PM_{15}) and 2.5 μm ($PM_{2.5}$) aerodynamic diameter, ozone (O_3), and nitrogen dioxide (NO_2). The ranges of pollutants measured during this study were lower than those reported during the 1974-1977 study. The ranges of 12-month mean pollution levels, in $\mu g/m^3$, were as follows: TSP (34 to 71), SO_2 (3 to 28), FSO_4 (4 to 14), PM_{15} (20 to 59), $PM_{2.5}$ (12 to 37), O_3 (18 to 38), and NO_2 (6 to 23). Reported rates of chronic cough, bronchitis, and chest illness were positively associated with the mean annual pollution measurements, except for O_3. Children with a history of wheeze or asthma had a higher prevalence of respiratory symptoms and a stronger relationship between pollution levels and respiratory symptoms, than did children without a history of wheeze or asthma. No associations were found between pollutant concentrations and pulmonary function.

Another arm of the Six Cities Study involved serial pulmonary function testing in a group of 200 children, aged 6 to 8 years, in Steubenville, Ohio (12). Four separate evaluations were completed in 1978, 1979 (2 evaluations), and 1980. Researchers performed testing before, during, and after pollution episodes when daily TSP levels rose as high as 422 $\mu g/m^3$ and daily SO_2 levels rose as high as 439 $\mu g/m^3$. When the four studies were grouped together, they demonstrated a 1% to 2% decline in FEV_1 associated with increases in daily levels of TSP or SO_2 near or above the 24-hour ambient air standards (260 $\mu g/m^3$ for TSP and 365 $\mu g/m^3$ for SO_2).

The Six Cities Study also collected data on 8,480 adults, aged 25 to 74 years (13). These subjects completed a respiratory questionnaire and pulmonary function testing. Symptom rates were adjusted for age and smoking status, but not for environmental tobacco smoke exposure or socioeconomic status. All pulmonary symptoms were associated with cigarette smoking. There was no clear relationship between pulmonary symptoms and air pollution. Pulmonary function was likewise related to tobacco smoking, with more of an effect noted in men than in women.

A mortality study examined daily non-accidental deaths in the Steubenville, Ohio, metropolitan area over the 11-year period from 1974 to 1984 (14). Daily deaths were correlated with weather factors (temperature and relative humidity), season, particulate pollution, and sulfur dioxide pollution. Deaths increased in winter and spring, and in association with hot and humid weather. In addition, an increase in particulate pollution of 100 $\mu g/m^3$ was associated with a 4% increase in mortality on the following day.

A second mortality study followed 8,111 adults that were part of the original Six Cities study group for 14 to 16 years (15). Researchers found, after adjusting for smoking status (at the time on entry into the study), educational level, age, body-mass index, and occupational exposure the mortality rate ratio between the most polluted city and the least polluted city was 1.26 (95 percent confidence interval, 1.08 to 1.47). The association between air pollution and mortality was seen for lung cancer deaths and cardiopulmonary deaths, but not for other deaths. Mortality was most strongly associated with fine particulate and sulfate levels. A weakness of this analysis is the lack of data on smoking histories of the subjects after they entered the study. This could have altered the results if there were regional differences in smoking cessation rates.

The Six City Studies were well designed, had good measurements of pollutants, and considered all important confounders in most analyses. The studies demonstrated a consistent relationship between pollution levels and respiratory symptoms in children, but not between pollution levels and respiratory symptoms in adults or pulmonary function. A weakness of these studies, similar to the Chestnut Ridge studies, is the lack of personal exposure data, which could differ from community exposure data.

New Cumberland, West Virginia, was the site of a 7-month study of a panel of asthmatics (16). This town of 2,100 is located very close to a coal-fired power plant. Researchers recruited 43 subjects, both children and adults, with a history of intermittent wheezing to record their pulmonary symptoms daily for 7 months. Fourteen subjects dropped out of the study and only 20 of the 29 remaining subjects reported more than one attack of wheezing during the study period. Levels of the outdoor pollutants SO_2, TSP, SO_4 and nitrates were recorded daily. Mean daily SO_2 measurements ranged from 100 to 750 $\mu g/m^3$, and TSP ranged from 35 to 790 $\mu g/m^3$ during the study period. Attacks of wheezing were most strongly associated with minimum temperature, although the attacks were also associated with pollutant levels, including SO_4, SO_2, TSP, and nitrates.

INTERNATIONAL STUDIES

Health effects in people living close to power plants has also been studied internationally. One study was done in 1983 and 1984 in the Gardanne coal-basin, located near Marseille, France. Three coal-fired power plants release approximately 420 metric ton of SO_2 daily. Over a four month period, 450 children, aged 9 to 11 years, completed pulmonary symptom diaries (17). The study was done in eight communities, five of which had high pollution levels, and three of which had low levels. Researchers measured SO_2 and respirable particles daily. Mean monthly SO_2 levels ranged from 23 to 156 $\mu g/m^3$ in the high-pollution community and from 14 to 30 $\mu g/m^3$ in the low-pollution community. Subjects were asked whether they had "pulmonary symptoms" (morning cough or wheezing), or "ear, nose, and throat (ENT) Symptoms" (eye irritation, runny nose, or sneezing) on each day of the study. Partial correlation coefficients between daily mean symptoms scores and daily SO_2 level showed a significant correlation with cough, eye irritation, runny nose, and wheezing in two of the most highly polluted communities, but not in the other communities. Daily mean temperature was correlated with symptoms in both the high- and low-pollution communities. A separate analysis evaluated the two-week prevalence of symptoms during a period of high pollution compared with a period of low pollution, when mean minimum temperature and relative humidity were similar. In each of the five communities, both pulmonary symptoms ($p < 0.02$) and ENT symptoms ($p < 0.05$) were 10% to 20% higher during the more polluted period.

This study did not adjust for significant confounders, such as temperature and humidity, in its overall analysis of respiratory symptoms and pollution measurements. Some of the analyses only looked at one month of data, although four months of data were collected. The one month selected for these analyses had the highest pollution levels of the four months that were studied.

Finnish researchers have completed studies that deal with fossil fuel-related air pollution. One study took place in three cities, comparing the frequency of respiratory illnesses in groups of children in these cities (18). The ages of children studied were 14 to 18 months old and 6 years old. The polluted city had a higher daily maximum SO_2 (57 to 83 $\mu g/m^3$) than did the reference cities (37 to 56 $\mu g/m^3$ and 53 $\mu g/m^3$). Levels of particulates were similar in all three cities. The odds ratio for respiratory illnesses, adjusted for atopy, environmental tobacco smoke exposure, socioeconomic status, and number of daily contacts, was elevated in the polluted city for both 14- to 18-month old children (2.0, 95% confidence interval 1.3-3.2), and 6-year old children (1.6, 95% confidence interval 1.2-2.1).

Another Finnish study evaluated the relationship between air pollution and symptoms over a 4 month period in 158 subjects who lived in the vicinity of a coal-fired power plant (19). Subjects completed a daily dairy recording symptoms and medication use. Daily mean ambient air measures of NO_2, SO_2, suspended particles, and soot were recorded, along with meteorologic data. Daily measurements of SO_2 ranged from 3.8 to 99.1 $\mu g/m^3$, and measurements of suspended particles ranged from 2 to 66 $\mu g/m^3$. This study demonstrated no clear relationship between ambient levels of air pollution and pulmonary symptoms.

A third Finnish study evaluated the relationship between weekly levels of air pollution and respiratory infection and absenteeism in children and adults (20). This study was undertaken in the city of Helsinki, which has a population of 492,000. The main pollution sources are road traffic, and 3 coal-fired and 6 oil-fired power plants. Researchers obtained health data (respiratory infection and tonsillitis) from 25 communal health centers and 14 day care centers. Absenteeism data were obtained from 210 schools, 14 day care centers, and 12 workplaces over a period of one year. Mean weekly SO_2 levels ranged from 9-62 $\mu g/m^3$ during the study. Upper respiratory infections and tonsillitis were correlated with mean weekly temperature and mean weekly SO_2 level ($p < 0.001$). After controlling for temperature, upper respiratory infection still correlated with mean weekly SO_2 ($p < 0.05$). Mean weekly absenteeism due to a febrile illness was correlated with mean weekly SO_2 and temperature in both the day care center children ($p < 0.05$) and adults ($p < 0.001$), but not in schoolchildren.

Australian researchers evaluated the relationship between pulmonary symptoms and air pollution in a panel of 99 asthmatic schoolchildren over a one year period (21). Part of the study population lived close to two coal-fired power stations, and the remainder lived in an unpolluted area. Maximum mean daily levels of SO_2 were 26 $\mu g/m^3$ in the polluted area and 11 $\mu g/m^3$ in the unpolluted area. This study did not find any correlation between air quality measurements and respiratory symptoms.

Japanese researchers completed a mortality study in Yokkaichi, Japan (21). This city contains an oil-fired power station and an oil refinery. Researchers obtained death certificates that listed asthma or chronic obstructive pulmonary disease as the underlying cause of death. The researchers divided the city into a polluted and a nonpolluted region, on the basis of SO_x precipitation plates. SO_x levels in the polluted area ranged from 2 to 3 mg/day, and levels in the nonpolluted area ranged from 0.1 to 0.3 mg/day. Over the 21-year study period, death rates for both asthma and chronic obstructive pulmonary disease (COPD) were

TABLE 1
Summary of Studies Included in This Review, Listed Chronologically

First Author	Year	Number of Subjects	Age Range	Sulfur Dioxide Levels	Particulate Levels	Positive findings Inconclusive Findings	p value
Cohen (16)	1972	20	NR	100-750 $\mu g/m^3$ (daily mean)	35-790 $\mu g/m^3$ (daily mean)	Attacks of wheeze weakly associated with SO_2 and TSP Attacks of wheeze not associated with suspended nitrates or suspended sulfates.	p<0.01
Morris (5)	1976	835	30->75	0.6-3.7 mg ($SO_3/100cm^2/day$)	109-151 $\mu g/m^3$ (daily mean)	Excess risk of air pollution for death in non smokers approximately 10% of excess risk for heavy smokers in non polluted area. Evidence of additive interaction of air pollution and smoking.	NR
Ferris (13)	1979	8,480	25-74	10-100 $\mu g/m^3$ (annual mean)	22-180 $\mu g/m^3$ (annual mean)	No poositive findings FEV_1, when stratified by smoking status and gender, not associated with air pollutants. Wheeze prevalence or phlegm prevalence, when stratified by smoking status and gender, not associated with air pollutants.	NR
Schenker (9)	1983	5,557	17-74	17-74 $\mu g/m^3$ (annual mean)	62-99 $\mu g/m^3$ (annual mean)	Wheezing in nonsmokers associated with high SO_2 exposure areas. Chronic cough, chronic phlegm, and dyspnea (grade 3), not associated with high SO_2 in nonsmokers. Chronic cough, chronic phlegm, dyspnea (grade 3), and wheezing not associated with high SO_2 in smokers.	p<0.02
Ferris (12)	1983	200	6-8	18-439 $\mu g/m^3$ (daily mean)	20-422 $\mu g/m^3$ (daily mean)	Increase in TSP to 260 $\mu g/m^3$ and SO_2 to 365 $\mu g/m^3$ associated with 1% to 2% decline in FEV_1. None reported.	
Pershagen (19)	1984	158	15-64	3.8-99.1 $\mu g/m^3$ (daily mean)	2-66 $\mu g/m^3$ (daily mean)	Temperature negatively correlated with COPD exacerbations. Ambient air soot associated with COPD exacerbations when adjusted for temperature. Ambient SO_2, and NO_2 not associated with cough, irritative symptoms or COPD exacerbations.	p<0.01 p<0.01
Imai (22)	1986	128,000	0->60	0.1-3 mg SO_x/day (annual mean)		Death from asthma and COPD were increased in the polluted area. No negative findings.	p<0.03
Schenker (7)	1986	4,071	6-11	65-91 $\mu g/m^3$ (annual mean)		Increase prevalence of respiratory illness prior to age 2 in the most polluted areas. Chronic cough or phlegm, persistent wheeze, or chest illness in the prior year not associated with air pollutant levels. Pulmonary function, after adjustment for height, gender, age, doctor diagnosed asthma, chest illness prior to age 2, prolonged hospitalization at birth, parental smoking, gas stove use, and socieoeconomic status, not associated with air pollutant levels.	p<0.05
Ware (10)	1986	8,380	6-11	3.4-68.3 $\mu g/m^3$ (daily mean)	19-114 $\mu g/m^3$ (daily mean)	Increased rates of cough, bronchitis, and chest illness associated with high pollutant levels. Polmonary function not associatedd with air pollutant levels.	p<0.05

Vedal (8)	1987	351	6-11	18-176 μg/m³		Decrease in temperature positively associated with increased rate of lower and upper respiratory illnesses. Lower respiratory illness, upper respiratory illness, or wheeze not associated with sulfur dioxide or coefficient of haze.	$p<0.05$
Charpin (17)	1988	450		14-156 μg/m³ (monthly mean)		Cough, eye irritation, runny nose, and wheezing associated with increased pollution in two communities with mean monthly sulfur dioxide level of 146μg/m³ Cough, eye irritation, runny nose, and wheezing not associated with increased pollution in six communities with mean monthly sulfur dioxide level of less than 98μg/m³	$p<0.05$ $p<0.05$
Dockery (11)	1989	5,422	10-12	3-28 μg/m³ (annual mean)	34-71 μg/m³ (annual mean)	Increased rates of cough, bronchitis, and lower respiratory illness associated with pollutants. Asthma, persistent wheeze, hay fever, or nonrespiratory illness not associated with pollutants. Pulmonary function not associated with pollutants.	$p<0.05$
Ponka (20)	1990	492,000	Children and Adults	9-62 μg/m³ (weekly mean)	48-123 μg/m³ (annual mean)	Upper respiratory infection and tonsillitis associated with SO_2 and temperature in the entire population. After controlling for temperature, relationship between SO^2 and upper respiratory infection still significant. Upper respiratory infection, tonsillitis, otitis, or lower respiratory infection not associated with SO_2 or NO_2 in a group of 1532 children from day care centers.	$p<0.05$
Ponka (20)	1990	83,987	Children and Adults	9-62 μg/m³ (weekly mean)	48-123 μg/m³ (annual mean)	Absenteeism and SO_2 levels correlated in adults. Absenteeism and SO_2 levels not correlated in children.	$p<0.01$
Henry (21)	1991	59	School-children	11-26 μg/m³ (daily maximum)		No positive findings. Cough, wheezing, or breathlessness not associated with SO_2 or NO_2 levels.	$p>0.05$
Jaakola (18)	1991	1,438	1-6	53-83 μg/m³		Increased rates of respiratory illness associated with exposure to higher pollutant levels. No negative findings reported.	$p<0.05$
Schwartz (14)	1992	163,099	NR	18-143 μg/m³ (daily, 10th to 90th percentile)	36-209 μg/m³ (daily, 10th to 90th percentile)	100μg/m³ increase in particulate pollution associated with 4% increase in mortality during the following day. Sulfur dioxide, after adjustment for particulate pollution, not associated with mortality.	$p<0.05$
Dockery (15)	1993	8,111	25-74	5-64 μg/m³ (mean over 14 yrs.)	32-89 μg/m³ (mean over 14 yrs.)	Mortality rate-ratio 1.26, comparing most polluted city to least polluted city. Mortality most strongly associated with fine particulates and sulfates.	$p<0.05$

consistently higher in the polluted area compared with the nonpolluted area. Because of the nature of the data, corrections for smoking status, socioeconomic status, and residence history could not be made.

DISCUSSION

Air pollution exposure has been linked to adverse health effects for many years. Power plants that burn fossil fuels are an important source of sulfur-containing air pollutants and have been central to the studies reviewed here. In order of decreasing importance, the spectrum of biologic response to pollutant exposure ranges from increased mortality to mere annoyance with odors (4). Summaries of the studies included in this review are shown in Table 1.

Four studies looked at mortality related to fossil fuel-burning pollution. One found a slightly higher age-adjusted mortality for asthma and COPD in the polluted area compared with the nonpolluted area (22). This study was not able to control for other significant predictors of asthma and COPD mortality, such as occupational exposure, socioeconomic status, and tobacco smoking. The second study, which evaluated death as an "acute" event, was a 10-year time-series analysis of deaths in Steubenville, Ohio, and found a correlation between non-accidental deaths and particulate pollution on the previous day (14). In studies that investigate death as an "acute" event, deaths typically occur in individuals with severe underlying cardiorespiratory diseases. The third study (15) followed a population over a 14 to 16 year period and found a 26% increase in mortality when the most polluted city was compared to the least polluted study. Although this study controlled for smoking status in the subjects at the time of entry into the study, it did not have data on smoking cessation or smoking initiation. The fourth study examined mortality in a community over a period of 13 years, and found a nonsignificant excess risk for death in nonsmokers of approximately one tenth of the excess risk for smokers (5). These studies all suggest increased mortality related to air pollutant exposure, although they all lacked data on potential exposure to indoor pollutants such as environmental tobacco smoke.

The next most significant adverse respiratory health effect from air pollution, according to the American Thoracic Society (ATS) paradigm (4), is increased incidence of cancer. No study evaluated this endpoint of exposure to fossil fuel-related air pollution, although one study reported an increased lung cancer mortality in the most polluted city when compared to the least polluted city (15).

The next three adverse health effects in the ATS paradigm are increased frequency of symptomatic asthmatic attacks, increased incidence of LRI, and increased exacerbations in persons with COPD. Asthma attacks were the focus of one panel study where a weak correlation between exacerbations and pollution was found (16). Another panel study did not find any significant relationship between air pollutants and asthma exacerbations, when temperature was considered

(8). A third panel study did not demonstrate a relationship between power plant-generated air pollution and asthma symptoms, although this study was hindered by the very low levels of pollutants detected (21). LRI and chest illness were evaluated in several studies of children. TSP and TSO_4 were associated with LRI but not with chest illness in one study (10). In another study, chest illness was not significantly correlated with pollutant measures (11). Only one study looked at COPD exacerbations and did not find any relationship between air pollutants and exacerbations (18). Overall, there does not seem to be an association between air pollution from power plants and these illnesses.

The next adverse health effect ATS considers important is decreased pulmonary function. Three studies found no relationship between air pollutants and lung function when other variables were controlled for (7, 10, 13). Another study (12) looked at serial changes in spirometry in children following pollution episodes and found a 1% to 2% decrement in FEV_1 following exposures to pollutants at levels near or above the 24-hour ambient air standards.

The next group of significant adverse health effects include increased prevalence of wheezing, chest tightness, and bronchitis. Five studies found at least one relationship between a pollutant measure and these symptoms (7, 9, 10, 11, 17). The most consistent relationship was between bronchitis and particulate pollutants (7, 10, 11).

The next adverse health effect in the ATS paradigm is URI. This was evaluated in one study where living in a more polluted area was associated with an increased odds ratio for URI in both 14- to 18-month old children and 6-year old children (18). In another study (20), upper respiratory infection was associated with SO_2, even after adjusting for temperature. In this study absenteeism from work was also associated with elevated SO_2 levels in adults.

The final adverse health effects are ENT irritation and odors. ENT irritation was evaluated in one study, which found a higher two-week prevalence of symptoms during a high-pollution period compared with a low-pollution period, when temperature and humidity were similar (17).

A weakness in all of these studies is an absence of personal exposure measures or biomarkers for exposure. Other studies have demonstrated a poor relationship between ambient outdoor air levels of pollutants and indoor air levels of pollutants. This finding is especially important because people spend most of their time indoors (23). Inaccurate exposure data could potentially bias the findings in all of the studies we reviewed. There exists a clear need to develop new methods for assessing personal exposure and response to air pollution to incorporate these methods into epidemiologic studies. We also need to develop useful physiologic, biochemical, and morphologic markers of the subclinical effects of air pollution exposure (24).

Potential biomarkers might be found in the blood, which interfaces extensively with pollutants present in alveolar air. Because the lifespan of red blood cells is 120 days, biomarkers that appear in red blood cells might indicate chronic (4- to 10-week) exposures.

CONCLUSION

Pollutants related to power plants have been studied in several different communities in this country and around the world. Most studies have demonstrated a relationship between high levels of pollutants and at least one pulmonary symptom, but not between these high levels and lung function. No consistent health effect was noted across the studies reviewed, although the studies varied in the populations studied and in the intensity of pollutant exposure. A need exists for more extensive evaluation of individuals with COPD who live in the vicinity of power plants that burn fossil fuels, as this group would be expected to have the most severe reaction to high pollutant levels. A need also exists for the development of better personal exposure monitors and biomarkers.

REFERENCES

1. Shy, C.M. 1989. Review, discussion, and summary of epidemiological studies. Environ Health Perspect. 79:187-190.
2. Abramson, M. and T. Voigt. 1991. Ambient air pollution and respiratory disease. Med J Aust. 154:543-553.
3. U.S. Environmental Protection Agency. 1990. National air quality and emissions trends report, 1990. Officc of Air Quality Planning and Standards. Research Triangle Park, N.C., EPA.
4. American Thoracic Society. 1985. Guidelines as to what constitutes an adverse respiratory health effect, with special reference to epidemiologic studies of air pollution. Am Rev Respir Dis. 131:666-668.
5. Morris, S.C., Shapiro, M.A. and J.H. Waller. 1976. Adult mortality in two communities with widely different air pollution levels. Arch Environ Health. 38:248-254.
6. U.S. Environmental Protection Agency. 1982. Air quality criteria for particulate matter and sulfur oxides. Research Triangle Park, N.C.: Environmental Criteria and Assessment Office, 1982b. EPA-600/8-82-029 (PB 54-156801/reb).
7. Schenker, M.B., Vedal, S., Batterman, S., Samet, J and F.E. Speizer. 1986. Health effects of air pollution due to coal combustion in the Chestnut Ridge region of Pennsylvania: cross-section survey of children. Arch Environ Health. 41:104-108.
8. Vedal, S., Schenker, M.B., Munoz, A., Samet, J.M., Batterman, S. and F.E. Speizer. 1987. Daily air pollution effects on children's respiratory symptoms and peak expiratory flow. Am J Public Health. 77:694-698.
9. Schenker, M.B., Speizer, F.E., Samet, J.M., Gruhl, J. and S. Batterman. 1983. Health effects of air pollution due to coal combustion in the Chestnut Ridge Region of Pennsylvania: results of cross-sectional analysis in adults. Arch Environ Health. 38:325-330.
10. Ware, J.H., Ferris, B.G. Jr., Dockery, D.W., Spengler, J.D., Stram, D.O. and

F.E. Speizer. 1986. Effects of ambient sulfur oxides and suspended particles on respiratory health of preadolescent children. Am Rev Respir Dis. 133:834-842.

11. Dockery, D.W., Speizer, F.E., Stram, D.O., Ware, J.H., Spengler, J.D. and B.G. Ferris, Jr. 1989. Effects of inhalable particles on respiratory health of children. Am Rev Respir Dis. 139:587-594.
12. Ferris, B.G. Jr., Dockery, D.W., Ware, J.H., Speizer, F.E. and R. Spiro, 3rd. 1983. The Six-City Study: examples of problems in analysis of the data. Environ Health Perspectives. 52:115-123.
13. Ferris, B.G. Jr., Speizer, F.E. and J.D. Spengler et al. 1979. Effects of sulfur oxides and respirable particles on human health. Methodology and demography of populations in study. Am Rev Respir Dis. 120:767-779.
14. Schwartz, J. and D.W. Dockery. 1992. Particulate air pollution and daily mortality in Steubenville, Ohio. Am J. Epidemiol. 135:12-19.
15. Dockery, D.W., Pope, C.A. and X Xu, et al. 1993. An association between air pollution and mortality in six U.S. cities. NEJM. 329:1753-1759.
16. Cohen, A.A., Bromberg, S., Buechley, R.W., Heiderscheit, L.T. and C.M. Shy. 1972. Asthma and air pollution from a coal-fueled power plant. Am J Public Health. 62:1181-1188.
17. Charpin, D., Kleisbauer, J.P., Fondarai, J., Graland, B., Viala, A. and F. Gouezo. 1988. Respiratory symptoms and air pollution changes in children: the Gardanne Coal-Basin Study. Arch Environ Health. 43:22-27.
18. Jaakkola, J.J., Paunio, M., Virtanen, M. and O.P. Heinonen. 1991. Low-level air pollution and upper respiratory infections in children. Am J Public Health. 81:1060-1063.
19. Pershagen, G., Hrubec, Z., Lorich, U. and P. Ronnqvist. 1984. Acute respiratory symptoms in patients with chronic obstructive pulmonary disease and in other subjects living near a coal-fired plant. Arch Environ Health. 39:27-33.
20. Ponka, A. 1990. Absenteeism and respiratory disease among children and adults in Helsinki in relation to low-level air pollution and temperature. Environ Research. 52:34-36.
21. Henry, R.L., Bridgman, H.A., Wlodarczyk, J., Abramson, R., Adler, J.A and M.J. Hensley. 1991. Asthma in the vicinity of power plants: II. Outdoor air quality and symptoms. Pediatr Pulmonol. 11:134-140.
22. Imai, M., Yoshida, K. and M. Kitabatake. 1986. Mortality from asthma and chronic bronchitis associated with changes in sulfur oxides air pollution. Arch Environ Health. 41.29-35.
23. Szalai, A., editor. The use of time: daily activities of urban and suburban populations in twelve countries. The Hague: Mouuton, 1972.
24. National Research Council, Committee on the Epidemiology of Air Pollutants. 1985. Epidemiology and air pollution. Washington, D.C.: National Academy of Sciences.

Environmental Contaminants, Ecosystems and Human Health. Edited by S.K. Majumdar, E.W. Miller, and Fred J. Brenner.

Chapter Fourteen

CLIMATE CHANGE AND HUMAN HEALTH

CASEY CRUMP and RUTH A. ETZEL
Air Pollution and Respiratory Health Branch
Division of Environmental Hazards and Health Effects
National Center for Environmental Health
Centers for Disease Control and Prevention
Atlanta, GA 30341-3224

INTRODUCTION

Ongoing climatic changes as the result of human activity include global warming caused by the greenhouse effect and changes caused by stratospheric ozone depletion and rising levels of tropospheric pollutants. These changes are interrelated in several ways: they all result mainly from anthropogenic emissions, they share common biochemical processes, and their effects on living organisms may be synergistic. These changes have the potential to affect human health in unprecedented ways. In this chapter, we will explore these issues by reviewing the literature on climate change and discussing the potential impact of such changes on human health.

HISTORICAL BACKGROUND

The physical process by which radiatively active gases can elevate air temperatures and alter the global climate has been known for at least a century. In 1827, Jean-Baptiste Fourier raised important questions about the variability of global temperatures and suggested that human activity can modify climate (Fourier 1827). The formal foundation for the greenhouse theory of climate change was laid in 1896, when Svante Arrhenius studied the climatic effects of changes in atmospheric carbon dioxide (Arrhenius 1896). On the basis of his finding that a doubling of atmospheric CO_2 would warm the planet by 5 K, Arrhenius concluded that past glacial epochs may have occurred largely because of a reduction in atmospheric CO_2. Trewartha coined the term "greenhouse effect" in 1937 to describe how atmospheric gases in the troposphere trap infrared radiation and thus maintain a warmer temperature at the Earth's surface (Last 1991). At about the same time, Callendar concluded that industrial and other human activities were causing an increase in atmospheric CO_2 and that this increase could initiate global warming. Quantitative evidence for CO_2-induced global warming was first provided in 1967 by Manabe and Wetherald, who used a forecasting model based on the concepts of global energy balance and radiative-convective equilibrium. The theory of climate change involving variations in levels of atmospheric CO_2 is now one of the major mechanisms for understanding climate variations of the past, as well as for predicting future climate changes from anthropogenic sources.

SCIENTIFIC BACKGROUND FOR CLIMATE CHANGE

The greenhouse effect is a well-established theory of atmospheric science. The planet Venus, for example, with its dense CO_2 atmosphere, has surface temperatures near 700 K; Mars, with a very thin CO_2 atmosphere, has temperatures of only 220 K. This difference is due to the effect of gases and particles that preferentially allow the visible and near-infrared wavelengths of sunlight to filter through to the surface of the planet relative to the amount of longer, infrared wavelengths emitted by the planet that the atmosphere allows to escape into space. The greater the concentration of such gases, the less infrared energy that can escape, and the greater the "greenhouse" effect. Hence, increasing the amount of greenhouse gases increases the planet's surface temperature by increasing the amount of heat that is trapped in the lower atmosphere. Exactly how much the earth's surface temperature will rise given a certain increase in a trace greenhouse gas such as CO_2 is a critical and uncertain issue (Schneider 1989).

Because of its abundance, water vapor is the gas that contributes most to the greenhouse effect (Leggett 1990). Carbon dioxide, the second most important greenhouse gas, is added to the atmosphere both naturally, by volcanic eruptions and natural organic recycling pathways, and unnaturally, through fossil fuel combustion and deforestation. During the past decade, carbon dioxide emission from fossil fuels and deforestation accounted for more than half the total anthropogenic contribution to the greenhouse effect (Leggett 1990). Indeed, carbon dioxide is our society's single largest waste product, with an estimated world production in 1988 of 5 billion metric tons (Leaf 1989). Numerous other trace gases further contribute to the greenhouse effect, including, in descending order of relative contribution, chlorofluorocarbons, methane, nitrous oxide, and tropospheric ozone. Chlorofluorocarbons (CFCs) and related synthetic gases are produced for a variety of industrial uses such as refrigeration, foam blowing, and use as solvents. CFCs are particularly efficient absorbers of infrared radiation. The addition of one molecule of CFC-11 ($CFCl_3$) or CFC-12 (CF_2Cl_2) to the atmosphere, for instance, can have the same greenhouse effect as the addition of 10^4 molecules of CO_2 (Ramanathan 1988). Chlorofluorocarbons also have detrimental effects on the stratospheric ozone layer. Methane (CH_4) is contributed significantly to the atmosphere by rice paddies, enteric fermentation, and leakage from natural gas production. Nitrous oxide (N_2O) is contributed by biomass burning and fertilizer use, as well as by fossil fuel combustion (Leggett 1990). Tropospheric ozone, formed by chemical interactions between uncombusted hydrocarbons and nitrogen oxides in the presence of sunlight, is an additional contributor to the greenhouse effect.

Analyses of antarctic ice cores suggest that during the past several thousand years, atmospheric levels of carbon dioxide have been reasonably close to the preindustrial value of 280 ppm (Schneider 1989). Interestingly, data from these analyses reveal a close correspondence between the CO_2 concentration of gas bubbles trapped in the ice cores and the inferred surface temperature at Antarctica at the time the ice was formed (Barnola 1987). Methane concentrations in ice cores also show a close relationship with climate during the past 150,000 years (MacDonald 1988). Since 1850, atmospheric CO_2 levels have increased by about 25%. Half the total carbon dioxide produced during this period remains in the atmosphere today, the rest having been incorporated into living matter or sequestered more definitively in the oceans. The record at Mauna Loa observatory shows that atmospheric concentrations of CO_2 have increased from about 310 ppm in 1958 to 353 ppm in 1988, levels that are substantially higher than at any time in the past 160,000 years (Leggett 1990).

Accompanying the increase in atmospheric carbon dioxide has been an increase of 0.6°C in the earth's mean surface temperature in the past hundred years. The 7 warmest years during that period have occurred since 1980 (Bolin 1991). Over the same period, the average global sea-level has risen by 10 to 20 cm. These increases have neither been regular in time nor uniform over the globe. The extent of this warming is broadly consistent with the predictions of climate models, but it is also of the same magnitude as natural periodic climatic variability. It has been pointed out that this background variability of global temperature may still hide, to a considerable degree, the changes caused by rising greenhouse gas concentrations (Wigley 1991). Not until the observed changes of the global mean temperature have increased beyond about 1°C will it be possible to establish the sensitivity of the global climate system to changes in greenhouse gas concentrations with more certainty. Such a cumulative 1°C increase is not likely to occur until the first decade of the 21st century; furthermore, any change in the global climate may not have major effects until a decade or two later (Bolin 1991).

The increasing use of fossil fuels for economic development worldwide, concomitant with further deforestation, makes a continuing rise in the carbon dioxide level inevitable. According to most projections for future CO_2 levels, which are based on a 0.5% to 2% annual growth range for fossil fuel use, CO_2 concentrations will increase to 600 ppm sometime between the years 2030 and 2080, and other trace greenhouse gases will continue to rise at even faster rates (Schneider 1989). For example, during the past decade, the atmospheric concentration of CFCs increased at an annual rate of about 4%, eight times the annual increase in the concentration of carbon dioxide. Exactly how closely these increases in greenhouse gases will correspond to future temperature changes is unknown.

Mathematical models have been used to simulate the present climate, including the effects of volcanic eruptions, as well as to explain the evolution of past climates, such as those of the ice ages and the Cretaceous period. The accuracy of these paleoclimatic simulations lends credibility to projections derived from the same models of how future climates will be affected by greenhouse gases and other pollutants. The precise composition and complexity of such models depend on how long a period of time they are designed to address, the spatial resolution involved, the number of dimensions simulated, and the amount of spatial detail included. An example of an extremely simple model is one that calculates only the average temperature of the earth, independent of time, as an energy balance arising from the earth's average reflectivity and the average "greenhouse" properties of the atmosphere. Such a model is zero-dimensional: it collapses the real temperature distribution on the earth to a single point, a global average. In contrast, three-dimensional climate models reproduce the way temperature varies with latitude, longitude, and altitude. The most sophisticated of them are known as general-circulation models. They predict changes not only of temperature but also of humidity, wind speed and direction, soil moisture, and other climatic variables (Schneider 1987).

Most general-circulation models have suggested that if atmospheric CO_2 were to double in the next century as predicted, the earth's surface temperature would eventually increase between 1.5° and 5°C (Schneider 1989). For comparison, the global mean surface temperature during the Ice Age extreme 18,000 years ago was only about 5°C colder than today. A sustained global increase of more than 2°C above the present level would be unprecedented. The projected increase would be 10 to 60 times faster than the natural average rate of temperature change that occurred from the end of the last Ice Age to the present warm period (Schneider 1989).

In 1990, the Intergovernmental Panel on Climate Change (IPCC) began an assessment of the likelihood and expectable characteristics of a human-induced change in the world's climate (Bolin 1991). On the basis of current general circulation models, the IPCC projected a rate of increase during the next century of 0.2° to 0.5°C per decade. Abandoning the use of CFC gases would reduce this rate of increase by 15 to 20%. They projected that land surfaces will warm more rapidly than the oceans, and that high-northern latitudes will warm more in winter than will the global mean. Furthermore, at present rates of greenhouse gas emission, the global mean sea-level will rise by an average of about 6 cm per decade during the next century (with a mean uncertainty range of 3-10 cm per decade), mainly because of the thermal expansion of the oceans and the melting of some land-ice. The IPCC predicted that ocean levels will rise about 20 cm by 2030 and 65 cm by the end of the next century (Bolin 1991). However, many uncertainties exist in these projections, including an incomplete understanding of the sources and sinks of greenhouse gases, as well as multiple potential feedback mechanisms. The effects of clouds, oceans, and polar ice-sheets on the magnitude of temperature change and sea-level rise are also poorly understood. For instance, the amount, height, or brightness, of clouds may change, substantially altering the climatic response to CO_2 and other gases. In addition, a CO_2-induced warming would likely decrease the regions of Earth covered by snow and ice and decrease the global reflective power, and hence the amount of solar energy reflected.

Numerous scenarios have been hypothesized regarding the impact of the projected climate changes on physical and biological processes at the regional level. The scope of plausible effects on agriculture, forestry, and water resources is extensive, and the biosociological implications of such effects are critical. Alterations in the availability of water and the duration of the growing season, together with more frequent storms and other extreme events, could have adverse effects on agricultural production. The world's current food production system has developed within a fairly narrow climatic region. The distribution and timing of a 2°or 3°C warming could cause drought and desertification in areas that are now fertile and move arable areas into higher latitudes. Because of increased carbon dioxide levels and a prolonged growing season, the potential productivity of high and middle latitudes may increase. Studies have not yet determined whether, on average, global agricultural potential will increase or decrease. On the basis of the results of controlled laboratory experiments, researchers have determined that elevated atmospheric carbon dioxide levels could increase the yield of some food crops, such as wheat, rye, soybeans, rice, and barley (Haines 1991). Some weeds may also show increases in growth. But it is not yet clear whether the possible increases in photosynthesis and growth will persist or to what extent they will occur in natural ecosystems. Increased desertification will undoubtedly occur in many regions, with severe effects felt in some areas of particular vulnerability to declines in production, including much of Brazil, Peru, the Sahel region of Africa, Southeast Asia, Central Asia, and China (Bolin 1991). Patterns of agricultural trade could be altered by decreased cereal production in some of the current high-production areas, such as Western Europe, the southern United States, parts of South America, and Western Australia. For instance, wheat yields in Europe and North America could decline by between 3 and 17% with a 2°C rise in temperature (Haines 1991). Such a decline would have worldwide implications, considering that the United States alone provides 90% of the world's grain surplus. Although advances in agricultural technology and biotechnology may serve to offset some of the negative effects of climate change, the rate and degree of change could potentially have very serious effects on the availability of food for the world's population.

Projected changes in temperature and precipitation suggest that climatic zones could shift several hundred kilometers toward the poles over the next 50 years (Bolin 1991). Although ecosystems would also gradually shift toward the poles, the movement of flora and fauna would lag behind these climatic shifts. Some species could be lost owing to increased ecological stress, leading to a reduction in global biological diversity. Because the rotation period of forests is long, plantations and current young forests will mature and decline in climates to which they are increasingly more poorly adapted. The actual impact of global warming will depend on the physiological adaptability of trees and their host-parasite relationships. The most sensitive areas will be where species are close to their biological limits in terms of temperature and moisture, for example, in semi-arid areas. Losses from wildfires will be increasingly extensive. The expected increasing and nonsustainable uses of forest will progressively place more pressure on forest investments, conservation, and management.

Relatively small climate changes can cause large water resource problems in many areas. Although about 9000 km^3 of water is available for human exploitation, in theory enough to sustain around 20 billion people, the local availability of water varies widely (Haines 1991). Nearly three-quarters of the water used is for agricultural purposes, and the area of irrigated land is increasing steadily each year. Although current models of climate change are inadequate to give clear predictions of local change, there is relatively high confidence in predicted levels of change at a continental level. For instance, by 2020 in central North America, temperature is predicted to increase 3-4°C in winter and 2-3°C in summer, rainfall is predicted to increase by up to 20% in winter and decrease by between 5 and 10% in summer (Haines 1991). This could result in a 10 to 15% reduction of soil moisture in summer. In Southern Europe, a decrease in summer precipitation is also predicted, but in Southeast Asia, precipitation is predicted to increase by 5-15% in the summer. The patterns of tropical and monsoon rains that are critical to vegetation and agriculture in Africa, Asia, and South America would also likely be altered.

Global warming could cause the sea level to rise 0.5 to 2 meters by 2100 (Haines 1991). This rise would lead to increased salinization of rivers, wetlands, and aquifers, causing harm to some aquatic plants and animals and posing a threat to the use of water by humans. The most serious implications are for those densely populated parts of the world that are vulnerable to small rises in sea level, such as the Nile Delta in Egypt and the Ganges Delta in Bangladesh, where the homes and livelihoods of 46 million people are potentially threatened. In addition to raising the sea level, global warming will modify ocean circulation and change marine ecosystems, with considerable socioeconomic consequences.

OTHER ANTHROPOGENIC EFFECTS ON CLIMATE

Ozone in the upper atmosphere is important in modulating the amount of solar and infrared radiation that reaches the troposphere. Chlorofluorocarbons are not only the most effective greenhouse gases, but they have detrimental effects on stratospheric ozone. CFCs have no known important sinks in the troposphere (Ramanathan 1988). They are ultimately broken down in the stratosphere above 25 km by photolysis, which frees the highly reactive chlorine. The chlorine competes with nitrogen oxides for odd oxygen species, including O_3, and the result is an efficient catalytic destruction of O_3 by chlorine until it is removed from the system as HCl. Atmospheric chemistry models predict that, at present rates of emission of CFCs, severe O_3 reduction ranging from 15% near 30 km to 40% near 45 km will result.

The protection conferred by the ozone layer is selective, with shorter wavelengths absorbed preferentially. Thus, energy in the shorter (200-290 nm) ultraviolet wavelengths, UV-C, is completely absorbed by the ozone layer; the UV-B (290-320 nm) is only partially absorbed; and the UV-A (320-400 nm) is not absorbed at all. With a depletion of the stratospheric concentration of ozone caused by the production and release of CFCs, it has been predicted that more UV-B, particularly in the shorter wavelengths, will reach the earth. The amount of UV-A will remain unchanged, and UV-C is still expected to be completely absorbed (Longstreth 1991). Both UV-B and UV-A are biologically active; however, UV-B, especially the shorter wavelengths, is generally more damaging and increased levels of UV-B radition on the earth's surface will have deleterious effects on human health. These health effects will be discussed later.

Ozone in the troposphere undergoes photolysis to produce an oxygen radical that reacts with water to release hydroxyl (OH) radicals. The highly reactive OH is the primary sink for many tropospheric gases and pollutants including O_3, CH_4, CO, and NO. Thus, the increases in CH_4 levels that occurred during the last century may have caused a substantial (20-40%) reduction in OH levels, which in turn, could cause tropospheric O_3 levels to increase by as much as 20% (Ramanathan 1988). An increase in average temperature may likewise lead to increased concentrations of tropospheric ozone because of faster reaction rates. In addition, the production of ozone may be increased by the higher natural and man-made emissions of hydrocarbons, which could also occur as a result of elevated mean temperatures. The same is true for acidic materials such as sulfates because sulfur and nitrogen oxides would be oxidized more rapidly (Longstreth 1990). A preliminary analysis in the San Francisco Bay area of the possible regional impacts of such increases showed that a 4°C temperature increase during certain portions of the summer could result in the maximum ozone concentrations rising by about 20%, approximately doubling the area in which National Ambient Air Quality Standards (NAAQS) are exceeded and approximately tripling the person-hours of exposure. Similar extimates for the Midwest indicated that on certain days during the summer, a 4°C rise in temperature would cause area in which the NAAQS would be exceeded to almost triple (Longstreth 1990).

HEALTH EFFECTS OF CLIMATE CHANGE

The climatic changes that have been projected could have a number of primary, secondary, and tertiary effects on human health. An immediate difficulty in estimating the health impact of climate change is that there are few empirical precedents for most of the ongoing or anticipated environmental changes. For example, it is unknown how a 3°C rise in average global temperature would affect the frequency of heatwaves, agricultural productivity, or cyclonic activity. Likewise, a sustained 10-20% loss in stratospheric ozone has never occurred within our recorded existence. Some potential health effects may be inferred from previously studied relationships between health effects and warming, such as generalized heat-related morbidity or the effects of temperature on communicable disease epidemiology. In the context of environmental and demographic changes projected, these problems may be compounded by widespread malnutrition in a burgeoning world population. Regional impacts of global warming on agricultural production could have adverse consequences for food supplies worldwide. Droughts or floods could severely affect crop yields, and the same conditions that could potentially enhance the effects of vector-borne human diseases could result in increases in plant and animal pests, thereby contributing to additional loss of food supplies. Particularly for populations already underfed, the consequent increase in malnutrition could be severe.

HEAT-RELATED ILLNESS

One of the most studied effects of climate on human health is the effect of temperature on generalized mortality. Changes in climate are most likely to affect those who are already vulnerable for other reasons, in particular the elderly (Kilbourne 1990). A variety of studies have derived relationships between monthly mean temperatures and either generalized mortality or mortality from specific diseases such as coronary heart disease or stroke (Longstreth 1991). Data from these studies show an inverse relationship between mortality rates and temperature for temperatures between about -5°C and 25°C, but at temperatures above and below this range, mortality rates increase sharply.

To determine the potential impact of global warming on human mortality, the U.S. Environmental Protection Agency (EPA) commissioned a series of studies to evaluate past mortality trends in major urban areas across the United States and then to use this information to assess the potential impact of global warming on U.S. mortality rates (Kalkstein 1989). The primary finding of these studies was that, historically, mortality rates increase once temperatures exceed or fall below certain thresholds and that each city has its own characteristic thresholds. Furthermore, mortality rates are to a large degree dependent on the duration of periods in which the temperature thresholds are breached. These studies indicated that people over age 65 years were the most susceptible population (although infants less than a year old were excluded from the analysis). Interestingly, these studies also indicated that the effects of weather on mortality rates vary by region, with the heat-related impacts greater in the North (which is unaccustomed to hot weather) and the cold-related impacts greater in the South (which is unaccustomed to cold weather). The time of year during which a heat wave or cold spell occurred also made a difference, with heat waves in spring or early summer and cold spells in fall or early winter having more of an impact (Kalkstein 1986).

The most reasonable hypothesis for these relationships is that populations have the ability to adapt to extremes in temperature given a sufficient period of time. Whether the adaptation is physiological on an individual basis, or institutional on a population basis (for instance, because of changes in housing structure) is unknown. This series of EPA-commissioned studies covered 15 cities, and the results were extremely dependent on the degree of adaptation hypothesized; with an acclimatization equivalent to that seen in southern cities, this analysis indicated that global warming was likely to result in the number of temperature-related deaths increasing from 1399 to 2114 in these 15 cities (Kalkstein 1989). This projected increase included a large increase in the number of summer deaths and a decrease in the number of winter deaths. When the effects of acclimatization were not factored in, the projected net increase in mortality for these 15 cities was more than 5-fold (from 1399 to 7454).

These outcomes should be interpreted with caution for a number of reasons. First, there is a large degree of variability in the projections depending on the degree of acclimatization, and little is known about the factors that contribute to acclimitization and how they will be affected by climate change. Second, the central concept of threshold temperatures focuses the analysis on deaths occurring once these thresholds have been breached. Other researchers, however, have found that the inverse relationship between temperature and both mortality and morbidity extends over a wider range of temperatures above and below these threshold values (Longstreth 1991).

Other researchers using U.S. data show that under conditions of doubled CO_2 concentrations and in the absence of acclimatization to higher temperatures, the number of heat-related and temperature-related deaths might rise from an estimated current total of 1150 to around 7400, with deaths among those aged 65 and over accounting for about 60% (Haines 1991). The increasing proportion of very elderly people in the population of many Western countries will probably increase the number of people at risk from heat-related death, because the main causes of such deaths are likely to be cardiovascular and respiratory diseases. Even with full acclimatization, the rise in the number of summer deaths associated with increased temperature is likely to exceed the reduction in the number of winter deaths. Using mortality data from New York City, an apparent threshold temperature of 92°F (33°C) appeared to exist above which total mortality increased. In a study of mortality during three heatwaves in Los Angeles, in which daily maximum temperatures averaged above 41°C, researchers found mortality at all ages to be between 172 and 445% of that expected. Among those people 85 years and older, the peak mortality was between 257 and 810% of that expected (Oeschli 1970).

COMMUNICABLE DISEASES

Global changes in climate could also have major implications for a variety of communicable diseases. Of particular concern are changes in the epidemiology of vector-borne diseases. In addition, a variety of other communicable diseases associated with poor hygiene and crowded conditions could be exacerbated should global climate change place additional strain on public health resources.

The present distributions of malaria, trypanosomiasis, leishmaniasis, amoebiasis, bancroftian and brugian filariasis, onchocerciasis, schistosomiasis (except for Schistosoma japonica), hookworm, guinea worm, and even to a large extent tapeworm and roundworm infections, are closely correlated to temperature (Haines 1991). Variations in temperature and humidity can directly affect the multiplication rates of the infectious agent or the vector; the biting rate of the vector; the geographic distribution of the intermediate animal hosts; and the amount of time that intermediate hosts or human hosts are exposed to the vector. In addition, changes in climate may indirectly affect the relationships between these components by leading to changes, either beneficial or adverse, in their habitats. For instance, more forage could result in a population explosion in the intermediate host, or warming could lead to a different distribution of plants so that the relative abundance of forest versus meadow is altered, leading to a different distribution of the vector.

In an examination of the potential effects that climate change may have on vector-borne diseases, Gillett (1981) found that insects and the pathogens they transmit are ectothermic; their patterns of development and multiplication are temperature dependent. For instance, the extrinsic incubation period for yellow fever virus varies from 3 days to a few weeks, depending entirely on the mean ambient temperature. The extrinsic incubation periods of the four malarial species affecting humans have also been shown to vary inversely with temperature. Another temperature effect on parasites is manifested in Egypt, which is partly in the hot zone and partly in the temperate zone: water snails tend to lose their schistosome infections during January, February, and March each year, when the temperature is lower (Gillett 1981). A warming of the climate will lead to expanded distributions of most vector-borne diseases, increased rates of vector and parasite multiplication, and increased human exposure to vectors and their parasites. Vector populations may recover more quickly from control programs, and problems of drug and insecticide resistance could arise more quickly in parasites and vectors. These changes could enhance the transmission of infection leading to heavier parasitic loads in the human population (Haines 1991).

Although vector-borne diseases are relatively less common in industrialized countries, there are still a number of significant conditions in these countries that could be affected by climate change. Computer simulations of various climate scenarios for the United States have been used to study the effect of global warming on vector-borne disease epidemiology. Such simulations have indicated that the distribution of tick-borne diseases (for example, Rocky Mountain spotted fever and Lyme disease) would expand northwards. Models for the relationship of malaria or Rocky Mountain spotted fever to various weather and climate parmeters show that the southern portion of the United States would become no more hospitable than it already is for malaria, and that the distribution of the ixodid tick, vector for Rocky Mountain spotted fever, would likely spread into Canada (Longstreth 1991). Other important factors that were, however, not included in the analysis include the size of forests and of the small mammal population (Haines 1991). Five mosquito-borne diseases have been considered to be potential risks to the United States consequent to a climate warming: of those five, malaria, dengue fever, and arbovirus-induced encephalitides were considered significant risks, and yellow fever and Rift Valley fever possible risks.

Other communicable diseases could be affected by climate changes, including water-borne diseases and those diseases to which people's susceptibility increases with malnutrition. A rise in sea level could compromise sewage treatment capabilities, threatening water supplies and facilitating the water-borne spread of diseases present in fecal material; it could also result in mass movements of people, putting additional strain on the infrastructure of those areas to which such refugees flee. Diseases such as cholera and typhoid, eminently controllable with adequate sanitation, could reappear in some areas and become worse in areas where they currently exist. To the extent that regional agricultural productivity is disrupted, the nutritional state of populations may be further compromised, making them more susceptible to communicable diseases such as measles and infant diarrhea (Longstreth 1990).

RESPIRATORY ILLNESS

Global warming is expected to increase urban air pollution by causing increased concentrations of tropospheric ozone and other photochemical oxidants. Increased ozone levels are of major concern because the concentrations of ozone required to induce respiratory function deficits after 6- to 8- hour exposures may be at or below 0.12 ppm (Longstreth 1990). Increased ozone levels exacerbated by elevated temperatures could have an impact on asthmatics and on people with acute or chronic bronchitis (Haines 1990). Such people have heightened sensitivity to the toxic effects of ozone; increased levels can trigger acute asthma attacks and also lead to recurrent episodes of bronchitis.

Another effect of increased global mean temperatures could be changes in seasonal patterns of respiratory illness. Such seasonal patterns have been well studied. Patients with chronic obstructive airways disease are more likely to experience exacerbations in the winter, and pneumonia and bronchitis are also more common in winter months. Asthma and hay fever tend to peak in incidence in the spring months, with a second peak of acute asthma in autumn. Hospital admissions are generally more common in the last quarter of the year, an increase probably related to viral infections (Longstreth 1990). Increased winter temperatures from global warming could result in declines in the seasonal patterns of bronchitis and perhaps pneumonia. However, warmer temperatures and higher humidity could also lead to greater production and distribution of pollens and molds; the resultant longer and more intense allergy season and more diverse set of allergens could result in both increased incidence and severity of asthma and allergies.

EFFECTS OF STRATOSPHERIC OZONE DEPLETION

At least two of the potential effects of stratospheric ozone depletion are likely to intensify the adverse consequences of global warming on human health. These are the increase in ultraviolet radiation and the increase in tropospheric ozone levels. There is good experimental evidence that ultraviolet B radiation (UV-B) at doses equivalent to those received ambiently by humans causes a local immunosuppression and at higher doses causes systemic immunosuppression (Longstreth 1990). Most of the early work in this area has revealed that the immunosuppressive effects of UV-B consist of the impairment of the antigen-presentation step accompanied by the appearance of suppressor T lymphocytes specific for the antigen administered. Subsequent work in a number of animal models of infectious disease has shown that UV-B can suppress immune responses to infections such as tuberculosis, leishmaniasis, candidiasis, herpes virus infections, and possibly malaria. All of these diseases could potentially increase in incidence and distribution with global climate change, and the severity of such infections could conceivably be exacerbated by increases in UV-B.

The incidence of all varieties of skin cancer among white populations will increase with more exposure to ultraviolet B. The incidence of melanoma has already increased; in the United States, it has risen 83% over the past 7 years (Leaf 1989). Unlike basal-cell and squamous-cell carcinoma, which seem to increase in proportion to total exposure to ultraviolet radiation, melanoma appears to be associated with acute exposure, as in the case of severe sunburns. The U.S. Environmental Protection Agency (EPA) estimates that ozone depletion will lead to an additional 31,000 to 126,000 cases of melanoma among U.S. whites born before 2075 and an additional 7,000 to 30,000 fatalities (Leaf 1989).

An increased incidence of cataracts may be a more widespread health effect of ultraviolet B radiation than cancer because it will affect all populations. Cataracts are currently the third-largest cause of preventable blindness in the United States (Leaf 1989). According to the same EPA predictions, an additional 550,000 to 2.8 million Americans born before 2075 will have cataracts. To a considerable extent, the loss of vision due to cataracts is correctable today with surgery and the implantation of prosthetic lenses, but a large increase in the number of cataracts in younger people would nevertheless place an additional burden on the health care system.

CONCLUSIONS

The projections of climate change and its effects on human health will change as knowledge increases and more observations about ongoing changes accumulate. While the global mean surface-air temperature has increased by 0.3° to 0.6°C during the past century, it should be emphasized that the magnitude of this observed warming is consistent with natural periodic climatic variability. The atmospheric concentration of carbon dioxide, however, has climbed to levels that are unprecedented, from a relatively stable preindustrial value of 280 ppm to more than 350 ppm today. The eventual effect of this increase on global temperatures will depend on complex interrelationships of factors of which we are not yet fully aware. It has been argued that a projected rise in global temperatures of 1.5° to 4.5°C over the next century consequent to a CO_2 doubling is an overestimate, because the important feedback effects of cloud formation and cloud dissipation have not been adequately described or factored into the projection (Lindzen 1990). It has not conclusively been shown, for example, whether higher CO_2 levels would result in an increase or decrease in the amount of clouds, or exactly what effect such increases or decreases in cloud cover would have on temperatures.

In the future, as less developed countries continue to develop industrially, they will account for larger proportion of total greenhouse gas emissions. Data from 1994 show that Western industrialized countries now emit less than half of the emission of carbon dioxide from industrial sources.

Because of the general inertia of the climate system and of human society, there will be considerable lag time between when preventive measures for stabilizing or reducing anthropogenic impacts on climate are instituted and any appreciable effects are made. The IPCC reported that the emissions of carbon dioxide, CFCs, and nitrous oxide would have to be reduced by more than 60% from their current levels in order to stabilize their atmospheric concentrations (Bolin 1991). Moreover, it would take a century or more for the concentrations of these gases to become reduced substantially, even if the emissions were stopped immediately. Some have argued that the proposed policies designed to prevent global warming will have virtually no effect. Given the logarithmic dependence of warming on CO_2 levels, for example, the elimination of America's 20% contribution to CO_2 production would merely reduce warming by a fraction of a degree for a few years (Lindzen 1990).

The best current projections of climate change for the next century have important implications for the global environment and human health. If these projections are correct, climate change will affect world agricultural production, rates of heat-related disease, the epidemiology of infectious diseases, and other areas pertinent to human health and survival. Clearly, much additional research is needed to reach an adequate understanding of global climate change, its human health effects, and appropriate preventive or adaptive measures.

REFERENCES

Arrhenius, S. 1896. *Philos. Mag.* 41:237.

Barnola, J.M., Raynaud, D., Korotkevich, Y.S. and C. Lorius. 1987. *Nature* 329:403.

Bolin, Bert. 1991. Man-induced Global Change of Climate: The IPCC Findings and Continuing Uncertainty Regarding Preventive Action. *Environmental Conservation* 18:297-303.

Brown, Katrina and Neil Adger. 1993. Estimating National Greenhouse Gas Emissions Under The Climate Change Convention. *Global Environmental Change* 3:149-158.

Cooperstock, M. and R.A. Wolfe. 1986. Seasonality of Preterm Birth in the Collaborative Perinatal Project: Demographic Factors. *Am. J. Epidemiol.* 124:234-241.

Cox, Louis A., Jr. and Paolo F. Ricci. 1992. Dealing With Uncertainty: From Health Risk Assessment to Environmental Decision Making. *J. Energy Eng.* 118:77-94.

Fourier, Jean-Baptiste. 1827. *Mem. Acad. Sci. Inst. Fr.* 7:569.

Fouts, James R. and Warren T. Piver 1992. Human Health Effects and Global Climate Change. *MTS Journal* 25:38-44.

Gillett, J.D. 1981. Increased Atmospheric Carbon Dioxide and the Spread of Parasitic Disease. *Soc. Protozo. Spec. Publ.* 1:106-111.

Gillett, J.D. 1985. The Behaviour of Homo Sapiens, the Forgotten Factor in the Transmission of Tropical Disease. *Royal Soc. Trop. Med. Hyg.* 79:12-20.

Gillett, J.D. 1974. Direct and Indirect Influences of Temperature on the Transmission of Parasites from Insects to Man, pp. 79-95. In: A.E.R. Taylor and R. Muller (Ed.) *The Effects of Meteorological Factors Upon Parasites.*

Haines, Andrew. 1991. Global Warming and Health. *Brit. Med. J.* 302:669-670.

Haines, Andrew and Chris Fuchs. 1991. Potential Impacts on Health of Atmospheric Change. *J. Public Health Med.* 13:69-80.

Kalkstein, L.S. 1989. The Impact of CO_2 and Trace Gas-Induced Climate Change upon Human Mortality. Chapter 1 in: J.B. Smith and D. Tirpak (Ed.) *The Potential Effects of Global Climate Change on the United States, Appendix G-Health.* U.S. Environmental Protection Agency, EPA-230-05-89-057, Washington, D.C.

Kalkstein, L.S., Davis, R.E., Skindlov, J.A. and K.M. Valimont. 1986. The Impact of Human-Induced Climatic Warming upon Human Mortality: a New York case study. *Effects of Changes in Stratospheric Ozone and Global*

Climate. U.S. Environmental Protection Agency, United Nations Environment Programme, Washington, D.C.

Keller, C.A. and R.P. Nugent. 1983. Seasonal Patterns in Perinatal Mortality and Preterm Delivery. *Am. J. Epidemiol.* 118:689-698.

Kilbourne, E.M. 1990. Atmospheric Change and Temperature-Related Health Effects. In: White, Wagner and Beal: *Global Atmospheric Change And Public Health.*

Kutschenreuter, Paul H. 1959. A Study of the Effect of Weather on Mortality. *New York Acad. Sci.* 22:126-138.

Last, John M. 1993. Global Change: Ozone Depletion, Greenhouse Warming, and Public Health. *Annu. Rev. Public Health* 14:115-136.

Last, John and Tee L. Guidotti. 1991. Implications for Human Health of Global Ecological Changes. *Public Health Reviews* 18:49-67.

Last, John M. 1989. A Vision of Health in the 21st Century: Medical Response to the Greenhouse Effect. *Can. Med. Assoc. J.* 140:1277-1279.

Leaf, Alexander. 1989. Potential Health Effects of Global Climatic and Environmental Changes. *New Engl. J. Med.* 321:1577-1583.

Leggett, Jeremy. 1990. *The Nature of the Greenhouse Threat* pp. 15-43.

Liebetrau, Albert M. and Michael J. Scott. 1991. Strategies for Modeling the Uncertain Impacts of Climate Change. *Journal of Policy Modeling* 13:185-204.

Lindzen, Richard S. 1990. Some Remarks on Global Warming. *Environ. Sci. Technol.* 24:424-426.

Longstreth, Janice. 1990. Anticipated Public Health Consequences of Global Climate Change. *Environ. Health Perspectives* 96:139-144.

Longstreth, Janice D. 1991. Global Warming: Clues to Potential Health Effects. *Environ. Carcino. Revs.* C8:139-169.

McMichael, A.J. 1993. Global Environmental Change and Human Population Health: A Conceptual and Scientific Challenge for Epidemiology. *Int. J. Epidemiol.* 22:1-8.

Manabe, S. and R.T. Wetherald. 1967. *J. Atmos. Sci.* 24:241.

Morison, Warwick L. 1989. Effects of Ultraviolet Radiation on the Immune System in Humans. *Photochem. and Photobiol.* 50:515-524.

Oeschli, F.W. and R.W. Buckley. 1970. Excess Mortality Associated with Three Los Angeles September Hot Spells. *Environ. Res.* 3:277-284.

Pitcher, Hugh M. and Janice D. Longstreth. 1991. Melanoma Mortality and Exposure to Ultraviolet Radiation: An Empirical Relationship. *Environment International* 17:7-21.

Pittock, A. Berrie. 1987. The Greenhouse Effect and Future Climatic Change. *Engineers Australia* 6 February 1987: 40-43.

Pollard, Douglas. 1989. Climate Change and its Effects on Forests. *Canadian Forest Industries* Vol #:56-61.

Ramanathan, V. 1988. The Greenhouse Theory of Climate Change: A Test by an Inadvertent Global Experiment. *Science* 240:293-299.

Schlesinger, Michael E. and John F.B. Mitchell. 1987. Reviews of Geophysics

25:760-798.
Schneider, Stephen H. 1987. Climate modeling. *Science* 256:72-80.
Schneider, Stephen H. 1989. The Greenhouse Effect: Science and Policy. *Science* 243:771-781.
Shope, Robert. 1991. Global Climate Change and Infectious Diseases. *Environ. Health Perspect.* 96:171-174.
Tromp, S.W. and Janneke J. Bouma. 1982. The Influence of Weather and Climate on Asthma and Bronchitis, pp. 147-156. In: A.R. Rees and H.J. Purcell (Ed.) *Disease and the Environment*. John Wiley and Sons, Chichester.
Wigley, T.M.L. and S.C.B. Raper. 1991. Detection of the Enhanced Greenhouse Effect on Climate, pp. 231-242. In: Proceedings of the Second World Climate Conference, Cambridge University Press, Cambridge, England, U.K.

Environmental Contaminants, Ecosystems and Human Health. Edited by S.K. Majumdar, E.W. Miller, and Fred J. Brenner. © 1995, The Pennsylvania Academy of Science.

Chapter Fifteen

ACIDIC AEROSOL EXPOSURES AND HEALTH

HELEN H. SUH and PETROS KOUTRAKIS

Environmental Science and Engineering Program
Department of Environmental Health
Harvard University, School of Public Health
Boston, MA 02115

INTRODUCTION

Acid aerosols are secondary pollutants that are formed through atmospheric reactions. A major constituent of acid aerosols are sulfates (SO_4^{2-}), which are formed through the oxidation of sulfur dioxide (SO_2), a gas generated during the combustion of coal or other sulfur-containing fuels. Oxidation of SO_2 occurs through a series of homogeneous or heterogeneous reactions that convert SO_2 to sulfuric acid (H_2SO_4) particles. Once formed, H_2SO_4 particles may remain as H_2SO_4 or may react with ammonia (NH_3) to form the salts, ammonium sulfate ($(NH_4)_2SO_4$), ammonium bisulfate (NH_4HSO_4), or letovicite ($(NH_4)_3H(SO_4)_2$). Concentrations of these sulfate species usually are expressed in terms of the total SO_4^{2-}, with the acidic fraction expressed in terms of the free or titratable H^+, which typically is referred to as the aerosol strong acidity.

Acidic sulfate aerosols have been associated with a variety of adverse effects to the environment and the human health. Effects to the environment, which include acidification of lakes and soil as well as reductions in fish populations and visibility, have been studied extensively and are well documented. Less is known about their impact on human health, but available evidence from both laboratory and epidemiologic studies suggests that acidic sulfate aerosols may adversely affect the human respiratory system (1, 2).

In light of this evidence, considerable effort has been made to characterize exposures to acidic sulfate aerosols. Although these efforts originally focused on outdoor concentrations, more recent investigations also have begun to examine their indoor concentrations and personal exposures. Together, these efforts have significantly enhanced our understanding of acidic sulfate aerosol exposures. Through its application to future health effect studies, this understanding may substantially enhance our ability to assess, and if necessary reduce, the risks posed by exposures to acid aerosols.

This chapter focuses on acidic sulfate aerosols and its major constituents: sulfates (SO_4^{2-}) and aerosol strong acidity (H^+) and presents a brief overview of the findings from health effect studies of acid aerosols followed by a review of studies of acid aerosol exposures. The review of acid aerosol exposures is divided into three sections. The first section discusses outdoor concentrations and includes information about the diurnal and spatial variability in SO_4^{2-} and H^+ concentrations. The second and third sections discuss indoor SO_4^{2-} and H^+ concentrations and personal SO_4^{2-} and H^+ exposures, respectively. Since the indoor concentration and personal exposure databases are dominated by measurements obtained in a study conducted in State College, Pennsylvania (3, 4), these discussions are based primarily on findings from this study.

HEALTH EFFECTS

Perhaps the first indication of the potential health risks posed by acid sulfate aerosols was provided by the severe air pollution episodes of the mid-1900s. The most serious of these episodes occurred in December 1952, in London, England, during which approximately four thousand excess deaths occurred (5). Mortality was closely associated with British Smoke (BS), a measure of particle blackness, and to a lesser degree with sulfur dioxide (SO_2) (6). Subsequent reanalysis of the 1952 London pollution and mortality records further suggested particle acidity (H^+) to the toxic agent (7). Other severe air pollution events in The Meuse Valley, Belgium, and Donora, Pennsylvania, support the hypothesis that exposures to H^+ lead to excess mortality and increased hospital admissions (5). Since no measurements of H^+ concentrations were made during these episodes, however, direct evidence of the association between H^+ concentrations and excess mortality could not be obtained.

In the United States, air pollution episodes of the severity of that in London no longer occur. Significant advances in air pollution control technology have succeeded in substantially reducing concentrations of both particulate matter and SO_2. At the same time, however, these technologies have changed the spatial and temporal patterns of air pollution. Today, electric utilities use tall smoke stacks to emit SO_2 high in the troposphere, where acid aerosols are formed and transported over long distances with the prevailing winds. In North America, these winds tend to transport acid aerosols from the Ohio River Valley to areas throughout the northeastern United States and Canada, potentially exposing millions of people to their effects.

Whether current levels of acid aerosol exposure present a danger to public health is not known. Many controlled laboratory studies of adult and adolescent asthmatics have demonstrated decreased pulmonary function and mucociliary clearance after acute exposures to sulfuric acid (H_2SO_4) concentrations of 250 $\mu g \cdot m^{-3}$ or greater (8-12). Adverse health effects also were found for lower H_2SO_4 exposures in a study conducted by Koenig et al. (1), as adolescent asthmatics showed significant decrements in pulmonary function after acute exposures to 100 $\mu g \cdot m^{-3}$ of H_2SO_4. The magnitude of these effects has been found to increase with the acidity of the SO_4^{2-} particle, with H_2SO_4 being the most potent, followed by ammonium bisulfate (NH_4HSO_4), and ammonium sulfate ($(NH_4)_2SO_4$) (13, 14). In addition, the magnitude of these effects is likely to be greatest in persons with asthma or exercise-induced bronchospasm (EIB), a group estimated to include at least 10% of the U.S. general population or 25 million people (1, 6).

Concentrations administered in these controlled laboratory studies are substantially higher than those found in the ambient environment. Extrapolations to "real world" exposures, however, may be possible, since the cumulative dose administered to these study participants is comparable to those experienced today by individuals exercising outdoors on a polluted day (6). [Controlled laboratory studies generally administered exposures of 10-30 minute durations, resulting in a cumulative dose of approximately 30 μg (assuming 20 breaths$\cdot$min^{-1} of 500 ml). Exercising individuals on a polluted day, on the other hand, may be exposed to concentrations of 30 $\mu g \cdot m^{-3}$ for three hours, for a cumulative dose of 54 μg.] These extrapolations suggest that exposures to acidic sulfate aerosols at current levels may result in adverse effects to the human respiratory system.

Further evidence is provided by results from epidemiologic studies of acid aerosols, which suggest that similar doses administered over slightly longer time periods may result in similar adverse effects. Recent epidemiologic studies have demonstrated significant associations between outdoor SO_4^{2-} and H^+ concentrations and a variety of adverse respiratory system effects. Among the most important of these studies is the 24-city study, which was specifically designed to examine the health effects resulting from acidic sulfate exposures (2, 15-16). As part of this study, health questionnaires were completed by the parents of over fifteen thousand children (ages 8-12) living in 24 North American communities, while pulmonary function measurements were obtained for over ten thousand children living in 22 of these communities. Preliminary analyses of this data show that both bronchitic symptoms and decrements in pulmonary function (FVC and FEV_1) are significantly associated with yearly outdoor H^+ concentrations.

These results have been supported by findings from other studies. Raizenne et al. (17), Neas et al. (18), and Thurston et al. (19) found similar associations between H^+ concentrations and decrements in pulmonary function, while Speizer (2) and Damokosh et al. (20) reported significant associations between H^+ concentrations and increased incidence of bronchitis. In addition, H^+ and/or SO_4^{2-} concentrations have been linked to other adverse health effects, such as increased mortality (21) and increased hospital admissions as well (22-26). These effects have been found in studies conducted in a number of communities, both rural and urban.

Despite these findings, questions remain about the adverse effects of SO_4^{2-} and H^+ exposures. Many of the adverse effects found in these studies were at the same time associated with other pollutant measures, such as particulate mass and ozone (O_3). Furthermore, results from recent mortality analyses suggest that particles (with an aerodynamic diameter less than 2.5 microns or $PM_{2.5}$) themselves are the causative agent, and not their acidic components (27-29). These results demonstrate the difficulty in separating the specific effects of acidic sulfate aerosols from those of $PM_{2.5}$, O_3, and other strongly correlated pollutants. As a result, definitive conclusions about the effects of SO_4^{2-} and H^+ on human health cannot yet be made. Nevertheless, results from these epidemiologic studies, in combination with those from controlled laboratory studies, provide considerable evidence that exposures to SO_4^{2-} and H^+ may result in adverse health effects.

ACID AEROSOL EXPOSURES

The extent to which acid aerosols pose a public health danger depends in large part on the magnitude and duration of the acid aerosol exposures experienced by the general population. Since acid aerosols are produced primarily by outdoor sources, their concentrations outdoors are of obvious interest and are important to characterize and assess. In addition, since people spend the majority of their time inside (30), indoor acid aerosol concentrations are also important to characterize and assess.

Outdoor Concentrations

Outdoor SO_4^{2-} and H^+ concentrations have been measured routinely as part of several studies of acid aerosols. Among the first of these studies was conducted by Pierson et al. (31) in the Allegheny Mountains in Western Pennsylvania. In this study, SO_4^{2-} of varying degrees of acidity was found to be the principal component of fine particle pollution, with regional transport an important determinant of both SO_4^{2-} and H^+ concentrations outdoors. The ability of SO_4^{2-} and H^+ to accumulate far from the sources of the precursors is detailed in a 1985 Office of Technology Assessment (OTA) report (32). In this report, OTA estimated that approximately one-third of the total sulfur compounds (including both SO_2 and SO_4^{2-}) deposited over the northeastern U.S. originated from sources over 500 kilometers (km) away from the region in which they are deposited, one-third from sources between 200 and 500 km away, and one-third from sources within 200 km.

Acidic sulfate aerosols generally are less than 2.5 (μm in diameter, with a mass median aerodynamic diameter between 0.2 and 1.0 μm (33-36). In North America, outdoor SO_4^{2-} and H^+ concentrations tend to be elevated throughout the northeast United States and Canada, with the highest concentrations found in communities located near the Ohio River Valley (37). In communities with elevated acid aerosol levels, outdoor SO_4^{2-} and H^+ concentrations are highest during the summer months (May through September), when approximately two-thirds of the total yearly acidity may be present (38).

Among the highest SO_4^{2-} and H^+ concentrations have been measured in Uniontown, Pennsylvania, a small semi-rural community located in the southwest corner of the state (Table 1). In Uniontown, mean summertime SO_4^{2-} and H^+

TABLE 1

Maximum SO_4^{2-} and H^+ concentrations measured in North American Cities. H^+ concentrations expressed as "H_2SO_4" equivalents. "SC" indicates semi-continuous measurements. Table adapted with permission from reference (6).

		Maximum Concentration		
Location	Sample Duration (h)	SO_4^{2-} ($\mu m \cdot m^{-3}$)	H_2SO_4 ($\mu m \cdot m^{-3}$)	Reference
Lennox, CA	2-8	18	0.1	50
Smoky Mountains	12	17	10	51
High Point, NJ	6	37	18	52
Brookhaven, NY	3	24	10	52
Tuxedo, NY	1-12	41	9	53
St. Louis, MO	SC	25	7	54
St. Louis, MO	SC	43	34	55
Los Angeles, CA	12	10	3	56
Harriman, TN	SC	47	18	57
Watertown, MA	SC	31	14	57
Fairview Lake, NJ	SC, 4	27	12	52
Warren, MI	24	37	9	58
Whiteface Mt., NY	24	59	14	59
Toronto, Ontario, Canada	8, 16	75	19	40
Allegheny Mt., PA	7, 10	45	31	60
Laurel Mt., PA	7, 10	56	42	60
Harriman, TN	24	28	14	61
St. Louis, MO	24	40	6	61
Topeka, KA	24	14	3	62
Watertown, MA	24	23	9	62
Stebenville, OH	24	56	18	37
Portage, WI	24	33	4	37
Kanawha Valley, WV	24	46	22	63
Dunville, Ontario, Canada	24	31	15	64
Hendersonville, TN	24	23	11	64
Livermore, CA	24	9	2	64
Morehead, KY	24	23	14	64
Monroeville, PA	24	42	18	64
Pembroke, ON, Canada	24	29	14	64
Springdale, AR	24	11	2	64
Newtown, CT	24	26	8	65
Allegheny Mt., PA	12	33	20	31
Uniontown, PA	12, 24	52	39	42
State College, PA	12	47	25	3
Philadelphia, PA	24	39	9	48
Pittsburgh, PA	6, 24	27	15	39

levels measured over 6-hr. (10am-4pm) periods have been found to equal 150 and 175 nmoles·m^{-3} (39). For 6-hr. measurements made during the same summer in Pittsburgh, a metropolitan area located approximately 60 kilometers northwest from Uniontown, outdoor SO_4^{2-} concentrations were found to be comparable to those in Untiontown, with a summer time average of 125 nmoles·m^{-3} (39). For H^+, however, outdoor concentrations measured in Pittsburgh were substantially lower, averaging only 80 nmoles·m^{-3} over 6-hr. periods (39).

Diurnal Variability. Outdoor SO_4^{2-} and H^+ concentrations vary diurnally, with concentrations higher during the day as compared to the night (40-42). Peak SO_4^{2-} and H^+ concentrations generally occur between 12noon and 6pm, but the peak acid aerosol period will vary with radiative heat input and wind speed (41). Geometric mean concentrations for daytime (8am-8pm) and following nighttime (8pm-8am) SO_4^{2-} concentrations in State College, PA, were found to equal 90 and 64 nmoles·m^{-3}, respectively during the summer of 1991 (3). For H^+ concentrations, the diurnal variation was more pronounced, as geometric mean concentrations equaled 72 nmoles·m^{-3} during the day and 26 nmoles·m^{-3} during the night (3).

Diurnal variation in SO_4^{2-} and H^+ levels probably is due to the effects of atmospheric mixing, where air containing high concentrations of SO_4^{2-} and H^+ is mixed downwards during the daytime hours (41). For H^+, this effect is enhanced by ammonia (NH_3), where at night H^+ is neutralized below the inversion layer by NH_3 emitted from the ground (41). The observed diurnal variation patterns for SO_4^{2-} and H^+ indicate that children are most likely to be outside during peak acid aerosol periods.

Spatial Variation. Outdoor SO_4^{2-} exhibits little spatial variability, as outdoor concentrations tend to be uniform across large areas. Within rural and suburban communities, little to no spatial variability in SO_4^{2-} concentrations typically is found (3, 42). Similarly, no spatial variation in outdoor SO_4^{2-} concentrations has been observed within urban environments. In a study conducted by Waldman et al. (40) in metropolitan Toronto, outdoor SO_4^{2-} concentrations were found to be similar in three sites, located 33 km apart and with different population densities. These results were supported by findings from a subsequent monitoring study conducted by Suh et al. (43), where SO_4^{2-} concentrations were found to be uniform across eight sites located within the metropolitan Philadelphia area. Uniformity in SO_4^{2-} concentrations also has been demonstrated for communities separated by large distances. A comparison of the outdoor SO_4^{2-} concentrations of Uniontown, State College, and Pittsburgh, Pennsylvania—three communities separated by large distances (at least 60 km) and by vastly different population densities showed comparable SO_4^{2-} concentrations (39).

As was the case for SO_4^{2-}, outdoor H^+ exhibits little spatial variation within rural and suburban communities (3, 42). Spatial variation, however, has been observed within urban communities (40, 43-44) and across communities with different population densities (3) or separated by large distances (39). This spatial variation may be substantial, as H^+ concentrations were found to differ by approximately 30% in metropolitan Toronto (40) and by as much as 50-90%

within metropolitan Philadelphia (43). The extent of spatial variation in H^+ concentrations does not appear to depend on the overall H^+ concentration (43). It, however, is strongly associated with local NH_3 concentrations.

Evidence for the importance of outdoor NH_3 is given by results from a study conducted in State College, Pennsylvania (3). In this study, outdoor concentrations were measured at a total of 47 homes within residential State College and at a stationary ambient monitoring (SAM) site located in a heavily wooded area, only six kilometers outside the residential community. Outdoor H^+ concentrations at the home sites were found to be significantly lower than those measured at the SAM site (Figure 1), with home site concentrations 31% or 29.3 nmoles·m^{-3} lower on average than the SAM site levels. Lower home site concentrations were related to NH_3, as outdoor NH_3 concentrations were significantly higher at the home as compared to the SAM site (Figure 1).

The strong association between H^+ spatial variation and local NH_3 concentrations

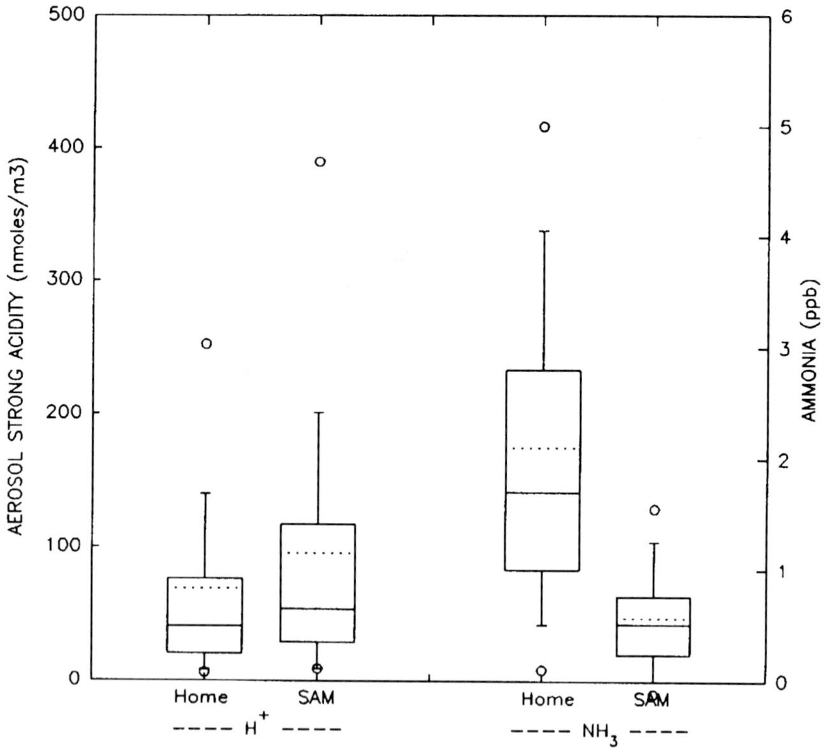

FIGURE 1. Boxplot of outdoor H^+ and outdoor NH_3 concentrations measured within a suburban community and at a rural stationary ambient monitoring (SAM) site. Concentrations were collected over 24-h (8am-8am) periods. All data from a study conducted in State College, Pennsylvania, during the summer of 1991. Adapted from references (3, 42).

is not surprising, since NH_3 readily neutralizes H^+. Thus, areas with high NH_3 concentrations would be expected to have lower H^+ concentrations as compared to other areas. Differences in NH_3 concentrations may be attributed to a number of factors. Of these factors, population density may be the most important (43). Since people and their activities are primary sources of NH_3, population density is strongly associated with NH_3 concentrations. As a result, differences in population density will result in corresponding differences in NH_3 concentrations, and thus in H^+ concentrations as well. These differences may be particularly significant in urban environments, where population density may vary substantially over short distances.

Indoor Concentrations

Residential Buildings. Characterization of indoor SO_4^{2-} and H^+ concentrations has been performed primarily for detached, single family homes located in rural and suburban communities (3, 42-43, 45). As discussed previously, the majority of acid aerosol data collected inside homes was collected by a study conducted in State College, Pennsylvania, during the summer of 1991, where multiple 12-h (8am-8pm) acid aerosol measurements were made inside and immediately outside the homes of 47 children (3).

Sulfate. SO_4^{2-} concentrations inside homes comprise a substantial fraction of outdoor levels, with air conditioner use an important determinant of indoor levels (3, 43, 45). In non-air conditioned homes, indoor SO_4^{2-} concentrations are comparable to, and sometimes higher than, corresponding outdoor levels. For air conditioned homes, differences between indoor and outdoor SO_4^{2-} concentrations tend to be more pronounced and tend to exhibit greater inter-home variability.

The importance of air conditioning to indoor SO_4^{2-} concentrations is the result of air conditioning's effect on air exchange rates. Air conditioned homes generally have lower air exchange rates as compared to non-air conditioned homes (3), which is expected since air conditioned homes were characterized by closed windows and doors. As a result of these lower air exchange rates, the residence time of air inside air conditioned homes is longer, allowing more time for SO_4^{2-} to deposit inside these homes. This theory is supported by results from the State College study (3), which show (a) that indoor/outdoor SO_4^{2-} ratios increase with air exchange rates and (b) that this relationship is consistent for air and non-air conditioned homes, even when air conditioned homes have uncharacteristically high air exchange rates (>1 exchange$\cdot$hr^{-1}).

Air exchange rates may also affect the resuspension of SO_4^{2-} particles indoors (46). Since air exchange rates may be indicative of indoor air flow rates, SO_4^{2-} resuspension, and thus indoor/outdoor ratios, may be greater inside homes with high air exchange rates. Thus, higher SO_4^{2-} concentrations in non-air conditioned homes also occurs as the result of increased SO_4^{2-} resuspension. Using mass-balance techniques, resuspension rates have been estimated to equal approximately 20 nmoles$\cdot$m$^{-3}\cdot$hr^{-1} in State College (3). Similarly, in a study conducted in 394 homes in New York, source strengths were estimated to equal 26% of the indoor SO_4^{2-} concentration (46). Direct comparison of these estimated values is not possible, since their units are not equivalent. Using the geometric mean SO_4^{2-} concentration found in State College, however, the resuspension rate found in the New York study suggests a mean resuspension rate of 18.7 nmoles·m^{-3}, which is comparable to that found in State College.

Penetration efficiencies for SO_4^{2-} also have been estimated using mass balance techniques, equalling 84% in the State College study (3) and 74% in the New York study (46). These estimates are similar, with the lower penetration efficiency in the New York study attributable to the fact that it was conducted during the winter when homes are tightly sealed.

Aerosol Strong Acidity. Indoor H^+ concentrations are substantially lower than corresponding outdoor levels (3, 42, 45). In homes using air conditioners, indoor H^+ concentrations generally are near or below the LOD, while for non-air conditioned homes, H^+ levels tend to be higher, with more pronounced inter-home variability. In State College, H^+ concentrations were found to reach a maximum of 125 nmoles·m^{-3} (3). Differences between indoor and outdoor H^+ concentrations are substantially larger than those for SO_4^{2-}. Since virtually all H^+ is associated with SO_4^{2-}, larger indoor/outdoor concentration differences demonstrate the presence of other removal mechanisms for H^+ indoors. As was the case outdoors, the most important of these is the neutralization of H^+ by NH_3.

The importance of indoor NH_3 is demonstrated by the observed negative association between indoor H^+ and indoor NH_3 concentrations. As shown on Figure 2, NH_3 concentrations greater than approximately 50 ppb were sufficient to neutralize H^+ indoors completely. [The apparent coexistence of H^+ and NH_3 indoors may be an artifact of the 12-hr. sample durations, which may hide temporal variations in indoor H^+ and indoor NH_3 concentrations.] This threshold probably results from the influence of air exchange rates, which were also associated with indoor H^+ concentrations. Results from State College show that air exchange rates lower than one exchange·hr.$^{-1}$ are associated with little to no H^+, but high NH_3, concentrations indoors. These results indicate that long indoor air residence times allow for sufficient NH_3 to accumulate inside these homes to neutralize all of the entering H^+.

Non-Residential Buildings

Studies characterizing SO_4^{2-} and H^+ concentrations inside non-residential buildings are sparse; however, the studies conducted to date do provide further evidence of the importance of air exchange rates and ventilation conditions as a determinant of indoor SO_4^{2-} and H^+ concentrations. In a study of child and elderly care facilities, indoor/outdoor SO_4^{2-} and H^+ ratios were compared to those measured inside residential buildings (47). Air conditioner use and indoor NH_3 were again identified as important determinants of indoor SO_4^{2-} and H^+. In contrast, telephone equipment buildings, which used large air handling systems to control ventilation, had substantially lower indoor/outdoor SO_4^{2-} ratios as compared to residential buildings (48). Filter efficiency and fresh air recirculation were identified as important determinants of indoor SO_4^{2-} concentrations.

PERSONAL EXPOSURES

As was the case for indoor concentrations, the largest study characterizing personal acid aerosol exposures was conducted in State College, Pennsylvania, during the summer of 1991, where the personal exposures of 47 children were measured over multiple daytime (8am-8pm) periods (4).

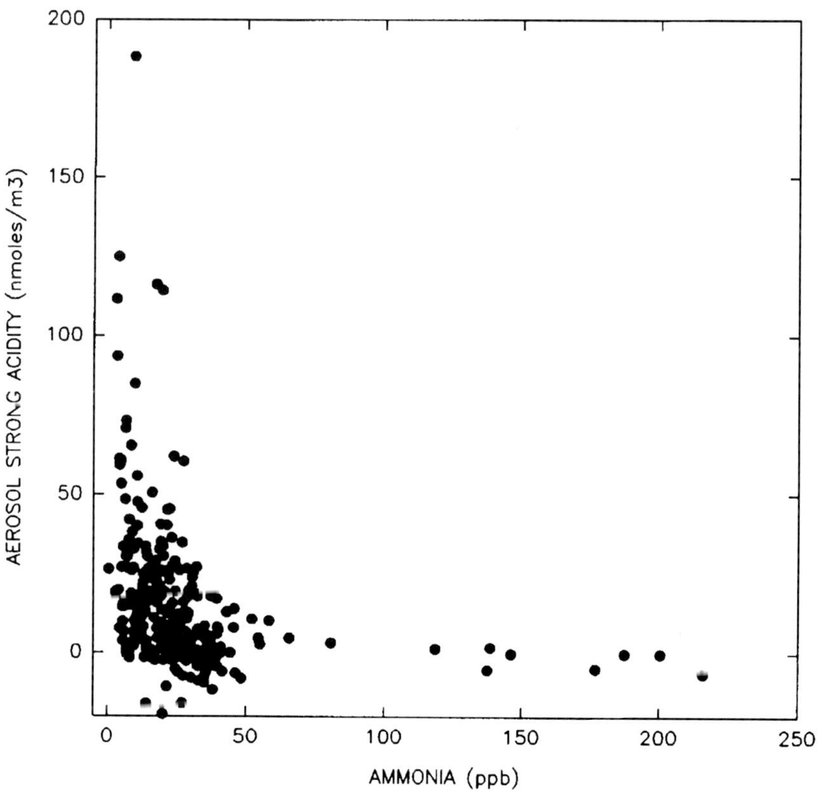

FIGURE 2. Indoor H^+ vs. Indoor NH_3 concentrations. Data collected over 12-hr. daytime (8am-8pm) and nighttime (8pm-8am) periods in Uniontown, PA, during the summer of 1990 and over 12-h daytime (8am-8pm) periods in State College, PA during the summer of 1991. Adapted from references (3, 42).

Levels

Personal SO_4^{2-} exposures are consistently lower than corresponding outdoor levels and tend to be higher than indoor concentrations. Personal SO_4^{2-} exposures exhibit considerable interpersonal variability, differing by as much as 200% on days with similar outdoor concentrations (4). This interpersonal variability is due in part to differences in the characteristics of the children's homes, where children living in air conditioned homes tend to have lower personal SO_4^{2-} exposures as compared to children living in non-air conditioned homes. [Indoor SO_4^{2-}- concentrations were lower inside air conditioned homes.] Interpersonal variability also may result from differences in the amount of time the children spent outdoors, as children who spent more time outdoors are likely to have higher personal exposures.

These factors also may contribute to the observed interpersonal variability in personal H^+ exposures. As was the case for SO_4^{2-}, personal H^+ exposures tend to be lower than corresponding outdoor concentrations and higher than corresponding indoor levels (4, 42, 49). The disparity between personal H^+ exposures and indoor and outdoor concentrations is generally more pronounced than that for SO_4^{2-}. Personal H^+ exposures exhibit substantial interpersonal variability, also as the result of differences in housing characteristics and time/activity patterns. In State College, personal H^+ exposures were found to vary by as much as 300% on days with similar outdoor concentrations (4). As a result of this interpersonal variability, both outdoor and indoor H^+ concentrations tend to be poor predictors of personal H^+ exposures (4).

Personal Exposure Models

Personal exposures to both SO_4^{2-} and H^+ may be estimated using microenvironmental exposure-based models, which have been developed and validated for children living in rural and suburban environments (4, 42). Their validation makes their use possible in future health effect or exposure assessment studies. Both models rely on concentrations measured in two microenvironments: outdoors at a stationary ambient monitoring (SAM) site and indoors at a home or other residence. In addition, both models require information about the individuals' time/activity patterns.

For SO_4^{2-}, the personal exposure model incorporates an expression for the microenvironmental concentration and also a correction term:

$$E = \underbrace{(f_iC_i+f_oC_o)}_{\text{time-weighted concentration}} - \underbrace{k_p \cdot (f_iC_i+f_oC_o)}_{\text{correction term}}$$

where f_i represents the fraction of time spent indoors, f_o the fraction of time spent outdoors, C_i the indoor concentration, C_o the outdoor concentration, and k_p the personal correction factor. The value of the personal correction factor is equal to

0.115 and reflects either the sampling error or the presence of particle concentration gradients (42). Results from the State College study show that this model predicted personal SO_4^{2-} exposures in State College with near perfect accuracy and precision, accounting for 95% of the variability in measured personal exposures (4).

The personal exposure model for H^+ also includes expressions for the time-weighted concentration and the correction term. The H^+ model includes an H^+-NH_3 reaction term as well:

$$E = \underbrace{(f_iC_i+f_oC_o)}_{\text{time-weighted concentration}} - \underbrace{k_p*(f_iC_i+f_oC_o)}_{\text{correction term}} - \underbrace{(f_iC_i+f_oC_o)*[NH_3]_p*k_{NH3}}_{NH_3\text{ reaction term}}$$

where $[NH_3]_p$ is the personal NH_3 concentration, k_{NH3} the rate constant for the H^+-NH_3 reaction, and $[NH_3]_pk_{NH3}$ the reaction rate of H^+ with NH_3. This model also includes the correction factor from the SO_4^{2-} model (k_p=0.115), since virtually all H_+ is associated with SO_4^{2-}. The H^+-NH_3 reaction rate represents (a) the number of H^+-NH_3 collisions that will occur and (b) the fraction of these collisions that will result in a neutralization reaction. The value for k_{NH3} equals 0.006 ppb^{-1}. This model was found to predict personal H^+ exposures in State College reasonably well, as it accounted for 63% of the variability in measured H^+ exposures, a substantial improvement over outdoor concentrations alone (4). The demonstrated high performance of both models shows that the models' correction and neutralization terms are stable and that the models may be used to estimate personal SO_4^{2-} and H^+ exposures for children living in non-urban environments.

DISCUSSION AND CONCLUSIONS

Studies conducted to date have advanced our understanding of indoor, outdoor, and personal acid aerosol concentrations substantially and have provided detail information on their distribution and magnitude. With the application of these findings to future health effect studies, the risk posed by acid aerosol exposures on human health may be determined more readily, as may the identification of populations at risk and the adoption of effective acid aerosol control strategies.

Studies have established that 24-hour integrated (8am-8am) outdoor SO_4^{2-} concentrations are uniform across large areas, exhibiting no spatial variation within and across urban and non-urban communities. Similarly, findings demonstrate that outdoor H^+ levels do not vary spatially within non-urban environments. They, however, may vary spatially within urban environments. Since this variation is related to outdoor NH_3 concentrations, outdoor H^+ will depend on population and traffic density, both indicators of outdoor NH_3. As was observed in State College, this effect may be dramatic, as outdoor H^+ concentrations were found to differ for sites separated by only six kilometers. These results demonstrate that a single outdoor monitor is sufficient to estimate outdoor SO_4^{2-} and H^+

concentrations for non-urban populations, provided that the monitor is located within the community to be studied. For studies of H^+ in urban environments, however, several outdoor monitors or air pollution models will be necessary to estimate outdoor concentrations accurately for a study community.

Acid aerosol studies also have demonstrated that indoor SO_4^{2-} and H^+ concentrations measured over 12-hour (8am-8pm and 8pm-8am) periods typically are lower than corresponding outdoor levels and may vary substantially from home-to-home. For SO_4^{2-}, indoor concentrations were found to comprise a substantial fraction of those outdoors, with indoor concentrations highest inside non-air conditioned homes. Indoor H^+ concentrations also were higher inside non-air conditioned homes as compared to air conditioned homes; however, indoor levels generally were low. Indoor SO_4^{2-} and H^+ concentrations were highly correlated to outdoor levels, suggesting that outdoor air is the primary source for SO_4^{2-} and H^+ indoors.

The studies also showed that daytime (8am-8pm) personal SO_4^{2-} and H^+ exposures are lower than corresponding outdoor levels, with substantial inter-personal variability observed in personal SO_4^{2-} and H^+ exposures. Since indoor concentrations are an important component of personal exposures, housing factors were found to influence personal SO_4^{2-} and H^+ exposures. Activity patterns also were found to be an important determinant of personal exposures, since they may affect the amount of time spent indoors during the day. In addition, personal NH_3 concentrations affected personal H^+ exposures through neutralization of H^+ around the individuals.

Personal exposure models present considerable promise as a method to assess SO_4^{2-} and H^+ exposures. Models which included a correction term for SO_4^{2-} and a correction and H^+-NH_3 reaction term for H^+ were found to predict personal exposures with a high degree of accuracy and precision. Both the correction term and the reaction terms were stable, indicating the models could be used to estimate SO_4^{2-} and H^+ exposures for children living in non-urban environments.

Although these models are promising, they are not yet suitable for widespread use, as several key issues governing the behavior of SO_4^{2-} and H^+ in indoor and personal environments remain unresolved. To date, little or no acid aerosol data have been collected in urban environments or for other population sub-groups, making the characterization of personal acid aerosol exposures for urban populations and for adult and senior citizens difficult. Similarly, little to no information is available to resolve questions raised during the development of the personal exposure models, such as those concerning the existence of particle concentration gradients around individuals and the interaction of H^+ and NH_3 in the indoor and personal environments.

With the resolution of these issues, indoor and personal exposure models for SO_4^{2-} and H^+ can be refined and improved. The improved accuracy of these models would benefit epidemiologic, exposure assessment, and regulatory studies, as they could better determine selection criteria for "exposed" and "non-exposed" study participants and better identify parameters that are important determinants of personal acid aerosol exposures. Ultimately, improved exposure estimates for acid aerosols can be used to determine to what extent acid aerosol pose risks to human health. If the risks are significant, the improved exposure estimates could be used to reduce these risks through identification of populations at risk and through the establishment and evaluation of regulatory control programs for acid aerosols.

REFERENCES

1. Koening, J.Q., Covert, D.S. and W.E. Pierson. 1989. Effects of inhalation of acidic compounds on pulmonary function in allergic adolescent subjects. *Environ. Health Perspect.* 79:173-178.
2. Speizer, F.E. 1989. Studies of acid aerosols in six cities and in a new multicity investigation: design issues. *Environ. Health Perspect.* 79:61-67.
3. Suh, H.H., Koutrakis, P. and J.D. Spengler. 1994. The relationship between airborne acidity and ammonia in indoor environments. *J. Exp. Anal. Environ. Epi.* 4:1-22.
4. Suh, H.H., Koutrakis, P. and J.D. Spengler. 1993. Validation of personal exposure models for sulfate and aerosol strongly acidity. *JAWMA* 43:845.
5. Lippman, M. 1989. Background on health effects of acid aerosols. *Environ. Health Perspect.* 79:3-6.
6. Spengler, J.D., Brauer, M. and P. Koutrakis. 1990. Acid air and health. *Environ. Sci. Technol.* 24:946-956.
7. Thurston, G.D., Ito, K., Lippmann, M. and C. Hayes. 1989. Reexamination of London, England, mortality in relation to exposure to acidic aerosols during 1963-1972 winters. *Environ. Health Perspect.* 79:73-82.
8. Utell, M.J., Morrow, P.E., Speers, D.M., Darling, J. and R.W. Hyde. 1983. Airway responses to sulfate and sulfuric acid aerosols in asthmatics: an exposure-response relationship. *Am. Rev. Respir. Dis.* 128:444-450.
9. Spektor, E.M., Leikauf, G.D., Albert, R.E. and M. Lippmann. 1985. Effects of submicrometer sulfuric acid aerosols on mucociliary transport and respiratory mechanics in asymptomatic asthmatis. *Environ. Res.* 37:174-191.
10. Avol, E.L., Linn, W.S., Anderson, K.R., Shamoo, D.A., Velencia, L.M., Little, D.E. and J.D. Hackney. 1988. Respiratory dose-response study of normal and asthmatic volunteers exposed to sulfuric acid aerosol in the submicrometer size range. *Toxicol. Indust. Health* 4:173-184.
11. Utell, M.J., Mariglio, J.A., Morrow, P.E., Gibb, F.R. and D.M. Speers. 1989. Effects of inhaled acid aerosols on respiratory function: the role of endogeneous ammonia. *J. Aerosol Med.* 2:141-147.
12. Spektor, D.M., Yen, B.M. and M. Lippmann. 1989. Effect of concentration and cumulative exposure of inhaled sulfuric acid on tracheobronchial particle clearance in healthy humans. *Environ. Health Perspect.* 79:167-172.
13. Amdur, M.O., Bayles, J., Ugro, V. and D.W. Underhill. 1978. Comparative irritant potency of sulfate salts. *Environ. Res.* 16:1-8.

14. Utell, M.J., Morrow, P.E. and Hyde, R.W. 1982. Comparison of normal and asthmatic subjects' responses to sulphate pollutant aerosols. *Ann. Occup. Hyg.* 26:691-697.
15. Dockery, D.W., Damokosh, A.I., Neas, L.M., Raizenne, M., Spengler, J.D., Koutrakis, P., Ware, J.H. and F.E. Speizer. 1993. Health effects of acid aerosols on North American children: respiratory symptoms and illness. *Am. Rev. Resp. Dis.* 147:A633.
16. Raizenne, M., Neas, L., Damokosh, A., Dockery, D., Spengler, J., Koutrakis, P., Ware, J. and F. Speizer. 1993. Health effects of acid aerosols on North American children: pulmonary function. *Am. Rev. Respir. Dis.* 147:A635.
17. Raizenne, M.E., Burnett, R.T., Stern, B., Franklin, C.A. and J.D. Spengler. 1989. Acute lung function responses to ambient acid aerosol exposures in children. *Environ. Health Perspect.* 79:179.
18. Neas, L.M., Dockery, D.W., Koutrakis, P., Spengler, J.D., Tollerud, D.J. and F.E. Speizer. 1994. The association of ambient air pollution with twice daily peak expiratory flow rate measurements in children. *Am. J. Epi.* in press.
19. Thurston, G., Lippmann, M., Bartoszek, M. and J. Fine. 1993. Air pollution associations with asthma exacerbations, peak flow changes, and respiratory symptoms in children at a summer asthma camp. *Amer. Rev. Respir. Dis.* 147:A632.
20. Damokosh, A.I., Spengler, J.D., Dockery, D.W., Ware, J.H. and F.E. Speizer. 1993. Effects of acidic particles on respiratory symptoms in seven US communities. *Am. Rev. Resp. Dis.* 147:A632.
21. Özkaynak, H. and G.D. Thurston. 1987. Associations between 1980 U.S. mortality rates and alternative measures of airborne particle concentrations. *Risk Analysis* 7:448-461.
22. Bates, D.V. and R. Sizto. 1989. The Ontario Air Pollution Study: identification of the causative agent. Environ. Health Perspect. 79:69.
23. Thurston, G.D., Ito, K., Kinney, P.L. and M. Lippmann. 1992. A multi-year study of air pollution and respiratory hospital admissions in three New York state metropolitan areas: results from 1988 and 1989 summers. *J. Expos. Anal. Environ. Epidem.* 2:429-450.
24. G., Ito, K., Lippmann, M. and D.V. Bates. 1993. Summertime haze air pollution and asthma hospital admissions in Toronto, Ontario. *Amer. Rev. Respir. Dis.* 147, A634.
25. Ayres, J., Fleming, D., Williams, M. and G. McInnes. 1988. Measurements of respiratory morbidity in general practice in the United Kingdom during the acid transport event of January 1985. *Environ. Health Perspect.* 79:83.
26. Bates, D.V., Baker-Anderson, M. and R. Sizto. 1990. Asthma attack periodicity: a study of hospital emergency visits in Vancouver. *Environ. Res.* 51:51.
27. Pope, C.A., Dockery, D.W., Spengler, J.D. and M.E. Raizenne. 1991. Respiratory health and PM_{10} pollution. *Am. Rev. Respir. Dis.* 144:668-674.

28. Dockery, D.W., Schwartz, J. and J.D. Spengler. 1992. Air pollution and daily mortality: associations with particulates and acid aerosols. *Environ. Res.* 59:362-373.
29. Dockery, D.W., Pope, C.A., Xu, X., Spengler, J.D., Ware, J.H., Fay, M.E., Ferris, B.G., Speizer, F.E. 1993. An association between air pollution and mortality in six U.S. cities. *NEJM* 329:1753-1808.
30. Spengler, J.D. and K. Sexton. 1983. Indoor air pollution: a public health perspective. *Science* 221:9-17.
31. Pierson, W.R., Brachaczek, W.W., Truex, T.J., Butler, J.W. and T.J. Korniski. 1980. Ambient sulfate measurements on Allegheny Mountain and the question of atmospheric sulfate in the northeastern United States. *Ann. N.Y. Acad. Sci.* 338:145-173.
32. Office of Technology Assessment, U.S. Congress. 1985. *Acid Rain and Transported Pollutants: Implications for Public Policy.* Unpub: New York.
33. Milford, J.B. and C.I. Davidson. 1987. The sizes of particulate sulfate and nitrate in the atmosphere: a review. *JAPCA* 37:125-134.
34. Martonen, T.B., Barnett, A.E. and F.J. Miller. 1988. Ambient sulfate aerosol deposition in man: modeling - the influence of hygroscopicity. *Environ. Health Perspect.* 63:11-24.
35. Lioy, P.J. and J.M. Waldman. 1989. Acid sulfate aerosols: characterization and exposure. *Environ. Health Perspect.* 79:15-34.
36. Koutrakis, P. and B.P. Kelly. 1993. Equilibrium size of atmospheric aerosol sulfates as a function of particle acidity and ambient relative humidity. *J. Geophys. Res.* 98:7141-7147.
37. Spengler, J.D., Keeler, G.J., Koutrakis, P., Ryan, P.B., Raizenne, M. and C.A. Franklin. 1989. Exposures to acidic aerosols. *Environ. Health Perspect.* 79:43,
38. Thompson, K.M., Koutrakis, P., Brauer, M., Spengler, J.D., Wilson, W.E. and R.M. Burton. 1991. Measurements of aerosol acidity: sampling frequency, seasonal variability, and spatial variation. In: *Proceedings of the AWMA Conference,* Paper No. 91-89.5.
39. Burton, R.M., Wilson, W.E., Koutrakis, P. and L.S. Liu. 1992. Comparison of aerosol acidity in urban and semi-rural environments. In: *Proceedings of the EPA-AWMA Conference on the Measurement of Toxic & Related Air Pollutants,* Air and Waste Management Association, Pittsburgh, PA, pp. 1051.
40. Waldman, J.M., Lioy, P.J., Thurston, G.D. and M. Lippmann. 1990. Spatial and temporal patterns in summertime sulfate aerosol acidity and neutralization within a metropolitan area. *Atmos. Environ.* 24B:115.
41. Wilson, W.E., Koutrakis, P. and J.D. Spengler. 1991. Diurnal variations of aerosol acidity, sulfate, and ammonia in the atmosphere. In: *Proceedings of the AWMA Conference,* Paper No. 91-89.9.
42. Suh, H.H., Spengler, J.D. and P. Doutrakis. 1992. Personal exposures to acid aerosols and ammonia. *Environ. Sci. Technol.* 26:2507.

43. Suh, H.H., Allen, G.A., Koutrakis, P. and R.M. Burton. 1994. Spatial variation in acidic sulfate concentrations within metropolitan Philadelphia. To be submitted to *JAWMA*.
44. Allen, G.A., Koutrakis, P., Wilson, W., Burton, R., Waldman, J. and F. Hauptman. 1993. Spatial variation of strong aerosol acidity in the Philadelphia metropolitan area. Presented at the 1993 EPA-AWMA Conference on the Movement of Toxic & Related Air Pollutants, Durham, North Carolina, May 1993.
45. Brauer, M., Koutrakis, P., Keeler, G.J. and J.D. Spengler. 1991. Indoor and outdoor concentrations of inorganic acidic aerosols and gases. *J. Air Waste Manage. Assoc.* 41:171.
46. Koutrakis, P., Briggs, S.L.K. and B.P. Leaderer. 1992. Source apportionment of indoor aerosols in Suffolk and Onondaga Counties, New York. *Environ. Sci. Technol.* 26:521.
47. Liang, C.S.K. and J.M. Waldman. 1992. Indoor exposures to acid aerosols at child and elderly care facilities. *Indoor Air* 2:196-207.
48. Sinclair, J.D., Psota-Kelty, L.A. and C.J. Weschler. 1988. Indoor/outdoor ratios and indoor surface accumulations of ionic substances at Newark, New Jersey. *Atmos. Environ.* 22:461.
49. Brauer, M., Koutrakis, P. and J.D. Spengler. 1989. Personal exposures to acidic aerosols and gases. *Environ. Sci. Technol.* 23:1408.
50. Appel, B.R., Hoffer, E.M., Tokiwa, Y. and E.L. Kothny. 1982. Measurement of sulfuric acid and particulate strong acidity in the Los Angeles Basin. *Atmos. Environ.* 16:589-93.
51. Stevens, R.K., Dzubay, T.G., Shaw, R.W., McClenny, W.A., Lewis, C.W. and W.E. Wilson. 1980. Characterization of the aerosol in the Great Smoky Mountains. *Environ. Sci. Technol.* 14:1491-1498.
52. Lioy, P.J., Samson, P.J., Tanner, R.L., Leaderer, B.P., Minnich, T. and W. Lyons. 1980. The distribution and transport of sulfate "species" in the New York metropolitan area during the 1977 summer aerosol study. *Atmos. Environ.* 14:1391-1407.
53. Morandi, M.T., Kneip, T.J., Cobourn, W.G., Husar, R.B. and P.J. Lioy. 1983. The measurement of H_2SO_4 and other sulfate species at Tuxedo, New York with a thermal analysis flame photometric detector and simultaneously collected quartz filter samples. *Atmos. Environ.* 17:843-848.
54. Huntzicker, J.J., Hoffmann, R.S. and R.A. Cary. 1984. Aerosol sulfur episodes in St. Louis, Missouri. *Environ. Sci. Technol.* 18:962-967.
55. Ferris, B.G. and Spengler, J.D. 1985. Problems in the estimation of human exposure to components of acid precipitation precursors. *Environ. Health Perspect.* 63:5-9.
56. John, W., Wall, S.M. and J.L. Ondo. 1985. *Dry Acid Deposition on Materials and Vegetation: Concentration in Ambient Air.* Report CA/DOH/A1HL/SP-34. Resources Board, Sacramento, CA.
57. Spengler, J.D. 1986. In: Lee, S.D. (Ed.) *Proceedings of 2nd Annual*

U.S.-Dutch International Symposium on Aerosols. Lewis Publishers, Chelsea, MI, pp. 107-120.
58. Cadle, S.H. 1985. Seasonal variations in nitric acid, nitrate, strong aerosol acidity, and ammonia in an urban area. *Atmos. Environ.* 19:181-188.
59. Kelly, T.J., McLaren, S.E., Kadlecek, J.A. 1989. Seasonal variations in atmospheric SO_x and NO_y species in the Adirondacks. Atmos. Environ. 23:1315-1332.
60. Pierson, W.R., Brachaczek, W.W., Gorse, R.A., Japar, S.M., Norbeck, J.M. and G.J. Keeler. 1989. Atmospheric acidity measurements on Allegheny Mountain and the origins of ambient acidity in the northeastern United States. *Atmos. Environ.* 23:431-459.
61. Koutrakis, P., Wolfson, J.M. and J.D. Spengler. 1988. An improved method for measuring aerosol strong acidity: results from a nine-month study in St. Louis, Missouri and Kingston, Tennessee. *Atmos. Environ.* 22:157-162.
62. Koutrakis, P. and P. Mueller. 1989. In: *Proceedings of the 82nd Annual Meeting of the Air Pollution Control Association,* Anaheim, CA. Paper 89-71.4.
63. Cohen, M.A. 1990. Assessment of community exposures to volatile organic compounds. Ph.D. Thesis, School of Public Health, Harvard University: Cambridge, MA.
64. Keeler, G.J., Koutrakis, P.E. and J.D. Spengler. 1989. Atmospheric acidity in rural eastern North America. Presented at 1989 Meeting of the American Association for Aerosol Research, Reno, NV.
65. Keeler, G.J., Spengler, J.D. and R.A. Castillo. 1991. Acid aerosol measurements at a suburban Connecticut site. *Atmos. Environ.* 25A:681-690.

Environmental Contaminants, Ecosystems and Human Health. Edited by S.K. Majumdar, E.W. Miller, and Fred J. Brenner. © 1995, The Pennsylvania Academy of Science.

Chapter Sixteen

PESTICIDE CONTAMINANTS AND HUMAN HEALTH

M.K.K. PILLAI and AMIT NAIR
Department of Zoology
University of Delhi
Delhi-110007
India

INTRODUCTION

In the contemporary world, use of pesticides has become an indispensable input to agriculture, forestry and public health. Since the introduction of DDT in 1943, the last five decades have witnessed introduction of a variety of pesticides and a spectacular rise in their production. The global pesticide use increased by 1,300 percent between 1945 and 1980 (1). An estimate of world market in pesticides in 1990 reached a peak of 18.5 billion US $ which included 33 percent for insecticides, 39 percent for herbicides and 21 percent for fungicides. The pattern of use of pesticides, however, varied in different countries. For instance, in North America and Western Europe herbicides constituted 70 percent of the pesticides consumed, in Western Europe fungicides alone accounted for 43 percent and in developing countries insecticides claimed 62 percent of the total pesticides used.

Pesticides, being toxic, can become a potential hazard to the manufacturer, the user, the public at large and the environment. Extensive deliberate use of pesticides has resulted in contamination of our vital supplies, air, water and food. The risk to humans may be short-term or long-term depending on the persistence of the pesticides and the exposure period. In USA, it is estimated that 300,000 farm workers annually suffer from pesticide poisoning (2). In the present scenario, long term effect of chronic poisoning by pesticide contaminants is of paramount importance. A multidisciplinary study involving toxicology, epidemiology, physiology and behavioral sciences is essential for the proper assessment of human health hazards related to pesticide exposure. The presence of insecticide residues especially of DDT in man is well recognized but its long term effects are still elusive. Public concern about many pesticides turning out as etiological agents for many major and minor human diseases is still unabated. However, there is enough scientific data based on epidemiological studies conducted among occupationally exposed people such as pesticide manufacturers, applicators and farmers to assess the potential human risk from pesticide contaminants. In this chapter pesticide-related human hazards are critically examined.

ACUTE POISONING SYMPTOMS OF PESTICIDES IN MAN

The main groups of pesticides of our concern are insecticides, herbicides, fungicides and certain soil fumigants (Table 1). Organochlorine (OCl) insecticides, due to their inherent chemical stability coupled with environmental persistence, are considered to be the main pesticide contaminants in the environment, their residues being dynamic in any ecosystem as they pass through food chains of different trophic levels (3). These compounds are generally cumulative poisons. DDT, a nerve poison, though less toxic to man, can cause at high doses symptoms such as paresthesias of the tongue, lips and face, apprehension, irritability, dizziness, tremors and convulsions. Acute poisoning due to DDT is rare in man as in many accidental cases recovery has been reported. Lindane (γ-isomer of HCH) is more toxic than DDT to man and poisoning stimulates nervous system characterized by increase in blood pressure, bradycardia and irregular encephalogram. Though very high level of HCH is required to cause mortality in

TABLE 1
Classification of Pesticides with Examples

Class	Chemical Type	Examples
Insecticide	DDT analogues	DDT
	BHC isomers	BHC (HCH)
	Cyclodienes	aldrin, dieldrin, endrin, chlordane, heptachlor
	Organophosphates	parathion, malathion, TOCP
	Carbamate	carbaryl
	Dinitrophenol	DNOC
	Inorganic	arsenicals, fluorides
	Botanicals	rotenone, nicotine
	Pyrethroids	permethrin, deltamethrin, fenvalerate
Nematocide	Brominated compound	DBCP
Fungicide	Dithiocarbamate	maneb
	Chlorinated aromatic	Pentachlorophenol,HCB
Herbicide	Phenoxy	2,4-D; 2,4,5,-T,MCPA
	Dinitrophenol	DNOC
	Bipyridyl	paraquat
	Triazine	atrazine
Rodenticide	Anticoagulants	Warfarin
	Thiourea	ANTU (α-naphthylthiourea)

man, children are more susceptible and may succumb to fatal poisoning even with very low doses. Until recently aldrin and dieldrin were the extensively used cyclodiene insecticides. The symptoms of poisoning due to aldrin and dieldrin resemble those of HCH as both are nerve stimulants. These symptoms include headache, nausea, vomiting, dizziness, general weakness and convulsions in severe poisoning.

Both organophosphorus (OP) and carbamate pesticides, most extensively used pesticides nowadays, are not cumulative but highly toxic to man as compared to OCl insecticides. Detailed poisoning symptoms are given in Table 2.

Acute poisoning due to phenoxy herbicides results in depression of central nervous system, bradycardia, hypotension, skeletal muscle myotonia and myoglobinuria. Their major target sites are central nervous system and cardiovascular system. Survivors exhibit peripheral neuropathy. Paraquat, a non-selective herbicide, on ingestion causes edema, burns, ulceration in the mucosa of the mouth and the digestive tract, damage of liver, kidney and skeletal muscle. It causes a progressive irreversible pulmonary fibrosis. Acute poisoning may cause death within few days from cardiovascular collapse, and subacute doses result in pulmonary fibrosis. Diquat resembles paraquat in poisoning symptoms except that it does not produce fibrosis of the lung.

Acute poisoning with pentachlorophenol, a fungicide causes irritation of eyes, respiratory tract, bronchitis and chest tightness. Other symptoms are tachycardia, nausea and vomiting. Target organs for toxicity are liver, kidney, haematopoietic, pulmonary and nervous system. Severe poisoning by inhalation causes pulmonary edema and death. Chronic poisoning leads to liver injury. The fungicides, dithiocarbamates generally cause irritation of eyes, skin and mucosa of gastric tract, nausea, dizziness, headache, confusion, ataxia, vomiting, diarrhea and flaccid paralysis. Ingestion of large quantities may lead to death.

TABLE 2

Clinical Symptoms of Acute Poisoning with OP Pesticides

Target Site	Clinical manifestations
Muscarinic synapses	Respiratory difficulty, increased bronchial secretion, dyspnea, lung edema, hypotention, bradycardia, heart spasms, vomiting, abdominal pain, cramps, increased salivation, gastrointestinal hypermotility, diarrhea, miosis of pupils, blurred vision, increased lacrimation, excessive perspiration, involuntary defecation and urination.
Nicotinic synapses	Voluntary muscles - cramps, muscular twitches, fasciculations and paralysis. Tachycardia, hypertension
Central nervous system	Restlessness, irritability, anxiety, confusion, dizziness, tension, insomnia, ataxia, coma, convulsions, respiratory depression, cardiac block and arrest.

ACUTE PESTICIDE POISONING AND MORTALITY

Accidental, occupational or intentional (suicide, homicide and unspecific forensic cases) exposure to pesticides and the consequent mortality has been the primary cause for many isolated episodes of acute pesticide poisoning all over the World. Though approximately 80 percent of the total agrochemicals are used in developed countries, deaths due to acute poisoning are almost negligible. However, in developing countries it is a serious problem and a prioity public health issue at present. For example, in Sri Lanka pesticide poisoning cases registered an annual rate of 50,000 resulting in over 2,000 deaths (4). Levine on the basis of data from 34 countries for 15 years, from 1970 to 1984 arrived at an estimate of 1,111,000 incidents with 20,000 deaths annually based on mortality method and 1,528,000 cases with 28,000 deaths based on morbidity method (5). Using a case fatality rate of 0.5 percent for the developing countries and 0.25 percent for the developed countries, global deaths due to acute pesticide poisoning has been estimated to be 80,000 per annum (5).

Non-occupational pesticide poisoning is a major public health problem in developing countries such as Sri Lanka, Indonesia, Malaysia, Thailand and Philippines as indicated by WHO surveys (6). Non-accidental poisoning especially self-inflicted suicides using pesticides are very high as it accounted for 53 percent of the unreported cases in Sri Lanka and 64 percent of 603 deaths reported between 1980-87 in Philippines (7). In Kenya, out of 5 million agricultural workers, 350,000 are poisoned annually by pesticides (8).

Many episodes of acute pesticide poisoning which resulted in high mortality were mainly due to food contamination alone (Table 3) (9). Organophosphorus pesticides accounted for majority of the poisoning incidents. In Japan, China and Nicaragua acute poisoning were mainly due to methyl parathion. In China, out of 573 cases of acute poisoning, exposure to synthetic pyrethoids resulted in only 5 deaths which included 2 deaths due to occupational exposure to deltamethrin and 3 accidental ingestive cases of fenvalerate (10). Herbicides, though not highly toxic, were found to be responsible for many poisoning epidemics in Fiji, Papua New Guinea, Columbia, Western Samoa and in Malaysia (11). Out of 569 deaths recorded in 1987 in Malaysia 310 were due to the herbicide, paraquat (11). High rate of acute pesticide poisoning in developing countries is due to either lack of government regulatory processes or not strictly following safety regulatory measures.

CHRONIC POISONING AND MAJOR HUMAN DISEASES

A survey of literature reveals the fact that the majority of papers published on human hazards attributing to chronic exposure to pesticide contaminants pertain to various types of human cancers. But incrimination of pesticides as etiological agents for human cancer is being hotly debated as the validity of many studies showing positive association of pesticides with human cancers is doubtful. This is on the basis of the methodologies adopted, the lack of precise exposure data and absence of rigorous statistical methods (12). Occupational exposure studies are usually performed on farmers, pesticide manufacturers, formulators and applicators. In the case of farmers, data on linkage with many types of cancers is generally found to be weak, as data on the identity of pesticide and quantitative exposure are insufficient. Infact, a dose-dependent effect is presented only in a few cases. Though carcinogenic effects of many pesticides have been established in animals, such data cannot be applied to man. Often in animals very high levels of pesticides were used, as compared to occupational exposures in man. Also, mathematical models based on high-low dose extrapolation are not completely reliable in evaluating the risk factor.

Epidemiological studies of human risk assessment associated with pesticide exposure are usually carried out on occupationally exposed people by morbidity and mortality evaluation essentially using case control, cohort, proportionate, or cross-sectional studies. Inspite of inherent limitations, these studies merit serious consideration as they are useful in evaluating pesticides as human carcinogenic agents. Linkage of some of the suspected cases of human cancers is discussed below.

Non-Hodgkin's Lymphoma (NHL)

Recently there has been an increased public concern on the association of

TABLE 3

Acute Pesticide Poisoning and Mortality Reported from Different Countries

Country	Year	Nature of Pesticide Exposure	No. of Cases	No. of Deaths	Pesticides Involved
Sri Lanka*	1975-1980	1,2,3	13,000	1,000	mainly OP pesticides
Sri Lanka*	1988	1,2,3	50,000	2,000	mainly OP pesticides
Philippines	1980-1987	1,2,3	4,031	603	mainly OP pesticides
Pakistan	1976	1,2,3	2,800	5	iso-malathion
Thailand	1987	1,2,3	4,633	5	parathion
Malaysia	1978-1986	1,2,3	569	310	paraquat
Kenya**	1981	1,2,3	721	13	OP pesticides
Columbia	1984	1	600	88	parathion, paraquat
Brazil	1984	1	173	nil	OP pesticides
China	1983-1988	1	573	5	synthetic pyrethroids
Qatar	1967	2a	691	24	endrin
India	1958	2a	360	102	parathion
Mexico	1968	2a	559	16	parathion
Turkey	1955-1959	2a	3,000-5,000	3 to 11 per cent of the total cases	hexachlorobenzene
U.S.A.	1985	2a	1,175	nil	aldicarb

*Annual average, **Records from Kenyatta National Hospital

1 - occupational, 2 - accidental, 2a - accidental food consumption, 3 - intentional

pesticide contaminants with NHL among many types of human leukemia. In USA, 50 percent increase in NHL incidence was observed during 1950 to 1980 predominantly in agricultural areas. Epidemiological studies in Sweden, USA, Canada, UK, Italy, New Zealand and Australia confirm higher risk among farmers exposed to pesticides as 34 case control studies were found to be positive and 16 negative, while 16 cohort studies were positive and 8 negative (12). A firm relationship between phenoxy herbicides especially, 2,4-D with 2 to 8-fold ratios for NHL in farmers was established from different countries. Studies also revealed that 2,4-D used had no contamination with dioxins, such as TCDD, which is a known carcinogen (13). A Swedish study showed a 6-fold increase, while in case of Kansas a 7-fold increase in risk was observed among persons who used 2,4-D for more than 21 days per year (14, 15). It is concluded that a large increase in the use of phenoxy herbicides in USA accounts for the increased incidence of NHL. Besides phenoxy herbicides, Kansas and Nebraska studies indicate positive relation of NHL with triazine herbicides and fungicides (15, 16). Pesticide handling, mixing and application showed elevated risk of NHL. A recent study in Iowa and Minnesota, USA suggests involvement of several insecticides such as carbaryl, chlordane, DDT, diazinon, dichlorvos, lindane, malathion, nicotine and toxaphene with NHL and the association was stronger for those exposed prior to 1965 (17). This study, however, needs confirmation. There is also a report implicating a single episode of high exposure of a soil fumigant (nematocide) 1,3-dichloropropene causing NHL in farmers, six years after exposure (18).

Soft Tissue Sarcoma (STS)

The first report on the association of phenoxy herbicide exposure with STS was by Hardell (19), who reported a statistically significant risk ratio of 1.8 to 6.8 among herbicide applicators exposed for 10 to 20 years. Similarly, out of 10 cases of STS reported from USA, 7 were associated with workers engaged in the manufacture of 2,4,5-T and related compounds and 3 deaths occurred among Vietnam veterans exposed to Agent Orange, a phenoxy herbicide formulation (12). Epidemiological cohort studies of workers exposed to MCPA in a Danish study observed a risk ratio of 2.72 (20). Similar linkage of MCPA and STS was found in a large cohort of manufacturers and sprayers, exposed mainly to MCPA in UK (21). However, a case of statistically non-significant 2 fold excess risk was observed (22) from an International cohort involving 12,492 manufacturers and 5,898 sprayers (22).

Case control studies in Sweden showed statistically significant association with STS and exposure to phenoxy herbicides (23, 24, 25). A similar effect was found among forestry workers from New Zealand and among female rice weeders in Italy exposed to phenoxy herbicides (26, 27). Though a Kansas study in USA could not find in farmers, any association with phenoxy herbicide exposure and STS (15), there appears from other available evidences, a strong indication in favor of linkage of phenoxy herbicides with STS.

Brain Cancer

An association between farming and brain cancer among children has been reported in many studies. Prenatal and postnatal exposure to chlordane increased cases of brain cancer among children (28). In Maryland, USA, brain tumor incidence in children was associated with exposure to insecticides used in the household with an odds ratio of 2.3 (29). Similarly, a significant relationship was reported with maternal use of aerosol pesticides during pregnancy and after birth and brain cancer in children (30). A case control study indicates association between childhood brain cancer and pesticides such as dichlorovos, chlordane, lindane and carbaryl, though the study is not conclusive (31).

Cohort studies revealed linkage of brain cancer among pesticide applicators (32) and white male farm owners from California (33). Out of the 12 mortality studies associated with pesticide exposure and brain cancer, only 5 were found to be positive and significant (12).

In a case control study in Italy, farmers associated with increased use of organophosphorous and organochlorine pesticides and fungicides showed significant increased risk for brain glioma, the risk being greater among farmers exposed for 10 or more years (34, 35). Out of 9 case control studies from Sweden, USA, Canada, Italy and New Zealand, 6 studies showed positive association between farming and brain cancer, though data on exposure to specific pesticides was lacking to establish a dose dependent relationship (12). The Milan study during 1983-84 found association of insecticides and fungicides with brain cancer showing significant increase in relative risk (35). However, fungicides used in Italy contained alkyl urea, a known precursor of N-nitroso alkyl urea capable of inducing neurogenic tumors in animals. Further epidemiological studies are warranted to unequivocally establish causal relationship between pesticides and brain cancer (35).

Lung Cancer

A significant increase in mortality due to lung cancer was found among pesticide applicators, the increase correlated with longer exposure (32). Similar positive association between agricultural workers and lung cancer mortality was reported from East Germany (36). Lung cancer mortality among workers in a plant in Maryland, USA was associated with exposure to inorganic arsenicals (37). Similarly, increase in lung cancer was evident among agricultural workers after adjusting for smoking in Shanghai, China (38). Also in Finland, wood workers exposed to pesticides such as inorganic arsenic, aldrin and lindane showed increased risk for lung cancer (39). A relative risk of 2.4 for lung cancer was detected among sugarcane workers in Louisiana, USA (40). However, farmers from Saskatchewan, Canada did not show any correlation between lung cancer and occupational exposure to pesticides. Overall evidences indicate a positive correlation between lung cancer and occupational exposure to inorganic arsenicals.

Association of lung cancer and farming in many instances, though showing a significant relationship, the precise nature of exposure to pesticides is wanting. A serious criticism levelled against many epidemiological studies of lung cancer is that the involvement of an important confounder, viz, smoking, is not excluded in all such studies.

Other Human Cancers

Liver cancer: In USA, there was a general decline in the incidence of liver cancer during 1930 to 1972 inspite of extensive use of organochlorine pesticides such as DDT, aldrin, chlordane, heptachlor, etc. (41). Also in most of the studies, no association between liver cancer and farming was evident in USA. No association of liver cancer with DDT could be established in a study in Sweden (42). Implication of hexachlorobenzene (HCB) in liver cancer does not seem to be unequivocal (43). However, there are many reports that organochlorine pesticides cause liver tumors in mice.

Cancer of digestive system: The important source of entry of etiological agents, in digestive tract cancer is through food and water. A possible risk of stomach cancer is linked to dietary intake of nitrates and nitrites which can be converted to nitrosamides *in vivo* in stomach causing cancer (44).

There are few reports indicating significant increase in mortality due to cancers of pancreas, colon, stomach, esophagus among farmers in USA and Canada (45, 33). However, there are many studies where linkage between occupational exposure to pesticides and digestive tract cancers could not be established. Among the case control studies, statistically significant odds ratio of 1.32 was evident in the case of stomach cancer associated with farmers in Iowa, USA but similar studies on colon and stomach cancers were negative (46, 51). On the whole, evidences for pesticide exposure being responsible for digestive tract cancer are not fully established.

Urinary bladder cancer: Occupational exposure to rodenticide, ANTU (α-naphthylthiourea) in UK, and chlordane and heptachlor in USA caused a statistically significant higher mortality due to bladder cancer (47, 48). However, there are many reports of exposure to insecticides and herbicides not contributing to any increase in bladder cancer.

Kidney cancer: Alavanja in two separate studies in USA, involving agricultural workers, forest and soil conservationists exposed to various insecticides and herbicides, showed significant increase in kidney cancer (45, 49). However, MCPA and other phenoxy herbicide manufacturers, formulators and applicators in UK did not show any increase in the incidence of kidney cancer (21).

Prostate cancer: Majority of the studies conducted in USA, UK and Canada (7 out of 10) showed significant positive association between occupational exposures to different insecticides, MCPA and other phenoxy herbicides and prostate cancer (12, 45, 49, 21, 33). Two case control studies from USA confirmed significant excess risk of pesticide exposure and prostate cancer (50, 51). Among the urinary tract cancers, an increased risk is clearly evident among many farmers and similar occupationally exposed people, but the identity of a particular pesticide is not clear.

Cancer of testis and ovary: Testicular cancer is less common in man. A significant increase in testicular cancer was observed among people associated with farming, forestry and fishing in Texas, USA (52). A similar significant increase in relative risks for testicular cancer during 1974 to 1979 in Swedish pesticide applicators was attributed to pesticide exposure (53).

A case control study in Italy showed an increased risk for ovarian tumors associated with exposure to triazine herbicides (54). Further, a population based case-referent study in the same area by Donna et al. found a 2 to 3 fold increase in risk of epithelial ovarian cancer among women exposed to triazine herbicides (55).

MUTAGENIC EFFECTS

Mutagenicity tests indicate the rate of mutations as a result of genetic impairment and the mutagens as potential carcinogens. Mutagenic tests have limitations as the tests may not be sensitive to very low levels of chronic exposure to environmental contaminants. However, with high levels of exposures and possible exclusion of other factors, these tests will be useful for epidemiological studies. Mutagenicity of pesticides and contaminants can be evaluated in human epidemiological studies. Mutagenicity of pesticides and contaminants can be evaluated in human peripheral blood lymphocytes by determining chromosomal aberrations and sister-chromatid exchanges. It is established that high frequency of chromosomal aberrations indicates an increased risk for cancer but the biological significance of sister-chromatid exchanges is not clear.

Most of the cytogenetic studies are conducted with pesticide applicators subjected to multiple exposures. Increase in somatic chromosomal aberrations/sister-chromatid exchanges was observed in pesticide workers/applicators exposed to OP, carbamates and OCl pesticides and the fumigant, phosphine (12). Work from DDT factories in Brazil showed significant dose dependent linear increase in the frequency of chromosomal aberrations (56). However, in other positive studies, precise exposure data is lacking and exposure to many pesticides makes it difficult to assess the genotoxic effects or risk factor of each insecticide separately. The available data seem to be inadequate to determine genotoxic effects of pesticide exposure and their epidemiological significance.

HUMAN PATHOLOGY AND PESTICIDE EXPOSURE

Effects on Skin

Irritation and allergic dermatitis: Contact with dithiocarbamate fungicide, zineb by agricultural workers resulted in irritation of the skin and chronic allergic dermatitis, characterized by diffuse erythrodermia with erythemia and appearance of localized eczematoidic epidermis. Similarly, tobacco workers developed acute and chronic allergic lesions of the skin especially at the upper extremities upon contact with tobacco treated with maneb and zineb. Exposure to Naled (Dibrom) caused dermatitis with mild irritation and maculopopular patchy eruption among female workers of a chrysanthemum grower's farm (57).

Porphyria cutanea tarda: Accidental consumption of wheat grains treated with the fungicide, hexachlorobenzene (HCB) in Turkey during 1955-1959 caused porphyria cutanea tarda in more than 3,000 people (58). Approximately 0.05 to 0.2 g of HCB was ingested by the victims per day for a long period. The disease was prevalent more in males and most patients were in the age group of 4-14 years. Children born to mothers who had consumed HCB-treated grain exhibited pembe yara or pink sore and died within 2 years. Porphyria cutanea tarda had early clinical symptoms which induced photosensitive skin lesions with vesicles containing clear fluid and hyperpigmentation of the skin. These studies confirmed HCB as the etiological agent for porphyria cutanea tarda (59).

Paresthesia: Occupational exposure to synthetic pyrethoroids in China, Sweden and USA resulted in episodes of transient paresthesia, characterized by abnormal facial sensations with symptoms of burning, itching or tingling coupled with irritation of the upper respiratory tract. Erythema occurred only in a few cases. Fenvalerate produced a 4-fold increased paresthesia as compared to permethrin. With deltamethrin, dermal symptoms were different as it caused tenacious and painful puruitus by local burning, erythema and desquamation (60). In China, 547 cotton growers exposed to pyrethroids for 5 years experienced paresthesia, dizziness, headache or nausea after each spraying but recovered within 2 days with no sequelae indicating that pyrethroids have no chronic toxicity to humans (10).

Prolonged supernormal period of median nerve excitability was found in farmers spraying deltamethrin for 3 days in China. This indicates that the facial reactions are not an instance of skin irritation but a sign of involvement of peripheral nerves (61). These symptoms are caused by repetitive firing of sensory nerve endings. The effects depend on pyrethroid structure and vary with formulations and environmental factors. Cyanopyrethroids are more effective in this respect.

Effects on the Reproductive System

In animals, pesticides are known to cause adverse effects on reproduction but their relevance to man is uncertain. A well established case of adverse reproductive effects is reported to be due to exposure to a nematocide, DBCP (1,2-Dibromo-3-chloropropane). Ecological, cohort and cross sectional studies involving occupationally exposed people associated with manufacture, formulation and application showed suppression of spermatogenesis leading to oligospermia, azoospermia and infertility coupled with higher levels of hormones such as FSH and LH in the blood (62, 63, 64, 65). This effect was found to be reversible in oligospermic cases when exposure is discontinued (65).

Paternal exposure to DBCP with male fertility effect resulted in reducing the male infant proportion probably due to damage of Y-containing sperms (66). Similarly, workers exposed to chlordecone showed oligospermia with abnormal and non-motile sperm (67). Also impotence in spraymen exposed to dieldrin was found to be reversible with treatment by methyltestosterone (60).

Neurobehavioral Effects

Neural effects: Long term exposure to organophosphorus pesticides can lead to neurophysiological and neurobehavioral impairments. These pesticides being potent inhibitors of acetylcholinesterase (AChE) cause accumulation of acetyl choline in muscarinic and nicotinic synapses eliciting a spectrum of abnormal physiological effects in different systems and organs (Table 2). In addition, acute intoxication may lead to dysfunction of central nervous system with several clinical symptoms. Acute OP poisoning results in depression of activity of AChE activity in blood plasma and RBC and this effect lasts 103 weeks in plasma and as long as 12 weeks in RBC. The neurological symptoms disappear once the AChE level is brought back to normal. Exposure to organophosphorus pesticides also elicit subtle changes in electromyography (60). Determination of blood AChE level, insecticide level in serum and their residues/metabolites in urine and electromyography test can be used to detect OP poisoning.

Polyneuropathy: OP pesticide poisoning may also produce polyneuropathy which is not related to the inhibition of AChE. Neuropathy usually occurs 1-2 weeks after the exposure and it is characterized by symptoms such as muscular cramps and progressive weakness, initially of legs and later of the arms. This syndrome was earlier called as 'ginger paralysis' as it was first detected in people who took Jamaican ginger adulterated with TOCP (triorthocresyl phosphate). Polyneuropathy in man has been reported after acute accidental or occupational exposure to TOCP, trichlorphon, trichloronate and methamidophos. It is initiated by the inhibition of neuropathy target esterase (NTE) in the nervous system by certain OP compounds (68). NTE activity in peripheral lymphocytes can be used as a biochemical diagnostic test for polyneuropathy.

Behavioral effects: Chronic occupational exposure to OP pesticides may result in neurobehavioral changes such as impaired concentration and vigilance, slowing of psychomotor speed and information processing, defective memory, impaired language function, confusion, depression, anxiety and irritability (69). Behavioral abnormalities include a wide range of neuropsychological variables. A persistent reduction in neuropsychological functioning was noticed among agricultural workers poisoned with OP pesticides in Nicaragua (70). Transient abnormalities in electroencephalogram (EEG) similar to epileptic patients were noticed among occupationally exposed workers (69).

Polyneuritis: Chronic exposure to DDT, aldrin and endrin may produce polyneuritis characterized by weakness of hands, legs, paresthesia and numbness of distal part of the extremities. It may also cause muscle weakness, difficulty in walking leading to ataxia. Polyneuritis has been recorded among malaria workers who sprayed DDT for 4 years in India (71). However, all the symptoms are reversible after the exposure is terminated.

MISCELLANEOUS MINOR EFFECTS

Parkinson's Disease (PD)

Association of Parkinson's disease with occupational exposure to pesticides especially herbicides has been indicated in few studies with agricultural workers in rural areas. Two epidemiological studies conducted in Canada showed a significant linkage of PD between workers engaged in orchard, crop farming, grain farming and use of herbicides (72, 73). Both univariate and multivariate conditional regression analyses clearly implicated occupational use of herbicides as a significant risk factor for PD (73). Though there are few studies with negative results, the available evidences point to the herbicide, paraquat as the probable etiological agent for PD.

Lung Diseases

High levels of exposure to herbicide, paraquat causes irreversible chronic obstructive pulmonary disease leading to degenerative lung fibrosis, lethal lesions and finally loss of lung capacity and function. Paraquat, due to its cationic nature produces generation of superoxide anion and depletion of cellular reducing equivalents. A Columbian study revealed a significantly higher incidence of chronic obstructive pulmonary disease among agricultural workers exposed to paraquat (74).

A new disease of the respiratory tract named as "Vineyard sprayer's lung" was prevalent among vineyard sprayers in France engaged in spraying Bordeaux mixture containing copper sulfate (75). This disease is characterized by histiocytic granulomas with nodular and fibrohyaline scars having plenty of deposits of copper. These conditions either may regress or develop into diffuse pulmonary fibrosis (75).

Hyperthermia

Herbicides, dinitrophenol (DNP), dinitrocresol (DNOC) and pentachlorophenol and fungicide, chlorophenol disrupt ATP synthesis in the mitochondria as they function as uncouplers of oxidative phosphorylation, thereby enhancing oxygen consumption and heat production. DNOC acts as a cumulative poison and acute exposure alters heat regulatory system of the body resulting in overheating or hyperthermia. The clinical symptoms are high respiratory rate, increased heart beat rate, sweating, nausea and finally comatous condition. The symptoms appear at a rapid rate. Severe poisoning may lead to fatal hyperthermia, otherwise recovery generally occurs within 24 to 48 hrs.

Liver Diseases

Vineyard workers in Portugal exposed to copper sulfate having "Vineyard sprayer's lung" developed liver diseases with varied symptoms such as proliferation of kupffer cells, histocytic granulomata, fibrosis and in some cases angiosarcoma of the liver. Affected liver contained deposits of copper establishing etiological relationship between exposure to copper sulfate and liver lesions (76). Similar liver disease was also found to be associated with vineyard growers in France exposed to arsenic. Workers exposed to DDT showed hyperlipidemia.

Manufacturers of the insecticide, chlordecone (Kepone) with high levels in blood and adipose tissue showed induction of drug metabolizing microsomal enzymes in the liver characterized by proliferation of smooth endoplasmic reticulum (77). Factory workers engaged in the production of aldrin, dieldrin and endrin showed enhanced liver function on the basis of SGPT and SGOT tests (78). A similar induction of many enzymes resulting in rapid antipyrine clearance was observed in workers with chronic exposure to OCl pesticides. These changes do not cause permanent damage to the liver. These effects are found to be reversible when exposure to insecticide is discontinued. Induction of enzymes may bring about abnormalities related to hormonal imbalance.

Cardiovascular Diseases

Elevated levels of DDT and DDE in serum were found to be associated with hypertension, arteriosclerosis and possibly diabetes in due course, in occupationally exposed people in USA (79).

Immunosuppression

Occupationally exposed people such as green house workers, farmers, general population residing in agricultural regions and people using household pesticides showed changes in humoral factors responsible for immunity. Pesticides can cause impairment of lymphoid organs thereby disturbing the immune responses. Pesticide induced immunosuppression resulting in loss of resistance has been found to be the cause of development of allergy, respiratory diseases and general morbidity among pesticide users (60). An epidemiological study in Oregon, USA suggests the potential of Bacillus thuringiensis, a safe microbial pesticide, causing disease in immuno-compromised persons (80).

Vision

Chronic exposure to pesticide can cause eye impairment and visual abnormalities. Pesticide production workers showed loss of color discrimination and dark adaptation (60). OCl pesticide workers generally showed allergic conjunctivitis and retinal angiopathy and agricultural workers, with chronic exposure to

OCl and OP pesticides, become susceptible to myopia, retinal degeneration, damage of optic nerve, accommodation spasm and glaucoma (60). More studies are required to evaluate the association of visual impairment with chronic exposures to pesticides.

CONCLUSIONS

Steady increase in global consumption of pesticides may potentially enhance global pesticide contamination and subsequent ecological perturbations with direct and indirect effects on human health. Main pesticide contaminants are OCl pesticides which is exemplified by the fact that DDT residues have become ubiquitous in our environment. High burden of DDT residues in human tissues and newly born babies in many developing countries (81, 82, 83) pose a serious problem, though their possible health hazards are still uncertain. No serious health hazard is evident, even though OCl insecticides have been extensively used for the last half a century.

OP pesticides though easily biodegradable, due to its high toxicity has been the source of acute accidental poisoning in many developing countries. Such cases are avoidable, provided pesticide regulations are strictly followed.

Among the pesticides, phenoxy and triazine herbicides and inorganic arsenicals have been well established as etiological agents for human cancers. However, linkage of OCl, OP and carbamate insecticides with cancers needs further confirmation. However, their potential in this respect cannot be completely ruled out. It is apparent that carefully planned and statistical valid epidemiological studies are essential to clarify the status of deleterious effects of pesticide contaminants on human health.

Association of pesticide with general morbidity due to minor diseases/ pathological conditions also deserve attention so that such effects can be totally eliminated or minimized. In general, this can be possibly achieved by phasing out dangerous pesticides, strict adherence of pesticide regulations and an eco-friendly approach in pest control programs by minimizing their use and maximizing their beneficial effects.

ACKNOWLEDGMENTS

We are grateful to Prof. K. Muralidhar, Dr. Umesh Chandra Ojha, Dr. Suguna Yesodharan and Dr. Arunima Sahgal for the help rendered in the preparation of the manuscript.

REFERENCES

1. Marcello, Lotti. 1987. Production and use of pesticides, pp. 11-17. In: L.G. Costa, C.L. Galli and S.D. Murphy (Ed.) *Toxicology of Pesticides: Experimental, Clinical and Regulatory Perspectives.* Springer-Verlag.
2. Labonte, R.N. 1989. Pesticides and healthy public policy. *Can. J. Public Health.* 80:238-242.
3. Pillai, M.K.K. 1986. Pesticide pollution in soil, water and air in Delhi area, India. *Sci. Total Environ.* 55:321-328.
4. Fernando, R. 1990. National poisons information centre in a developing Asian country: the first year's experience. *Human Exptl. Toxicol.* 9: 161-164.
5. Levine, R.S. 1986. Assessment of mortality and morbidity due to unintentional pesticide poisonings. *WHO/VBC*/86.929.
6. Jayaratnam, J. 1985. Health problem of pesticide usage in the third world. *Brit. J. Ind. Med.* 42:505-506.
7. Lum, K.Y., Mamat, M.J., Cheah, U.B., Castaneda, C.P., Rola, A.C. and P. Sinhaseni. 1990. Pesticide research for public health and safety in Malaysia, the Philippines and Thailand, pp. 31-48. In: G. Forget, T. Goodman, and A. de Villiers (Ed.) *Impact of PEsticide Use on Health in Developing Countries,* IDRC, Canada, pp. 335.
8. Choudhury, A.W. and Levy, B.S. 1988. Morbidity estimates of occupational illness and injuries in Kenya: human and economic costs. Paper presented at the 9th Annual Medical Scientific Conference, 1-5 February 1988. Kenya Medical Research Institute and Kenya Trypanosomiasis Research Institute, Nairobi, Kenya.
9. Ferrer, A. and J.P.R. Cabral. 1989. Epidemics due to pesticide contamination of food. *Food Additives. Contaminants.* 6:S95-S98.
10. He, F., Wang, S., Lium, L., Chen, S., Zhang, Z. and J. Sun. 1989. Clinical manifestations and diagnosis of acute pytrethroid poisoning. *Arch. Toxicol.* 63:54-58.
11. Forget, G. 1990. Balancing the need for pesticides with the risk to human health. pp. 2-16. In: G. Forget, T. Goodman, and A. de Villiers (Ed.). *Impact of Pesticide Use on Health in Developing Countries.* IDRC. Canada, pp. 335.
12. Maroni, M. and Fait, A. 1993. Health effects in man from long-term exposure to pesticides. *Toxicol.* 78:1-180.
13. Zahm, S.A. and A. Blair. 1992. Pesticides and Non-Hodgkin's lymphoma. *Cancer Res.* 52:5485-5488.
14. Hardell, L. 1979. Malignant lymphoma of histiocytic type and exposure to phenoxyacetic acids of chlorophenols. *Lancet.* 1:55-56.
15. Hoar-Zahm, S., Blair, A., Holmes, F.F., Boysen, C.D., Robel., R.J., Hoover, R. and J.F. Fraumeni. 1986. Agricultural herbicide use and risk of lymphoma and soft-tissue sarcoma. *J. Am. Med. Assoc.* 256:1141-1147.

16. Hoar-Zahm, S., Weisenburger, D.D. and P.A. Babbitt. 1990. A case control study of non-Hodgkin's lymphoma and the herbicide 2,4-dichlorophenoxyacetic acid (2,4-D) in eastern Nebraska. *Epidemiol.* 1:349-356.
17. Cantor, K.P., Blair, A., Everett, G., Gibson, R., Burmeister, L.F., Brown, L.M., Schuman, L. and F.R. Dick. 1992. Pesticides and other agricultural risk factors for non-Hodgkin's lymphoma among men in Iowa and Minnesota. *Cancer. Res.* 52:2447-2455.
18. Markovitz, A. and W.H. Crosby. 1984. Chemical carcinogenesis: a soil fumigant, 1,3-dichloropropene, as possible cause of haematologic malignancies. *Arch. Int. Med.* 144:1409-1411.
19. Hardell, L. 1977. Soft tissue sarcomas and exposure to phenoxyacids: A clinical observation. *Lakartidningen.* 74:2753-2754.
20. Lynge, E. 1985. A follow-up study of cancer incidence among workers in manufacture of phenoxyherbicides in Denmark. *Brit. J. Cancer.* 52:259-270.
21. Coggon, D., Pannett, B., Winter, P.D., Acheson, E.D. and J. Bonsall. 1986. Mortality of workers exposed to 2 methyl-4-chlorophenoxyacetic acid. *Scand. J. Environ. Health.* 12:448-454.
22. Saracci, R., Kogevinas, M., Bertazzi, P.A., Bueno de Mesquita, B.H., Coggon, D., Green, L.M., Kauppinen, T., L'Abbe, K.A., Littorin, M., Lynge, E., Mathews, J.D., Neuberger, M., Osman, J., Pearce, N. and R. Winkelman. 1991. Cancer mortality in workers exposed to chlorophenols. *Lancet.* 338:1027-1032.
23. Eriksson, M., Hardell, L., Berg, N.O., Moller, T. and O. Axelson. 1981. Soft-tissue sarcomas and exposure to chemical substances: a case-referent study. *Brit. J. Ind. Med.* 38:27-33.
24. Eriksson, M., Hardell, L. and H.O. Adami. 1990. Exposure to dioxins as a risk factor for soft tissue sarcoma: a population based case control study. *J. Natl. Cancer. Inst.* 82:486-490.
25. Hardell, L. and M. Eriksson. 1988. The association between soft tissue sarcomas and exposure to phenoxyacetic acids: a new case-referent study. *Cancer.* 62:652-656.
26. Reif, J., Pearce, N., Kawachi, I. and J. Fraser. 1989. Soft-tissue sarcoma, non-Hodgkin's lymphoma and other cancers in New Zealand forestry workers. *Intern. J. Cancer.* 43:49-54.
27. Vineis, P., Terracini, B., Ciccone, G., Cignetti, A., Colombo, E., Donna, A., Maffi,. L., Pisa, R., Ricci, P., Zanini, E. and P. Comba. 1986. Phenoxy herbicides and soft tissue sarcomas in female rice weeders: a population-based case-referent study. *Scand. J. Work Environ. Health.* 13:9-17.
28. Infante, P.F., Epstein, S.S. and W.A. Newton. 1978. Blood dyscrasias and childhood tumors and exposure to chlordane and heptachlor. *Scand. J. Work. Environ. Health.* 4:137-150. IARC Monogr. on the evaluation o the carcinogenic Risk of Chemicals to Humans. World Health Organization.
29. Gold, E., Gordis, L., Tonascia, J. and M. Szklo. 1979. Risk factors for brain tumors in children. *Am. J. Epidemiol.* 109:309-319.

30. Sinks, T.H., Jr. 1985. N-nitroso compounds, pesticides and parental exposures in the work place as risk factors for childhood brain cancer. A case-control study. *Dissertation Abstr. Int.* 46:1886-B.
31. Davies, J.R., Brownson, R.C., Garcia, R., Bentz, B.J. and A. Turner. 1993. Family pesticide use and childhood brain cncer. *Arch. Environ. Contam. Toxicol.* 24:87-92.
32. Blair, A., Grauman, D.J., Lubin, J.H. and J.F. Fraumeni. 1983. Lung cancer and other causes of death among licensed pesticide applicators. *J. Natl. Cancer Inst.* 71:31-37.
33. Stubbs, H.A., Harris, J. and R.C. Spear. 1984. A proportionate mortality analysis of California agricultural workers, 1978-1979. *Am. J. Ind. Med.* 6:305-320.
34. Musicco, M., Filippini, G., Bordo, B.M., Melotto, A., Morello, G. and F. Berrino. 1982. Gliomas and occupational exposure to carcinogens: a case-control study. *Am. J. Epidemiol.* 116:782-790.
35. Muscico, M., Sant, M., Molinari, S., Filippini, G., Gatta, G. and F. Berrino. 1988. A case-control study of brain gliomas and occupational exposure to chemical carcinogens: the risk of farmers. *Am. J. Epidemiol.* 128:778-785.
36. Barthel, E. 1981. Increased risk of lung cancer in pesticide-exposed male agricultural workers. *J. Toxic. Health.* 8:1027-1044.
37. Mabuchi, K., Lilienfeld, A.M. and L.M. Snell. 1979. Lung cancer among pesticide workers exposed to inorganic arsenicals. *Arch. Environ. Health.* 312-320.
38. Levin, L.I., Zheng, W., Blot, W.J., Gao, Y. and J.F. Fruameni. 1988. Occupation and lung cancer in Shanghai: a case-control study. *Brit. J. Ind. Med.* 45:450-458.
39. Kauppinen, T.P., Partanden, T.J., Nurminen, M.M., Nickels, J.I., Hernberg, S.G., Hakulinen, T.R., Pukkala, E.I. and E.T. Savonen. 1986. Respiratory cancers and chemical exposures in wood industry: a nested case-control study. *Brit. J. Ind. Med.* 43:84-90.
40. Rothschild, H. and J.J. Mulvey. 1982. An increased risk for lung cancer mortality associated with sugarcane farming. *J. Natl. Cancer Inst.* 68: 755-760.
41. Deichmann, W.B. and W.E. MacDonald. 1976. Liver cancer deaths in continental USA from 1930-1972. *Am. Ind. Hyg. Assoc. J.* 495-498.
42. Hardell, L., Bengtsson, N.O., Jonsson, U., Eriksson, S. and L.G. Larsson. 1984. Aetiologial aspects on primary liver cancer with special regard to alcohol, oganic solvents and acute intermittent porphyria - an epidemiological investigation. *Brit. J. Cancer.* 50:389-397.
43. Axelson, O. 1986. A review of porphyria and cancer and the missing link with human exposure to hexachlorobenzene pp. 585-589. In: C.R. Morris and J.R.P. Cabral (Ed.). *Hexachlorobenzene: Proceedings of an International Symposium.* IARC, Lyon, pp. 691.
44. WHO-EHC (IPCS) 1977. Nitrate, Nitrites and N-Nitroso Compounds.

45. Alavanja, M.C.R., Blair, A., Merkle, S., Teske, J. and B. Eaton. 1989. Mortality among forest and soil conservationists. *Arch. Environ. Health.* 44:94-101.
46. Burmeister, L.F., Van Lier, S.F. and P. Isacson. 1982. Leukemia and farm practices in Iowa. *Am. J. Epidemiol.* 115:720-728.
47. Davies, J.M., Thomas, H.F. and D. Manson. 1982. Bladder tumors among rodent operatives handling ANTU. *Brit. Med. J.* 285:927-931.
48. Wang, H.H. and B. MacMohan. 1979. Mortality of pesticide appliators. *J. Occup. Med.* 21:741-744.
49. Alavanja, M.C.R., Blair, A., Merkle, S., Teske, and B. Eaton. 1988. Mortality among agricultural extension agents. *Am. J. Ind. Med.* 14:167-176.
50. Brownson, R.C., Chang, J.C., Davis, J.R. and J.R. Bagby. 1988. Occupational risk of prostate cancer: a cancer registry-based study. *J. Occup. Med.* 30:523-526.
51. Burmeister, L.F., Everett, G.D., Van Lier, S.F. and P. Isacson. 1983. Selected cancer mortality and farm practices in Iowa. *Am. J. Epidemiol.* 118:72-77.
52. Mills, P.K., Newell, G.R. and D.E. Johnson. 1984. Testicular cancer associated with employment in agriculture and oil and natural gas extraction. Lancet. I:207-210.
53. Wiklund, K., Dieh, J. and L.E. Holm. 1986. Testicular cancer among agricultural workers and licensed pesticide applicators in Sweden. *Scand. J. Work. Environ. Health.* 12:630-631.
54. Donna, A., Betta, P.G., Robutti, F., Crosignani, P., Berrino, F. and D. Bellingeri. 1984. Ovarian mesothelial tumos and herbicides: a case-control study. *Carcinogenesis.* 5:941-942.
55. Donna, A., Crosignani, P., Robutti, F., Betta, P.G., Bocca, R., Mariani, N., Ferrario, F., Fissi, R. and F. Berrino. 1989. Triazine herbicides and ovarian epithelial neoplasms. *Scand. J. Work. Environ. Health.* 15:47-53.
56. Rabello, M.N., Becar, W., De Almedia, W.F., Pigati, P., Ungaro, M.T., Murota, T. and C.A.B. Pereira. 1975. Cytogenetic study on individuals occupationally exposed to DDT. *Mutation. Res.* 28:449-454.
57. Edmundson, W.F. and J. Davies. 1967. Occupational dermatitis from naled. A clinical report. *Arch. Environ. Health.* 15:89.
58. Cam, C. and G. Nigogosyan. 1963. Acquired toxic porphyria cutanea tarda due to hexachlorobenzene. *J. Am. Med. Assoc.* 183:88-91.
59. Peter's, H.A., Cripps, D.J., Lambrecht, R.W., Erturk, E., Morris, C.R. and G.T. Byran. 1986. History and Geography of hexachlorobenzene poisoning in southeastern Turkey, pp. 131-137. In: C.R. Morris and J.R.P. Cabral (Ed.) *Hexachlorobenzene: Proceedings of an International Symposium.* IARC Scientific Publication, pp. 691.
60. Kaloyanova, F.P. and M.A. El. Batawi. 1991. *Human Toxicology of Pesticides.* CRC Press, pp. 196.
61. Vijverberg, H.P.M. and J. van den Berken. 1990. Neurotoxicological

effects and the mode of action of pyrethroid insecticides. Crit. Rev. Toxicol. 21:105-126.

62. Whorton, M.D., Krauss, R.M., Marshall, S. and T.H. Milby. 1977. Infertility in male pesticide workers. Lancet. ii:1259-1261.
63. Whorton, M.D. and T.H. Milby. 1980.Recovery of testicular function among DBCP workers. *J. Occup. Med.* 22:117-119.
64. Milby, T.H. and M.D. Whorton. 1980. Epidemilogical assessment of occupationally related, chemically induced sperm count suppression. *J. Occup. Med.* 22:77-82.
65. Lantz, G.D., Cunningham, G.R., Huckins, C. and L.I. Lipschultz. 1981. Recovery from severe oligospermia after exposure to dibromochloropropane. *Fertil. Steril.* 35:46-53.
66. Potashnik, G., Goldsmith, J. and V. Insler. 1984. Dibromochloropropane - induced reduction of the sex-ratio in man. *Andrologia.* 16:213-218.
67. Simic, B. and J. Kniewald. 1980. Effects of pesticides on the reproductive system. *Acta Pharm. Yugoslavia.* 30:59.
68. Johnson, M.K. 1977. Improved assay of neurotoxic esterase for screening organophosphates for delayed neurotoxicity potential. *Arch. Toxicol.* 37:113-115.
69. Savage, E.P., Keefe, J.A., Mounce, L.M., Heaton, R.K., Lewis, J.A. and P.J. Burcar. 1988. Chronic neurological sequelae of acute organophosphate pesticide poisoning. *Arch. Environ. Health.* 43:38-45.
70. Rosenstock, L., Keifer, M., Daniell, W.E., McConnel, R. and K. Claypoole. 1991. Chronic central nervous system effects of acute organophosphate pesticide intoxication. *Lancet.* 338:223-227.
71. Chakravarti, H. 1965. Polyneuritis due to DDT. *J. Ind. Med. Assoc.* 45:598. Hertzmann, C., Wiens, M., Bowering, D., Snow, B. and D. Calne. 1990. Parkinson's disease: a case-control study of occupational and environmental risk factors. *Am. J. Ind. Med.* 17:349-355.
72. Semchuk, K.M., Edgar, J.L. and R.G. Lee. 1992. Parkinson's disease and exposure to agricultural work and pesticide chemicals. *Neurology.* 42:1328-1335.
73. Arroyave, M.E. 1990. Pulmonary obstructive disease in a population using paraquat in Columbia. pp. 85-93. In: G. Forget, T. Goodman and A. de Villiers (Ed.) *Impact of Pesticide Use on Health in Developing Countries.* IDRC, Canada, pp. 335.
74. Pimentel, J.C. and F. Marques. 1969. "Vineyard sprayer's lung". A new occupational disease. *Thorax.* 24:678-688.
75. Pimentel, J.C. and A.P. Menezes. 1977. Liver disease in vineyard sprayers. *Gastroenterology.* 72:126-140.
76. Guzelian, P.S., Vranian, G., Boiland, J.J., Cohn, W.J. and R.V. Blanke. 1990. Liver structure and function in patients poisoned with chlordecone (Kepone). *Gastroenterology.* 78:206-213.
77. Jager, K.W., Roberts, D.V. and A. Wilson. 1970. Neuromuscular function

in pesticide workers. Brit. J. Ind. Med. 27:273-278.

79. Morgan, D.P., Lin, L.I. and H.H. Saikaly. 1980. Morbidity and mortality in workers occupationally exposed to pesticides. *Arch. Environ. Contam. Toxicol.* 9:349-382.
80. Green, M., Heumann, M., Sokolow, R., Foster, L.R., Bryant, R. and M. Skeels. 1990. Public health implications of the microbial pesticide Bacillus thuringiensis. An epidemiological study, Oregon, 1985-86. *Am. J. Public Health.* 80:848-852.
81. Nair, A. and M.K.K. Pillai. 1992. Trends in ambient levels of DDT and HCH residues in humans and the environment of Delhi, India. Sci. *Total Environ.* 121:145-157.
82. Nair, A., Dureja, P. and M.K.K. Pillai. 1992. Aldrin and dieldrin residues in human fat, milk and blood serum collected from Delhi. *Hum. Exptl. Toxicol.* 11:43-45.
83. Dureja, P., Nair, A. and M.K.K. Pillai. 1991. Aldrin, and dieldrin in maternal serum, cord serum and breast milk in human samples from Delhi, India. *Intern. J. Environ. Anal. Chem.* 44:253-256.

Environmental Contaminants, Ecosystems and Human Health. Edited by S.K. Majumdar, E.W. Miller, and Fred J. Brenner. © 1995, The Pennsylvania Academy of Science.

Chapter Seventeen

BIOLOGICAL CONSEQUENCES OF METALLIC POLLUTION IN AN URBAN ECOSYSTEM AS ENVIRONMENTAL CONTAMINANTS AND THEIR IMPACT ON HUMAN HEALTH

B. BHATTACHARYA* and S.K. SARKAR**

*Department of Metallurgical Engineering
Jadavpur University
Calcutta 700 032, India
and
**Department of Marine Science
Calcutta University, Calcutta 700 019, India

INTRODUCTION

Calcutta, the tenth most polluted city in the world with an area of 1,030 sq. km., is the hub of economic activities of the entire eastern and north-eastern India. The urban climates of this congested city are a complex assemblages of thousands of overlapping local climates. This microclimatic condition is deteriorating day by day to become alarming due to increasing generation and persistence of substances usually resulting from different activities of man. This poor environmental condition in the city is due to a variety of factors. These include unplanned growth and tremendous influx of population, poor socio-economic conditions, the location of industries within the city, an overloaded transport system with restricted roadways, and deteriorating air and water quality affecting the health of most of the city dwellers.

The present worsening picture can be rightly justified by the following facts:

(a) The unplanned expansion of the city in a linear orientation on the north-south axis along the bank of Hoogly river.

(b) Most populous city in the country and the fourth largest in the world. City's population is more than 12 million, with an average density 33,000 per sq. km. In North Calcutta, there are four wards where the population density is above 175,000 per sq. km. the highest in the world. The WHO standards for a healthy city is 2,500 per sq. km.

(c) The tree density in Calcutta is 21 per sq. km. against the International standard of 100.

(d) Growth of vehicular population (5 lakhs registered vehicles in 1992) with extremely inadequate motorable road space (only 6% of the total city area against average 10 - 15% in other metropolitan cities in the country).

(e) Existence of 11,516 legally registered factories in and around Calcutta, including 40 lead factories. Lethal residue of Pb along with Cd, Cu, Ni, As and Hg are a threat to the residents of this city.

(f) The suspended particulate matter (SPM) in Calcutta air was 1,000 micrograms per cubic meter in December 1993 (The Statesman 22 April, 1994), compared to 254.7 $\mu g/m^3$ in Athens, 361 $\mu g/m^3$ in Tehran and 22.8 $\mu g/m^3$ in London. The matter contained several toxic substances, including large amounts of the carcinogenic compound, benzopyrene, a polynuclear aromatic hydrocarbon. According to WHO, SPM beyond 90 $\mu g/m^3$ can pose a serious environmental problem..

The above-mentioned points clearly indicate the sordid picture of environmental degradation of Calcutta through industrial, vehicular, noise, and other types of pollution. Mr. M.N. Buch, the former Director of the National Institute of Urban Affairs, New Delhi, India has rightly warned that 'the urban ecosystem is in a crisis which will increase in geometric progression as urbanization accelerated and the availability of financial resources for urban development declines'. The present investigation aims to analyze the metal levels in blood of normal and disease-affected persons residing at various corners of this metropolitan city with the following objectives:

(a) To establish some clinical phenomenon with the abnormal level of particular metals.

(b) Monitoring of environmental pollution and occupational exposure.

NATURE OF TOXICITY

It is very difficult to demarcate toxic metals from essential ones. In fact, any metal could become toxic if ingested in sufficiently large amounts (Kneip, 1974). A toxic metal is one which is neither essential nor beneficial but exhibits a positive catastrophic effect on normal metabolic function even when present in small amounts, and may, at any time, be responsible for permanent disorders or manufacturing or organ systems leading finally to death (Browning, 1969).

The toxicity of a metal, as of any other substance, depends on a number of factors, such as (I) physical state (ii) nature of its compound (iii) mode of its entry into the human body and (iv) susceptibility of the exposure (Browning, 1969).

MATERIALS AND METHODS

Apparatus

A Perkin-Elmer Atomic Absorption Spectrophotometer (Model, 2380) equipped with BG corrector and MHS - 10 and Orion ion selective electrodes, Eppendrof automatic pipette were used for determining metallic contents of blood.

Reagents and Glassware

Nitric acid (suprapur), perchloric acid 70% GR.E. Merck, speckpure metals of Cu, Hg, Se and Te from Johnson and Matthey, UK., standard solutions of K, Li and Na from Sigma, USA have been used throughout the experiment.

All the glasswares and plasticwares were cleaned by soaking for 72 hours in 25% (v/v) nitric acid. After cleaning, all containers were thoroughly rinsed with water and dried in a dust-free environment.

Collection and Preparation of Samples

Samples of various blood were taken from high risk and normal healthy persons with disposable PVC syringe under hospital conditions using sodium heparine as additive. The samples were preserved at -4°C and analyzed within 24 hours of collection.

The mineralization of the blood samples was done by taking 5 ml of whole blood and 5 ml of nitric acid. After digestion on water bath for 30 minutes the samples were treated with 2 ml of perchloric acid and digested again for 30 minutes. The samples were then diluted to 10 ml with 15% nitric acid solution.

Copper, cadmium, lithium, sodium, potassium and zinc were analyzed with flame atomic absorption spectroscopy using acetylene as fuel and air as oxidant using suitable dilution.

Mercury, selenium and tellurium were determined by cold vapor technique and hydride technique respectively using 3% $NaBH_4$ (E. Merck) in 1% NaOH (E. Merck) solution. High purity argon gas was used as purge gas. Peak areas were calculated using integrator for quantification of mercury, selenium and tellurium.

Sodium and potassium were determined by diluting the sample with 1,000 ppm lithium chloride to control ionization interference. Lithium was similarly determined by diluting with 1,000 ppm potassium chloride solution.

LEAD IN BLOOD

The most important constituent emitted through automobile exhaust gases are tiny particles of lead compounds. These are present in petrol in both organic and inorganic form. Organic lead used as an anti-knock reagent in gasoline is emitted into the atmosphere as lead oxide when gasoline is burned in automobile engines. A major portion of these Pb compounds remain airborne for a long time and enter the human system through inhalation. Although the major route of entry of lead into human beings is through food and drink, contributions of airborne lead to the people living near congested areas is also significant. While the permissible limit is 30 μg/100 ml of blood, an average citizen of this city has a lead level between 10 - 40 μg/100 ml of blood. High lead exposure is required to bring about clinical physiological symptoms of lead toxicity. A recent survey reveals that respiratory as well as kidney diseases have increased significantly in Calcutta owing to a sharp rise in the pollution (The Statesman, May 1, 1994). It is reported that the children having lead levels of 30 - 40 μg/100 ml possess low IQ, irritability and hyper-activity in comparison to the children bearing lower levels of blood lead.

MERCURY IN BLOOD

The mercury content of human blood both in fish eating cases and pure vegetarians are shown in Table 1a and 1b respectively. Average mercury content in

TABLE 1A

Concentration of mercury in the blood samples of normal people consuming fish available from local markets in Calcutta (expressed in μg/100 ml)

RSD:4.5%							DL: 0.00047 μg/ml
Sample No.	Sex	Age	Hg concentration	Sample No	Sex	Age	Hg concentration
1	M	52	0.85	19	M	50	0.75
2	M	48	0.58	20	F	46	0.29
3	M	18	0.62	21	M	27	0.80
4	F	40	0.82	22	M	10	0.49
5	M	58	0.95	23	M	35	0.64
6	M	50	0.68	24	M	49	0.48
7	M	18	0.72	25	M	42	0.88
8	F	31	0.98	26	F	50	0.19
9	F	30	0.88	27	M	48	0.93
10	M	58	0.48	28	M	35	0.85
11	M	38	0.80	29	M	30	0.16
12	F	42	0.82	30	M	31	0.40
13	M	32	0.69	31	M	25	0.55
14	M	65	0.55	32	M	48	0.95
15	M	19	0.68	33	M	55	0.90
16	M	35	0.19	34	M	35	0.19
17	F	50	0.82	35	F	15	0.75
18	M	55	0.16				

RSD: Relative standard deviation for 10 determinations DL: Detection limit of metal the 35 blood samples analyzed for fish eaters was 0.64 μg/100 ml while average mercury content in blood samples of 22 vegetarian people was 0.20 μg/100. The ratio of increase in mercury content in blood of fish eaters showed 3.2 times greater than non-fish eaters. The results warrant that the release of mercury through food stock (mainly fishes) has to be monitored and there should be legislation establishing a minimum for the heavy metals; especially for mercury which is a severe neurological toxic metal.

CADMIUM IN BLOOD

Table 2 depicts the Cd, Cu, Se, Te, Na, K and Li values of various disease-affected people of Calcutta. In all cases Cd was below the flame atomic absorption detection limit. The presence of Cd has been reported in fruits, vegetables, food grains and rice irrigated with Cd contaminated water coming from metal mines upstream (Nogawa et al., 1978). Cd ingestion by human beings may also occur through drinking Cd contaminated water and drinks. The daily intake of Cd through food and beverages, in absence of pollution, has been found to be 10 μg to 50 μg/day for a 70 kg man in U.K., 10-20 μg/day in Sweden and comparatively higher 20-60 μg/day by an adult (20 year old) in the USA (Mahaffey, 1975).

Until now it apparently seems that the environment of Calcutta is not getting polluted with Cd. Petering et al., (1974) established that when the amount of Zn is marginal, toxic effects of Cd increase and signs of Zn deficiency occur.

COPPER IN BLOOD

The level of Cu in blood, and in plasma, is affected principally by diet and diseases. It is also under hormonal influence (Underwood, 1962). Under normal

TABLE 1B

Concentration of mercury in the blood samples of normal vegetarian people in Calcutta (expressed in μg/100 ml)

RSD:4.5% DL: 0.00047 μg/ml

Sample No.	Sex	Age	Hg concentration	Sample No	Sex	Age	Hg concentration
V_1	M	40	0.21	V_{12}	M	38	0.15
V_2	M	28	0.28	V_{13}	M	40	0.18
V_3	M	42	0.19	V_{14}	M	31	0.04
V_4	F	35	0.31	V_{15}	M	29	0.13
V_5	M	30	0.49	V_{16}	F	30	0.20
V_6	M	28	0.01	V_{17}	M	32	0.18
V_7	M	26	0.18	V_{18}	F	28	0.27
V_8	M	32	0.09	V_{19}	M	30	0.20
V_9	M	35	0.33	V_{20}	M	26	0.33
V_{10}	M	30	0.40	V_{21}	F	53	0.15
V_{11}	F	27	0.12	V_{22}	M	44	0.19

RSD: Relative standard deviation for 10 determinations DL: Detection limit of metal conditions ingested Cu is absorbed from the stomach and upper part of the small intestine. The element passes into the blood stream and is relatively loosely bound, probably to albumin, and passes to the liver (Hill, 1977). The Cu content of the whole blood for human beings is 0.89 - 0.99 μg/100 ml. Wilson's disease represents a special case of Cu toxicosis. This condition is an inborn error of human metabolism in which the liver, kidney, brain and cornea accumulate large amounts of Cu.

The present study shows a comparatively elevated value of Cu in cases of hypertensive and diabetic patients. Mckenzie and Kay (1973) also presented a report of increased Cu in urine of hypertensive patients.

The protective effects of Cu against Cd poisoning are not as conspicuous as those of Zn. It has been reported that Cu with appropriate amounts of Zn and Mn significantly decreased the accumulation of Cd (Fox, 1976). These findings have been interpreted to show that together they protect against the adverse effects of Cd.

TABLE 2
Metal levels in disease affected persons
(expressed in μg/ml)

Disease concerned	Sex	Age	Cd	Cu	Se	Te	Na	K	Li
RSD			1.5%	1.2%	4.5%	4.5%	1.2%	1.2%	1.2%
DL (μg/ml)			0.0028	0.0077	0.0005	0.0005	0.02	0.02	0.02
HTN	M	74	BDL	0.10	0.02	0.12	3100	140	4.10
HTN	F	50	BDL	0.10	0.01	0.14	2900	130	4.00
HTN	M	62	BDL	0.10	0.01	0.12	3200	145	4.20
HTN	M	42	BDL	0.80	BLD	0.19	3300	142	4.30
HTN	M	55	BDL	0.40	0.01	0.30	3350	150	5.20
DM	M	60	BDL	0.12	0.04	0.10	3050	110	1.50
DM	M	75	BDL	0.80	0.02	0.06	2900	135	4.30
DM	M	45	BDL	0.10	0.02	0.10	3100	142	4.10
DM	M	16	BDL	0.80	0.04	0.18	2800	135	4.00
DN	M	70	BDL	0.80	0.28	0.12	3400	150	4.80
DN	M	45	BDL	0.80	0.02	0.14	3200	145	4.50
DN	F	50	BDL	0.80	0.01	0.10	2900	135	3.80
DN	M	63	BDL	0.80	0.02	0.16	3100	140	4.20
DN	F	40	BDL	0.16	0.44	0.12	3000	145	4.10
DN	M	44	BDL	0.80	0.20	0.18	3200	150	4.50
CN	M	25	BDL	0.14	0.10	0.12	3100	140	4.50
CIN	M	60	BDL	0.10	0.50	0.14	3300	145	5.10
CGN	F	35	BDL	0.80	0.12	0.10	3000	135	4.20
CGN	M	42	BDL	0.80	0.01	0.22	3100	140	4.50
CGN	M	50	BDL	0.80	0.04	0.42	3200	1.45	4.80
CGN	M	40	BDL	0.80	0.01	0.12	3300	150	4.40
PK	F	60	BDL	0.10	0.01	0.04	3100	145	3.80
RPGN	M	47	BDL	0.80	0.02	0.06	3200	140	4.20
Gout	M	47	BDL	0.60	0.10	0.40	3100	140	4.50

BDL - Below detection limit. RSD - Relative standard deviation for 10 determinations. DL - Detection limit of metals. HTN - Hypertension; DM - Diabetic mellitus; DN - Diabetic nephritis. RPGN - Rapidly progressive glomerular nephritis. PK - polycystic kidney. CGN - Chronic glomerular nephritis. CN - Chronic nephritis. CIN - Chronic intestinal nephritis.

SELENIUM IN BLOOD

Shamberger et al., (1973 a, b) measured blood Se in normal males and females in different age groups and in patients with cancer and other diseases. Normal values for 41 males and females were 22.9 ± 3.52 (SD) μg/100 ml. Patients with gastrointestinal cancer or metastases to gastrointestinal cancer or metastases to gastrointestinal organs had significantly lower ($p < 0.001$) values. Patients with Hodgkins disease, cirrhosis and hepatitis also had lower blood selenium. In carcinocoma of the breast, most blood cancers, Crohn's disease, other gastrointestinal disorders, certain sarcomas, lipomas, most patients with rectal cancer, and patients with several other diseases, Se values were normal.

The present study reveals normal and below normal values of Se in the whole blood of different patients excepting a few cases with higher values showing (Table 2). It was suggested by Shamberger et al., 1973a that measurements of blood Se may be consistent with epidemology studies.

Because of the striking inhibitory effects of Se in animals against a variety of carcinogens and the possibility that Se at higher levels might be enhancing carcinogenesis, it is intriguing to determine whether Se occurrence might be correlated with human cancer mortality. Shamberger and Frost (1969) described a very high negative correlation ($R = -0.96$; $p < 0.001$) between the Se level of the blood of adult males in the United States and human cancer death rates. This correlation was based on the values for 10 cities or counties with populations secured by Allway et al., (1968) with normal blood Se values ranging from 10 to 34 μg/100 ml.

TELLURIUM IN BLOOD

The use of Te and its derivatives is considered minor and not hazardous in terms of direct industrial use exposure. Although Te and its compounds possess low orders of acute toxicity, relatively little is known regarding their chronic effects in low doses (Fishbein, 1977).

SODIUM, POTASSIUM AND LITHIUM IN BLOOD

Sodium, potassium and lithium in blood serum were analyzed in 24 disease affected persons to correlate cation transport of Na^+ - K^+ pump, Na+ - K^+ cotransport system and Na^+ - Li^+ countertransport system. No significant correlation was found between the cation transport pathways and trace metal contaminants. Normal serum level of Na, K and Li is 3151-3381 μg/ml, 136-195 μg/ml and 4.2-9.1 μg/ml respectively. The results obtained on disease affected persons reveals normal and below normal values. Further research is necessary in order to establish a conclusive statement about accumulation of toxic metals in hair and nails, and effect on ion transport pathways.

REMEDIAL MEASURES

At the end of the 20th century, the people of Calcutta must be aware of its alarming microclimatic condition. The time has come for serious thinking to reduce the level of automobile emissions as well as industrial effluents in this urban ecosystem. This can be effectively achieved through the following measures:

a. Strict enforcement of law for proper maintenance of vehicles and treatment of sewage and industrial wastes.

b. Create awareness among the people to protect the environment; different scientific mass media should be installed to raise public consciousness about the ill effects of pollution.

c. Introduction of more scientific traffic management and emission control systems.

d. Decongestion of the busy streets and intersections.

e. Alternate day use of private cars should be instituted.

f. Implementation of pollution control devices.

ACKNOWLEDGMENTS

We express our gratitude to Dr. K.J. Goswami for collection of blood samples and University Science Instrumentation Centre, Jadavpur University for providing laboratory facilities.

REFERENCES

Allway, W.H., Kubota, J., Losee, F. and M. Roth. 1968. Selenium, molydenum and vanadium in human blood. *Arch. Environ. Health.* 16:342-350.

Browning, E. 1969. *Toxicity of industrial metals* (Ed.), Butter Worth, London.

Fishbein, L. 1977. Toxicology of selenium and Tellorium. In *Advances in Modern Toxicology,* Vol. 2, R.A. Goyer and M.A. Mehlman (Eds.), pp. 191-240, John Wiley and Sons.

Fox, M.R.S. 1976. In *Trace elements in human health and disease,* Vol. II. A.S. Prasad (Ed.) pp. 401-406, Academic Press, N.Y.

Hill, C.H. 1977. Toxicology of Copper. In *Advances in Modern Toxicology.* Vol. 2, R.A. Goyer and M.A. Mehlman (Eds.), pp. 123-127.

Kenip. T.J., Underwood, E.J. and R.E. Hazen. 1974. Metals in nature. In *Preventive medicine and metal toxicity* (Eds. J.S. Edmonds and B.W. Raston), Academic Press.

Mckenzine, J.M. and D.L. Kay. 1973. *N.Z. Med. J.*, 78:68-70.

Mahaffey, K.R. 1975. *Environ. Health Perspec.*, 12:68.

Nogawa, K., Ishizaki, A. and S. Kawano. 1978. *Environ. Res.*, 15(2), 185.

Petering, H.G., Johnson, M.A. and K.L. Stemmer. 1971. *Arch. Environ. Health.* 23:93-101.

Perry, H.M. Jr. and M.W. Erlanger. 1974. *J. Lab. Clin. Med.,* 83:510-515.

Shamberger, R.J. and D.V. Frost. 1969. Possible protective effect of selenium against human cancer. *Can. Med. Assoc. J.,* 100:682.

Shamberger, R.J. and C.E. Willis. 1971. Selenium distribution and human cancer mortality. *Crit. Rev. Clin. Lab. Sci.*, 2:211-221.

Shamberger, R.J. Rukovena, E., Longfield, A.K., Tytko, S.A., Doedhar, S. and C.E. Willis. 1973b. Antioxidants and cancer. I. selenium in the blood of normal and cancer patients. *J. Natn. Cancer Inst.,* 50:863-870.

Shamberger, R.J., Tytko, S. and C.E. Willis. 1973a. Selenium in the blood of normals, cancer patients and patients with other diseases. *Clin. Chem*., 19:672-675.

Environmental Contaminants, Ecosystems and Human Health. Edited by S.K. Majumdar, E.W. Miller, and Fred J. Brenner. © 1995, The Pennsylvania Academy of Science.

Chapter Eighteen

WATER AND HEALTH PROBLEMS IN PUERTO RICO

SONIA I. ARBONA
Department of Geography
Shippensburg University
Shippensburg, PA 17257

INTRODUCTION

At the global scale the predominant health impacts resulting from the interaction between water and health have followed a sequential pattern in which water pollution concerns have shifted from biodegradable organic wastes to contamination with toxic chemicals. This change has been associated with changes in demographic patterns and economic activities. For the most part, with higher economic development comes the virtual elimination of the microbiological type of pollution from potable supplies. In turn, water contamination with synthetic organic chemicals emerges and replaces microbiological pollution as the most pressing water contamination problem. This is, however, an oversimplified pattern since economically advanced societies must rely on a sophisticated water purification system that has to work properly all the time to avoid contamination of potable supplies (1). Occasional failures in the system have the potential to expose the population to sub-standard water quality. In addition to the biological pollution, less developed societies have chemical contamination of their water, usually resulting from development strategies in which new and foreign technologies have been adopted.

Human health and development are related in complex interactions, and water related health problems in Puerto Rico offer a demonstration of this association. In the last 30 years, the growth of Puerto Rico's economy and population has been among the most rapid in the Caribbean region. This growth was initiated largely by a program of industrialization to convert the island's economy from an agricultural to an industrial base. Urban and industrial growth occurred rapidly, but a lag arose in the provision of the infrastructure necessary to allow for a balanced expansion. Moreover, although a legal framework to control pollution exists, the financial resources for adequate implementation and enforcement of the laws is lacking. Today water quality analyses in Puerto Rico have indicated pollution of ground and surface waters (2, 3, 4). Contributions to water quality deterioration come from practices at the individual, community and industrial level and cover a wide spectrum of water contaminants.

The increasing contamination of surface waters make them hazardous to human health (5). Groundwater, a major resource for water supply, must often be utilized to satisfy demand. However, percolation of pollutants into aquifers threatens contamination of this water source, too (6). The public health implications of these high risk situations are acute and long term.

After describing the general characteristics of Puerto Rico briefly, this chapter proceeds to identify the current nature and extent of water pollution on the island. A discussion on the impacts of development approaches on the quality of the water and the infrastructure for the treatment, purification and supply is then presented. Although very little epidemiological evidence exists on the connection between water quality and human health, the current and potential health threats associated with poor water quality are also discussed.

PUERTO RICO - GENERAL CHARACTERISTICS

The island of Puerto Rico is located at about 18°N and 66°W. The total land area is approximately 3,470 square miles including Vieques and Culebra, its two offshore islands. Average population density is a little over 1,000 persons per square mile; however, population is concentrated in the four major urban areas of San Juan, Ponce, Mayagüez, and Arecibo. In general, the coastal plains, especially the northern one, have the great majority of the population (Figure 1). It is in the coastal plains where most of the urban and industrial growth has occurred. The population in 1990 was 3.5 million and estimates of the Puerto Rico Planning Board indicate that it will increase to approximately 3.9 million by the year 2,000. The quality of the waters in Puerto Rico has been affected substantially by the island's population growth.

Although small in size, the topography and geology of Puerto Rico are diverse and the water resources abundant. The diversity is caused mostly by the topographic relief and its effect on rainfall distribution more than by any other factor. A complex central ridge system forms a barrier to the northeast trade winds with the effect of producing greater precipitation in the northern part of the island, which averages about 80 inches per year compared to 35 inches in the southern part.

The principal rivers, of which there are 100, flow from the central ridge to the sea. There are no natural lakes, but 28 reservoirs have been constructed for water supply, irrigation and hydroelectric power. Surface water is the major source of freshwater and provides close to three-quarters of the need for this natural resource (7).

The aquifer system is one of the prime sources of water supply and constitutes a substantial proportion of the total water used in Puerto Rico. In the last 25 years, development of groundwater resources for public supply and industrial uses has increased significantly. The demand and dependence on groundwater varies throughout the island, but the northern limestone and the southern alluvial aquifers are important water resources for their regions. In the northern plains, the limestone aquifer provides 20% of the water potable supply to north-coast communities (8).

THE EXTENT OF WATER CONTAMINATION

The biennial water-quality reports to the Environmental Protection Agency (EPA) and the U.S. Congress produced by the Puerto Rico Environmental Quality Board (PREQB) provide a general scenario of the quality of the waters in Puerto Rico. In this report, water quality is evaluated according to its designated use: swimming (primary contact recreation), secondary contact recreation (such as fishing and boating), drinking water and preservation of aquatic life. In the 1992-1993 report, the majority of the surface waters evaluated were classified as not supporting the designated uses (Table 1). Moreover, a substantial proportion of those waters that supported the intended use are threatened by pollution. The principal causes for less than full support were organic enrichment, pesticides, metals, and pathogen indicators. Among the sources of pollution

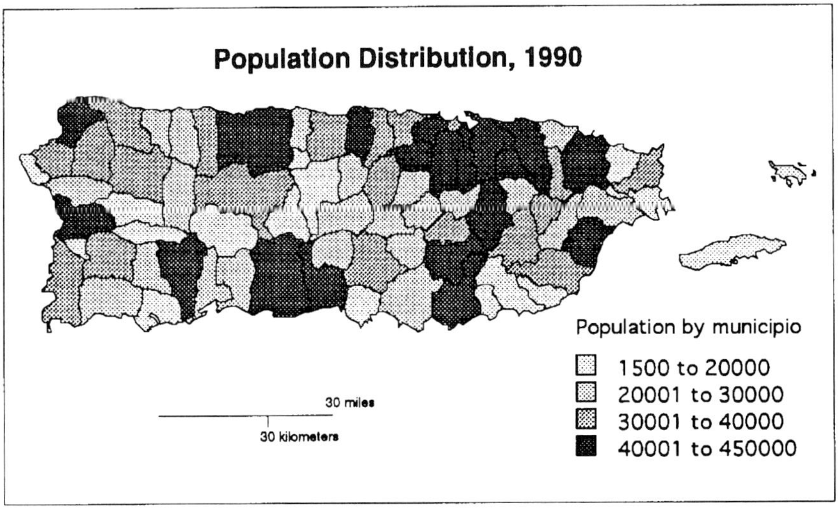

FIGURE 1. Population distribution in 1990 by municipio, the local government area. (Source: U.S. Bureau of the Census, 1990).

identified most often were land disposal (especially wastewater), urban runoff, agriculture (especially livestock enterprises), and municipal point source discharges from sewage treatment plants (9).

Regarding groundwater, the major sources of contamination were septic tanks, livestock enterprises, agriculture, storage tanks and landfills. In a groundwater priority system included in the biennial document, the presence of pollutants and the magnitude of the contamination in groundwater supplies are indicated. For example, the presence of pollutants in the groundwater reservoir is reported as being detected in 37 of the 78 municipalities of Puerto Rico (10). Wells monitored for organic and inorganic contaminants did not exceed drinking water standard, but pollutant levels were in high enough concentrations to be detected. It is also reported that due to contamination problems, the Puerto Rico Aqueduct and Sewer Authority (PRASA) stopped the pumping of drinking water at various wells tested. Well closings have become a repetitive action taken by the water authority in order to avoid exposure of the population to contaminants. Typically, another well is sunk in order to satisfy demand, but migration of contaminants, although poorly understood, is suspected to be taking place.

After appropriate treatments, the waters of Puerto Rico meet the drinking water standards established by the Puerto Rico Department of Health and the PREQB. However, the nature and extent of water contamination has changed over the years, making it increasingly difficult to supply potable water.

FROM SCHISOSOMIASIS TO CANCER RISK

During the first four decades of this century, agriculture, mainly the cultivation of sugarcane, tobacco and coffee, was the predominant economic activity in Puerto Rico. The development of large projects to acquire water supply that began early in this century made the growth of the sugarcane industry possible, particularly in the drier southern coastal area. At that time, the extensive use of irrigation had an unexpected side effect of prompting the emergence of high infection rates of schistosomiasis, a tropical disease caused by the schistosome parasite and spread by infected waters.

S. mansoni, the species found in Puerto Rico, is transmitted in feces. When human waste reaches water, the water-dependent transmission cycle is initiated.

TABLE 1
Percentage of Surface Waters Supporting Designated Use

Designated Use of Surface Water Surface Water	Drinking Water		Swimming		Secondary Contact		Aquatic Life	
	S	T	S	T	S	T	S	T
Rivers and streams	8.5	9.6	13.6	7.2	19.2	6.6	8.1	9.4
Lakes	35.3	22.3	45.5	20.2	80.3	9.3	7	32.8

S = supporting T = threatened

People do not need to drink contaminated water to contract schistosomiasis; all is needed is water contact like swimming, bathing, doing the laundry, or working in close contact with contaminated water.

Although the disease became distributed widely on the island during the first decades of this century, prevalence was spatially variable. By the 1930s, the southeastern part of the island was considered the most important focus of schistosomiasis, and the presence of this focus was attributed to exposure among irrigation workers (11). Efforts to control the disease focused on the control of the intermediary snail necessary to complete the schistosome life cycle. These efforts were reinforced by the program started in 1945 by PRASA with the objective to improve sewer systems and the provision of safe water supplies. It has been suggested that the migratory movement of population within Puerto Rico after World War II has implications for the control of schistosomiasis as well (12). This was attributed to the principal migratory pattern being movement to the capital city of San Juan and other major cities characterized by being well-sewered urban areas. These urbanized areas provided the infrastructure for the proper disposal of human wastes.

Migration to urban centers was promoted by a shift toward industrialization. In the 1950s, during the first years of industrialization, firms were attracted to Puerto Rico by lower labor and production costs, and later by tax exemption. In those early years, emphasis was placed upon labor-intensive industries such as food processing and related activities including rum distilling, textiles, apparel, and tobacco products. The manufacture of ceramics and glass was also important. These industries were established in the principal cities, especially in San Juan and neighboring municipalities.

With rapid growth, the provision of facilities and necessary services for the new city inhabitants became difficult, including the provision of safe drinking water supplies and the adequate disposal and treatment of wastewaters. In an attempt to divert the flow of population toward the capital, a program of industrial decentralization was adopted. The program was based on providing better incentives to those industries that were to be located in less developed municipalities. Nevertheless, urban and industrial concentrations persisted in and around San Juan.

By the mid 1960s, the establishment of chemical and petrochemical industries exhibited rapid expansion. These activities are more capital-intensive and require more specialized labor but fewer number of workers. The emphasis on this type of capital intensive industry continued although its success was not as much as had been hoped, mainly due to the lack of a cheap source of petroleum on which it was dependent. The oil crisis in the 1970s demanded still another shift in industrial development strategies.

The latest stage of industrialization has been dominated by pharmaceutical, chemical, and electronic industries. The establishment of industries in Puerto Rico continues to be attractive due to duty-free entry into the United States and tax exemption on the island for 10 to 25 years, depending on location. The concentration of these industries in the northern coastal plains is pronounced. The abundance of underground water reservoirs, used intensively by the chemical industry, has been an attractive location feature. Population concentrations in this area have also increased in association with employment opportunities, services and general infrastructure development.

Today, the north coast limestone region, which occupies 20% of the island's surface area, is inhabited by 48% of its total population and contains 55% of the manufacturing plants. Numerous wells have been drilled for industrial water supply near the coast and for public water supply further inland. This has put the aquifer under great pumping stress, as well as being threatened by industrial discharges, pollution by landfills, and saltwater intrusions.

The most evident results of the impact of industrial and urban growth have been a deteriorated rural landscape, the development of industrial nuclei, and a highly commercialized urban landscape. Water pollution has become a by-product of this process and constitutes the most serious environmental problem of the island.

Fecal coliforms, nitrates, and synthetic organic compounds with potential carcinogenic effects are among the contaminants detected in the aquifers of the northern and southern coasts. The sources of synthetic compounds are past agricultural practices, especially the use of pesticides such as DDT and dieldrin. Solvents used in industrial activities such as trichloroethylene and tetrachloroethylene reported in above standard concentrations are also implicated. Halogenated volatile compounds such as these may persist in the underground environment which makes them a constant threat to groundwater supplies. It may take years for polluted water in one location to appear in another because groundwater generally moves slowly through an aquifer, rendering monitoring and detection difficult.

Nitrate pollution of groundwaters, the cause of closure of many wells, might be indicative of pollution through disposal of sewage or organic waste. Typically, rural areas are exposed to diverse sources of nitrate; among them are nitrate fertilizer applications, barnyards, leaking septic tanks and cesspools. However, the sources of organic nitrogen include not only natural materials, but also numerous organic forms of the element likely to be associated with industrial processes. The main concern with this type of pollution is that in the gastrointestinal tract of infants, nitrates metabolize to nitrates, which can reduce the oxygen carrying capacity of the red blood cells. But there is no public health data to substantiate this health threat to the infant population of Puerto Rico.

Although bacteriological pollution has been reduced substantially and chemical pollution has become the most serious concern to water quality, the first one has nonetheless persisted as a public health threat and remains as the principal surface water pollution problem. Bacteriological pollution has been associated in the rural areas with leaking cesspools, barnyards and direct discharge of body wastes in the environment. Moreover, those areas, rural or urban, connected to a sewerage system now face its deterioration and the accompanying contamination of waters and potable supplies. Inadequate sewage treatment plants are contaminant sources themselves, and the aging water purification plants are overburdened by the increasing demand and the contamination of waters.

The Water System Infrastructure

As mentioned above, during the 1940s and 1950s, PRASA extended the aqueduct and sewer system to almost all of the island. Today, 100% of the urban areas and about 75% of the rural areas are covered by this system. However, there are serious problems with the sewage treatment plants. In 1978, the majority of the plants that were evaluated by a monitor from the EPA did not comply with the National Pollutants Discharge Elimination System (NPDES) permits, and a Federal Court Order in which the United States Supreme Court assumed jurisdiction over those treatment facilities was issued. Inclusion in the court order forbids new connections to the system and involves a heavy fine.

In the first report by the monitor, only three sewage treatment plants were working properly, seven were marginally satisfactory, and seventy-eight were evaluated as unsatisfactory. Over the years, plants operating properly have been taken out of the court order list; in addition some plants have been eliminated by regional facilities. In 1994, a total of 41 out of 78 sewage treatment plants remain under the court order. Moreover, periodical evaluations of sewage treatment plants to identify problems and observe progress in them include only those under the federal court order. Thus, it is not to be assumed that plants not included in the list are necessarily in compliance with environmental laws (Dr. Rafael Ríos, EPA monitor, pers. com., 5-2-94).

As a result of the sub-standard performance of sewage treatment plants, water bodies receive discharges that reduce their quality substantially. This in turn poses a threat to health when plants used for the purification of potable supplies are overburdened. They must work above design capacity, particularly in urban and industrial areas, and purify waters of decreasing quality. Ruptures and other faults in the water pipe system further hamper the quality of the potable supply delivered, even if the processing is of high quality at the plant.

Examination of geographical patterns of bacteriological contamination of waters are associated strongly with urban areas. In a report by the United States Geological Survey (USGS) in 1982, it is indicated that four wells in the aquifer of the heavily populated northern plain were contaminated with fecal bacteria 40 to 100 percent of the time (13). This study also indicated that the highest concentrations of fecal coliforms and streptococcal bacteria occur in surface waters of the San Juan metropolitan area which uses surface water reservoirs rather than groundwater as the main source of its drinking supply.

The limited epidemiological connection between water pollution and health has been examined using data reported in 1987 by the Puerto Rico Department of Health on the number of cases of acute gastroenteritis. An interpretation of these data concluded that the San Juan metro area has water supplies which are the best protected from bacteriological contamination because it is serviced by more sewage treatment plants than other areas. On the other hand, the non-metropolitan northern limestone region, heavily dependent on groundwater, had a mean gastroenteritis rate equal to the rest of Puerto Rico. This suggest a general lack of aquifer protection against biological contamination in that region (14).

According to public statements made by head officials of the Puerto Rico Department of Health, the weekly average number of cases of gastroenteritis in Puerto Rico fluctuates around one thousand, thus a substantial increase of cases above that number is required for official recognition and reporting of an outbreak. Despite this restriction and the fact that cases of gastroenteritis treated by private physicians do not reach the register of the department of health, the health agency has reported gastroenteritis as the most frequent transmissible disease on the island during recent years. Laboratory analyses and epidemiological studies have suggested that there is an association between the contamination of potable water with fecal coliforms and its improper chlorination with disease transmission in many of the outbreaks. Transmission can take place by consumption of food or drink or by person-to-person contact, but cumulative evidence points toward contaminated water as a major culprit in the island-wide outbreaks.

Giardiasis, another water-related infection producing gastrointestinal symptoms, has been mentioned as another possible cause in the gastroenteritis outbreaks. However, the position of PRASA officials is that *Giardia lamblia,* the pathogenic agent of giardiasis, is resistant to chlorination, and therefore its presence in potable supplies does not imply an inefficient water purification system. Nevertheless, the cysts of this protozoan can be removed through the process of coagulation and sedimentation at the treatment plant, provided the system is working properly at the volume or capacity for which it was designed.

In Puerto Rico many of the existing water purification plants work 10 to 20% above intended capacity. Purification plants can withstand this overload, but under those circumstances, additional monetary and energy resources are required to preserve the quality of the water served for public consumption and to maintain the equipment to function properly. In addition to the increasing overload, many of the purification plants have been in use for decades; when examination in combination with the frequent and multiple cases of pipe ruptures, the scenario of an antiquated system that needs replacement and reparation is evident. Although improvement and expansion of the water purification and distribution system are needed urgently, the crisis in the sewage treatment plants that became apparent in the 1980s required the capital investment of PRASA in that sector of its infrastructure. This action diverted the attention and monies from the system of purification and distribution of the potable supply.

The cumulative sum of the fine paid by PRASA since 1980 surpasses $3,000,000, and with more than half of the sewage treatment plants still not complying with environmental regulations, the depletion of PRASA's budget continues. The negative impact on drinking water is likely to continue as long as water treatment plants must process larger volumes of raw water of diminishing quality.

The Solid Waste Disposal and the Superfund Sites

The Superfund National Priorities List is an inventory of contaminated sites that represent an imminent threat to human health and the environment and require removal and remedial actions. Superfund sites are a special case in the legacy of environment contamination. In Puerto Rico, most of these sites involve water resources and the population depending on them.

In the early years of industrialization, in the United States, as well as Puerto Rico, it was general practice to dump industrial wastes in sanitary landfills that were not designed to meet the conditions necessary to receive that type of waste. The disposal of industrial wastes in this manner is now prohibited, but toxic substances remain buried under tons of sanitary landfill, and illegal dumping of industrial waste is still in practice. The discharge of industrial pollutants and garbage in urban areas is increasing and only a small part is being recycled.

Toxic wastes in Puerto Rico have been generated mainly by oil refineries, petrochemical companies, electronic companies, chemical and medical product industries, and thermoelectric manufacturers. These industries produce and use chemical solvents, many of the polychlorinated type, and heavy metals. Moreover, although small scale agriculture has been almost eliminated from the island, commercial agriculture uses a wide array of pesticides.

The federal government has designated nine sites in Puerto Rico under Superfund. All nine sites represent a threat to surface or groundwater resources. Five of the sites listed below are located over the northern limestone aquifer which, as previously noted, is the largest on the island and the most important source of underground potable supply (Figure 2). The history of these sites

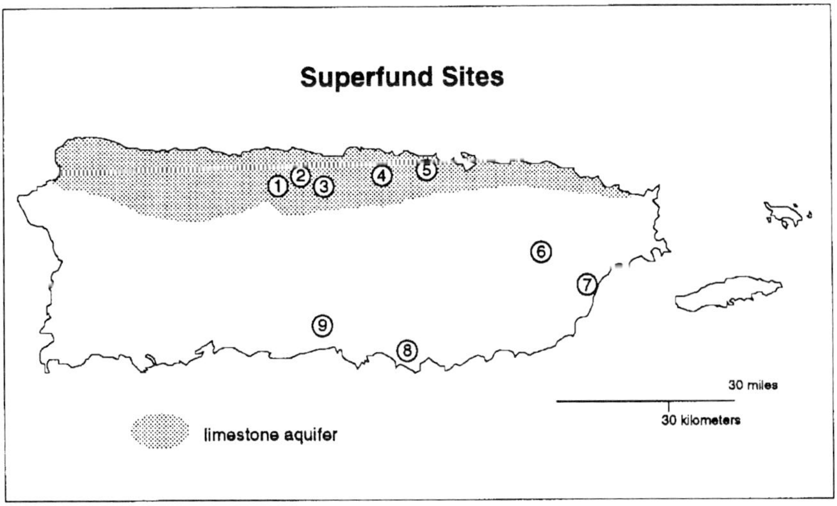

FIGURE 2. Location of existing Superfund sites. (Source: Puerto Rico, Environmental Quality Board, 1994).

represents both the results of actions taken before the creation of environmental laws to protect the environment and human health, and leniency in the enforcement of regulations once they were created.

DESCRIPTION OF SUPERFUND SITES

Numbering coincides with the map (Figure 2):

1. The UpJohn Facility site contains a pharmaceutical manufacturing plant. In 1982, approximately 15,300 gallons of waste material, including carbon tetrachloride, leaked from an underground storage tank. Six wells were sampled for contamination and four were taken out of service. UpJohn conducted remedial action which removed large areas of contamination from soil and groundwater. However, concern exists as to whether the total extent of the pollution plume has been eliminated. The approximately 12,000 people residing within a 3-mile radius of the site depend on groundwater for their potable supply.

2. RCA del Caribe in operation from 1971 through 1987 manufactured masks for television screens. Spent ferric solution from the manufacturer's operations was stored in four lined surface lagoons. These lagoons were breached due to sinkhole development, which caused the discharge of approximately one million gallons of the contaminant into the sinkholes. Among the contaminants are heavy metals including chromium, selenium, and iron from former manufacturing processes. This site is a short distance from the first one, thus the same population is at risk.

3. The Barceloneta landfill in a limestone formation continues to accept municipal and industrial waste. About 300 tons of hazardous wastes have been placed in sinkholes. No artificial or natural barrier exists to keep wastes from moving into groundwater. Preliminary sampling results have identified various heavy metals and volatile organic compounds (VOCs) in sludges. Sampling of surface water disclosed toluene and heavy metals in surface runoff. The site is located relatively short distance from the first two sites above.

4. In Vega Alta, public water supply wells were tested in 1983 and found contaminated with volatile organic compounds due to industrial practices. Six of the fifteen wells at this site were closed and new ones were constructed to compensate for the shortage. Close to 27,600 people live near the site.

5. The Naval Security Group Activity site is a naval communication station. From the early 1950s through 1979, the Operation's Public Works Department deposited all waste generated at the station at various areas. Materials included paints, solvents, waste oil, and battery acid. Pesticides were also spilled. Soil samples show elevated levels of arsenic, lead and chlordane. Rain could wash soil contaminants through a drainage ditch to a marsh, and the fractured limestone bedrock may allow pollutants to move into groundwater. Approximately 47,000 people living in and around the station obtain drinking water from public wells within 3 miles of the site.

6. In the Juncos landfill, thermometers containing mercury from the pharmaceutical company Becton Dickinson were dumped. This is one of three sites contaminated with the heavy metal. The landfill was closed in 1984 but leachate seeps and soil erosion were evident during inspections conducted by the EPA. Groundwater is contaminated with heavy metals and chloroform. Several streams flow through this site into the Gurabo River, which drains into Lake Loíza, the most important surface water reservoir for the potable supply of the San Juan metropolitan area. The population at most immediate risk include 10,000 people living within a 3-mile radius of the site. An epidemiological study is being designed to ascertain the magnitude of the effects on the population.

7. Frontera Creek received industrial waste discharges including mercury and the pesticide lindane. The public became concerned about the creek's possible contamination in 1977 following the death of thirty cows that had grazed in the area. Adjacent to the creek are two large fresh-water lagoons which are connected directly to the creek and that local residents used for fishing and recreation. The 500 residents of the housing development Ciudad Cristiana, which was built along the creek in 1979, began to complain of health problems within a year after their arrival. Blood and urine samples of the residents, obtained by the Puerto Rico Department of Health, showed above normal concentrations of mercury. Residents were evacuated permanently from the housing development and an epidemiological study is underway.

8. The Fibers Public Supply Wells site serves as a stand-by water supply for the municipality of Guayama with more than 41,000 inhabitants. Four of the five wells were closed in 1982 due to contamination by halogenated solvents believed to have originated from a nearby fiber manufacturer. Solvents were discharged through the drainage system into lagoons and seeped into the soil.

9. The General Electric Wiring Devices manufactured mercury light switches from 1957 until 1969. Groundwater, soil, and debris are contaminated with organic mercury that was discarded into a waste area at the site. There are approximately 10,000 people living within 3 miles of the site. Groundwater in the area is used as a source of drinking water; one of the wells used for that purpose is located approximately 1,500 feet of the site.

Currently the EPA has identified the parties responsible for eight of the nine hazardous sites. Under the Superfund Law, those responsible will be the ones required to cover the cost of the clean ups. Remedial design or clean up activities have begun at all except the Barceloneta landfill and the General Electric Wiring Devices sites (Miguel A. Maldonado, PREQB officer, pers. com., 5-4-94). Health effects on the exposed population remain to be qualified and quantified.

It must be pointed out that since Superfund sites are only those toxic sites that represent an immediate threat to the population and the environment, the number of sites in the National Priorities List is much more numerous. The list include many of the municipal landfills on the island. In fact, as mentioned above, any sanitary landfill on the island may be suspected of containing toxic wastes. Most of the landfills are located on highly permeable soils that allow contamination to spread by seepage and drainage far beyond the landfills, including the groundwater reservoir. The problem is magnified by the pattern of clandestine and illegal dumping found everywhere in Puerto Rico.

Nevertheless, after years of little activity, the Solid Waste Management Authority was reorganized in 1988 and is now playing a more active role in the search for solutions to the solid waste problem faced by municipalities in Puerto Rico. Water quality can only benefit from this. In 1993, approximately 45% of the 60 municipal landfills had serious deficiencies and were closed. Ideally, landfill closures will be accompanied by the introduction of new technologies, including transfer stations and incinerators.

CONCLUSIONS

From the foregoing discussion of the water-related health problems within the context of Puerto Rico, it is clear that reported cases of acute gastroenteritis are the most obvious connection between disease and water pollution, and thus have heightened community concern about contamination of the natural resource. However, the absence of rigorous epidemiological studies and the possible multiple causes of gastroenteritis hamper the determination of a cause and effect relationship.

When it comes to chemical contamination and human health effects, the relationship is much more difficult to establish. The effects of chemical pollution are silent and long-lasting, and are thus more likely to go undetected. The problem of latency complicates the situation further because most of the health effects of concern appear only after chronic exposure. The clinical signs and symptoms may become evident after the exposure has ceased to exist.

However, it is not necessary to have occurrences of large numbers of people becoming ill simultaneously to have endemic levels of waterborne disease occurring continually. Under those circumstances, the health of the entire population is under threat, but special sectors of the population, such as the elderly, the very young, pregnant women, AIDS victims and all those with a weak immunological system, are at a greater risk.

Some of the persistent organic compounds present in water supplies today are objectionable because of known harmful effects on humans, fish, and/or wildlife. Others are considered objectionable because of adverse effects that are suspected but whose nature and extent are unclear. This problem is different in character and dimensions from the concerns about oxygen demand and bacteriological pollution in water bodies, which presented the principal pollution control objective until recently.

Most of the monitoring done to assess the effectiveness of water quality programs is usually directed to gross pollution parameters, such as turbidity, color, biochemical oxygen demand, and suspended solids. Only in some circumstances are monitoring activities conducted for specific constituents, such as metals and organic compounds. Generally these are done in connection with specific industrial discharges, or in water supplies where there are special grounds for concern. As more specific chemicals are defined in water quality standards, there is an increased need for developing monitoring methods for those substances. But no water treatment system in Puerto Rico has been designed for the removal of toxic materials, and monitoring of such materials is not generally performed.

After appropriate treatment, the quality of the waters in Puerto Rico meets the drinking-water standards but, as pointed out, due to a lack of investment in the infrastructure for water, the system is in decadence and relies on technology not designed to remove new and potentially dangerous contaminants. These have become more abundant after the process of quick industrialization in which safe disposal of industrial waste was not part of the development scheme.

The particular case of groundwater is one in which chlorination is the only purification treatment applied prior to distribution for public consumption. Although the present quality of the groundwater reservoir meets drinking water standards, some of the contaminants that have been observed are not regulated, and thus are not considered in the laboratory analyses to determine water potability. The potential health hazards are unknown, particularly when pollutants occur in combination, in which not only additive but possible synergistic effects must be considered. Furthermore, chlorination treatment, while disinfecting water, may also react with pollutants present in the water to produce volatile organic compounds with carcinogenic effects.

In absolute terms, water-related morbidity and mortality have been reduced over the last four decades, but the gains of the past are still insufficient to protect public health from water-related diseases. Moreover, past progress is now undermined not only by the introduction of synthetic contaminants but also by the presence of pathogenic microbiology. This is the result of a decaying aqueduct and sewer infrastructure, and of inadequate land use activities that continue to encroach surface and groundwater reservoirs.

REFERENCES

1. Page, G. William. 1987. Water and Health. In: M.R. Greenberg (Ed.) Public Health and Environment, the United States Experience. The Guilford Press, New York, pp. 105-138.
2. United States Geological Survey. 1987. *Planning Report for the Caribbean Islands Regional Aquifer System Analysis Project.* Water Resources Investigations, Report 86-4074, Washington, D.C.: United States Geological Survey.
3. ______. 1985. *Reconnaissance of Trace Organic Compounds in Ground Water Throughout Puerto Rico.* Open File Data Report 84-810. Washington, D.C.: United States Geological Survey.

4. ______. 1982. *Reconnaissance of Ground-Water Quality Throughout Puerto Rico, September-October 1981.* Water Resources Division, Open-File Report 82-334. Washington, D.C.: United States Geological Survey.
5. Arbona, Sonia I. and John M. Hunter. 1993. The Geography of Land Use and Water Quality in Puerto Rico. *Conference of Latin Americanist Geographers, Year Book 1993*. 19:93-111.
6. ______. Economic Development Threatens Groundwater Quality in Puerto Rico: Results of a Preliminary Field Study. ms.
7. Vachier, Ricardo J. 1993. *Puerto Rico Stream Water Quality.* Water-Supply Paper 2400. Washington, D.C.: United States Geological Survey. 471-476.
8. Zack, Allen, Rodríguez Alonso, Teresita and Angel Román Mas. 1987. *Puerto Rico, Ground-Water Quality.* Water-Supply Paper 2325. Washington, D.C.: United States Geological Survey. 437-441.
9. Environmental Quality Board. 1994. *Goals and Progress of Statewide Water Quality Management Planning, Puerto Rico 1992-1993.* Government of Puerto Rico, San Juan.
10. Ibid.
11. Haddock, Kenneth, C. 1981. Control of Schistosomiasis: The Puerto Rican Experience. *Social Science and Medicine.* 15D: 501-514.
12. Jobin, W.R. 1980. Sugar and Snails: The Ecology of Bilharziasis Related to Agriculture in Puerto Rico. *American Journal of Tropical Medicine and Hygiene.* 29:91.
13. United States Geological Survey. 1982. *Reconnaissance of Ground-Water Quality Throughout Puerto Rico, September-October 1981.* Water Resources Division, Open-File Report 82-334. Washington, D.C.: United States Geological Survey.
14. John M. Hunter and Sonia I. Arbona. Paradise Lost, in press.

Environmental Contaminants, Ecosystems and Human Health. Edited by S.K. Majumdar, E.W. Miller, and Fred J. Brenner.

Chapter Nineteen

ENVIRONMENTAL LINKS TO CHILDHOOD CANCER

DONA SCHNEIDER
Associate Professor
Department of Urban Studies and Community Health
Rutgers, The State University of New Jersey
New Brunswick, NJ 08903

INTRODUCTION

Cancer is a dreaded disease, linked to chronic health risk behaviors such as smoking, occupational exposures such as asbestos, and advancing age. Yet cancer also strikes children. In Western nations, the incidence of childhood cancer is estimated at 110-140 per million children (under age 15) per year (1-5). The largest number of new cases of cancer in children under age 15 occur as leukemias, followed by central nervous system (brain and spinal cord) cancers, and lymphomas.

There are significant differences in the incidence of childhood cancer among Western nations. Rates in Scandinavia, for example, are 30 percent higher than in the United States (2). Overall, rates in Western nations are higher than those in Asia (50% higher than rates in China, India, Japan, and Singapore) (2, 4, 6). These overall patterns suggest that the environment and/or differences in genetic stocks play a role in the development of these diseases.

The incidence of cancer in U.S. children, primarily leukemias, fell between 1950 and 1965, but overall rates have inched slightly upwards since 1970, possibly the result of better reporting practices (1, 5, 7). In 1991, the Surveillance Epidemiology and End Results (SEER) Program of the National Cancer Institute estimated about 7,500 new cases of cancer in children under age 15 in the United States in that year (5). While the chances of surviving childhood cancer have improved significantly since 1970 due to improved treatments, surgical interventions, and support services (4-5), SEER estimated that in 1991, cancer would take the lives of about 1,500 American children ages 0-14 years (6).

When a diagnosis of childhood cancer is made, it can devastate a family, causing severe physical, psychological and financial hardships. Parents seek answers about whether they may have caused their child to develop cancer with their own behaviors (such as exposing their child to side-stream smoke). They question whether their own occupational exposures might be the cause (such as prenatal exposure to solvents or radiation), or whether environmental exposures such as living near high tension wires or a toxic waste dump might be the culprit. These are hard questions, and researchers have literally launched thousands of research projects designed to examine potential links between childhood cancer and the environment. This chapter reviews the epidemiologic evidence linking physical, chemical, and biologic agents to cancers in childhood.

PHYSICAL AGENTS

Solar Radiation

Multiple studies have shown a strong relationship between sun exposure and skin cancers in adults. Specifically, malignant melanoma is associated with blistering sunburns (5 or more) at young ages (prior to 20 years), latitude of residence at between 15 and 20 years of age (low latitudes carry a higher risk), and cumulative sun exposure. Squamous-cell and basal-cell carcinomas are primarily associated with cumulative sun exposure (8-10).

While sun exposure plays a major role in the development of skin cancers, these diseases are rarely seen in children because there is a significant lag time between the initiating events and the development of the disease. When skin cancers do develop in children, there is usually a predisposing factor. For example, malignant melanoma may be acquired transplacentally from an affected mother (11). It is also more likely to develop if a genetic or developmental disorder is present. These genetic disorders are rare, however, and do not contribute significantly to childhood cancer incidence. Males and females younger than 20 years share the same risk for skin cancer incidence, 3.8 per million per year for whites. Skin cancers are exceedingly rare among black youths (12).

Ionizing Radiation

The link between ionizing radiation and childhood cancer has been firmly established. Survivors of the atomic bombing of Hiroshima and Nagasaki in 1945 suffered a leukemia peak within 5 years of exposure. This latency period was so short that it was possible to document that children under 10 years of age at the time of exposure had two to three times the relative risk of developing radiation-induced leukemia as those older than 10 years (13). Survivors exposed in utero had an even higher risk of developing leukemia than children less than ten years old at time of exposure (14). Children conceived by survivors after exposure had no excess risk regardless of parental radiation dose (15).

In the 1950s, children were exposed to ionizing radiation from a variety of sources. Prenatal exposures were common as physicians, especially in Britain, performed routine pelvimetry during the third trimester of pregnancy. Shoe-fitting was done with fluoroscopic examinations, where children would have fun looking at the bones of their feet in the fitting box at the shoe store. Physicians used radiation to treat children with benign diseases such as acne, enlarged thymus or tonsils, and tinea capitis (ringworm). Chest and dental x-rays were routine and done at far higher doses than they are today. Any child with a positive tuberculin test would be followed-up with a chest x-ray and these examinations would often be repeated every 6 months.

American and British expert panels issued warnings about the indiscriminate use of radiation in 1956 (13). Negative publicity quickly eliminated fluoroscopes from the marketplace and regulations were tightened to restrict access to x-ray machines. In that same year, the first study appeared linking prenatal diagnostic radiation with leukemia and other childhood cancers (16). Dose reduction was achieved almost instantaneously, and pelvimetry was abandoned. Some obstetric radiography continued, but by 1970 it was essentially replaced by sonography (17). More than a dozen studies appeared before 1985 linking prenatal x-ray exposure to a variety of childhood cancers. Mole (17) reviewed the literature on both atomic bomb survivors and prenatal X-rays and concluded there is no threshold for radiation exposure in pregnancy. Exposure within the first few weeks post-conception cannot be linked to childhood cancer, but this is probably because radiation carcinogenesis during this period of rapid cell division results in spontaneous abortion. Later exposure, however, is clearly linked to increased childhood cancers, regardless of dose or period of gestation (17).

With prenatal radiation exposure drastically reduced by the late 1950s, leukemia mortality in children under 5 years dropped significantly by 1965, the period before chemotherapy was successful in increasing survival (18). For radiation exposures in children after the prenatal period, cancers took longer to develop. Hempelmann (19) followed 3,000 children treated with radiation for an enlarged thymus and 5,000 of their nonirradiated siblings. By age 50, 30 of the irradiated group and only 1 of the nonirradiated group developed thyroid cancer. The shortest latency period for a tumor to arise was 6 years. The longest was 35 years (20).

Studies also show that children who had radiation for full mouth dental examinations before 1945 (21), treatment of enlarged tonsils (22), and tinea capitis (23), had increased rates of brain and central nervous system cancer as adults. Today, these exposures are no longer a concern. Instead, we worry about background and small radiation exposures. The greatest epidemiologic concern has been exposure from nuclear power plant accidents.

In 1979, the Three-Mile Island nuclear power facility in Pennsylvania triggered releases of radioactivity into the environment. Exposures to the population living within 5 miles (8 km) of the plant were calculated by dosimetry experts as whole-blood gamma exposures of 0.1-0.25 millisieverts (mSv) (24), a fraction of the average yearly background gamma radiation exposure (0.8-1 mSv) of a U.S. resident (25). Public furor caused epidemiologic investigations of health effects, but none were shown. Neither adult nor childhood cancers were significantly increased five years after the event (26). Hatch and Susser (27) reported a slight increase in childhood cancers around the TMI site, but this was based on residing within 10 miles of the nuclear plant between 1975 and 1985, not on residing there after the release. A third study of populations living near U.S. nuclear facilities found no excess cancer risk for either children or adults (28).

The U.S. findings on cancer risks from residing near a nuclear facility run counter to findings in the United Kingdom. Studies have documented an increased incidence of childhood leukemia around two nuclear fuel reprocessing plants, the Sellafield plant in the northwest of England (29), and the Dournreay plant in northern Scotland (30). In southern England, the Aldermaston and Burghfield plants, producers of nuclear weapons, also showed an excess number of childhood leukemia cases (30). Studies examining paternal occupational exposures to radiation have been done in both Canada and the United Kingdom and do not support the contention that radiation exposures before conception increase the risk of childhood cancer (31-32). These later studies support the findings from studies of atomic bomb survivors-that is, there is no increased incidence of childhood cancer due to preconceptual exposure to radiation.

It is difficult to study the relationship between childhood cancer and residential proximity to nuclear installations because the numbers of cases are small, the levels of exposure are close to background, and confounders are difficult to control. The studies that have been done have yielded conflicting results and, taken together, do not support the contention that low level radiation exposure is a causative factor for childhood cancer. It is possible that low-level radiation interacts with other carcinogens in the environment or with behavioral or genetic risk factors that we have not yet identified. These relationships remain to be studied. Nevertheless, the studies of childhood cancer incidence and nuclear power stations such as those near Sellafield have stirred major political controversies.

In the instance of large scale nuclear accidents, such as the one at Chernobyl in 1986, we should be able to detect results somewhere between those of the atomic bomb blast and that of background. As soon as the Chernobyl accident occurred, an international effort to measure exposure (direct and from fallout) and health effects went into effect. The European Childhood Leukemia-Lymphoma Incidence Study (ECLIS) was launched, monitoring the cancer registries of 20 European countries. The first report of ECLIS was released in 1993 (33), and covered the period from 8-32 months after the accident. While it is still too early to see effects, the group estimates an overall leukemia increase of 0.8 percent in areas reporting to the ECLIS, with the highest increase in Belarus (5.8%) (33). If the group is correct, a change in childhood cancers in regions near the accident should eventually be apparent, but an increase in childhood cancers in Europe as a whole might not be detectable.

Radon

Radon gas, a colorless, inert radioactive decay product of uranium, is a common indoor air pollutant. It enters homes through soil, water, building materials, and natural gas. As radon gas undergoes spontaneous decay, it produces progeny (polonium-218 and polonium-214) which are minute solid particles rather than gas. These radon daughters are responsible for the adverse health effects of radon exposure. The particles are inhaled and are retained in the lungs. They emit alpha radiation to the immediately surrounding tissues, but do not disperse to the rest of the body because of their rapid radioactive decay.

One Swedish study showed no relationship between developing childhood cancer and living in a home with a high radon concentration (34). Henshaw et al. (35) found that childhood cancers of the brain and spine, leukemia, osteosarcoma, and melanoma show a significant relationship with radon concentrations, and suggest that some of the clusters of childhood leukemia in the United Kingdom might be related to radon.

Although most epidemiologic studies of radon have focused on lung cancer as the outcome of interest, one ecological study of radon concentration in drinking water in North Carolina found that overall childhood cancer mortality was increased in counties with high and medium radon levels (36). These findings were general in nature and were subject to confounding and other biases. Irregardless, the American Academy of Pediatrics is taking no chances. Their Committee on Environmental Hazards has declared radon exposure a hazard to children and recommends that pediatricians recommend remediation for all homes with radon levels greater than 4pCi/L (37). This recommendation is based on the fact that lifetime exposure to radon is of concern-that is, the effects of radon, like sunlight, are cumulative.

Asbestos

Asbestos is a causative agent for developing malignant mesothelioma. As the disease has a latency period of between 10 and 40 years, it is extremely rare in children (38). Cases of childhood malignant mesothelioma have linked the disease to exposure to asbestos fibers on parents' contaminated work clothing (39). More commonly, childhood exposures from playing on asbestos dumps have been linked to the disease in adults (40).

Between 1950 and 1973, about 9,000 schools in the United States had their ceilings sprayed with asbestos. It was also used for floor tiles, insulation wrap for pipes and ducts, and as surfacing material on girders and walls (41). In 1987, the Environmental Protection Agency issued a ruling that required school districts to identify asbestos-containing materials in school buildings and develop a management plan. As parents across the country reacted with fear and outrage to the news that their school contained asbestos, the management plan too often selected was removal (42). The choice to remove asbestos was a bad one for many schools for two reasons. First, the costs were staggering. Second, poorly trained contractors made asbestos exposures worse by slip-shod removal techniques. Greenberg and Schwartz (42) suggest that the median cost to schools ($100,000) for asbestos removal is not justified, especially since risk was actually increased in about half of asbestos removal jobs. Given that childhood cancers have never been linked to asbestos exposures in schools (there has been no increase in mesothelioma among non-occupationally exposed individuals), they are probably correct in their assessment.

Electromagnetic Fields

More than 40 studies have attempted to link cancers with low-frequency alternating electromagnetic fields (EMF). Given the number of studies done, it is not surprising that some have found a relationship between childhood cancer and EMF. Most, however, have not. EMF studies are difficult to compare because exposure is rarely estimated in the same fashion. Some studies estimate exposure based on residential wiring configurations. Others use self-reported use of appliances. Still others use actual measurements within the home or estimate exposure by residential proximity to high-voltage power lines. To further confuse the matter, some studies found relationships with leukemia but not other cancers, whereas other studies found relationships with brain tumor but not leukemias. Taken as a body of literature, then, EMF studies are confusing and inconclusive.

Michaelson (43) notes the lack of studies showing a biological link between EMF and cancer in living cells. To date, there has been no link between EMF and chromosome damage or DNA repair mechanisms. No laboratory experiments link cell changes to electric or magnetic field exposures (43). Given the lack of a biologic link, we need to consider whether the exposure measures used to estimate EMF may actually be surrogates for another exposure we have not yet identified. In the meantime, it can be assumed that EMF, if it is a risk for childhood cancer, is not a large one.

CHEMICAL AGENTS

Therapeutic Drugs

Some drugs have been specifically linked to childhood cancer. Diethylstilbestrol (DES) is a known transplacental carcinogen linked to rare vaginal cancers in the daughters of women given the drug during pregnancy (4, 13). Case reports linking the use of phenytoin, used for the treatment of epilepsy, with neuroblastoma spurred large scale studies that showed no association (4). Case-control studies linking analgesics, antibiotics, antihistamines, antipyretics, barbiturates, and diuretics with childhood cancers have never been replicated on a large scale and cannot be considered conclusive (4).

Intramuscular vitamin-K has been statistically associated with an increased risk of childhood leukemia in the United Kingdom, but these findings have been strongly challenged (44). Given that almost all newborns in the United States are given intramuscular vitamin K at birth, and given that there has been no increase in childhood leukemia since the practice was instituted, this finding is incompatible with the U.S. experience.

Cytotoxic agents, such as cyclophosphamide, chlorambucil and phenylalanine mustard, are used as cancer treatments. Their use has been linked to the development of non-lymphocytic leukemias as second cancers in persons treated for cancer in childhood (13). Immunosuppressants used for organ transplantation can cause reticulum cell sarcoma and lymphoma, especially in those less than 20 years of age. Since the numbers of transplant cases are small, the contribution of immunosuppressants to the overall incidence of childhood cancer is insignificant (4, 13).

Other Chemicals and Environmental Contaminants

There are no studies linking air pollution to childhood cancer. Nor has residing near a landfill or toxic waste dump been shown to be associated with an increased childhood cancer risk (45). Even residing near Love Canal, with its 82 chemicals buried on 16 acres, is not statistically linked to an increase in the number of cancer cases (46). How can this be?

Studies of cancer clusters usually do not show an association with environmental contamination for two reasons. First, the numbers of cases are usually small and, secondly, few contaminants have been shown to be causal for cancer. One exception is the childhood leukemia cluster of Woburn, Massachusetts. Lagakos et al. (47) documented a statistical association between community well water (contaminated with trichloroethylene, tetrachloroethylene, and chloroform) and childhood leukemia in Woburn. Few other cluster studies, however, have yielded significant results.

Multiple case-control studies exist linking maternal use of hair dyes, incense, marijuana, and pesticides in the prenatal period with an increased risk of childhood brain tumors and leukemias. These findings are not consistent and the exposures have not been confirmed in large scale studies. Stiller (4) suggests these associations might be linked to constituents in the products, such as aldehydes or polycyclic aromatic hydrocarbons, but this hypothesis has yet to be tested.

Smoking

More than a dozen studies of parental smoking in both the prenatal period and from birth to age 14 have been undertaken to examine the link between exposure to tobacco smoke and childhood cancers, especially brain tumors and leukemias. Small scale studies tend to find relationships, but large scale, population-based studies fail to confirm these findings, suggesting selection bias in the small scale studies. While prenatal smoking does predispose an infant to fetal tobacco syndrome and parental smoking does predispose children to respiratory infections, evidence for a carcinogenic effect in children is inconclusive. Long-term effects, however, should be considered. Exposure to tobacco smoke, including sidestream smoke, is known to increase the risk of developing lung cancer. This means exposure in childhood can increase life-time risk for developing the disease.

Parental Occupations

O'Leary et al. (48) reviewed 32 studies of parental occupation and childhood cancer. The authors concluded that a preponderance of evidence links parental occupational exposures to childhood malignancies. Specifically, parental occupational exposures to paints, petroleum products, solvents (especially chlorinated hydrocarbons) and pesticides, and to metal exposure, were statistically linked to multiple types of cancers, including leukemias, brain tumors, Wilm's tumor, and hepatoblastoma.

While these studies may be statistically significant, they are hampered by the lack of exposure data. Occupation is taken as a surrogate for exposure and this may simply be an incorrect assumption. For example, studies linking parental occupation in the nuclear power industry with childhood leukemia assume paternal gonadal exposure to irradiation is a predisposing actor for the development of the disease. This assumption, however, runs counter to what we know about the role of parental gonadal exposure before conception from studies of atomic bomb survivors. Paternal occupation in the nuclear power industry, then, may be a risk factor for childhood leukemia, but gonadal irradiation may not be the correct exposure. We simply need better, and possibly different exposure data and larger studies to draw conclusions about these links.

BIOLOGICAL AGENTS

Infectious Diseases

Viral agents have been statistically linked to some childhood and adult cancers and they offer biologic plausibility for explaining why some children develop cancers and others do not. Stiller (4) provides an excellent discussion about the relationship between viral agents and childhood cancers. Specifically, hepatitis B infection is associated with hepatocellular carcinoma, human immunodeficiency virus (HIV) is associated with non-Hodgkin's lymphoma, and Epstein-Barr virus (EBV) is associated with Burkitt's lymphoma. EBV is also suspected of playing a role in the development of Hodgkin's disease (7, 49) and cancers of the uterine cervix are associated with human papilloma virus (HPV) (13).

Multiple studies have investigated the relationship of viral illnesses with the development of acute lymphocytic leukemia (ALL) in childhood. In the 1960s it was thought that murine and feline leukemia were the counterpart to leukemia in humans, but this theory was discounted as the diseases were not transmissible either vertically or laterally (13).

Kinlen (50) proposed that childhood leukemia is a rare response to a widespread viral infection. He suggested that high rates of childhood leukemia would be found where there were high levels of population mixing, situations that would yield more chances for exposures. This is one explanation offered for the clusters of childhood leukemias around nuclear power plants in the United Kingdom. Population influx due to job opportunities at the plants may have provided the opportunity for widespread viral infections as persons not previously exposed were exposed to new viral agents.

Kinlen's theory has been challenged by MacMahon (51), who notes that no increase of childhood leukemia occurred during the massive evacuation of mothers and children from British cities during World War II. If the development of ALL in childhood is a rare response to viral infections, claims MacMahon, it must be independent of population mixing.

The childhood cancer being a rare response to infection theory is supported in the literature by at least one leukemia cluster study. Eight cases of childhood leukemia occurred in Niles, Illinois, between 1957 and 1960. All eight children attended the same parochial school and all were accompanied by a parallel "rheumatic-like" illness that occurred among other students attending the same school (52). This now classic cluster investigation of childhood leukemia strongly suggests an infectious link to at least some childhood leukemias.

Genetics

Genetic influences clearly play a role in the development of childhood cancers. Indeed, some cancers are now recognized as requiring an environmental agent and a genetic trait for the development of the disease. For instance, children with hereditary dysplastic nevus syndrome may develop melanoma if they are exposed to ultraviolet radiation. Children with Fanconi's pancytopenia may develop hepatoma if exposed to androgens (53).

Some genetic and congenital disorders predispose a child or adolescent to cancer, even in the absence of environmental insults. Chromosomes may be translocated, absent or present in excess. Genes may interact, with no one gene or environmental factor being the triggering factor for the development of the disease. Some cancers such as retinoblastoma, Wilm's tumor and neuroblastoma-are now recognized as having two typologies, that is, they have a heritable form and a sporadic one. In the hertibale form, one mutation has already been established at birth and the second mutation occurs sporadically later. In the sporadic form, both mutations occur after birth, perhaps in response to environmental insults, perhaps spontaneously (53).

It is important that we keep in mind that living beings mutate. This is natural and to be expected. Unfortunately, not all mutations are compatible with survival.

While this fact may seem fatalistic to some who have staked their careers on finding environmental links to childhood cancers, it seems eminently sensible to recognize that genetics cannot be ignored when considering a disease linked to rapid cell division.

DISCUSSION

The popular perception that 80 percent of cancers are due to environmental factors (including lifestyle factors such as diet and smoking) simply does not hold for childhood cancers. In fact, few environmental exposures can be statistically linked, much less causally linked, to cancer in childhood. Some environmental exposures, such as solar radiation and radon, have lifetime cumulative effects and should be prudently monitored. Others, such as exposure to viral illnesses, may plan a role in the development of cancer in certain individuals. Yet others, such as EMF, may actually be surrogates for childhood cancer risks we have not yet identified.

Given that the evidence linking childhood cancers to environmental risks is mixed, and given that the worldwide incidence of childhood cancers remains fairly stable, we need to look at other factors to explain why some children get cancer and others do not. Genetic theories of cancer involving multiple mutations seem to hold the key. Living beings have unique genetic codes, and some are more or less prone to mutation than others. In other words, chance (heritable and sporadic mutations) plays a significant role in whether a child will develop cancer. This knowledge should help assuage the guilt of parents who have a child with cancer because it removes the idea that something they did or did not do caused the disease. It should also be useful knowledge for the rest of us. Knowing that genetics plays a major role in the development of childhood cancer should reduce our fears. This, in turn, should allow us to focus our attention on policies (such as nuclear power regulations) designed to protect all Americans, including children, from true cancer risks. It can also protect us from spending enormous amounts of money and energy trying to protect children from unproven low-level and background exposures (such as residing near a landfill, nuclear facility, or power line) that do not pose significant risks for developing the disease.

REFERENCES

1. Young, J.L., Ries, L.G. and E. Silverberg, et al. 1986. Cancer incidence, survival, and mortality for children younger than age 15 years. *Cancer* 58:598-602.
2. Parkin, D.M., Bieber, A. and G.J. Draper, et al. 1988. *International Incidence of Childhood Cancer*. International Agency for Research on Cancer Scientific Publications (No. 87), Lyon, France. pp. 401.

3. Levi, F., La Veccia, C. and F. Lucchini, et al. 1992. Patterns of childhood cancer incidence and mortality in Europe. *European Journal of Cancer* 28A:2028-49.
4. Stiller, C.A. 1994. Malignancies, pp 439-472. In: I.B. Pless (Ed.) *The Epidemiology of Childhood Disorders.* New York: Oxford University Press. pp. 531.
5. Bleyer, A. 1990. The impact of childhood cancer on the United States and the world. *CA-A Cancer Journal for Clinicians* 40:355-67.
6. Breslow, N.E. and B. Langholz. 1983. Childhood cancer incidence: geographical and temporal variations. *International Journal of Cancer* 32:703-16.
7. Robison, L.L., Mertens, A. and J.P. Neglia. 1991. Epidemiology and etiology of childhood cancer, pp. 11-28. In: D.J. Fernbach and T.J. Vietti (Eds.). *Clinical Pediatric Oncology* 4th ed. Mosby Year Books, St. Louis, MO, pp. 720.
8. Marks, R., Jolley, D. and S. Lectsas, et al. 1990. The role of childhood exposure to sunlight in the development of solar keratoses and non-melanocytic skin cancer. *Medical Journal of Australia* 152:62-6.
9. Stern, R.S., Weinstein, M.C. and S.G. Baker. 1986. Risk reduction of non-melanoma skin cancer with childhood sunscreen use. *Archives of Dermatology* 122:537-45.
10. M.A. Weinstock, Colditz, G.A. and W.C. Willet, et al. 1989. Nonfamilial cutaneous melanoma incidence in women associated with sun exposure before 20 years of age. *Pediatrics* 84:199-204.
11. Trozak, D.T., Rowland, W.D. and F. Hu. 1975. Metastatic malignant melanoma in prepubertal children. *Pediatrics* 55:191-204.
12. Cangir, A. 1991. Miscellaneous childhood tumors, pp. 627-45. In: D.J. Fernbach and T.J. Vietti (Eds.). *Clinical Pediatric Oncology* 4th ed. Mosby Year Books, St. Louis, MO, pp. 720.
13. Miller, R.W. 1989. Frequency and environmental epidemiology of childhood cancer, pp. 3-18. In: P.A. Pizzo and D.G. Poplack (Eds.). *Principals and Practice of Pediatric Oncology.* J.B. Lippincott Company, Philadelphia, PA. pp. 1129.
14. Kato, H., Yoshimoto, Y. and W.J. Schull. 1989. Risk of cancer among children exposed to atomic bomb radiation in utero: a review. pp. 365-74. In: N.P. Napalkow, J.M. Rice, L. Tomasaki, et al. (Ed.). *Perinatal and Multigeneration Carcinogenesis.* International Agency for Research on Cancer: Lyon, France. pp. 436.
15. Yoshimoto, Y. 1990. Cancer risk among children of atomic bomb survivors: a review of RERF epidemiologic studies. *JAMA* 264:596-600.
16. Bithell, J. and A.M. Stewart. 1975. Pre-natal irradiation and childhood malignancy: a review of British data from the Oxford Study. *British Journal of Cancer* 31:271-87.
17. Mole, R.H. 1990. Childhood cancer after prenatal exposure to diagnostic X-ray examinations in Britain. *British Journal of Cancer* 62:152-68.
18. Miller, R.W. and F.W. McKay. 1984. Decline in US childhood cancer

mortality 1950 through 1980. Journal of the American Medical Association 251:1567-70.
19. Hempelmann, L.H., Hall, W.J. and M. Phillips, et al. 1975. Neoplasms in persons treated with x-rays in infancy: Fourth survey in 20 years. *Journal of the National Cancer Institute* 55:519-30.
20. Shore, R.E., Woodward, E. and N. Hildreth, et al. 1985. Thyroid tumors following thymus irradiation. *Journal of the National Cancer Institute* 74:1177-84.
21. Preston-Martin, S., Paganini-Hill, A. and B.E. Henderson, et al. 1980. Case-control study of intracranial meningiomas in women in Los Angeles County, California. *Journal of the National Cancer Institute* 65:67-73.
22. Shore-Freedman, E., Abrahams, C. and W. Recant, et al. 1983. Neurilemomas and salivary gland tumors of the head and neck following childhood irradiation. *Cancer* 51:2159-63.
23. Ron, E., Modan, B. and J.D. Boice, et al. 1988. Tumors of the brain and nervous system after radiotherapy in childhood. *New England Journal of Medicine* 319:1033-9.
24. Gur D., Good, W.F. and G.K. Tokuhataka, et al. 1983. Radiation dose assignment to individuals residing near the Three Mile Island nuclear plant. *Proceedings of the Pennsylvania Academy of Science* 57:99-102.
25. Upton, A.C. 1982. The biological effects of low-level ionizing radiation. *Scientific American* 246:41-9.
26. Hatch, M.C., Beyea, J. and J.W. Nieves, et al. 1990. Cancer near the Three Mile Island nuclear plant: radiation emissions. *American Journal of Epidemiology* 132:397-412.
27. Hatch, M and M. Susser. 1990. Background gamma radiation and childhood cancers within ten miles of a US nuclear plant. *International Journal of Epidemiology* 19:546:52.
28. Jablon, S., Hrubec, Z. and J.D. Boice. 1991. Cancer in populations living near nuclear facilities: a survey of mortality nationwide and incidence in two states. *JAMA* 265:1403-8.
29. Draper, G.J., Stiller, C.A. and R.A. Cartwright, et al. 1993. Cancer in Cumbria and in the vicinity of the Sellafield nuclear installation, 1963-90. *BMJ* 306:89-94.
30. Draper, G.J. 1992. An overview of reports and current research concerning childhood leukemia and cancer around nuclear installations in the U.K. *The Science of the Total Environment* 127:9-12.
31. Evans, H.J. 1990. Ionizing radiations from nuclear establishments and childhood leukemias - an enigma. *BioEssays* 12:541-49.
32. McLaughlin , J.R., King, W.D. and T.W. Anderson, et al. Paternal radiation exposure and leukemia in offspring: the Ontario case-control study. *BMJ* 307:959-66.
33. Parkin, D.M., Cardis, E. and E. Masuyer, et al. 1993. Childhood leukemia following the Chernobyl accident: the European Childhood Leukemia-

Lymphoma Incidence Study (ECLIS). European Journal of Cancer 29A:87-95.

34. Stjernfeldt, M., Samuelsson, S. and J. Ludvigsson. 1987. Radiation in dwellings and cancer in children. *Pediatric Hematology and Oncology* 4:55-61.
35. Henshaw, D.L., Eatough, J.P. and R.B. Richardson. 1990. Radon as a causative factor in induction of myeloid leukemia and other cancers. *Lancet* 335:1008-12.
36. Collman, G.W., Loomis, D.P. and D.P. Sandler. 1991. Childhood cancer mortality and radon concentration in drinking water in North Carolina. *British Journal of Cancer* 63:626-9.
37. Committee on Environmental Hazards, American Academy of Pediatrics. 1989. Radon exposure: a hazard to children. *Pediatrics* 83:799-802.
38. Brenner, J., Sordillo, P.P. and G.B. Magill. 1981. Malignant mesothelioma in children: report of seven cases and review of the literature. *Medical and Pediatric Oncology* 9:367-73.
39. Antman, K., Shemin, R. and L. Ryan, et al. 1988. Malignant mesothelioma: prognostic variables in a registry of 180 patients. The Dana-Farber Cancer Institute and Brigham and Women's Hospital experience over two decades. *Journal of Clinical Oncology* 6:147-53.
40. Wagner, J.C., Sleggs, C.A. and P. Marchand. 1960. Diffuse pleural mesothelioma and asbestos exposure in the North Western Cape Province. *British Journal of Industrial Medicine* 17:260-71.
41. Committee on Environmental Hazards, American Academy of Pediatrics. 1987. Asbestos exposure in schools. *Pediatrics* 79:301-5.
42. Greenberg, M.R. 1990. Asbestos: Is removal the best approach? *Forum for Applied Research and Public Policy* (winter):82-6.
43. Michaelson, S.M. 1991. Household magnetic fields and childhood leukemia: A critical analysis. *Pediatrics* 88:630-5.
44. Klebanoff, M.A., Read, J.S. and J.L. Mills, et al. 1993. The risk of childhood cancer after neonatal exposure to vitamin K. *New England Journal of Medicine* 329:905-8.
45. Muir, K.R., Hill, J.P., S.E. Parkes, et al. 1990. Landfill waste disposal: an environmental cause of childhood cancer? *Pediatric and Perinatal Epidemiology* 4:484-5.
46. Levine, A.G. 1982. *Love Canal: Science, Politics, and People.* Lexington Books, Lexington, MA, pp. 263.
47. Lagakos, S.W., Wessen, B.J. and M. Zelen. 1986. An analysis of contaminated well water and health effects in Woburn, Massachusetts. *Journal of the American Statistical Association* 81:583-614.
48. O'Leary, L.M., Hicks, A.M. and J.M. Peters, et al. 1991. Parental occupational exposures and risk of childhood cancer: a review. *American Journal of Industrial Medicine* 20:17-35.
49. Armstrong, A.A., Alexander, F.E. and R. Pinto Paes, et al. 1993.

Association of Epstein-Barr virus with pediatric Hodgkin's disease. American Journal of Pathology 142:1683-8.
50. Kinle, L. 1988. Evidence for an infective cause of childhood leukemia: comparison of a Scottish New Town with nuclear reprocessing sites in Britain. *Lancet* 3:1323-7.
51. MacMahon, B. 1992. Is acute lymphoblastic leukemia in children virus-related? *American Journal of Epidemiology* 136:916-24.
52. Heath, C.W. and R.J. Hasterlik. 1990. Leukemia among children in a sub-urban community. *CA-A Cancer Journal for Clinicians* 40:29-50.
53. Mulvihill, J.J. Clinical genetics of pediatric cancer, pp. 19-37. In: P.A. Pizzo and D.G. Poplack (Eds.). *Principals and Practice of Pediatric Oncology.* J.B. Lippincott Company, Philadelphia, PA, pp. 1129.

Environmental Contaminants, Ecosystems and Human Health. Edited by S.K. Majumdar, E.W. Miller, and Fred J. Brenner. © 1995, The Pennsylvania Academy of Science.

Chapter Twenty

LEAD POISONING AND CHILDHOOD HEALTH

ADRIAN J. BAILEY
Department of Geography
Dartmouth College
6017 Sherman Fairchild Hall
Hanover, NH 03755-3571

INTRODUCTION

Lead poisoning in children and adults arises as a result of the ways human society has concentrated lead in its built environments. This chapter proposes that a geographic analysis of lead poisoning can shed light on the distribution of lead within and between these built environments. In particular, the discussion stresses the interdependent role of urbanization and industrialization in creating landscapes of lead poisoning.

Children are the focus of this chapter. Children are more likely to be lead poisoned because part of their normal behavior includes hand-to-mouth activities, which may introduce lead to the gastrointestinal tract. Indeed, children tend to absorb and retain about 50% of the lead they ingest, compared to 8% for adults (Hammond 1982). Children continue to suffer from widely documented neurobiological and psychological dysfunctions. Although lead poisoning is a global and historic phenomenon, examples are restricted to the recent United States experience.

DEFINING LEAD POISONING

Individuals are considered lead poisoned when they have an elevated blood lead level. Exactly what constitutes elevated is a topic of some debate. As cumulative research enhances our understanding of the causes and consequences of lead poisoning, particularly among children, the recognized level for lead toxicity has shifted downwards. While a blood lead level above 60 mcg was considered toxic in the mid 1960s (Chisholm and Harrison 1956), levels above 30 mcg were considered toxic a decade later. More recently the Centers for Disease Control (CDC) has advocated replacing any single definition of childhood lead poisoning with a five class system (Centers for Disease Control 1991, Mason 1991). This system considers all children with blood lead levels in excess of 10 mcg to be poisoned, and counsels pharmalogic treatment in cases above 20 mcg, and chelation therapy in cases where blood lead levels exceed 45 mcg.

LEAD POISONING AND CHILDHOOD HEALTH

A broad, and widening slate of neurobiological and psychological dysfunctions suggests that lead poisoning is a major insult to the health of children (Neggers 1986, Sciarillo 1992). Much of the early literature on the effects of lead poisoning focused on damage to the central nervous system and the kidneys. While lead was thought to be involved in chronic renal failure and severe encephalopathy, the levels at which these insults were triggered was less well understood (Bornschein et al. 1985b).

More recently research has linked health impacts with bands of lead absorption, as reflected by blood lead levels. Thus, in extreme cases, where blood lead levels have exceeded 80 mcg, coma, convulsions, and death can result. Lower lead levels impair heme synthesis (Piomelli 1980) and are associated with reduced levels of 1,25-dihydroxyvitamin D, the metabolite involved in the gastrointestinal absorption of calcium and phosphorus (Rosen et al. 1980). Other "low level" effects include decreased stature or growth, decreased hearing acuity, and decreased ability to maintain a steady posture (Centers for Disease Control 1991:9). In some individuals reduced birth weights have been linked to maternal and cord blood lead levels between 10 and 15 mcg (ATSDR 1988).

Evidence increasingly suggests links between elevated blood lead and psychological changes (including intelligence deficits) in children (see, for example, Shaheen 1984, Schroeder et al. 1985, Landsdown et al. 1986, Fulton et al. 1987, Needleman et al. 1990, Dyer 1993). Declines in auditory and speech processing, attention deficits, and reduced verbal intelligence quotients were described by Needleman et al. in 1979. Subsequent prospective studies (Bellinger et al. 1991), Dietrich et al. 1991) have further linked pre- and post-natal exposure to slower sensory motor and delayed early cognitive development. Bellinger et al. (1991) reported that, as the child ages, improving socioeconomic conditions and reduced lead exposure can diminish some of these effects. In general terms, and having controlled for intervening factors (of which socioeconomic status is crucial), intelligence test scores appear to decline as blood lead levels increase above 10 mcg (Centers for Disease Control 1991:10).

SOURCES OF LEAD IN THE BUILT ENVIRONMENT

Natural concentrations of lead in the physical environment are generally insufficient to constitute a health risk to children. Therefore human activity, most generally in the guise of industrialization and urbanization, is directly and indirectly responsible for assembling harmful concentrations of lead in the built environment. Before discussing the main environmental sources of lead it is important to note that multiple sources of relatively low-level lead exposure can be as harmful as a single high-level lead exposure, and that the duration of exposure is linked to the body's lead burden.

Lead-based paint remains the single most important source of lead in the child's built environment (Mahaffey 1983, Chisholm et al. 1985, Rabinowitz et al. 1985, Centers for Disease Control 1991). Before the 1950s most interior paint contained up to 50% lead, and although the use of lead-based paint has since declined an estimated 74% of the privately owned occupied housing units in the United States that were constructed prior to 1980 contain some lead-based paint. Additionally, exterior paint that contains lead has been applied to the living spaces of houses. Lead may enter the child's body from paint through pica (the direct ingestion of non-food substances) and from the ingestion of household dust that contains small particles of eroded paint. Generally, then, as the condition of the paint deteriorates the risk of exposure increases. Older homes are not only likely to contain more lead paint, but they are also more likely, ceteris paribus, to be deteriorating. Alternatively, renovation and remodeling exercises have also been linked to increased lead in household dust (Rabinowitz 1985, Chisholm 1985, Farfel 1990). The longevity of exposure to this source, and the developmental normalcy of the hand-to-mouth action which introduces lead into the body combine to make lead-based paint the single most important source of environmental lead exposure. Lead-based paint may also be present on wood that is burnt in woodstoves, and on many bridges and other structures. In the latter case repair workers may bring chips of old paint into their homes. Additionally, the soil near to decaying structures may contain high levels of lead, and airborne lead levels will be raised if such structures are demolished.

Gasoline is a second major source of lead in the environment. The practice of adding lead to gasoline as an anti-knocking agent began in 1923 (Rosner and Markowitz 1985) and continued until the 1980s, when successive legislation has reduced and will virtually eliminate the use of leaded gasoline. In many other regions of the world, however, gasoline continues to be a significant source of lead. Although automobile emissions are now less responsible for elevating levels of airborne lead in the United States, the accretion of over sixty years of traffic pollution means that the soil and dust near major highways are particularly likely to be important lead pathways.

Industry, and industrial processes represent a third source of lead (Seligman 1986, Baser and Marion 1990, Maizlish et al. 1990, Tepper 1992, Waller 1992).

These include the mining and smelting of lead ores, or ores which may be contaminated with lead salts (Steenland 1992). The process of manufacturing storage batteries can involve exposure to lead. Point pollution around smelters is widely documented (Hunter 1976, Landrigan et al. 1976, Moehr 1993). Evidence from California and New York suggests that workers in certain occupations have higher lead exposure than others (see Table 1). These workers may introduce lead into the home environment on their clothes.

Among other sources of lead, solder is frequently found in older domestic water distribution systems and, in concert with various qualities of the water, may result in increased lead levels in drinking water. Lead pipes, lead connectors, and lead containing faucets can have similar effects. Such systems are more likely to be present in pre-1920s developments. Some water storage devices (for example various cisterns) may have lead liners or may contain lead as a result of repairs. Lead solder is used in hobbies like casting ammunition, making fishing weights, sculpting toy soldiers, and model railroading. Lead may also enter the home on some food cans (particularly those that have been imported), as part of traditional folk remedies, in cosmetics, and in moonshine, which may be contaminated if its distilling has involved parts of automobile engines that themselves contained lead solder. Ceramic and pottery items may have glazes with lead content that can leach into food or drink.

TABLE 1

Occupations Associated with Elevated Blood Lead Levels

STANDARD INDUSTRIAL CLASSIFICATION NUMBER (SIC CODE)	INDUSTRY DESCRIPTION
1622	Bridge, tunnel, and elevated highway construction
2816	Inorganic pigments
2899	Chemical and chemical preparations
3231	Glass products made of purchased glass
3269	Pottery (nec)
3341	Secondary smelting and refining of nonferrous metals
3362	Brass/copper foundry
3432	Plumbing fixture fittings and trim (brass goods)
3494	Valve and pipe fittings (not SIC 3432)
3691	Storage batteries
3692	Primary batteries, dry and wet
3714	Motor vehicle parts and accessories
5084	Industrial machinery and equipment
7539	Automotive repair shops
7997, 9221	Firing range

SOURCE: Compiled by Centers for Disease Control (1991:22, Table 3.1) from Baser and Marion 1990, Maizlish et al. 1990.

LEAD PATHWAYS

The major lead pathways include soil and dust, water, air, and food. Lead enters soil and dust from flaking and chipping lead-based paint, leaded gasoline, and industrial pollution, both directly (for example, a paint chip) and indirectly (for example, aerosol lead). A large volume of literature links lead in soil and dust with elevated blood lead levels (see Mielke 1983, Rabonowitz et al. 1985, Bellinger et al. 1986). Lead does not dissipate, biodegrade, or decay. Therefore lead in soil and dust will continue to represent a significant exposure pathway to children. Moreover, understanding the risk of exposure to lead in soil and dust warrants a historic perspective which focuses on how tracts of land have been utilized as urbanization and industrialization has proceeded.

The micro-geographic distribution of lead in soil and dust appears to be systematic. In a comprehensive review Hunter (1976) notes that aerosol lead fallout tends to decrease with increasing distance from sources, which include buildings covered with lead-based paint, highways, and smelters. In some cases this distance decay relationship may be modified by prevailing wind patterns, highway slope or gradient, age of vehicles on highway, vehicle speeds and actions, and modal mix. In a study in Minneapolis Mielke et al. (1984) report that soil lead levels increased from 5 parts per million (ppm) or less in rural areas to between 100 and 1,000 ppm in the central business district of the city. They also note considerable variability of soil lead levels within the city confines. Vertical gradients for soil lead appear to be steep.

Water becomes contaminated with lead mainly as a result of the presence of lead in water distribution systems. However, the corrosiveness of water (pH, alkalinity, and mineral content), and the duration of standing water contact both influence lead contamination. Absorption rates from water are typically greater than 50% for children (ATSDR 1988). However, when lead in soil, paint and dust is controlled for statistically, lead in water has been found to be insignificantly correlated with blood lead (suggesting water contamination is, in a relative sense, a less important pathway, Rabinowitz et al. 1985).

In many parts of the world industrial pollution and leaded gasoline emissions combine to make aerosol lead a major exposure pathway. Historically, too, airborne lead has assaulted the human body. In 1932-33 a measurement gauge, ironically placed in the grounds of a hospital, recorded the annual fallout equivalent of 2.2 tons of lead per square mile in Newcastle, England (Hunter 1976). However, in the United States at least, this pathway is now less important than it was, thanks largely to environmental legislation. For example, between 1970 and 1976, Billick et al. (1979) noticed as much as a 50% decline in the average quarterly air lead level recorded on the west side of Manhattan, New York. Readings of airborne lead levels are extremely sensitive to local geographic variations (for example, micro-climatological), time of day, prevailing weather conditions, and season (Hunter 1976, 1978, Muschett 1981, Greenland and Yorty 1985). Earickson and Billick (1988) find that air lead measurements recorded in 1979 were twice as high in downtown Louisville compared with the suburbs. Further, air lead values in Detroit generally declined with distance from the downtown and the lake. In addition to these urban gradients high dose exposures may still characterize locations close to smelters and other industrial point sources.

In concert with reductions in aerosol lead, the lead content of food has shown recent declines. For example, the Federal Drug Administration (FDA) estimated a 2-year old child would intake 30 mcg lead per day in 1982, 13 mcg per day in 1985, ad 5 mcg per day by the late 1980s. These improvements accompanied reductions in the proportion of food and soft drink cans that were lead soldered (down from 47% in 1980 to 1.4% in 1989, Centers for Disease Control 1991). However, food containers made outside the United States often contain lead solder. Additionally, lead may be present in some vegetables that are grown in lead contaminated soil. Other documented food pathways include food supplements, storage in lead-glazed ceramic containers or leaded crystal.

Lead enters the child's body through the lungs, the skin, and in complex ways across the gastrointestinal tract. Rates of absorption depend on the form in which the lead assaults the body and a range of other factors which include the presence of other foodstuffs and the general physiological state of the individual. For example, dietary calcium tends to block lead absorption across the gastrointestinal tract (Barton et al. 1978, Blake et al. 1983). The presence of fatty foods is thought to increase absorption (Smith 1992).

CORRELATES OF CHILDHOOD LEAD POISONING

As the above discussion implies, not all children are at equal risk from exposure to, and absorption of lead. Rates of lead poisoning should therefore vary across population sub-groups. However, few studies have attempted to establish "baseline" blood lead levels for the general population. Instead, research has concentrated on describing the conditions of children in "high risk" situations, and this has led to a relative abundance of studies of children living in inner city environments. In general terms, and with some exceptions (for example Fine et al. 1972, Bailey et al. 1994), rural and suburban environments are less intensively studied. Childhood lead poisoning is then popularly regarded as an urban phenomenon. The closing sections of this chapter suggest that a geographic analysis of childhood lead poisoning highlights the simultaneous and interdependent role of both urbanization and industrialization.

The second National Health and Nutrition Examination Survey (NHANES II) derived descriptive estimates of levels of blood lead for the United States population between 1976 and 1980 (Mahaffey et al. 1982). Five major correlates were uncovered, and have been widely replicated (Walter et al. 1980, Bornschein et al. 1985a, Brown 1990). First, blood lead levels were highest for children under five years of age. Increasing age was then negatively associated with blood lead level until the early teens. Between these ages and late middle age blood lead levels again rose. Finally, lead levels declined in older persons. Second, an individual's sex was only significantly related to blood lead level in adults. Males recorded blood lead levels approximately 50 percent higher than females in the adult age range. Third, African Americans had significantly higher blood lead levels compared to whites in the sample. This finding was robust across all age groups. The strong race effect has been reported elsewhere for African Americans and other minority groups, including Latinos and Asian Americans (for example, Billick et al. 1979, Daniel et al. 1990).

The NHANES II study also documented associations between both family income and degree of urbanization and blood lead levels. Among children, blood lead levels increased as total family income (from all sources) decreased. Means of blood lead levels ranged from 14 mcg in rural areas, 16.5 mcg in small urban areas (urbanized areas with a total population less than one million), and 18 mcg in large urban areas. Lead levels were also higher in central city locations compared to urbanized non-central city areas. Across both these socioeconomic correlates African Americans in the sample recorded blood lead levels 4 to 6 mcg higher than whites.

This influential baseline study-emphasizing the urban, low income, and racial component of elevated blood lead-also made mention of rural conditions. Here, increased lead exposure was regarded as some function of the presence of point sources (Landrigan et al. 1976), the presence of housing with lead-based paint, and the presence of in-migrant populations. Essentially, then, rural manifestations of lead poisoning were understood in terms of how urban phenomena applied to rural landscapes. Little attention was given, for example, to how processes of rural industrialization may differ from processes of industrialization in urban areas.

Continuing efforts to link indicators of urban structure with blood lead levels have highlighted the role of housing stock. Residence in modern public housing and residence in owner-occupied housing has been associated with lower blood lead levels when compared to residence in rental housing (Clark et al. 1985). Family structure is thought to be a surrogate for levels of child supervision, which are themselves argued to impact a child's exposure to lead (Stark et al. 1982).

In summary, then, a number of demographic and socioeconomic correlates have been linked to increased levels of blood lead in children. These correlates include age; sex; race; family income; degree of urbanization; housing stock; presence of in-migrant groups; family structure. While not proven, these associations resonate with our neurobiological expectations about lead sources, lead pathways, exposure, and absorption. For example, the young age effect is consistent with the idea that children absorb more lead than adults. The middle age effect among males is consistent with the idea that males are more likely than females to be employed in occupations which increase their exposure to lead (see Table 1). African Americans, Latinos, and Asian Americans are more likely to live in old, deteriorating rental housing than whites. This housing is, ceteris paribus, more likely to contain lead-based paint (which may be flaking and chipping), and it is more likely to contain lead contaminated water. Further, these residences are geographically concentrated in parts of urban (and perhaps rural) environments which have experienced the longest passage of industrialization and the heaviest volumes of automobile traffic, and which have high levels of lead in the soil and dust. Lead aerosol levels are still highest in these neighborhoods (Earickson and Billick 1988). Additionally, African American children tend to have a low calcium diet which may contribute to increased lead absorption (when present calcium helps to block absorption). Income is a particularly complex correlate because it acts through a variety of other factors, for example race, residence, and family structure. Income may also affect diet and promote relatively greater absorption rates among low-income children.

TOWARD A GEOGRAPHIC UNDERSTANDING OF CHILDHOOD LEAD POISONING

A geographic understanding of childhood lead poisoning is crucial in light of the available treatment and amelioration strategies that are available to health care providers. In extreme cases, where the body's lead burden poses an immediate hazard, lead can be removed using chelation therapy. This involves the use of drugs which bind the metal more tightly and make it more water soluble, in turn facilitating the urinary excretion of lead. Such extreme toxicity and treatment can be avoided using appropriate intervention and prevention techniques. Indeed, in the majority of cases, prevention is the most cost-efficient alternative (Anderson and Clark 1974, Berwick and Komaroff 1982, Martin 1991, Bellows 1993, Binder 1993). Prevention turns on education and abatement and is an essentially local, state-coordinated project. A knowledge of where to target scarce public health dollars assists care-givers in screening communities and identifying individuals who are most likely to be at risk from elevated blood lead levels.

Issues of scale and interdependence are critical elements of this geographic understanding. The environmental signature of lead poisoning is dependent upon the scale of interest. The geography of the lead hazard not only looks different across different environmental landscapes (the home, the yard, the neighborhood, the community, the town, the region) but is influenced by scale-dependent factors. The need to be aware of potential problems arising from the ecological fallacy can be illustrated with reference to a hypothetical example. In this example there are two towns, one of which (town B) possesses a lead smelter. However, at the town scale, town A has a mean blood lead level twice that of town B. On this evidence alone, we may falsely believe that the presence of a smelter has no impact, or actually reduces blood lead levels! A more local analysis could show that, indeed, in town B, blood lead levels were greatly increased in the vacinity of the point source, but due to some other factors (perhaps income structure or racial component), these elevated lead levels were masked for the town as a whole.

This hypothetical example points to the importance of considering both scale of inquiry and the interdependent relations between factors that promote lead poisoning (for example, there is often some relation between past industrialization and current residential structure). Three levels of interdependence can be recognized. First, places possess environmental characteristics which contribute directly to the risk of lead poisoning. These include the stock of lead-based paint, lead in soil and dust, lead in water distribution systems, lead aerosol, and point sources. These factors constitute the proximate source of lead exposure in the built environment and are the essential elements that must be reduced in any prevention strategy. Further, these factors are contained in a dynamic system. For example, flaking paint may enter the living spaces of the house directly or be taken up by garden vegetables from lead contaminated soil.

Second, places have a range of compositional factors that give rise to a systematic geographic distribution of lead poisoning within, and between places. These compositional factors may be demographic (racial and ethnic populations, in-migrant and immigrant populations, age-sex mix), industrial (heritage of industrial and, more recently, post-industrial development, giving rise to the current industrial mix), and socioeconomic (income and occupational characteristics). Not only is there a great deal of interdependence among these factors (for example, towns with a certain manufacturing base may attract immigrants and be especially susceptible to economic downswings, which then undermine the socioeconomic standing of residents), but it is also the case that interdependence exists between compositional and proximate factors.

To better understand this third source of variation a set of contextual effects can be highlighted. Contextual effects are those locally specific factors which are responsible for compositional factors interacting with proximate factors in different ways in different places. These contextual effects may be topographic (local relief may promote temperature inversions and make aerosol lead more likely to enter the soil), hydrologic (water quality can affect lead concentrations in distribution systems), geologic (availability of local building supplies can influence building practices, including density outcomes), and institutional (a broad category covering access to housing, access to health care, and access to social services in general).

Implicit in this discussion is the notion that the most general promulgators of lead poisoning-urbanization and industrialization-are inseparable. To suggest that lead poisoned children will only be found in inner-city areas, or in rural areas with "urban-like" characteristics, is to deny the complexity of industrial and urban processes. Further, such beliefs may also contribute to the systematic geographic biasing of prevention programs away from certain risk-groups. Indeed, where data have been made available on blood lead levels for children living in a range of built environments (urban, suburban, and rural), analysis confirms that the geography of lead poisoning is much more sophisticated than one derived from a simple rural-urban dichotomy (Bailey et al. 1994).

CONCLUSION

This chapter has argued that, in the United States, childhood lead poisoning is a serious public health issue that can be greatly improved with appropriate intervention strategies. Although the neurobiological mechanisms of lead poisoning are increasingly well understood, and although certain population groups have been consistently identified as high-risk, our ability to identify which geographic places are at greatest risk needs to be sensitive to issues of scale and interdependence. Specifically, the geography of childhood lead poisoning is connected to both urbanization and industrialization, and simple urban-rural markers of risk may be less appropriate than sociodemographic indicators such as poverty.

Promoting a geographic awareness of lead poisoning seems timely. Public health resources are scarce, and those places that are least able to afford new expenditures are often those places that have the highest risk of lead poisoning (Gellert 1993). The use of Geographic Information Systems and multi-level modeling can help identify high risk places. Perhaps, too, the geography itself is shifting, as the population of the United States variously suburbanizes, non-metropolitanizes, and gentrifies, as immigrants further increase inner city residential densities, and as the level at which lead poisoning is defined falls.

BIBLIOGRAPHY

Anderson, D.G. and J.L. Clark. 1974. Neighborhood Screening in Communities Throughout the Nation for Children with Elevated Blood Lead Levels. Environmental Health Perspectives. May:3-6.

ATSDR (Agency for Toxic Substance and Disease Registry). 1988. The Nature and Extent of Lead Poisoning in Children in the United States: A Report to Congress. ATSDR, Atlanta, GA.

Bailey, A.J., Sargent, J., Goodman, D., Freeman, J. and M.J. Brown. 1994. Poisoned Landscapes: The Epidemiology of Environmental Lead Exposure in Massachusetts Children 1990-1991. Social Science and Medicine. 39,6:757-766.

Barton, J.C., Conrad, M.E., Harrison, L. and S. Nuby. 1978. Effects of Calcium on the Absorption and Retention of Lead. J Lab Clin Med. 91:366-76.

Baser, M.E. and D. Marion. 1990. A Statewide Case Registry for Surveillance of Occupational Heavy Metals Absorption. Am J Public Health. 80:162-4.

Bellinger, D., Leviton, A., Rabinowitz, M., Needleman, H. and C. Waternaux. 1986. Correlates of Low-Level Exposure in Urban Children at 2 Years of Age. Pediatrics. 77:826-833.

Bellinger, D., Sloman, J., Leviton, A., Rabinowitz, M., Needleman, H. and C. Waternaux. 1991. Low-level Exposure and Children's Cognitive Function in the Preschool Years. Pediatrics. 87:219-227.

Bellows, J. 1993. The Initial Impact of a Workplace Lead-Poisoning Prevention

Project. American Journal of Public Health. 83:406-410.
Bewick, D.M. and A.L. Komaroff. 1982. Cost Effectiveness of Lead Screening. The New England Journal of Medicine. 306:1392-1398.
Billick, I.H., Curran, A.S. and D. Shier. 1979. Analysis of Pediatric Blood Levels in New York City for 1970-1976. Environmental Health Perspectives. 31:183-190.
Binder, S. 1993. Childhood Lead Poisoning: the Impact of Prevention. The Journal of the American Medical Association. 269:1679-81.
Blake, K.C. and M. Mann. 1983. Effect of Calcium and Phosphorus on the Gastrointestinal Absorption of 203Pb in Man. Environ Res. 30:188-194.
Bornschein, R.L., Succop, P., Dietrich, K.N., Clark, C.S., Que Hee, S. and P.B. Hammond. 1985a. The Influence of Social and Environmental Factors on Dust Lead, Hand Lead, and Blood Lead Levels in Young Children. Environmental Research. 38:108-118.
Bornschein, R.L., Hammond, P.B., Dietrich, K.N., Succop, P., Krafft, K., Clark, S., Berger, O., Pearson, D. and S. Que Hee. 1985b. The Cincinnati Prospective Study of Low-level Lead Exposure and Its Effects on Child Development: Protocol and Status Report. Environmental Research. 38:4-18.
Brown, M.J. 1990. Lead Poisoning in Children of Different Ages. The New England Journal of Medicine. 323:135-6.
Centers for Disease Control. 1991. Preventing Lead Poisoning in Young Children. U.S. Department of Health and Human Services, Public Health Service, Centers for Disease Control, Atlanta, GA.
Chisholm, J.J. 1986. Removal of Lead Paint from Old Housing: the Need for a New Approach. American Journal of Public Health. 76:236-7.
Chisholm, J.J. and H.E. Harrison. 1956. The Exposure of Children to Lead Pediatrics. 18:934-955.
Chisholm, J.J., Mellits, E.D. and S.A. Quaskey. 1985. The Relationship Between the Level of Lead Absorption in Children and the Age, Type, and Condition of Housing. Environmental Research. 38:31-45.
Clark, C.S., Bornschein, R.L., Succop, P., Que Hee, S.S., Hammond, P.B. and B. Peace. 1985. Condition and Type of Housing as an Indicator of Potential Environmental Lead Exposure and Pediatric Blood Lead Levels. Environmental Research. 38:46-53.
Daniel, K., Sedis, M.H., Polk, L., Dowuona-Hammond, S. and B. McCants. 1990. Childhood Lead Poisoning, New York City, 1988. Morbidity and Mortality Weekly Report. 39:1-7.
Dietrich, K.N., Succop, P.A., Berger, O., Hammond, P. and R.L. Bornschein. 1991. Lead Exposure and Cognitive Development of Urban Preschool Children: The Cincinnati Lead Study Cohort at Age 4 Years. Neurotoxicology and Teratology. 13:203-211.
Dyer, F.J. 1993. Clinical Presentation of the Lead-Poisoned Child on Mental Ability Tests. Journal of Clinical Psychology. 49:94-101.
Earickson, R.J. and I.H. Billick. 1988. The Aerial Association of Urban Air

Pollutants and Residential Characteristics: Louisville and Detroit. Applied Geography. 8:5-23.

Farfel, M.R. 1990. Health and Environmental Outcomes of Traditional and Modified Practices for Abatement of Residential Lead-based Paint. American Journal of Public Health. 80:1240-5.

Fine, P.R., Thomas, C.W., Suhs, R.H., Cohnber, R.E. and B.A. Flashner. 1972. Pediatric Blood Lead Levels: A Study in 14 Illinois Cities of Intermediate Population. Journal of The American Medical Association. 221:1475-1479.

Fulton, M., Raab, G., Thompson, G., Laxen, D., Hunter, R. and W. Hepburn. 1987. Influence of Blood Lead on the Ability and Attainment of Children in Edinburgh. Lancet. I:1221-6.

Gellert, G.A. 1993. Lead Poisoning Among Low-Income Children in Orange County, California: A Need for Regionally Differentiated Policy. The Journal of the American Medical Association. 270:69-71.

Greenland, D. and R.A. Yorty. 1985. The Spatial Distribution of Particulate Concentrations in the Denver Metropolitan Area. Annals of the Association of American Geographers. 75:69-82.

Hammond, P.B. 1982. Metabolism of Lead. In Lead Absorption in Children. Chisholm J.J. and D.M. O'Hara Eds. Urban and Schwartzenberg, Baltimore, MD, 11-20.

Hunter, J.M. 1976. Aerosol and Roadside Lead as Environmental Hazard. Economic Geography. 52:147-160.

Hunter, J.M. 1978. The Summer Disease - Some Field Evidence on Seasonality in Childhood Lead Poisoning. Social Science and Medicine. 12:85-94.

Landigran, P.J., Baker, E.L. and R.G. Feldman. 1976. Increased Lead Absorption with Anemia and Slowed Nerve Conduction in Children Near a Lead Smelter. J Pediatr. 89:904-910.

Landsdown, R., Yule, W., Urbanowicz, M.A. and J. Hunter. 1986. The Relationship Between Blood Lead Concentrations, Intelligence, Attainment, and Behavior in a School Population: The Second Study. Int Arch Occup Environ Health 57:225-235.

Mahaffey, K.R. 1983. Sources of Lead in the Urban Environment. American Journal of Public Health. 73:1357-8.

Mahaffey, K.R., Annest, J.L., Roberts, J. and R.S. Murphy. 1982. National Estimates of Blood Lead Levels: United States, 1976-1980. The New England Journal of Medicine. 307:573-579.

Maizlish, N., Rudolph, L., Sutton, P., Jones, J.R. and K.W. Kizer. 1990. Elevated Blood Lead in California Adults, 1987: Results of a Statewide Surveillance Program Based on Laboratory Reports. Am J Public Health. 80:931-4.

Martin, D. 1991. Lead Poisoning In Children: An Analysis of the Causes and Proposals for Prevention. Journal of Environmental Health. 54:18-19.

Mason, J.O. 1991. US Public Health Service's Strategic Plan to Eliminate Childhood Lead Poisoning. The Journal of the American Medical Association. 265:2049.

Mielke, H.H. 1983. Lead Concentrations in Inner-city Soils as a Factor in the Child Lead Problem. American Journal of Public Health. 73:1366-9.
Mielke, H.W., Blake, B., Burroughs, S. and N. Hessinger. 1984. Urban Lead Levels in Minneapolis: The Case of the Hmong Children. Environmental Research. 34:64-76.
Moehr, A.D. 1993. Childhood Lead Poisoning Near Abandoned Lead Mining and Smelting Areas: A Case Study of Two Affected Households. Journal of Environmental Health. 56:20-23.
Muschett, F.D. 1981. Spatial Distributions of Urban Atmospheric Particulate Concentrations. Annals of the Association of American Geographers. 71:552-565.
Needleman, H.L., Gunoe, C., Leviton, A., Reed, R., Peresie, H. and C. Maher. 1979. Deficits in Psychologic and Classroom Performance of Children with Elevated Dentine Levels. New England Journal of Medicine. 300:689-95.
Needleman, H.L., Schell, A., Bellinger, D., Leviton, A. and E.N. Allread. 1990. The Long Term Effects of Exposure to Low Doses of Lead in Childhood: An 11-year Follow-up Report. New England Journal of Medicine. 322:83-8.
Neggers, Y.H. 1986. Effects of High Lead Intake in Children. Journal of the American Dietetic Association. 86:938-40.
Piomelli, S. 1980. The Effect of Low Level Lead Exposure on Heme Metabolism. In Needleman H.L. Ed Low Level Lead Exposure: The Clinical Implications of Current Research. New York: Raven Press, 67-74.
Rabinowitz, M. 1985. Home Refinishing, Lead Paint, and Infant Blood Lead Levels. American Journal of Public Health. 75:403-4.
Rabinowitz, M., Leviton, A., Needleman, H., Bellinger, D. and C. Waternaux. 1985. Environmental Correlates of Infant Blood Lead Levels in Boston. Environmental Research. 38:96-107
Rosen, J.F., Chesney, R.W., Hamstra, A., DeLuca, H.F. and K.R. Mahaffey. 1980. Reduction in 1,25-dihydroxyvitamin D in Children with Increased Lead Absorption. New England Journal of Medicine 302:1128-31.
Rosner, D. and G. Markowitz. 1985. A 'Gift of God'?: The Public Health Controversy over Leaded Gasoline during the 1920s. American Journal of Public Health. 75:344-352.
Schroeder, S.R., Hawk, B., Otto, D.A., Mushak, P. and R.E. Hicks. 1985. Separating the Effects of Lead and Social Factors on IQ. Environmental Research. 38:144-154.
Sciarillo, W.G. 1992. Lead Exposure and Child Behavior. American Journal of Public Health. 82:1356-60.
Seligman, P.J. 1986. Occupation Lead Poisoning in Ohio: Surveillance Workers' Compensation Data. American Journal of Public Health. 76:1299-302.
Shaheen, S.J. 1984. Neuromaturation and Behavior Development: the Case of Childhood Lead Poisoning. Developmental Psychology. 20:545-550.
Smith, R.P. 1992. A Primer of Environmental Toxicology. Lea and Febiger, Philadelphia, PA.

Stark, A.D., Fitch, Q.R., Wister Meigs, J. and E.R. Delouise. 1982. Relationship of Sociodemographic Factors to Blood Lead Concentrations in New Haven Children. Journal of Epidemiology and Community Health. 36:133-139.

Steenland, K. 1992. The Mortality of Lead Smelter Workers: An Update. American Journal of Public Health. 82:1641-4.

Tepper, A. 1992. Surveillance of Occupational Lead Exposure in New Jersey: 1986-1989. American Journal of Public Health. 82:275-7.

Waller, K. 1992. Lead Exposure in the Construction Industry: Results from the California Occupational Lead Registry, 1987 though 1989. American Journal of Public Health. 82:1669-71.

Walter, S.D., Yankel, A.J. and I.H. von Lindern. 1980. Age-Specific Risk Factors for Lead Absorption in Children. Archives of Environmental Health. 35:53-58.

Environmental Contaminants, Ecosystems and Human Health. Edited by S.K. Majumdar, E.W. Miller, and Fred J. Brenner.

Chapter Twenty-One

LEAD AND OTHER HEAVY METALS IN URBAN AREAS AND CONSEQUENCES FOR THE HEALTH OF THEIR INHABITANTS

BRIAN E. DAVIES
Environmental Geochemistry Research Unit
Department of Environmental Sciences
University of Bradford
Bradford, West Yorkshire
BD7 1DP, England

INTRODUCTION

Heavy metal is a useful, but not very precise term. It generally refers to elements of a metallic nature with densities greater than 6 g cm^{-3}. Often it is extended to metalloids, especially arsenic, because interest centers on the toxicity of heavy metals to animals and humans. Heavy metals are also "trace elements," elements which occur in only tiny amounts in soils or animal tissue (usually <0.1%). Some heavy metals are essential for life processes in the sense that if they are not present in sufficient amounts in an organism it will suffer ill health or even die. Copper and zinc fall into this class, being needed for metabolic enzymes. In contrast, all heavy metals are toxic when their concentrations become too high, the toxicity threshold being different for different species. Some are well known poisons, such as arsenic which has a long history of homicidal use. Yet, even arsenic is needed in tiny traces for our health (1).

When metals or their compounds are heated, pulverized or dissolved they have a potential for escaping to the environment as contaminants. They follow natural geochemical pathways until they reach a sink where they then accumulate. Soils and sediments are major sinks since metal residues are lost from them only very slowly by leaching. Cities contain many sources of metals including vehicle and industrial emissions. Cities are not entirely concrete jungles: exposed soil surfaces are found in parks and waste areas and adjacent to houses as gardens or yards.

In western, industrialized countries most people live in towns and cities. For example, in the USA in the years 1970 and 1990 the urban population was 74% of the total population. In Europe as a whole the relative urban populations for the two same dates were however, 67% and 73%, i.e., increasing. Increases in urbanization are also happening in underdeveloped countries (2). Globally, the human habitat is now urban not rural. This juxtaposition of people living together in areas of high population density and an increasing and widespread use of heavy metal compounds inevitably exposes the population to these metals and invokes the danger of ill health arising from that exposure. Moreover, attention has increasingly switched from the clinically obvious, acute metal poisoning to the more insidious "sub-clinical" symptoms where health or well-being may be impaired but in a way that the cause is difficult to establish. Lead has received greatest attention. This chapter is concerned with describing first what we no know of the distribution of heavy metals in urban areas (urban geochemistry) and then with health problems relating to urban heavy metal exposure.

URBAN GEOCHEMISTRY

Whatever their differences in size, shape, architecture or street layout, world-wide research suggests that all towns and cities now have in common a heavy metal contamination problem in their soils and street and house dusts.

Leaded Petrol and Its Consequences

In the early years of this century there were rapid improvements in the design of the motorcar engine but the poor quality of the available petrol hampered the introduction of new models. The uneven combustion of petrol in the engine cylinders led to the damaging effect of "knocking". The discovery that tetraethyl lead dissolved in petrol stopped engine knocking revolutionized motoring. The first leaded petrol was sold in the USA in 1923, and rapidly became used world-wide. It is estimated that between 1926 and 1985, seven million tons of lead was burned in car engines in the USA (3).

It was not until the 1960s however, that much attention was paid to the possible dangers of inhaling the lead-rich aerosols emitted from vehicle exhausts. The introduction of lead-free fuel and the stepwise reduction of lead contents in petrol in the USA in the 1970s was not however, a response to the growing evidence that inhaled lead was deleterious. Rather, it was because of the introduction of platinum-based catalytic converters to reduce photochemical smog in Los Angeles and other cities since lead poisoned the platinum catalyst. The consequent reduction in consumption of leaded-petrol was matched by a steady decline in airborne lead levels in the USA (3). For example, in Minnesota annual lead fluxes to the ground surface through precipitation decreased over the period 1979 to 1983: 2000 to 370 ng cm^{-2} at an urban site and 430 to 100 ng cm^{-2} at a rural site (4). In Europe, the mandatory reduction of lead in petrol took place in 1986. Over the period 1984 to 1986 airborne lead at monitoring sites in western Britain decreased by 52-61% (5).

Warren and Delavault (6) in Canada discovered that lead was accumulating in the roadside ecosystem and suggested that the contributions of lead from petrol exhaust fumes merited attention. Cannon and Bowles (7) demonstrated that grass was contaminated by lead within 500 feet downwind of highways in Denver, Colorado and the lead content of soil decreased exponentially away from the road. In England, Davies and Holmes (8) reported that this roadside effect was also observed for major roads near Birmingham. The concentration of lead in roadside soils depends on traffic densities, topography, climate and wind conditions. The distance-decline relationship away from the road is curvilinear and local background levels are reached by some 30 to 50 m away from the road (9, 10, 11).

Changes in roadside soil lead levels have been monitored. Construction of new roads has usually been followed by a progressive development of a contaminated strip along the highway. In England, the Watford and South Mimms sections of the London Orbital (M25) motorway opened in late 1986. At South Mimms, the surface soil lead content before opening was 20.4 $\mu g\ g^{-1}$; at 6, 12, 18 and 24 months the respective soil lead concentrations were 164.1, 290.6, 421.5 and 446.9 $\mu g\ g^{-1}$ (12). Some soil lead levels have fallen following the phasing out of leaded petrol. Byrd et al. (13) determined soil lead contents adjacent to Interstate 20 in northeast Louisiana. Concentrations increased from 1973 to 1974 but decreased from 1973 to 1979.

Contamination of the roadside environment by lead has now been reported by many authors in many countries and it must be counted a world wide consequence of the use of leaded petrol.

Other metals which accumulate include nickel (added to petrol and originating in the metallic compounds of the vehicle), zinc and cadmium (from antioxidants in lubricating oils, in tires and in the car body or engine metal) (14). Ward (15) has shown that, beside lead, eleven elements (V, Cr, Mn, Co, Ni, Cu, Zn, Br, Mo, Cd and Ce) are concentrated in roadside soils and contents increase with increasing mean traffic density. Similar increases in metals other than lead have been reported for other countries (e.g., Australia (16), 1975; Hong Kong (17); Belgium (18); Greece (19). The increasing use of catalytic converters is now leading to increases in platinum road dusts and urban river sediments (20).

CONCENTRATIONS AND ORIGINS OF LEAD IN URBAN AREAS

Towns and cities are foci for traffic routes. As a city center is approached the road density increases and therefore the area of vehicle-contaminated land must also increase. It is not surprising that urban soils and dusts are intensively and extensively contaminated by lead and other heavy metals. Meteorologists speak of cities as being "heat islands" since larger cities tend to be warmer than the surrounding countryside. Similarly, we can speak of cities as being "heavy metal islands". Most of us live and work in this metal-rich environment.

Although early environmental concern was with air lead, the dangers posed by lead-rich soil and dust became apparent by the 1970s. Greenfield (21) commented that lead-contaminated dirt and dust are, theoretically, as available for children to ingest as is peeling lead paint. Lead contaminated dirt and dust could not be ignored as significant contributors to child lead body burden. Sayre (22) cast doubt upon the significance of lead in paint as the main source of childhood lead exposure. They proposed that lead-rich dust on fingers was more important.

Purves (23) first recognized that levels of several trace elements are high in urban soils. A study of garden soils in Edinburgh and Dundee, Scotland, revealed contamination by copper and boron. Purves and Mackenzie (24) later reported elevated concentrations of lead in Scottish urban soils.

Purves (23) speculated that soot and domestic chimney smoke might make a substantial addition of these elements to garden soils. Open coal fires are now rare in houses in urban Britain. The urban smogs and consequent raised mortality rates of the early 1950s led to the banning of coal as a domestic fuel in towns and only smokeless fuels may now be burned. Before then the accretion of ash in soil caused widespread low level contamination by lead and other metals (25).

Culbard et al. (26) surveyed lead and other metals in 5,235 housedusts an 4,650 garden soils collected from 100 households in each of 53 representative towns and city boroughs in Britain. This was a major and comprehensive national survey which has provided the baseline data necessary for future urban studies in Britain. Lead concentration ranged from 12 to > 104 $\mu g\ g^{-1}$ (geometric mean 230 $\mu g\ g^{-1}$) in 3,550 garden soils from locations which excluded London and areas affected by mining and smelting. In the 579 gardens sampled in 7 London Boroughs the mean value was 647 μg Pb g^{-1}. Agricultural soils in Britain usually contain from 10-150 μg Pb g^{-1}. Lower lead values in soils from households of less than 30 years of age compared with those of 100 or more years of age were considered by the authors to reflect (a) voluntary reduction in lead content of household paints and (b) a shorter history of soil contamination from bonfires and fossil fuel residues from open coal fires and soot commonly used as soil improvers an ameliorants.

Research in other countries has generally confirmed that urban soils or dust contain lead and other heavy metal residues (27, 28, 29, 30).

There are other sources of heavy metals besides vehicle emissions. Most cities have some industrial areas which emit metal-enriched dust particles and aerosols. Sweet et al. (31), showed that industrial complexes (smelters) were the most significant sources of copper and cadmium in the urban air of East St. Louis, Illinois while in Chicago manganese and chromium were derived from steel mills. A household rubbish incinerator in London, England was a source of cadmium and lead contamination (32). In Taiwan, a battery recycling plant polluted the area around a kindergarten with lead (33) and at Arnhem in the Netherlands a secondary lead smelter caused cadmium and lead pollution (34). In central London concentrations of street dust exceeded 5,000 μg Pb g^{-1} and a newspaper print works was identified as the source (35).

Paint and pigmented road markings are sources of heavy metals. From the late eighteenth century until the late nineteenth century the only white pigments used in paint were lead compounds. Since the end of the nineteenth century paint formulations have included TiO_2, $BaSO_4$, $CaSO_4$ and $CaCO_3$. White paints have generally relied on three carbonates of lead, namely, normal lead carbonate, PbCO3, and the basic lead carbonates, $2PbCO_3.Pb(OH)_2$ and $4PbCO_3.2Pb(OH)_2.PbO$. Other white lead compounds in use include $PbSO_4$ and the silicate $3PbO.2SiO_2.H_20$. Red colors can be based on red lead (Pb_3O_4) and the chromates ($PbCrO_4$ and $PbCrO_4.PbO$) provide the yellow colors for street markings (36).

Solomon and Hartford (37) noted raised lead levels in soil near buildings in Champaign-Urbana, Illinois, and ascribed this in part to flaking paint. Preer et al. (38) described this same near-house effect in Washington, DC. Where roof runoff water goes direct to the soil it carries with it metals derived both from deposited dust and dissolved from roofing material (39).

Because young children are particularly susceptible to lead poisoning the environment of schools has been investigated to identify health hazards. Rundle and Duggan (40) studied seven schools in inner London which were due to be repainted externally during the summer of 1982. Before this work started, mean lead concentrations in the playground dust ranged from 280 to 4,600 μg Pb g^{-1} and during the work the range was 370-53,000 μg g^{-1}. One individual sample contained 30% lead. Later, Schwar and Alexander (41) showed that the problem of high lead-in-dust concentrations in London was associated overwhelmingly with pre-1939 schools.

We still know relatively little about the distributional patterns of metals in city soils and dust. The examples cited above demonstrate that high concentrations will occur around significant point sources. But, with the exception of lead, we do not know whether there is any common zonation pattern. Traffic density is greater in city centers and this is also where the older buildings with lead paint occur. This increase in concentration of sources of lead towards the downtown area is paralleled by an increase in soil/dust lead concentrations. Mielke et al. (42) studied the four Minnesota cities, Minneapolis, St. Paul, Duluth and Rochester. Soil lead build up was mainly a function of urban size. Although specific topographical factors, such as hills, influenced the pattern of soil lead content generally there was a steady increase from the suburbs to the center.

In London a study was made of a transect from Hyde Park (in the center of London) to the rural hinterland 60 km to the north (43). The combined lengths of main roads were estimated from maps for successive 12 x 6 km areas. It was assumed that, if there was open ground along side a road, then there would be a 30 m zone of lead accumulation on each side of the road. Figure 1(A) plots the proportion of the total area theoretically contaminated by vehicle exhaust fallout. Davies et al. (44) reported that park and garden soils in London were contaminated by lead and concentrations progressively declined along a northerly transect from the center (Hyde Park). Subsequently this same transect was resampled more intensively (43) and the results are plotted in Figure 1B. The similarity between distributions A and B is clear and supports the hypothesis that vehicle exhaust fumes are a major source of soil lead. Davies and Houghton (45) published a similar study done in Birmingham and showed a similar distribution of soil lead along a southeasterly transect from the city center. Probably the distribution of soil lead fits this curvilinear model for all major British cities.

Soil lead contamination is not confined to major towns or cities. Davies (25) reported the lead content (EDTA-extractable) of soils from 87 British gardens in both rural and urban environments: 63% of the soils were contaminated when compared with rural norms. Although the highest values were derived from gardens in lead mining areas and city gardens, older, rural gardens were also mildly contaminated.

There appears to be a general anthropogenic process, a "habitation" effect, leading to accumulations of lead in garden soils. In two villages in Devon and Cornwall significant correlations were found between the age of the house or garden (i.e., duration of occupancy) and the soil lead content (25). A study of three traffic-less islands (Aran and Tory, off-shore from Ireland, and Sark, one of the Channel Islands) revealed raised lead contents in garden soils (27 - 128 μg Pb g^{-1}) despite the remoteness of these islands (46).

INDOOR/OUTDOOR RELATIONSHIPS

Not only do most of us live and work in cities we also spend most of our day indoors. Thus, the indoor environment has a special importance for our health.

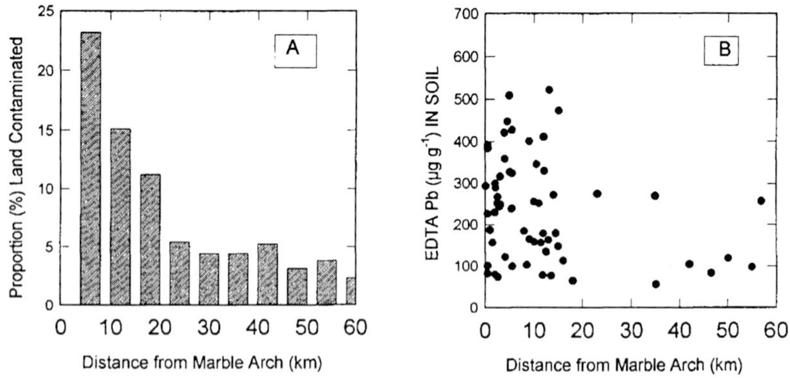

FIGURE 1.
Lead contamination in London
A: Estimated proportion of land contaminated by vehicle fumes in each zone.
B: EDTA-extractable lead in urban soils.

Where metals are concerned, greatest attention has again been paid to lead.

Lead inside a building may be derived from internal sources, for example, lead-based paints. Lead also comes in from outside by tracking in soil or dirt (47, 48) or as fine dusts in draughts (49).

Outdoor and indoor air exchange occurs continuously through draughts around doors and windows or intermittently when doors and windows are opened. Studies have shown that indoor air lead concentrations are less than those immediately outside the house (50, 49). Concentrations of lead in settled house dust correlate with contents in street dusts or garden soils but, as with air-borne levels, they may be lower than those outside (51, 52). In a study in north Wales, Davies et al. (48) concluded that about 30% of the lead in settled house dusts could be accounted for by lead in the garden soil outside. Culbard et al. (26) wrote "Pb in housedust is significantly related to the concentration of Pb in garden soil as well as the area of soil exposed".

Distribution of Lead Inside the Home

Redistribution within the home occurs as air-borne dusts settle and aerosols or very fine dust particles adhere to surfaces through impaction or electrostatic attraction. Settled dusts can temporarily be resuspended during normal household activities and redecoration may result in lead from old paint entering the house lead cycle. It is important to know something of the "geography of lead" inside the home in order to identify where young children might be at highest risk.

The studies of Thornton and co-workers provided early information on the distribution of lead within the home. In Birmingham the highest concentrations were found in door mats (79 - 15,000 μg Pb g^{-1}, geometric mean = 615 μg g^{-1}) compared with a range of 1 - 1,348 μg g^{-1} (geometric mean = 51 μg g^{-1}) in bedrooms (53). Houses undergoing redecoration at the time of sampling had almost double the lead concentrations in their floor dusts (26).

A recent study in England (54) has further clarified the in-house distribution of lead. Within a home, air lead concentrations were approximately the same upstairs and downstairs but there was less lead in settled floor dusts upstairs compared with downstairs and a front room downstairs had more lead in floor dust than a back room. Lead and dust loading on an entrance hall floor decreased away from the front door. Lead loadings on other surfaces were in the order window sills > furniture > walls > ceilings. Net curtains on windows effectively trapped lead aerosols. More lead was tracked into a hallway in wet weather and more lead entered through windows in summer months when windows were left open. Overall, lead loadings were higher for houses on busy streets near city centers. One unusual source of very high concentrations of lead was a leaded window (55). Such windows were popular last century and earlier this century in British houses. Corrosion of "cames", (i.e. the lead binding or decorative strips in these windows) is a previously unreported source of lead within older homes.

Surfaces close to the leaded windows were highly contaminated with lead but the content of lead in dust diminished rapidly within the first 2 m away from the window and was minimal after 3 m. The dust near the window was identified as mainly composed of basic lead carbonate. Condensed water on leaded windows contained appreciable lead whereas for unleaded windows it was negligible.

These data imply a complex interaction between the indoor and outdoor environment. Outside lead (soil, dust or airborne) enters the home and is subject to redistribution. The building itself acts as a collection surface for air-borne lead and thereby contributes to lead in garden soil as does any old, flaking paint. This interaction may explain the habitation effect discussed above.

Renovation of older houses can significantly increase the lead loading in the home environment. Marino et al. (56) reported the consequences of renovating a Victorian farmhouse in rural upstate New York. Over a 10-week period of 1987 hand sanders and blow torches were used to remove old paint from rooms throughout the house. The family was absent from their home during most of the work. On their return they found a thick dust had spread throughout the house. By mid-November, members of the family reported stomach aches and nausea. Blood lead levels proved to be so high that chelation therapy was needed i.e., patients were administered chelating agents to reduce body lead burdens. Chemical strippers should always be used during renovation, not sanding or torching.

Water

Finally, mention must be made of the quality of drinking water. Old houses may still have lead water pipes. Where the drinking water is soft (low in calcium salts and acid in pH) lead dissolves from the pipe wall (plumbosolvency) and first draw water, the water standing in the pipe overnight contains a significant dissolved lead burden (57). This water, when used for drinking, can significantly increase blood lead concentrations (58) and it is important to let the tap run for several minutes before filling a kettle or coffee pot. Solder in the plumbing system can give rise to the same problem (59). Corrosion of other metal pipes can increase ingested Cu, Fe and Zn (60) and even plastic (UPVC) pipes can cause problems (61).

HEALTH EFFECTS OF METALS

Frank cadmium and lead poisoning are adequately described in toxicological texts and reviews (62, 63). Clear cases of clinical toxicity are normally seen only in industrial exposure or after the accidental swallowing of metal-rich substances (metal salts, leaded paint chips, etc.). In Third World countries a widespread urban cottage industry is the recovery of gold and silver from jewelers' waste. In Sri Lanka, children exposed in this way had blood levels up to 56 $\mu g\ dl^{-1}$ (64). There are also occasional reports of some unusual sources of lead. A lead-based eye cosmetic (surma, khol) is traditional in Middle Eastern countries and can be hazardous (65). In Mexican-American children, lead poisoning has been encountered following the parental administration of *azarcon*, a folk remedy for colic which contains 86-93.5% Pb_3O_4 (66).

LEAD

Sub-clinical effects of environmental exposure have been the subject of continuous research for the last thirty years. Lead has received greatest attention and especially for children. Children are more susceptible to lead poisoning than adults and toddlers spend much of their time playing on the floor and sucking their fingers or licking toys and other objects. In some cases they develop a marked tendency to ingest soil or dust, a condition known as pica. Air, water and soil or dust are generally considered as the most important sources of lead in the urban environment. Lead in food can be a significant source, especially where people rely heavily on home grown vegetables in contaminated soils (67), but food usually comes from rural areas.

Blood Lead Levels: Limits and Changes

The relative ease of obtaining samples of blood for analysis have made it the tissue of choice for monitoring body lead burdens through determinations of blood-lead (68, 69). Blood lead concentrations are quoted as $\mu g/100$ ml or $\mu g\ dl^{-1}$ blood.

Historically, an acceptable blood lead concentration has been defined as the concentration below which adverse health effects are not likely to occur. Until about 1970 the limit for adults was 80 $\mu g\ dl^{-1}$ and for children, 50 - 60 $\mu g\ dl^{-1}$. The past twenty years have seen a steady lowering of the acceptable level in the USA and other countries (70). In 1975 the Centers for Disease Control (CDC) in Atlanta recommended 30 $\mu g\ dl^{-1}$ for children, reduced it to 25 $\mu g\ dl^{-1}$ in 1985 and then, in 1991, recommended a new limit of 10 $\mu g\ dl^{-1}$ as an action or intervention level for children (71). These USA limits can be compared with some values for remote populations. In the Himalayas, male blood lead (average value) has been reported as 3.4 $\mu g\ dl^{-1}$ and 2.9 $\mu g\ dl^{-1}$ for females and in Papua-New Guinea children, 5.2 $\mu g\ dl^{-1}$ (72, 73).

One of the symptoms of early chronic lead poisoning is anaemia. This is now known to be a consequence of lead's deactivation of the enzyme (-aminolaevulinic acid dehydratase or ALAD. This is one of the enzymes needed for the formation of haem (and hence, haemoglobin and red blood corpuscles) and is unusually sensitive to the presence of lead. Typical results are those of Roels et al. (74) who studied children in the 10-13 year age group in a rural, uncontaminated area and at two distances (<1 km and 2.5 km) from a lead smelter. Those closest to the smelter had a mean blood lead of 30.1 $\mu g\ dl^{-1}$ and ALAD activity 25.8; at 2.5 km blood lead was 21.1 $\mu g\ dl^{-1}$ and ALAD activity 38.3; in the rural area blood lead was 9.4 $\mu g\ dl^{-1}$ and ALAD activity was 53.6 (ALAD activity was measured as nmol per ml erythrocytes per minute).

Over recent years, blood lead levels have been declining in the general USA population according to analyses of data from the second National Health and Nutrition Examination Survey (NHANES II). Annest et al. (75) concluded that from 1976 to 1980 there was a reduction of 37% in average blood lead levels in the USA. Twenty two percent of the population had blood lead levels <10 μg d^{l-1} and 1.9% had blood lead >30 μg dl^{-1}.

Blood Lead, Air Lead and Leaded Petrol

The consensus view (76) is that as the concentration of lead in petrol has been reduced and as lead-free petrol is increasingly used, blood lead levels are declining (77). Annest examined the NHANES II data using regression analysis of blood lead levels on US EPA gasoline lead values. They suggested that the major contributory cause in changes in blood lead was the reduction of leaded gasoline use. The same trends are being observed in other countries. In South Africa, the permissible petrol lead content has been decreased from 0.836 g $^{l-1}$ in 1983 to 0.4 g $^{l-1}$ in 1989 and over this period the blood lead levels in a study population in Capetown dropped from 9.7 (g d^{l-1} to 7.2 (g d^{l-1} (78). In Helsinki, Finland (79) the average lead content of petrol has fallen from 0.55 g $^{l-1}$ in 1980 to 0.14 g $^{l-1}$ in 1988. The blood lead levels of children using a city center day care center fell from 4.8 (g d^{l-1} in 1983 to 3.0 (g d^{l-1} in 1988.

The best direct evidence is the Turin (Italy) Isotopic Lead Experiment which started in 1975 and ended in 1985 (80). Petrol was distributed containing lead derived only from the Australian Broken Hill Mine where the ore has a characteristic isotopic ratio of 296Pb/207Pb = 1.04. Before the start of the experiment the isotopic lead ratio in the petrol then being distributed was 1.19. Subsequent isotopic analyses of human blood (81) indicated that about 24% was contributed by petrol lead.

Conclusions concerning the consequences of reducing petrol-lead exposure have generally come from correlations, not from any direct evidence, and correlation is not necessarily the same as cause and effect. Elwood (82) was unconvinced by the reported associations between air lead and blood lead studies and preferred water lead as the main source of human exposure. Larbey (83) has also argued that the use of lead in petrol does not result in adverse health effects. These dissenting authors are however, in the minority.

Blood Lead and Water Lead

Plumbosolvency is well recognized as a cause of acute or chronic lead poisoning (58) although the symptoms can be confused with other diseases. Two case studies suffice to illustrate the problem. A patient in Yorkshire, England, complained of nausea and retching associated with colic and constipation. At first, he was treated for an intestinal obstruction but hospitalization failed to confirm this. Subsequently, his blood lead level was found to be 105 $\mu g\ dl^{-1}$ and his first-run tap water contained 6.0 mg Pb l^{-1} (84). In Scotland, a patient and his wife suffered abdominal discomfort and nausea intermittently over ten years. After an attack of vertigo and vomiting the quality of the water was suspected and a blood level of 418 $\mu g\ dl^{-1}$ was found. The water storage tank was lead lined and water analysis showed 0.32 mg Pb l^{-1} (85). Fortunately, it is a fast vanishing problem since the cure, replacing lead piping with plastic piping, is well known and limited only by an expense incurred.

Under more normal circumstances, the influence of drinking water on blood lead is difficult to quantify. Elwood et al. (86) studied women in areas of Wales (Britain) chosen to represent very different levels of exposure to traffic. They concluded that although air made a small, but significant contribution to blood lead water emerged as an important contributor to blood lead even though water lead was not high. In Glasgow, Lacey et al. (87) the contribution of intake of lead in water to blood lead levels of infants was difficult to quantify yet water did have an effect.

Lead in Blood and Dust or Soil

In the last ten to fifteen years attention has centered on soil and dust as important sources of lead exposure, especially for children. Barltrop et al. (88) studied the lead burden of children and mothers exposed to lead-contaminated soils and dusts in rural areas with minimal atmospheric pollution in England. They concluded that this source represented a small additional burden but in itself was insufficient to cause a hazard. In contrast Charney et al. (89) after a study of inner-city children in Rochester, New York, reported that lead contamination of house dust was a major factor in causing moderately elevated blood lead levels. In Cincinnati, Ohio, Bornschein et al. (90) clearly demonstrated a relationship between blood lead and dust lead for children up to 2 years of age. A major review of the problem was published by Duggan et al. (91). They summarized all the then available evidence from the international literature and paid special attention to the form of the suspended soil/dust - blood relationship. Their broad conclusion was that children's blood lead increases by 5 $\mu g\ dl^{-1}$ for every 1,000 $\mu g\ g^{-1}$ increment in dust lead.

Recently, Wixson and Davies (92) have summarized a report of a working party of the Society for Environmental Geochemistry and Health (93). A formula was proposed (Figure 2) which relates blood lead directly to soil lead assuming soil or dust (interior or exterior) represent the main exposure source. The formula allows the identification of target soil/dust lead levels which may then be used as a basis for remediation. Examples of the computational results are given in Table 1. It has been noted earlier that city soils quite commonly contain lead levels of around 300 - 500 g g^{-1} and the values in the Table suggest that most cities represent some degree of lead hazard.

Target soil dust lead guideline model

$$S = \frac{\left|\frac{T}{G^{n}} - B\right|}{\delta} \cdot 1000$$

where:

S is the soil or dust guideline, a geometric mean concentration in micrograms of lead per gram of dust ($\mu g\ g^{-1}$)

T is the blood lead guideline or target concentration ($\mu g\ dl^{-1}$) whole blood

G is the geometric standard deviation of the blood lead distribution, typically in the range of 1.3 to 1.5

B is the background or baseline blood lead concentration in the population from sources other than soil and dust

n is the number of standard deviations corresponding to the degree of protection required for the population at risk, and would normally follow from the way in which the blood lead guideline T was defined

δ is the slope or response of the blood lead-soil (dust) lead relationship and has the units of $\mu g\ dl^{-1}$ blood increase per 1000 $\mu g\ g^{-1}$ increment of soil or dust lead

FIGURE 2.

TABLE 1
Variation of Soil Lead Guideline with Target Blood Lead Concentration and Degree of Desired Protection[a].

Target PbB (μg/dL)	Soil lead standard for % of population, target PbB 95%	98%	99%	99.9%
10	880	500	300	-
15	2300	1860	1400	700
20	3750	4250	3700	2500
25	5200	4250	3700	2500

[a]Assumptions: δ = 2, background PbB = 4 μg dl-1, geometric standard deviation (GSD) = 1.4

General Health Effects Arising from Lead Exposure

Probably the most serious and contentious consequence of lead exposure for children is a possible effect on their mental development. Needleman (94) has argued that relatively low blood lead concentrations affect neuropsychological performance as demonstrated by psychometric intelligence, reaction time to stimuli and classroom behavior. In the discussion after his paper Needleman quoted work that suggested for a mean blood lead level of 13 $\mu g\ dl^{-1}$ there was still a 7 point IQ difference between children with higher and lower lead levels.

Needleman relied on psychological and other tests which some claim are inherently irreproducible. The work has also been criticized on the grounds that inclusion of socio-economic variables in the data analysis reduces any lead effect to below normal significance levels. Harvey et al. (95) reported a study in Birmingham, England with the clear statement that "the conclusion of this study is that there is no statistically significant association between blood lead and IQ after accounting for other factors in the child's environment". In contrast, Marecek et al. (96) worked with black inner-city children in Philadelphia and claimed that their results provided strong evidence that, even at moderate levels, lead intake results in decerements in neuropsychological functioning and lead in the environment represents a hazards to inner-city children. David et al. (97) went further. They treated hyperactive children (in Brooklyn, New York) with chelation therapy which reduced blood lead from a mean of 29.34 to 19.50 $\mu g\ dl^{-1}$ (there were also control groups receiving placebos) and resulted in statistically significant behavioral improvement. Goyer (71) has reviewed progress over twenty years in understanding the potential toxicity of lead. He states that "Taken together, the . . . studies demonstrate that lead has an adverse health effect on neurobehavioral and cognitive development at PbB levels in the 10 15 $\mu g/dL$ range in infants, children, and the fetus, and perhaps, at even lower levels."

In adults, the greatest interest concerning low-level lead exposure has been over cardiovascular disease. Schwartz (98) examined the NHANES II data and found a robust relationship between low-level lead exposure and blood pressure for adult men older than twenty grouped by age. He concluded that measures to reduce lead exposure in the U.S. population will lead to small, but important reductions in cardiovascular risk. Harlan (99) concurred that the NHANES II data confirmed the relationship of blood lead and blood pressure. In contrast, in the Netherlands (100) there was found no association between blood lead and coronary heart disease incidence and (101) in Wales, Great Britain (101). However, Goyer (71) concluded that an effect on blood pressure is now identified as the critical effect of lead in adults.

CADMIUM

Environmental cadmium first attracted attention in subsistence farmers in Japan. Osteoporosis in elderly, multiparous women was linked to excessive cadmium intake through fish living in and rice irrigated by cadmium-contaminated river water. The disease is now generally known as Itai-Itai disease but has not been identified outside Japan. Nomiyama (102) has reviewed that work. Other health problems which have been associated with environmental cadmium are hypertension, kidney damage and cancer (103).

Cadmium is absorbed both by inhalation and ingestion. Friberg (104) suggested that inhalation leads to 20 - 50% absorption and ingestion to 5% absorption. Cadmium is transported to the liver where it stimulates the synthesis of metallothionein and is then carried in the blood to the kidney. The critical kidney concentration is 200 $\mu g\ g^{-1}$ in the renal cortex at which level kidney damage is observed (104).

There is general agreement that vegetables comprise the main source of dietary cadmium. In addition, cigarette smoking is also important. Sartor et al. (105) reported that smoking 20 cigarettes per day increased cadmium body burden in Belgium by about 63% at age 50. Cadmium in vegetables was locally important and since 1985 people living in one area had been advised to refrain from eating vegetables grown in their own gardens.

There are rural locations where cadmium contents of the soil may be very high through pollution or geological causes. In Britain, the Somerset village of Shipham achieved some notoriety when it was discovered that soils there contained the highest cadmium contents then on record. However, the cadmium was not especially plant-available (106) and dietary intake was not regarded as excessive (107). Interestingly, the major soil cadmium problem in the USA and other countries arises from the application of sewage sludge to land, the contaminated sewage having come from cities with major industries and large populations (108).

Despite the evidence that city dusts and soils contain large residues of cadmium the implications have not been followed up. In the case of lead we have seen that the hand to mouth pathway is of greatest importance. There is no reason why cadmium should not reach the body by the same route.

CONCLUSIONS

There can now be no doubt that urban areas throughout the world contain significant accumulations of biologically active metals in their soils and dusts. The metal which has received greatest attention is lead because of the ubiquitous residues arising from the widespread use of petrol containing lead additives. Children are particularly susceptible to ill health effects arising from exposure to lead and, although still somewhat contentious, many workers believe that the main health effect is on mental development. This is an area of public health which undoubtedly will continue as a major focus of research in the future.

The latest evidence suggests that the levels of dust lead which are common in the urban environment can, through hand-to-mouth transfer, cause blood lead levels that are now regarded as hazardous. This exterior leaded-dust carries into the interior of homes and may be supplemented by other sources, especially paint. Most authorities now believe it is prudent to find effective means of abating lead, both inside and outside, despite the need for further studies to confirm the lead-mental development link in children.

The evidence for health effects in adults arising from the concentrations of lead commonly found in cities is less convincing. Cardiovascular disease is a major cause of death, especially in men, in many countries and although a role for lead in hypertension is suspected it appears small.

Cadmium is also of considerable interest in public health. Cigarettes and food appear to be the main sources and that in food arises from the use of contaminated sewage sludge as a manure. Sewage sludge is derived from urban areas and a reduction in environmental cadmium in cities would lead to a reduction in the build up of cadmium residues in rural soils. What is surprising is that a possible role for cadmium in health through hand-to-mouth transfer of city dust has been neglected.

There seems little likelihood that other metals in the city environment will be shown to have the same relatively direct effect on health which is seen for lead. Their effects on human health are more likely to be subtle. Nonetheless, the metal residues found in the urban environment are biologically active and can reach children and adults by the same routes as for lead. More needs to be done in comparing body burdens of these other heavy metals in rural and city dwellers.

REFERENCES

1. Uthus, E.O. 1992. Evidence for arsenic essentiality. Environ. Geoch. Hlth. 14:55-58.
2. UNEP. 1991. United Nations Environment Programme Environmental Data Report, 3rd ed. Oxford, England: Denis Blackwell.
3. Nriagu, J.O. 1990. The rise and fall of leaded gasoline. Sci. Tot. Environm. 92:13-28.
4. Eisenreich, S.J., Metzer, N.A., Urban, N.R. and J.A. Robbins. 1986. Response of atmospheric lead to decreased use of lead in gasoline. Environ. Sci. Technol. 20:171-174.
5. Page, R.A., Cawse, P.A. and S.J. Baker. 1988. The effect of reducing petrol lead on airborne lead in Wales, UK. Sci. Tot. Environ. 68:71-77.
6. Warren, H.V. and R.E. Delavault. 1960. Observations on the biogeochemistry of lead in Canada Trans. Roy. Soc. Canada. 54:11-20.
7. Cannon, H.L. and J.M. Bowles. 1962. Contamination of vegetation by tetraethyl lead. Science. 137:765-766.
8. Davies, B.E. and P.L. Holmes. 1972. Lead contamination of roadside soil and grass in Birmingham, England, in relation to naturally occurring levels. J. ag. sci. (Camb.) 79:479-484.
9. Mottom, H.L., Daines, R.H., Chilko, D.M. and C.K. Motto. 1970. Lead in

soils and plants: its relationship to traffic volume and proximity to highways. Environ. Sci. Technol. 4:231-238.
10. Smith, W.H. 1976. Lead contamination of roadside ecosystem. J. Air Poll. Control Assoc. 26:753-766.
11. Wheeler, G.L. and G.L. Rolfe. 1979. The relationship between daily traffic volume and the distribution of lead in roadside soil and vegetation. Environmental Pollution. 18:265-274.
12. Ward, N.I. 1990. Lead contamination of the London Orbital (M25) motorway (since its opening in 1986). Sci. Tot. Environ. 93:277-283.
13. Byrd, D.S., Gilmore, J.T. and R.H. Lea. 1983. Effect of decreased lead in gasoline on the soil of a highway. Environ. Sci. Technol. 17:121-123.
14. Falahi-Ardakani, A. 1984. Contamination of environment with heavy metals emitted from automotives. Ecotox. Env. Safety. 8:152-161.
15. Ward. N.I. 1990. Multielement contamination of British motorway environments. Sci. Tot. Environ. 93:393-401.
16. David, D.J. and C.H. Williams. 1975. Heavy metal contents of soils and plants adjacent to the Hume highway near Marulan, New South Wales. Aus. J. Exp. Agric. Animal Husb. 15:414-418.
17. Ho., Y.B. 1988. Monitoring lead, copper and manganese levels in soil, dust and vegetation at two roadside sites in Hong Kong from 1978 to 1987. Mem. Hong Kong Nat. Hist. Soc. 1988:41-46.
18. Albasel, N. and A. Cottenie. 1985. Heavy metal contamination near major highways, industrial and urban areas in Belgium grassland. Wat. Air Soil Poll. 24:103-109.
19. Fytianos, K., Vasilikiotos, G. and V. Saminidov. 1985. Contamination of roadside vegetation with lead, cadmium and zinc. Chemosphere. 14:271-277.
20. Wei, C. and G. Morrison. 1994. Platinum in road dusts and urban river sediments. Sci. Tot. Environ. 146/147:169-174.
21. Greenfield, S.M., Bridbord, K., Barth, D. and R. Engel. 1972. The changing perspectives regarding lead as an environmental pollutant. In Environmental Health Aspects of Lead, proceedings International Symposium, ed. D. Barth, A. Berlin, R. Engel, P. Recht, and J. Smeets. Luxembourg: Commission of the European Communities.
22. Sayre, J.W., Charney, E., Vostal, J. and I.B. Pless. 1974. House and hand dust as a potential source of childhood lead exposure. Am J Dis Child. 127:167-170.
23. Purves, D. 1966. Contamination of urban garden soils with copper and boron. Nature. 210:1077-1078.
24. Purves, D. and E.J. Mackenzie. 1970. Enhancement of trace-element content of cabbages grown in urban areas. Pl. Soil. 33:483-485.
25. Davies, B.E. 1978. Plant-available lead and other metals in British garden soils. Sci. Tot. Environ. 9:243-262.
26. Culbard, E.B., Thornton, I., Watt, J., Wheatley, M., Moorcroft, S. and M. Thompson. 1988. Metal contamination in British urban dusts and soils. J.

Environ. Qual. 17:226-234.
27. Klein, D.H. 1972. Mercury and other metals in urban soils. Environ. Sci. Technol. 6:560-562.
28. Beavington, F. 1973. Contamination of soil with zinc, copper, lead and cadmium in the Wollongong city area. Austral. J. Res. 11:27-31.
29. Fleming, G.A. and P.J. Parle. 1977. Heavy metals in soils, herbage and vegetables from an industrialized area west of Dublin city. Irish J. Agric. Res. 16:35-48.
30. Fergusson, J.E., Forbes, E.A., Schroeder, E.J. and D.E. Ryan. 1986. The elemental composition and sources of house dust and street dust. Sci. Tot. Environ. 50:217-221.
31. Sweet, C.W., Vermette, S.J. and S. Landsberger. 1993. Sources of toxic trace elements in urban air in Illinois. Env. Technol. 27:2502-2510.
32. Wadge, A. and M. Hutton. 1987. The cadmium and lead content of suspended particulate matter from a U.K. refuse incinerator. Sci. Tot. Environ. 67:91-95.
33. Wang, J.-D., Jang, C.-S., Hwang, Y.-H. and Z.-S. Chen. 1992. Lead contamination near a battery recycling plant. Bull. Environ. Contam. Toxicol. 49:23-30.
34. Keizer, M.G., Hooghiemsta-Tielbeek, M. and F.A.M. de Haan. 1982. Contamination of soil and street dust with lead and cadmium near a lead smelter at Arnhem, Netherlands. Netherl. J. agric. sci., 29:227-235.
35. Jensen, R.A. and D.P.H. Laxen. 1985. Sources of lead in urban dust: identification of a contribution from newspaper printworks. Sci. Tot. Environ. 46:19-27.
36. Dunn, E.J. 1975. Chapter 7: Lead Pigments. In Treatise on Coatings, vol. 3, ed. R.M. Myers and J.S. Long. New York: Marcel Dekker.
37. Solomon, R.L. and J.W. Hartford. 1976. Lead and cadmium in dusts and soils in a small urban community. Environ. Sci. Technol. 10:773-777.
38. Preer, J.R., Akintoe, J.O. and M.L. Martin. 1984. Metals in downtown Washington, DC gardens. Biol. Tr. Elements. 6:79-91.
39. Quek, U. and J. Forster. 1993. Trace metals in roof runoff. Wat. Air Soil. Poll. 68:373-389.
40. Rundle, J.A. and M.J. Duggan. 1986. Lead pollution from the external redecoration of old buildings. Sci. Tot. Environ. 57:181-190.
41. Schwar, M.J.R. and D.J. Alexander. 1988. Redecoration of external leaded paintwork and lead-in-dust concentrations in school playgrounds. Sci. Tot. Environ. 68:45-59.
42. Mielke, H.W., Burroughs, S., Wade, R., Yarrow, T. and P. Mielke. 1984/85 Urban lead in Minnesota: soil transect results of four cities. J. Minn. Acad Sci. 50:19-24.
43. Davies, B.E. 1995. Health and lead contaminated soils: the need for a globa audit? In Remediation of soils contaminated with metals, ed. I.K. Iskander, anc D.C. Adriano. Northwood, England: Science and Technology Letters. In Press

44. Davies, B.E., Conway, D. and S. Holt. 1979. Lead pollution of London soils: a potential restriction on their use for growing vegetables. J. ag. sci. (Camb.) 93:749-752.
45. Davies, B.E. and N.J. Houghton. 1984. Distance-decline patterns in heavy metal contamination of soils and plants in Birmingham, England. Urban Ecology. 8:285-294.
46. Elwood, P.C., Blaney, R., Robb, R.C., Essex-Carter, A.J. Davies, B.E. and C. Toothill. 1985. Lead levels on traffic-less islands. J. Epidem. Comm. Hlth. 39:256-258.
47. Diemel, J.A.L., Brunejree, B., Boleu, J.S.M., Biersteker, K. and S.J. Veenstra. 1981. The Arnhem lead study. II. Indoor pollution and indoor/outdoor relationships. Environ. Res. 25:449-456.
48. Davies, B.E., Elwood, P.C., Gallacher, J. and R.C. Ginnever. 1985. The relationships between heavy metals in garden soils and house dusts in an old lead mining area of north Wales, Great Britain. Environmental Pollution. 9:255-256.
49. Yocom, J.E. 1982. Indoor-outdoor air quality relationships; a critical review. J. Air. Poll. Control Assoc. 32:500-519.
50. Daines, R.H., Smith, D.W., Feliciano, A. and J.R. Trout. 1972. Air levels of lead inside and outside of homes. Ind. Med. 41:26-28.
51. Yocum, J.E., Clink, W.L. and Cole, W.A. 1971. Indoor/outdoor air quality relationships. J. Air Poll. Control Assoc. 21:251-259.
52. Schaefer, V.J., Mohnen, V.A. and V.R. Veirs. 1972. Air quality of American homes. Science. 175:173-175.
53. Davies, D.J.A., Watt, J.M. and I. Thornton. 1987. Lead levels in Birmingham dusts and soils. Sci. Tot. Environ. 67:177-185.
54. Al-Raday, A.S., Davies, B.E. and M.J. French. 1994. Distribution of lead inside the home: case studies in the north of England. Sci. Tot. Environ. 145:143-156.
55. Al-Raday, A.S., Davies, B.E. and M.J. French. 1993. Leaded windows as a source of lead within homes. Sci. Tot. Environ. 132:43-51.
56. Marino, P.E., Landrigan, P.J., Graef, J., Nussbaum, A., Bayan, G., Boch, K. and S. Boch. 1908. A case report of lead paint poisoning during renovation of a Victorian farmhouse. Am. J. Publ. Hlth. 80:1183-1185.
57. de Mora, S.J. and R.M. Harrison. 1984. Lead in tap water: contamination and chemistry. Chem. Brit. 20:900-906.
58. Moore, M.R. 1977. Lead in drinking water in soft water areas - health hazards. Sci. Tot. Environ. 7:109-115.
59. Benson, J.A. and H. Klein. 1983. Lead in drinking water. Investigation of a corrosive water supply. J. Environ. Hlth. 45:179-181.
60. Alam, I.A. and M. Sadiq. 1989. Metal contribution of drinking water from corrosion of distribution pipes. Environmental Pollution. 57:167-178.
61. Koh, L.L., Wong, M.K., Gan, L.M. and C.T. Yap. 1991. Factors affecting the leaching of lead from UPVC pipes. Env. Mon. Man. 19:203-214.

62. Green, V.A., Wise, G.W. and J. Callenbach. 1976. Lead poisoning. Clin. Toxic. 9:33-51.
63. Moore, M.R., Campbell, B.C. and A. Goldberg. 1977. Chapter Three: Lead. In the Chemical Environment, vol. Environment and Man: volume 6, ed. J. Lenihan, and W.W. Fletcher. London: Blackie.
64. Ramakrishna, R.S., Ponnampalam, M., Brooks, R.D. and D.E. Ryan. 1982. Blood lead levels in Sri Lankan families recovering gold and silver from jewelers' waste. Arch. Env. Hlth. 37:118-120.
65. Madany, I.M. and M.S. Akhter. 1992. Lead levels in some eye cosmetics used in Bahrein. J. Environ. Sci. Health. A27:1541-1547.
66. Bose, A., Vashistha, K. and B.J. O'Loughlin. 1983. Azarcon por Empacho - another cause of lead toxicity. Pediatrics. 72:106-108.
67. Gallacher, J.E.J., Elwood, P.C., Phillips, K.M., Davies, B.E., Ginnever, R.C., Toothill, C. and D.T. Jones. 1984. Vegetable consumption and blood lead concentrations. J. Epidem. Comm. Hlth. 38:173-176.
68. Roots, L.M. 1979. Tests available for assessing recent exposure to inorganic lead compounds and their use for screening purposes. Sci. Tot. Environ. 11:59-68.
69. Grunder, F.I. and A.E. Moffitt. 1982. Blood as a matrix for biological monitoring. Am. Ind. Hyg. Assoc. J. 43:271-274.
70. Schutz, A., Attewell, R. and S. Skerfving. 1989. Decreasing blood lead in Swedish children, 1978-1988. Arch. Env. Hlth. 44:391-394.
71. Goyer, R.A. 1993. Lead toxicity: current concerns. Env. Hlth. Persp. 100:177-187.
72. Piomelli, S., Corash, L., Corash, M.B., Seaman, C., Mushak, P., Glover, B. and R. Padgett. 1980. Blood lead concentrations in a remote Himalayan population. Science. 210:1135-1137.
73. Poole, C. and L.E. Smythe. 1980. Blood lead levels in Papua New Guinea children living in a remote area. Sci. Tot. Environ. 15:17-24.
74. Roels, H., Buchet, J., Lauwerys, R., Hubermont, G., Bruaux, P., Claeys-Thoreau, F., Lafontaine, A. and J. van Overschelde. 1976. Impact of air pollution by lead on the heme ciosynthetic pathway in school-age children. Arch. Env. Hlth. 31:310-316.
75. Annest, J.L., Pirkle, J.L., Makuc, D., Neese, J.W., Bayse, D.D. and M.G. Kovar. 1983. Chronological trend in blood lead levels between 1976 and 1980. N. Engl. J. Med. 308:1373-1377.
76. Fergusson, J.R. 1986. Lead: petrol lead in the environment and its contribution to human blood lead levels. Sci. Tot. Environ. 50:1-54.
77. Caplun, E., Petit, D. and E. Picciotto. 1984. Lead in petrol. Endeavour. 8:135-144.
78. Markesky, L.S. and S.R. Brobler. 1993. Effects of the reduction of petrol lead on blood lead levels of South Africans. Sci. Tot. Environ. 136:43-48.
79. Ponka, A., Salminen, E. and S. Ahonen. 1993. Lead in the ambient air and blood specimens of children in Helsinki. Sci. Tot. Environ. 138:301-308.

80. Facchetti, S. 1990. Isotopic lead experiment. Sci. Tot. Environ. 93: 537-538.
81. Facchetti, S. 1989. Lead in petrol. The isotopic lead experiment. Acc. Chem. Res. 22:370-374.
82. Elwood, P.C. 1986. The sources of lead in blood: a critical review. Sci. Tot. Environ. 52:1-23.
83. Larbey, R.J. 1994. Issues surrounding the use of lead in gasoline - energy, economic and environmental. Sci. Tot. Environ. 146/147:19-26.
84. Bacon, A.P.C., Froome, K., Gent, A.E., Cooke, T.K. and P. Sowerby. 1967. Lead poisoning from drinking soft water. Lancet. February 4:264-266.
85. Wilson, A.T. 1968. Acute vertigo and the lead content of food and dust. Practitioner. 200:282-288.
86. Elwood, P.C., Gallacher, J.E.J., Phillips, K.M., Davies, B.E. and C. Toothill. 1984. Greater contribution to blood lead from water than from air. Nature. 310:138-140.
87. Lacey, R.F., Moore, M.R. and W.N. Richards. 1985. Lead in water, infant diet and blood: the Glasgow Duplicate Lead study. Sci. Tot. Environ. 41:235-257.
88. Barltrop, D., Strehlow, C., Thornton, I. and J.S. Webb. 1975. Absorption of lead from dust and soil. Postgraduate Med. J. 51:801-804.
89. Charney, E., Sayre, J. and M. Coulter. 1980. Increased lead absorption in inner city children: where does the lead come from? Pediatrics. 65: 226-231.
90. Bornschein, R.L., Succop, P., Dietrich, K.N., Clark, C.S., Hee, S.Q. and P.B. Hammond. 1985. The influence of social and environmental factors on dust lead, hand lead, and blood lead levels in young children. Environ. Res. 38:108-118.
91. Duggan, M.J. and M.J. Inskip. 1985. Childhood exposure to lead in surface dust and soil: A community health problem. Public Health Rev. 13:1-54.
92. Wixson, B.G. and B.E. Davies. 1994. Guidelines for lead in soul. Environ. Sci. Technol. 28:26A-31A.
93. Wixson, B.G. and B.E. Davies. 1993. "Lead in Soil" Task Force: recommended guidelines. Northwood, England: Science Reviews.
94. Needleman, H.L. 1983. Low level lead exposure and neuropsychological performance. In Lead versus Health: sources and effects of low level lead exposure, ed. M. Rutter and R. Russel Jones. Chichester: John Wiley and Sons.
95. Harvey, P.G., Hamlin, M.W. and R. Kumar. 1984. Blood lead, behaviour and intelligence test performance in preschool children. Sci. Tot. Environ. 40:45-60.
96. Marececk , J., Shapiro, I.M., Burke, A., Katz, S.H. and M.L. Hediger. 1983. Low-level lead exposure in childhood influences neuropsychological performance. Arch. Env. Hlth. 38:355-359.
97. David, O.J., Hoffman, S.P., Clark, J.C., Grad, G. and J. Sverd. 1983. The

relationship of hyperactivity to moderately elevated lead levels. Arch. Env. Hlth. 38:341-346.

98. Schwartz, J. 1988. The relationship between blood lead ad blood pressure in the NHANES II survey. Env. Hlth. Persp. 78:15-22.
99. Harlan, W.R. 1988. The relationship between blood lead levels to blood pressure in the U.S. population. J. Environ. Hlth. 78:9-13.
101. Elwood, P.C., Davey-Smith, G., Oldham, P.D. and C. Toothill. 1988. Two Welsh surveys of blood lead and blood pressure. J. Environ. Hlth. 78:119-121.
102. Nomiyama, K. 1980. Recent progress and perspectives in cadmium health effects studies. Sci. Tot. Environ. 14:199-232.
103. Elinder, C.G. 1982. Cadmium in health: a survey. Intern J. Environmental Studies. 19:187-193.
104. Friberg, L. 1984. Cadmium and the kidney. Env. Hlth. Persp. 54:1-11.
105. Sartor, F.A., Rondia, D.J., Claeys, F.D., Staessen, J.A., Lauwerys, R.R., Bernard, A.M., Roels, H.A., Bruaux, P.J., Ducoffre, G.M., Linjnen, P.J., Lutgarde, B.T. and A.K. Amery. 1992. Impact of environmental cadmium pollution on cadmium exposure and body burden. Arch. Env. Hlth. 47:347-353.
106. Davies, B.E. and R.C. Ginnever. 1979. Trace metal contamination of soils an vegetables in Shipham, Somerset. J. ag. sci. (Camb.) 93:753-756.
107. Sherlock, J.C., Smart, G.A., Walters, B., Evans, W.H., McWeeny, D.J. and W. Cassidy. 1983. Dietary surveys on a population at Shipham, Somerset, United Kingdom. Sci. Tot. Environ. 29:121-142.
108. Naylor, L.M. and R.C. Loehr. 1981. Increase in dietary cadmium as a result of application of sewage sludge to land. Environ. Sci. Technol. 15:881-886.
109. Kromhout, D. 1988. Blood lead and coronary heart disease risk among elderly men in Zutphen, the Netherlands. Env. Hlth. Persp. 78:43-46.

Environmental Contaminants, Ecosystems and Human Health. Edited by S.K. Majumdar, E.W. Miller, and Fred J. Brenner.

Chapter Twenty-Two

MERCURY IN SEAFOOD AND HUMAN HEALTH

A. RENZONI

Universita degli Studi di Seina
Dipartimento di Biologia Ambientale
Via delle Cerchia, 3 53100 Siena, Italy

INTRODUCTION

Mercury is the only metal which is liquid at room temperature and below. Mercury volatilizes readily and is therefore distributed world-wide. In the atmosphere, mercury vapors are converted into insoluble forms that return to earth with precipitation. They enter water bodies (rivers, lakes, oceans) and undergo chemical modification in the water column, especially in sediments.

In the anaerobic layer of sediments, a large proportion is transformed into insoluble mercuric sulfide that remains trapped in the sediments. In the uppermost layer, most mercury is transformed into methylmercury and dimethylmercury: the latter is volatile and ascends the water column, evaporating back into the atmosphere; the former, after release into the aquatic environment, enters the food chain, accumulates in living organisms and is transferred to the highest levels of the chain, where it reaches maximum concentrations in top predators (tuna, swordfish, sharks, etc.).

Methylmercury is able to penetrate biological membranes. It is the most toxic short chain alkylmercurial. Once inside the cell, it interferes with cell metabolism and the physiological activity of the cell. Its great affinity for the central nervous system enables it to enter nerve cells, with severe consequences when accumulation is high.

Apart from the Minamata poisoning in Japan and several other incidents, the risk to which the adult population is normally exposed through is negligible. In pregnant women with a high fish consumption in the first months of pregnancy, the fetus may be at risk for neurological disorders due to placental transfer of organic mercury. Recent studies have shown that the frequency of mental retardation in children is highest in the group whose mothers are heavy consumers of seafood.

Selenium seems to counteract some of the toxic effects of mercury.

OCCURRENCE, PRODUCTION AND USE OF MERCURY

Both the mineral and the metal have been used for more than 2 millennia, and notwithstanding the many, true or suspected, deaths due to mercury poisoning, it has only been in the last 4 or 5 decades that documented cases have been described (Minamata Report 1968; D'Itri 1990; WHO 1990). Acute mercury poisoning has occurred in several regions (Iraq, New Mexico, Guatemala) with dozens of people dying after eating grain treated with organomercury seed-dressing or pork fed with dressed seeds (Bakir et al. 1974; Amin-Zaki et al. 1976; Kazantzis et al. 1976a; Kazantzis et al. 1976b; Greenwood 1985; Al-Mufti et al. 1986; Cox et al. 1989) intended for planting rather than human consumption. Chronic mercury poisoning, on the other hand, has occurred when mercury is released into the environment (even in small amounts) and does not affect the first living organisms encountered, but accumulates through one or more food chains to affect an organism at the top (Berglund and Berlin 1969; Fimreite 1970; Bernhard et al. 1982; Renzoni et al. 1986; Berg et al. 1986. André et al. 1990). The most studied case of this kind occurred almost 40 years ago in Japan at Minamata (Harada et al. 1977; Tsubaki et al. 1978; Tamashiro et al. 1984; Tsubaki and Takahashi 1986; Tamashiro et al. 1986; Tamarshiro et al. 1987). A large industrial plant (Chisso Corporation) producing acetaldehyde, vinylchloride and other chemicals was using a large quantity of mercury in its processes and eliminating it in its effluents. The wastes were carried by a stream into Minamata Bay, where several communities of fishermen relied on fish for their subsistence. The mercury, methylated in the bay sediment, entered the food chain and eventually ended up on the family dinner table. The high quantity of seafood in the diet of this fishermen population led to frequent instances of poisoning, with many cases of neurological impairment and even death.

A second and less localized episode of mercury poisoning occurred in Scandinavia, and others have probably also occurred elsewhere without ever being reported. The use of mercury-based fungicides for dressing forestry seeds in Scandinavia led to the almost total extermination of birds of prey in the area. How did this happen? Small rodents and seed-eating birds fed on the poisoned seed and accumulated high quantities of mercury, without suffering ill effects, as the period of availability of the seed was limited; birds of prey feeding on these animals accumulated much larger quantities of mercury in the liver, kidney and brain. The results was nervous system impairment, uncoordinated flight and movements, reduced reproductive capacity, decreased immune capacity and vulnerability to predators. The populations decreased sharply, almost reaching extinction in many cases. Since the prohibition of mercury dressing of seeds, the populations of these birds have somewhat recovered.

Mercury, as such, has been used (and in some cases is still being used) in the chlor-alkali industry, in temperature control systems, in electrical industries, in plastics, in the pulp and paper industry, in the extraction of gold, in fungicides and herbicides, in paints, in dental amalgam and so forth. Other anthropogenic sources of mercury include the smelting of sulfide ores (now reduced), the combustion of fossil fuel (ever increasing), cement production and waste incineration. However, a large proportion of the mercury present in various forms in the environment is of natural origin, e.g. weathering of the earth's crust, evaporation of large bodies of water, volcanic emissions. Mercury fluxes between the lithosphere and the atmosphere have been calculated by many authors (Weiss et al. 1971; Saha 1972; Korringa and Hagel 1974; Goldberg 1976; Berlin 1990), and values range from 30 to 150,000 tons per year. Most of the differences were probably due to the compartments taken into consideration by the authors; sometimes they refer to continents, sometimes to oceans, and sometimes to the earth as a whole. The most recent figures (Rasmussen 1994) are 21,000 tons per year from the oceans and 9,000 from the continents. The amount of mercury mined all over the world varies from year to year: in the many reports of the mining industry from 3,000 to 10,000 tons, but the trend is towards a progressive decrease due to its reduced use in industry and agriculture. Indeed several highly productive mines have recently been closed for example, Amiata in Italy and Idria in Slovenia.

MERCURY TRANSFORMATION IN NATURE

The natural release of mercury vapor plus man-made emissions of various forms of mercury are responsible for the presence of this heavy metal in air, water, sediments and biota. The global cycle of mercury probably involves mainly inorganic forms that do not accumulate (Bargagli et al. 1980; Minagawa et al. 1980) and hence do not pose problems for human health. Recent findings (Lindberg et al. 1987; Gaggi et al. 1991; Bacci et al. 1994) show that atmospheric mercury vapor may be taken up directly by plant foliage.

Mercury, like cadmium and lead, undergoes several abiotic and biotic modifications in the environment. Its transformation to organic forms has been extensively studied. After the discovery that methylmercury was responsible for the Minamata poisoning (the fraction eliminated as such by the factory and that produced in the bay from the inorganic mercury dumped by the plant), much research has been performed on how and where this process occurs. The most accepted opinion about this conversion is that it is performed by microbes in aquatic sediments (Figure 1). This conversion was first described by Wood et al. (1969) and almost at the same time by Jensen and Jernelöv (1969) in aquarium sediment. Much data subsequently suggested that sediments with microbial activity are the only compartment in which the process occurs, but other possible routes of conversion have been carefully considered and other more or less acceptable hypothesis have been put forward (Kiwimae et al. 1969, Rudd et al. 1980). Most recently new ideas in the process of mercury methylation and of methylmercury biodegradation too have been put forward (Baldi and Filippelli 1991; Baldi et al. 1993).

METHYLMERCURY IN SEAFOOD

Here seafood is intended to include fish, shellfish (crustaceans and molluscs), mammals, birds and even reptiles that are part of the human diet in various parts of the world.

Mercury from any source, in any form, sooner or later reaches our common sink, the oceans. There, dissolved in the water, absorbed by particles, incorporated

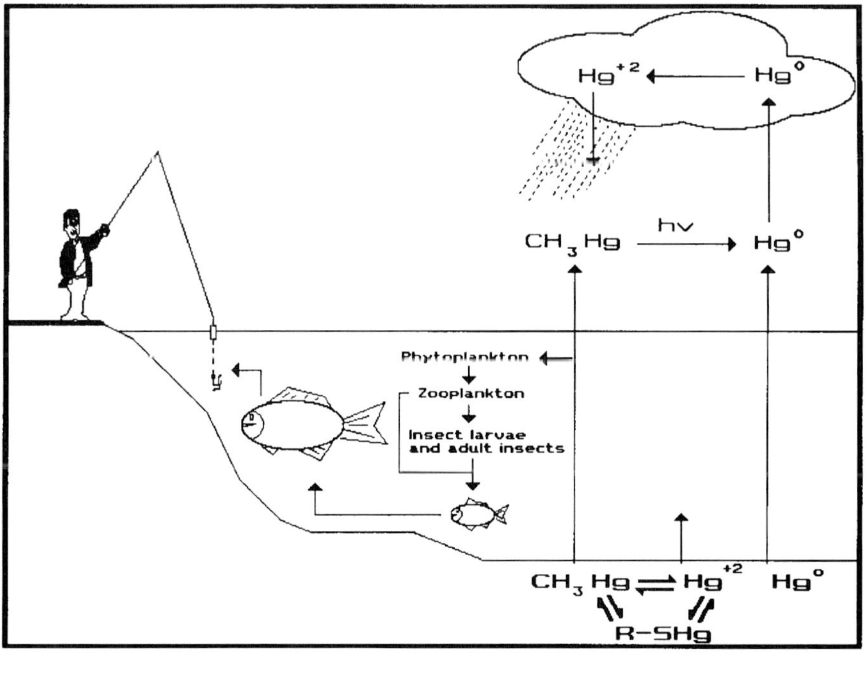

FIGURE 1. Simplified cycle of mercury in an aquatic environment.

in the detritus of dead plants and animals, it is absorbed by lower organisms such as bacteria, algae and aquatic plants, and enters animal organisms (Figure 1) principally through the gastrointestinal system (up to 90%) (Walsh, 1982).

In the water, a large portion of it is absorbed by microorganisms living in the upper part of the water column where it is transformed into methylmercury, the common organic compound responsible for most of the poisoning incidents in the last 40-50 years. This form of mercury is able to cross biological membranes and becomes available to all pelagic life forms feeding on microorganisms, whence it is passed on to fish and other animals at the top of the web, like tuna and swordfish.

Another large portion of mercury, however, descends the water column and reaches the upper layer of sediments where it undergoes different abiotic and biotic transformations. As in the pelagic environment, methylmercury is available to all benthic organisms in the various food chains. Shrimps, prawns, lobsters, crabs, squid and a large variety of benthic fish accumulate the bioavailable element. Typical biomagnifications occur from organisms at the lowest levels of the food chain (in both the pelagic and benthic environments) to those at the top of the food web and hence to man.

The concentration of mercury in aquatic animals is extremely variable and depends on many circumstances; the main differences, however, mostly depend on the animal's position in the food web. The levels of toxic substance bioaccumulated increase with the age of the animal, since uptake is faster than release. In time, levels may be reached that are dangerous for animals at the top of the food chains (e.g. cats in Japan) and for man (e.g. as seen in Japan, New Mexico and Iraq). In short-living organisms, such as most species of zooplankton and some organisms of the zoobenthos, bioaccumulation is negligible and differences between juveniles and adults are not detectable. In longer-living organisms the process of bioaccumulation has been repeatedly described in the marine (Cumont et al. 1975; Bernhard and Renzoni 1977; Gaskin et al. 1979; Honda et al. 1987; André et al. 1990) and freshwater environments (Johnels and Westermark 1969; Bligh 1970, 1971; Fimreite and Reynolds 1973); it is particularly evident in typically predatory fish such as tuna and swordfish (Miller et al. 1972; Menasveta and Siriyong 1977; Baldi et al. 1978; Renzoni et al. 1978; Buffoni et al. 1982). In most animals and man a parallel increase in selenium and mercury has been found (Capelli et al. 1989; Leonzio et al. 1991; Grandjean et al. 1992).

THE MEDITERRANEAN, AN ANOMALOUS BASIN

The Mediterranean Sea has been studied in great detail in the last 20 years, mostly under the auspices and with the financial support of the United Nations in the framework of the Environmental Programs of the WHO and FAO. An enormous amount of data has been collected and some of the results have contributed to a better understanding of the cycles, uptake and release of mercury in the biota and the risks for human health.

The data obtained during the 10-year project clearly indicates that organisms of the marine biota of this basin have higher mercury body burdens than the same (or similar) species in the Atlantic and other Oceans (Renzoni et al. 1978; Baldi et al. 1978; Schreiber 1983). This was confirmed in a study of bluefin tuna living their whole life in the Mediterranean and those entering in the Mediterranean from the Atlantic at adulthood for spawning. Specimens caught in the Atlantic in May, before entering the Mediterranean have the same body burden of Hg as Atlantic tuna caught in June in the Mediterranean in the same traps as Mediterranean tuna (Figure 2). However, the specimens resident in the Mediterranean for their entire life have mercury residues 3-4 times higher than those coming from the Atlantic only to spawn. A similar bioaccumulation pattern occurs even in tuna and fish of other oceans where the values reported (normalized according to length) are 2-3 times less than in Mediterranean fish (see Bernhard and Zattera 1975).

Some areas of the Mediterranean (e.g. the Tyrrhenian, where most of the data was gathered) show particularly high Hg levels in most species. High Hg residues have been reported in several crustaceans from this area, especially the Norwegian lobster which shows a clear correlation between mercury and body weight (Figure 3). This species lives in deep water in close contact with the upper

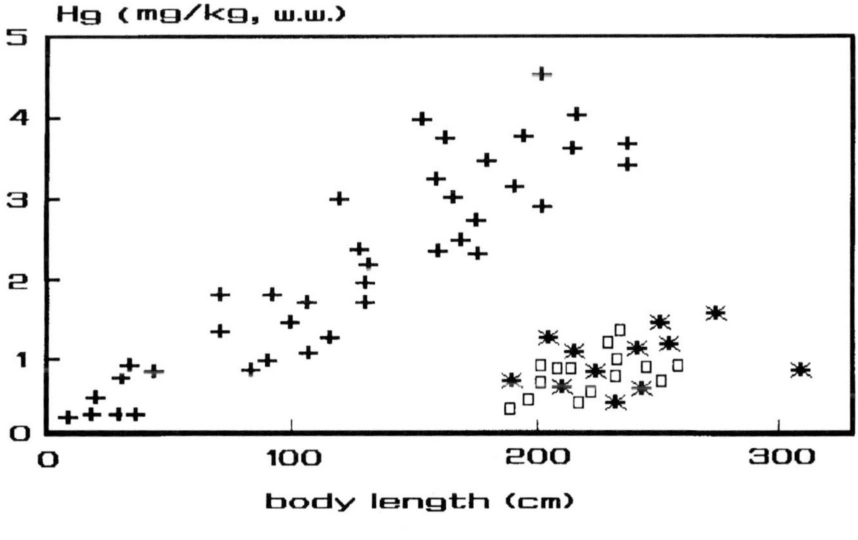

FIGURE 2. Total mercury concentration in bluefin tuna (*Thunnus thynnus thynnus*); + and * catch in Mediterranean traps: out of this group, those indicated with * are coming from the Atlantic for spawning in the Mediterranean; □ catch in the Atlantic before entering in the Mediterranean Sea.

layer of sediment, the site of mercury methylation. A positive correlation between mercury concentrations in sediment and mercury levels in benthic crustaceans and molluscs has been demonstrated in other parts of the Mediterranean (Hornung et al. 1984, Hornung 1986). The process of bioaccumulation, so evident in resident Mediterranean tuna, also occurs in representatives of the lower levels of the tuna food chain, particularly the anchovy, sardine and mackerel (Baldi et al. 1978); similar differences in the body burden of mercury between Mediterranean and Atlantic (or other oceans) species have been repeatedly demonstrated in organisms of the same species and in species with similar feeding habits or in the same food chain.

This is true for a marine bird, *Calonectris diomedea*, which lives in the Atlantic most of the year. Two groups, one nesting on Atlantic islands, and the other on some Mediterranean islands were studied recently (Renzoni et al. 1986; Renzoni & Massa 1989) in connection with research on mercury in other marine organisms. The eggs laid by the latter group after a period of 40-50 days in the Mediterranean, during which the birds fed on Mediterranean marine fauna, had a concentration of mercury 3-4 times higher than the eggs of the first group. Levels of mercury in the tissues of adult specimens of the second group showed the same pattern, reflecting the high levels of mercury in the birds' diet, composed of Mediterranean fauna. The differences between liver and kidney levels

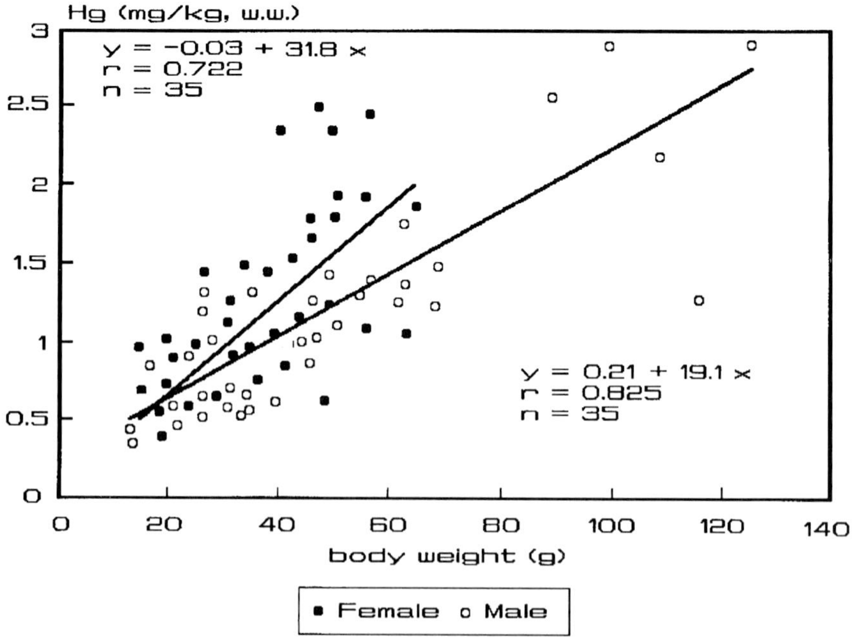

FIGURE 3. Total mercury in specimens of Norvegian Lobster (*Nephrops norvegicus*) catch in the Tyrrhenian Sea.

of Hg in groups breeding in the Atlantic and Mediterranean were quite high, whereas differences between brain and muscle levels were less evident (Renzoni et al. 1986). This suggests a rapid demethylation/excretion of Hg absorbed from Mediterranean prey. The high levels of mercury in Mediterranean biota have been the subject of several hypotheses in the last 20 years. Recent analysis of Mediterranean and other ocean water revealed wide ranges of concentrations, making it difficult to explain the high levels of Hg in Mediterranean benthic and pelagic biota. On the other hand, the fact that 65% of world mercury resources are located in the Mediterranean area (1% of the total surface area of the planet) and that the turnover of Mediterranean water through the Straits of Gibraltar is very low, explains the anomalous condition of this sea. As a consequence, Mediterranean marine organisms at the top of the food web show higher mercury residues than organisms from elsewhere.

RISKS FOR HUMAN HEALTH

The general population is exposed to mercury under various circumstances. Air, dental amalgam, water and food are the four main routes. Exposure to mercury vapors is the most frequent occupational exposure and it may occur in laboratories, factories and very polluted urban areas. Exposure to mercury of natural origin may occur near volcanoes (Siegel and Siegel 1979) and near active mines. Dental amalgam in persons with a large number of fillings may also be a significant continuous source of Hg. Vapor from dental amalgam and other sources is absorbed directly by the alveolar membrane of the lung. It is oxidized to mercury in the cells, reaching various body compartments where it may be retained for a variable period of time, interfering with cell and tissue metabolism. Excretion is mainly (Berlin 1990) performed by the kidney with the urine; other routes are feces, sweat and tears. In laboratory experiments it has been shown that selenium counteracts some of the toxic effects of mercury; this also seems to occur in nature since a correlation between levels of mercury and selenium in blood and other tissues has been repeatedly demonstrated (Ganther et al. 1972; Martoja and Viale 1977; Freeman et al. 1978; Chang and Suber 1982; Hansen et al. 1984; Capelli et al. 1989; Grandjean et al. 1992). Mercury occurs in very small concentrations in water; in drinking water and rain levels are of the order of 5-100 ng/l (average 20 ng/l); in the open ocean, 0.5-3 ng/l and in coastal waters, 2-15 ng/l. The contribution of water to man's body burden of Hg is almost negligible. There are no reports of methylmercury in drinking water. Except for people occupationally exposed to Hg in mines, smelting plants and industries using large quantities of mercury, the main source for the human population is food. Although concentrations of mercury in foodstuffs are very low and quite often below the detection limit, their contribution to Hg burden is not negligible (Goldwater 1971). However, the main contribution is from aquatic organisms in the diet. Without ignoring the importance of fresh-water organisms which could play an important role in mercury poisoning in certain continental areas, especially in northern Europe and North America (Mackeown-Eyssen and Ruedy 1983, Mackeown-Eyssen et al. 1983; Lindqvist et al. 1984), national and international agencies are concerned mainly with seafood in relation to human health.

The information in the literature has mainly been obtained from human observations conducted in Japan and Iraq and from experimental research in animals. From both sources it has been established that the effects, distribution and fate of Hg have similar patterns with minor variations between species. In most organisms studied, nerve tissue is the principal target. Observations of the effects of long - (Japan) and short-term (Iraq, New Mexico) mercury intake led to the definition of a series of considerations at a recent meeting of the WHO (WHO 1990) concerned with the review of published data on this subject:

"Methylmercury poisoning has several important features:
- a long latent period usually lasting several months;
- damage almost exclusively limited to the nervous system especially the central nervous system;
- areas of damage to the brain are highly localized (focal), e.g., in the visual cortex and the granular layer of the cerebellum, especially in the infolded regions (sulci);
- effects in severe cases are irreversible due to destruction of neuronal cells;
- the earliest effects are non-specific subjective complaints, such as paraesthesia, blurred vision, and malaise".

The distribution of mercury compounds in the organs and tissues of animals is well differentiated in the various species. In crustaceans and molluscs, Hg tends to concentrate in the viscera which include the digestive diverticula, the gastrointestinal systems and the kidney; smaller amounts are found in muscle. In fish, birds and mammals a substantial amount of the mercury burden is found in muscle, liver, kidneys and brain. Since most of the total amount of Hg in seafood is in the form of methylmercury, (Capelli et al. 1987), this is the form with which we are mainly concerned in this paper. Its possible degradation in the organs of the various species will only be mentioned for well documented data. Irrespective of its entry route into the body, the various chemical forms of mercury are distributed in different ways. In the blood, metallic and inorganic mercury is almost equally distributed between erythrocytes and the plasma, whereas methylmercury accumulates to a large extent in erythrocytes (Skerfving 1974, Berlin 1990). Most of the inorganic fraction passes into the liver, then the kidneys and from there is quickly eliminated; organic mercury (and particularly methylmercury) absorbed through biological membranes, is degraded much more slowly. The results of experiments conducted in the laboratory have been compared with data obtained in the field. Concentrations of total mercury and the methylmercury fraction, their biological half-life and distribution in the tissues are reported and discussed by several authors (Saha 1972, Miettinen et al. 1972,

Renzoni et al. 1973, Cunningham and Tripp 1975, Aston and Fowler et al. 1985, Bernard 1988, Clarkson 1992). The recent data of Capelli et al. (1985) clearly demonstrates that the inorganic fraction of the total mercury found in adult *Sarda sarda* is almost the same as in juveniles, whereas methylmercury increases with the size of the organism. Hence biomagnification through the food web and age/size-dependent bioaccumulation are mostly due to the accumulation of the methylmercury. Inorganic mercury seems to reach a certain level and remain constant. This model of accumulation needs to be verified in other species.

There is clear evidence that methylmercury is transferred from the mother to the fetus (Takeuki 1977, Harada and Moreiama 1977, Harada et al. 1977, Marsh et al. 1977, 1987). A close relationship between levels of total Hg in maternal hair and in the child has been demonstrated in the families of Seychelles fishermen (Matthweys 1983) and in mammals from several ocean regions (André et al. 1990).

However, recent analysis of the livers of newborn porpoises and common dolphin fetuses tend to exclude the existence of a major route of transfer of methylmercury from the mother to the embryo (Law et al. 1992).

In humans and animals, the developing nervous system is known to be much more sensitive to damage from methylmercury than the adult nervous system and other tissues and organs. In Minamata and elsewhere the children of mothers with long exposure to methylmercury in the first months of pregnancy, have been the subject of many studies. The symptoms and damage seem to be related to maternal plasma levels of mercury and the clinical picture may be indistinguishable from that of other diseases. In severe cases the whole brain is damaged, whereas in adults damage is more typically localized in a few areas. A wide range of damage (and hence consequences) has been observed in children prenatally exposed in the areas of major incidents (Minamata Report 1968, Amin-Zaki et al. 1974, Marsh et al. 1977, Clarkson and March 1982, Choi 1991, Clarkson 1992). WHO data (1990) indicates an increased incidence of mental retardation in children of mothers with hair mercury levels of over 10 (μ/g during pregnancy and a recent study in New Zealand (Kjellström et al. 1986) has shown "a statistically significant dose - response relationship between mean hair mercury during pregnancy and deficient Denver Test results in children".

CASE STUDY: A HOT SPOT AT ROSIGNANO SOLVAY, TYRRHENIAN SEA, ITALY

After various incidents such as Minamata, the findings of high levels of mercury in fish from many Canadian lakes and streams, and the banning of the sale of fish contaminated by mercury from about 40 lakes in Sweden, a survey was conducted (Renzoni et al. 1973) in the sea off the chlor-alkali plant at Rosignano Solvay, Tuscany, Italy. Coastal waters off the plant were found to have mercury levels one to two orders of magnitude higher than open waters. Analysis of marine sediment sampled near the plant showed levels one or two orders of magnitude higher than sediment from open waters. The sediment core sections corresponding to the years immediately after opening of the plant showed high levels of mercury. Seafood caught in the area had mercury residues 4-5 times higher than the same species caught a few miles north or south of the effluent discharge area. Up to 4.6 ppm was detected in the muscle of fish caught off the plant (Table 1). A small scale epidemiological survey was therefore performed among the few fishermen in the area. Urine, blood and hair wee analyzed for mercury residues. High levels of mercury were found in subjects having a diet high in seafood (Table 2) but since they were very few, no clinical examinations were performed. The local health authorities were informed and the factory staff

TABLE 1
Total mercury concentrations (mg/kg, w.w.) in tissues of 6 species of fish (on bracket, the number of the specimens).

Name	muscle $\overline{x} \pm$ S.D.	liver $\overline{x} \pm$ S.D.	kidney $\overline{x} \pm$ S.D.
Red mullet (*Mullus surmuletus* L.)	1.65±0.63 (6)	2.42±0.80 (6)	2.09±0.71 S.D.
Green wrasse (*Labrus turdus* L.)	1.89±0.57 (7)	7.31±3.07 (7)	8.14±4.06 (7)
Two-banded bream (*Diplodus vulgaris* Geoff.)	1.90±0.78 (19)	17.00±14.72 (10)	29.80 (2)
Small-scaled scorpion fish (*Scorpaena porcus* L.)	2.61±0.95 (17)	4.26±2.77 (17)	4.89±2.71 (15)
Rainbow wrasse (*Coris julis* L.)	2.77±0.65 (6)	16.12±5.31 (6)	12.23±2.80 (6)
Painted comber (*Serranus scriba* L.)	4.64±1.78 (13)	11.60±5.95 (9)	7.99±3.81 (6)

TABLE 2
Total mercury concentration in biological material from fishermen.

No. fish meals/week	Red blood cells (mg/g)	Plasma (mg/g)	Hair (μg/g)
0	$\overline{x}$ = 16 n = 2	$\overline{x}$ = 2 n = 2	$\overline{x}$ = 0.98 n = 2
1-3	$\overline{x}$ = 86.1 D.S. = 53.58 Range = 34-207 n = 10	$\overline{x}$ = 6.2 D.S. = 2.49 Range = 2-11 n = 10	$\overline{x}$ = 8.47 D.S. = 3.68 Range = 4-13.7 n = 9
4-6	$\overline{x}$ = 242.3 D.S. = 71.84 Range = 143-375 n = 10	$\overline{x}$ = 16.2 D.S. = 3.99 Range = 11-21 n = 10	$\overline{x}$ = 24.98 D.S. = 10.93 Range = 9.2-41.6 n = 8
>6	$\overline{x}$ = 320.83 D.S. = 227.48 Range = 102-912 n = 12	$\overline{x}$ = 22.83 D.S. = 12.10 Range = 12-4 n = 12	$\overline{x}$ = 45.47 D.S. = 47.36 Range = 10-109 n = 6

immediately modified the chemical process, obtaining a substantial reduction in mercury discharge. Species analyzed a few years later for mercury pollution showed a sharp decrease in total mercury residues (Figure 4). This decrease occurred much faster in the benthic herbivore molluscs *Patella* sp. than in the benthic predator fish *Scorpaena porcus*.

GENERAL COMMENTS

Clear evidence exists that organic mercury is accumulated more by aquatic than terrestrial organisms. After the Minamata outbreak, mercury pollution of freshwater organisms became the focus of attention in Canada and Scandinavia because of the large consumption of freshwater fish in these countries.

The danger of mercury poisoning due to habitual fish intake prompted the authorities to set maximum total mercury limits (0.5 or 0.7 or 1 mg/kg^{-1} fresh weight in edible parts of the organisms) for marketing and human consumption. However from a medical point of view, legal limits for maximum concentrations in seafood do not always guarantee the protection intended. In fact fishermen, fish vendors, and the staff of fish restaurants, for example, may eat so much fish, even containing mercury within legal limits, that their body burden reaches the damage threshold. International agencies, especially the WHO, have introduced and suggested a better system of prevention in the form of the PTWI (Provisional

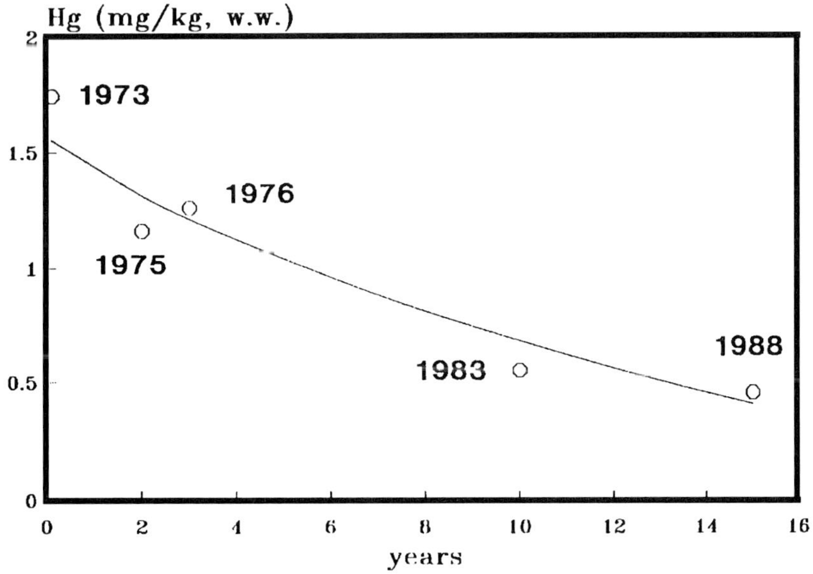

FIGURE 4. Decrease of total mercury concentration in muscle tissue of small-scaled scorpion fish (*Scorpaena porcus*).

Tolerable Weekly Intake) above which the body burden could reach the damage threshold. The WHO (1976) has estimated that the earliest symptoms of poisoning in the most sensitive groups appear after a long period with a daily intake of 200-500 (g Hg by a 70 kg person. The levels of mercury in the human body reached under these conditions are 200 to 500 (g/l in blood and 50 to 125 (g/g in hair. Since the first neurological problems may be observed at these levels, a safety factor of 10 was used, confirming the PTWI of 300 (g total Hg (200 (g of which as Me Hg) of the FAO/WHO (1976). As an example of correct administration, in the early seventies, the maximum weekly intake of fish from certain Scandinavian rivers and lakes was posted in signs along the banks.

TREND

The Japanese incidents and the episodes of mercury poisoning of man and wildlife by seed dressing had led to a better general understanding of the biogeochemical cycles and fate of many xenobiotics. At all levels, increased sensitivity to questions of environmental protection and management is evident. New techniques to recover mercury and other metals have been developed in the last 20 years. In industrialized countries where mercury consumption was exceptionally high between 1950 and 1980, methods to reduce its discharge into the environment have been implemented.

Clear evidence of its reduced use in industry is provided by the sharp reduction in demand and price. Production figures tend to reflect the amount of mercury marketed rather than the amount mined. Great quantities are held in storage for strategic purposes, and because of short demand by the industrial sector. Production is much less than 8.8 x 10^9 g/year mined in the late sixties (Weiss et al. 1968) when the production of mined mercury was at its maximum. The amount of mercury released in the burning of fossil fuels, especially coal, is still quite high, but new technologies have shown that it is possible to scrub most of the particulates, thus reducing their emission into the atmosphere. Mercury of anthropic origin does not seem to have affected the global content and cycle of mercury. The mercury content of museum specimens of deep sea fish is at the same high levels as today (Barber and Waling 1984), even for tuna (Miller et al. 1972). These studies show that elevated concentrations of mercury in ocean fish are a natural phenomenon and that the levels in fish have not changed in the last 100 years.

In his well documented review, “The significance of mercury in the environment”, (Saha 1972) underlined that “in the current controversy on the levels of mercury in fish and possible hazards from the consumption of contaminated fish, one question has not received much attention from scientists or the concerned public. This is background levels of mercury in fish”. Twenty years later the question is still unanswered and will probably remain so forever, because mercury concentrations vary widely in unpolluted waters.

REFERENCES

Al-Mufti, A.W., Copplestone, J.F., Kazantis, G., Mahmoud, R.M. and N.A. Majid. 1976. Epidemiology of organomercuric poisoning in Iraq. I) Incidence in a defined area and relationship to the eating of contaminated bread. Bull. World Health Organ. 53 (suppl.). 223-36.

Amin-Zki, L., Ebasshani, S., Majeed, M.A., Clarkson, T.W., Doherty, R.A., Greenwood, M. and T. Giovanoli-Jakubczak. 1976. Prenatal methylmercury poisoning in Iraq. Amer. J. Dis. Child. 130:1070-1076.

André, J.M., Ribeyre, F. and A. Boudou. 1990. Mercury contamination levels and distribution in tissues and organs of dolphinids (*Stenella attenuata*) from the eastern tropical Pacific, in relation to biological and ecological factors. Mar. Environ. Res. 30:43-72.

Aston, S.R. and S.W. Fowler. 1985. Mercury in the open Mediterranean: evidence of contamination. Sci. tot. Env. 43:13-26.

Bacci, E., Angotzi, G., Bralia, A., Lampariello, L. and E. Zanette. 1976. Etude sur une population humaine exposée au methylmercure par la consommation de poisson. Rev. Int. Ocean. Med. 41/42:127-141.

Bacci, E., Gaggi, C., Bargagli, R. and A. Renzoni. 1994. Mapping mercury vapors in an abandoned cinnabar mining area by azalea (*Azalea indica*) leaf trapping. Chemosphere 29:641-656.

Baldi, F., Renzoni, A. and M. Berhnard. 1978. Mercury concentration in pelagic fishes (anchovy, mackerel and sardine) from the Italian coast and the Strait of Gibraltar. IVes Journées d'études sur les pollutions marine en Méditerranée; Workshop on pollution of the Mediterranean; C.I.E.S.M. 251-254, Anthalya 1978.

Baldi, F. and M. Filippelli. 1991. New method for detection methylmercury by its enzymatic conversion to methane. Environ. Sci. Technol. 25:302-305.

Baldi, F., Pepi, M. and M. Filippelli. 1993. Methylmercury resistance in *Desulvibrio desulfuricans* strains in relation to methylmercury degradation. Appl. a. Environ. Microbiol 59:2479-2485.

Bakir, F., Damhyi, S.F., Amin-Zaki, L., Murthadha, M., Khalidi, A., Tikrita, S., Clarkson, T.W., Smith, J.C. and R.A. Doherty. 1973. Methylmercury poisoning in Iraq. An Interuniversity Report. Science 181:230-241.

Barber, R.T. and J.P. Whaling. 1984. Mercury in recent and century-old deep-sea fish. Envir. Sci. Technol. 18:552-555.

Bargalgi, R., Galiberti, A., Rossi, C. and L. Sarti. 1990. II giacimento musteriano di Montemileto (Avellino). Atti Soc. Tosc. Sci. Nat. Mem. Serie A., 86:341-403.

Berg, W., Johnels, A.G., Sjöstrand, B. and T. Westermark. 1966. Mercury content in feathers of Swedish birds from the past 100 years. Oikos 17:71-78.

Berglund, F. and M. Berlin. 1969. Risk of methyl-mercury accumulation in man and mammals and the relation between body burden of methylmercury and toxic effect. In: M.W. Miller and G.G. Berg (eds.); Chemical Fallout. C.C. Thomas Publ., Springfield, Illinois.

Beijer, J. and A. Jernelov. 1979. Methylation of mercury in aquatic environments. In: Nriagu, J.P. (ed.):The biogeochemistry of mercury in the environment. Amsterdam, Oxford, New York, Elsevier Science Publishers pp:235-238.

Berlin, M. 1990. Mercury. In: Friberg, L., Nordberg, G., Woulk, V.B. (eds.); Handbook on the toxicology of metals, II° Ed., Vol, II, Elsevier Amsterdam, N.Y., Oxford: Elsevier Science Publishers, 387-445.

Berhnard, M. and A. Zattera. 1975. Major pollutants in the marine environment. In: M. Pearson and E. Frangipane (Eds.) "Marine pollution and marine waste disposal" Pergamon Press, Oxford and New York 195-200.

Berhnard, M. and A. Renzoni. 1977. Mercury concentration in Mediterranean marine organisms and their environment: Natural or anthropogenic origin? Thal. Yugosl. 13, 265-300.

Berhnard, M., Buffoni, R. and A. Renzoni. 1982. Mercury in the Mediterranean tuna. Why are their levels higher than in Atlantic tuna? VII International Symposium "Chemistry in the Mediterranean" Primosten, Yugoslavia, 6-12 May.

Birke, G., Johnels, A.G., Plantin, B., Sjöstrand, B. and T. Westermark. 1967. Mercury poisoning through eating fish? Lakartidningen 64:3628.

Bligh, E.G. 1970. Mercury and the contamination of freshwater fish. Fish Res. Board Canada. Manuscript Rep. Series N. 1088, Winnipeg, Manitoba.

Bligh, E.G. 1971. Mercury levels in Canadian fish. Proc. Royal Soc. Can. Intern. Symposium "Mercury in man's environment", Ottawa 1971.

Bloom, G., Lundgren, K.D. and A. Swensson. 1955. Exposure and hazards from organic mercury compounds in connection with seed dressing on small farms. Nord. Hyg. T. 36:110.

Buffoni, G., Berhnard, M. and A. Renzoni. 1982. Mercury in Mediterranean tuna. Why is their level higher than Atlantic Tuna? A model. Thal. Yugos. 18:231-243.

Capelli, R., Minganti, V. and M. Bernhard. 1987. Total mercury, organic mercury, copper, manganese, selenium and zinc in *Sarda sarda* from the Gulf of Genoa. Sci. Total Environ. 63:83-99.

Capelli, R., Pellegrini, R., Minganti, V. and R. Poggi. 1989. Preliminary results on the presence on inorganic mercury and selenium in striped dolphin (*Stenella coeruleoalba*) from the Ligurian Sea: European Cetacean Society, 3rd Annual Conference 24-26 February.

Chang, L.W. and R. Suber. 1982. Protective effect of Selenium on methylmercury toxicity: a possible mechanism. Bull. Env. Cont. Toxic. 29:285-289.

Choi, B. 1991. Effects of methylmercury on the developing brain. In: T. Suzuki, M. Imura and T.W. Clarkson, (Eds.); Advances in mercury toxicology. Plenum Press, New York, 315-338.

Clarkson, T.W. 1992. Mercury: major issues in environmental health. Envir. Health Perspectives 100:31-38.

Cox, C., Clarkson, T.W., Marsh, D.O., Amin-Zaki, L., Tikriti, S. and G.G. Myers. 1989. Dose-response analysis of infants prenatally exposed to methylmercury.

An application of a single compartment model to single-standard hair analysis. Environ. Res. 49:318-332.
Cunningham, P.A. and M.R. Tripp. 1975. Factors affecting the accumulation and removal of mercury from tissues of the American oyster (*Crassostrea virginica*). Mar. Biol. 31:311-319.
D'Itri, F.M. 1990. Mercury contamination. What we have learned since Minamata. Env. Mon. Ass. 19, 165-182.
Engel, G.T. 1967. Mercury. In: Kirk-Othmer (Ed.), Encyclopedia of chemical technology. Second ed. 13, New York; Wiley.
Fimreite, N., Fyfe, R.W. and J.A. Keith. 1970. Mercury contamination of Canadian prairie seed eaters and their avian predators. Can. Field-Naturalist 84:269.
Fimreite, N. and J. Reynolds. 1973. Mercury contamination of fish in northwestern Ontario. Jour. Wld. Manag. 37:62-68.
Freeman, H.C., Shum, C. and J.F. Uthe. 1978. The selenium content in swordfish (*Xiphias gladius*) in relation to total mercury content. Jour. Environ. Sci. Health A 13:235-240.
Gaggi, C., Chemello, G. and E. Bacci. 1991. Mercury vapor accumulation in azalea leaves. Chemosphere 22:869-872.
Ganther, H.E., Gondie, C., Sunde, M.L., Kopecky, M.J., Wagner, P.O. and W.G. Hoekstra. 1972. Selenium: relation to decreased toxicity of methylmercury added to diets containing tuna. Science 175:1122-1124.
Gaskin, D.E., Stonefield, K.I., Suda, P. and R. Frank. 1979. Changes in mercury levels in harbour porpoises from the Bay of Fundy (Canada) and adjacent waters during 1969-77. Arch. Env. Contam. Toxicol. 8:733-762.
Goldberg, E.D. 1976. The health of oceans. The Unesco Press: pp. 170.
Goldwater, L.J. and M.B. Jacobs. 1964. Absorption and excretion of mercury in man. VII. Significance of mercury in blood. Arch. Environ. Health 9:735.
Grandjean, P., Weihe, P., Jorgensen, P.J., Clarkson, T., Cornichiari, E. and T. Videro. 1992. Impact of maternal seafood diet on fetal exposure to mercury, selenium and lead. Arch. Environ. Health 47:185-195.
Greenwood, M.R. 1975. Methylmercury poisoning in Iraq. An epidemiological study of the 1971-72 outbreak. Jour. Appl. Toxicol. 5(3):148-159.
Hammond, A. 1971. Mercury in the environment: natural and human factors. Science 171:788-789.
Hansen, J.C., Kromann, N., Wulf, H.C. and K. Alboge. 1984. Selenium and its interrelation with mercury in whole blood and hair in an east Greenlandic population. Sci. Total Environ. 38:33-40.
Harada, Y. and H. Moriyama. 1977. Study on the motor ability of school children in the Minamata district. Environ. Health Rep. 40:97-99 (in Japanese).
Harada, M., Fujino, T. and K. Kabaskima. 1977. Study of methylmercury concentration in the umbilical cords of the inhabitants born in Minamata area. Brain Dev. 9:79-84 (in Japanese).
Honda, K., Yamamoto, Y., Kato, H. and R. Tatsukuwa. 1987. Heavy metals

accumulation and their recent changes in southern Minke Whales *Balaenoptera acutorostrata.* Arch. Envir. Contam. Toxicol. 16:209-216.

Hornung, H. 1986. Assessment of mercury pollution in coastal marine sediments and in benthic organisms. In: FAO/UNEP/WHO/IOC/IAEA meeting of the biogeochemical cycle of mercury in the Mediterranean, FAO Fish. Rep. no. 325 (Suppl.), Rome, pp. 104-110.

Hornung, H., Krumholz, B.S. and Y. Cohen. 1984. Mercury pollution in sediments, benthic animals, fishes and sediments in Haifa Bay, Israel. Mar. Environ. Res. 12:191-208.

Jensen, S. and A. Jernelöv A. 1969. Biological methylation of mercury in aquatic organisms. Nature 223:753.

Johnels, A.G. and T. Wetermark. 1969. Mercury contamination of the environment in Sweden. In: M.W. Miller and G.G. Berg (Eds.): Chemical Fallout, p. 221, C.C. Thomas, Springfield, Illinois.

Kazantzis, G., Al-Mufti, A.W., Al-Jawad, A., Al-Shahwani, J., Majid, M.A., Mahmoud, R.M., Soufi, M., Tawfiq, K., Ibrahim, M.A. and H. Dabagh. 1976. Epidemiology of organomercury poisoning in Iraq: II Relationship of mercury levels in blood and hair to exposure and to clinical findings. Bull World Health Organ. 53 (suppl):37-48.

Kazantis, G., Al-Mufti, A.W., Copplestone, J.F., Majid, M.A. and R.M. Mahoud. 1976b. Epidemiology of organomercury poisoning in Iraq: III Clinical features and their changes with time. Bull. World Health Organ. 53 (suppl): 49-60.

Kjellström, T., Kennedy, P., Wallis, S. and C. Mantel. 1986. Physical and mental development of children with prenatal exposure to mercury from fish. Rep. 3080, National Swedish Environmental Protection Board, Solna, Sweden.

Korringa, P. and P. Hagel. 1974. Mercury in the Environment. In: Proceedings of the International symposium on the problems of contamination of man and his environment by mercury and cadmium. Luxembourg, 3-5 July 1973; CEC, Luxembourg p. 279.

Kiwimäe A., Swensson, A., Ulfarvson, V. and G. Westöö. 1969. Methylmercury compounds in eggs from hens after oral administration of mercury compounds. J. Agr. Food Chem. 17:1014-1019.

Ladd, C., Berkhout, P.G. and M. Jacob. 1964. Absorption and excretion of mercury in man. VII° Significance of mercury in blood. Arch. Environ. Health 9:735.

Law, R.J., Jones, R.B., Baker, J.R., Kennedy, S., Milne, R. and R.J. Morris. 1992. Trace metals in the livers of marine mammals from the Welsh coast and the Irish Sea. Mar. Poll. Bull. 24(6):296-304.

Leonzio, C., Focardi, S. and C. Fossi. 1991. Heavy metals and selenium in stranded dolphins of the North Tyrrhenian Sea (N.W. Mediterranean) Sci. Tot. Env. 119:77-84.

Lindberg, S., Stockes, P., Goldberg, E. and C. Wren. 1987. Group Report: Mercury. In: Hutchinson, T.W. and Meena, K.M. (Eds.). Lead, mercury, cadmium and arsenic in the environment. John Wiley and Sons, New York,

Chichester, Brisbane, Toronto:pp. 17-34.
Lindqvist, O., Jernelöv, A., Johansson, K. and R. Rodhe. 1984. Mercury in the Swedish environment: global and local sources, National Swedish Environment Protection Board (Rep. 1816):105 pp, Solna, Sweden.
Marsh, D.O., Mayers, G.J., Clarkson, T.W., Amin-Zaki, L. and S. Tikriti. 1977. Fetal methylmercury poisoning. New data on clinical and toxicological aspects. Trans. Am. Neurol. Assoc. 102:69-71.
Marsh, D.O., Turner, M.D., Smith, J., Choi, J.N. and T.W. Clarkson. 1974. Methylmercury in human populations eating large quantities of fish. II° American Samoa. Cannery and working fishermen. In: Proceeding of the First Int. Cong. on Mercury. Barcelona, May 6-10, Vol. II, 235-239.
Marsh, D.O., Clarkson, T.W., Cox, C., Myers, G.J., Amin-Zaki, L. and M. Al-Tikriti. 1987. Fetal methylmercury poisoning. Arch. Neurol. 44: 1017-1022.
Martoja, R. and D. Viale. 1977. Accumulation des granules de séleniure mercurique dans le foie d'Odontocétes (Mammifers, Cétacés): un mecanisme possible de détoxication du méthylmercure par le sélénium. C.R. Acad. Sci. Paris t2 Série D, 285:109-112.
Matthews, A.D. 1983. Mercury content of commercially important fish of the Seychelles and hair mercury levels of a selected part of the population. Environ. Res. 30:305-312.
McKeown-Eyssen, G.E. and J. Ruedy. 1983. Methylmercury exposure in Northern Quebec. I° Neurological finding in adults. Am. J. Epidem. 118:461-469.
McKeown-Eyssen, G.E., Ruedy, J. and A. Neims. 1983. Methylmercury exposure in Northern Quebec. II° Neurological findings in children. Am. J. Epidem. 118:470-479.
Menasveta, P. and R. Siriyong. 1977. Mercury content of several predacious fish in the Andaman Sea. Mar. Poll. Bull. 8:200-204.
Miettinen, J.K., Heyraud, M. and S. Kèckes. 1972. Mercury as hydrospheric pollutant. II° Biological half-time of methylmercury in four Mediterranean species: a fish, a crab, and two molluscs. Marine Pollution and Sea Life (Ruivo, M. ed.) pp. 295-298, FAO Fishing News, London 1972.
Miller, G.E., Grant, P.M., Kishore, R., Steinkruger, F.J., Rowland, F.S. and V.P. Gurnn. 1972. Mercury concentrations in Museum specimens of Tuna and Swordfish. Science 175:1121-1122.
Minagava, K., Sazaki, T., Takizawa, K., Tamura, R. and T. Oshina. 1980. Accumulation route and chemical form of mercury in mushroom species. Bull. Environ. Contam. Toxic. 25:382-388.
Minamata Report. 1968. In: Kutsuna (ed.) Minamata disease: Study Group of Minamata Disease, Kumamoto Univ., Japan.
Monteiro, L.R. and H.D. Lopes. 1990. Mercury content in Swordfish, *Xiphias gladius* in relation to length, weight, age and sex. Mar. Poll. Bull. 21(6):293-296.
Rasmussen, P.E. 1994. Current methods of estimation atmospheric mercury

fluxes in remote areas. Envir. Sci. Technol. 28:2233-2241.
Renzoni, A., Bacci, E. and L. Falciai. 1973. Mercury concentrations in the water, sediments and fauna of an area of the Tyrrhenian coast. Rev. Int. Ocean. Med. 31/32:17-45.
Renzoni, A., Berhnard, M., Sarà, R. and M. Stoeppler. 1978. Comparison between the Hg body burden of *Thunnus thynnus* from the Mediterranean and the Atlantic: CIESM, Antalya:255-260.
Renzoni, A., Focardi, S., Fossi, C., Leonzio, C. and J. Mayol. 1986. Comparison between concentration of mercury and other contaminants in eggs and tissues of Cory's Shearwater (*Calonectris diomedea*) collected on Atlantic and Mediterranean islands. Envir. Poll. (Series A.) 40:17-37.
Renzoni, A. and B. Massa. 1989. Contaminants in Mediterranean sea birds. 2nd Mediterranean Seabird Symposium, Calvià, 21-26 March, 207-212.
Rudd, J.W.M., Furutani, A. and M.A. Turner. 1980. Mercury methylation by fish intestinal contents. Appl. Envir. Microbiol. 40:77-82.
Saha, J.G. 1972. Significance of mercury in the environment. Residue Review 42:103-163.
Skerfving, S. 1979. Methylmercury exposure, mercury levels in blood and hair, and health status in Swedes consuming contaminated fish. Toxicology 2:3-23.
Takeuchi, T. 1977. Pathology of fetal Minamata disease. The effect of methylmercury on intrauterine life of human beings. Pediatrician, 6:69-87.
Takeuchi, T. and K. Eto. 1975. Minamata disease. Chronic occurrence from pathological viewpoints. In: Studies on the health effects of alkylmercury in Japan. Tokyo, Japan Environmental Agency, pp. 28-62.
Tamashiro, H., Akagi, H., Arakaki, M., Fatatsuka, M. and L.H. Roth. 1984. Causes of death in Minamata disease: analysis of death certificates. In: Arch. Occup. Environ. Health 54:181-185.
Tamashiro, H., Arakaki, M., Fatatsuka, M. and S. Lee. 1986. Methylmercury exposure and mortality in southern Japan: a close look at causes of death. Jour. Epidemiol. Community Health, 40:181-185.
Tamashiro, H., Fukutomi, K. and S. Lee. 1987. Methylmercury exposure and mortality in Japan: a life table analysis. Arch. Environ. Health, 42:100-107.
Turner, M.D., Marsh, D.O., Rubino, C.E., Chiriboga, J., Chiriboga, C.C., Smith, C.J. and T.W. Clarkson. 1974. Methylmercury in a population eating large quantities of marine fish. I° Northern Peru. In: Proceed. of the First Int. Cong. on Mercury, Barcelona, May 6-10, Vol. II:229-234.
Tsubaki, T. and K. Irukayma. Eds. 1977. Minamata disease. Amsterdam, Oxford, New York, Elsevier Science Publishers.
Tsubaki, T., Hirota, K., Shirakawa, K., Kondo, K. and T. Sato. 1978. Chemical, epidemiological and toxicological studies on methylmercury poisoning. In: Plaa, G.L. & Ducan, W.A.M. (Eds.) Proceedings of First International Congress on Toxicology. New York, London, San Francisco, Academic Press. pp. 339-357.

Tsubaki, T. and K. Takashashi. 1986. Recent advances in Minamata disease studies, Tokyo, Kodansha.
Walsh, C.T. 1982. The influence of age on the gastrointestinal absorption of mercuric chloride and methylmercury chloride in the rat. Environ. Res. 27:412-420.
Weiss, H.V., Kaide, M. and E.D. Goldberg. 1971. Mercury in a Greenland ice sheet: evidence of recent input by man. Science 174:692-694.
Westöo G. 1966. Determination of methylmercury compounds in foodstuffs. I° Methylmercury compounds in fish: identification and determination. Acta Chem. Scand. 20:2131.
WHO/IPCS. 1990. Environmental Health Criteria no. 101, Geneva.
Wood, J.M., Kennedy, F.S. and C.E. Rosen. 1969. Synthesis of methylmercury compounds by extracts of methanogenic bacteria. Nature 220:173-174.

Environmental Contaminants, Ecosystems and Human Health. Edited by S.K. Majumdar, E.W. Miller, and Fred J. Brenner.

Chapter Twenty-Three

ESTIMATING CHILD EXPOSURE TO ELECTROMAGNETIC FIELDS ASSOCIATED WITH ELECTRIC POWER TRANSMISSION LINES

ELLEN K. CROMLEY[1] and KEVIN P. JOY[2]

[1]Department of Geography
University of Connecticut
Storrs, CT 06265
and
[2]EnviroGraphics
Chester, CT

INTRODUCTION

Electromagnetic fields (EMFs) are an environment feature of increasing concern to researchers and policymakers since the mid 1960s (1) when Soviet researchers published reports of medical problems among switchyard workers (2). Over the past thirty years, biologists have explored the possible association between electric and magnetic field exposure and changes in the cells and systems of the human body (3). Although no studies to date have produced irrefutable evidence linking EMF exposure to adverse health effects, scientists have established that particular field strengths induce biological reactions and offered several explanations for a relationship between prolonged exposure to low-level electromagnetic fields and an increased risk of cancer (4, 5). Epidemiological research has focused attention on the electromagnetic fields associated with electrical power systems including generating stations, high voltage transmission lines, and primary distribution lines, and cancer, particularly childhood leukemia (6). In consideration of this research, at least seven states in the U.S. have enacted legislation or established guidelines for the management of future high voltage transmission lines and many utilities have explored and implemented EMF mitigation measures (7).

Environmental contamination from EMFs thus represents a particular kind of public health problem. Unlike contaminants emitted from single sources, EMFs arise from the operation of every electrical appliance, distributor, and generator, varying only in relative intensity, and are omnipresent in post-industrial societies (8). In contrast to contaminants like short fiber asbestos, EMFs have not been irrefutably linked to adverse human health impacts. Finally, particularly in the case of EMFs associated with electric power and distribution systems, proven mitigation techniques are available. Given these dimensions of EMFs as a public health problem, understanding geographical patterns of exposure to EMFs has important policy and research implications. On the public policy side significant reductions in total exposure are possible by focusing mitigation efforts on those areas where exposure is greatest. On the research side, epidemiological research can be advanced by better methods for estimating exposure fields. The purpose of this chapter is to demonstrate the use of geographic information system (GIS) technology for estimating child exposure by residential or school location to the potential environmental hazard of EMFs associated with electrical power transmission lines.

TEXT

Background

The electromagnetic field emanating from objects carrying electrical current has two components (9). The electric component pushes or pulls charged particles in the direction of the field. The magnetic component acts on moving charged particles and pushes them perpendicular to their direction of motion. The intensity of an electric field is directly proportional to the magnitude of force of the charged particle but the magnetic field is a function of the wire current and charge and velocity of the moving particle. A common unit of measurement for magnetic fields is the milligauss or mG. Intensity drops off rapidly with an increase in distance from the power source. Although the magnetic fields associated with many household appliances are often stronger than those found adjacent to power transmission and distribution lines, appliance fields typically exhibit dramatic distance decay (from several thousand mG adjacent to the appliance to less than 50 mG at a distance of a meter) and appliances are not considered a major contributor to time-averaged magnetic fields posing a possible health hazard to humans.

Epidemiological research has instead focused on the electrical power transmission system. In the U.S., electrical power is distributed to end users through a vast network of transmission and distribution circuits. The transmission segment of the system consists of high voltage transmission lines connecting power generating facilities with outlying substations. The lines are typically supported by metal or wooden structures reaching heights in excess of 120' and transmitting power at 69 to 765 kV. From the intermediate substations where power is transferred from high to low voltage, primary and secondary distribution lines deliver electricity to end users. These lines carry loads of up to 35 kV that are gradually stepped down to allow the safe operation of household electrical appliances.

The previously cited research by Wertheimer and Leeper conducted in Denver, Colorado, was one of the first controlled epidemiological studies of residential exposure to EMFs in the U.S. Children under age 18 from high-exposure homes were more than twice as likely to develop cancer as children living in non-exposed homes. Criticisms of this study were addressed in later research by Tomenius (10) and Savitz (11), among others, yielding similar results. A recent study conducted in Sweden calculated relative risk estimates for the incidence of cancer across levels of EMF exposure. A relative risk of 2.7 for children exposed to a 2 mG electromagnetic field for childhood leukemia was established with a 95 percent confidence interval (12). Given the growing body of evidence suggesting a potential link between exposure to EMFs and childhood cancer, techniques for better evaluating geographical patterns of exposure are needed.

Study Area

This research mapped child exposure to transmission line EMFs in excess of 2 mG in Hartford County, Connecticut. Although the analytical procedures outlined here could be replicated for exposure analysis in other communities across the U.S., several specific attributes of Hartford County made it an appropriate setting for this research. As a whole, Connecticut is a densely populated region. Estimated child population density of approximately 175 children per square mile based on 1990 Census data ranks Connecticut fourth behind New Jersey, Rhode Island, and Massachusetts. Hartford County, located in the north central part of the state had a 1990 population of 851,783 distributed across 29 towns. The population under age 18 was 192,343 or 23 percent of state's population in that age group.

The towns in Hartford County are diverse in their socioeconomic characteristics and settlement geography. Hartford, the state capital, has the largest population (139,739) of any town in the county and the largest minority population in the region. Other towns are older or emerging suburban communities. In the northern and eastern periphery of the county, the towns are predominantly rural. Selection of Hartford County as the study area enabled exploration of exposure patterns in various community settings.

Northeast Utilities, a major public utility in Connecticut and the New England region, is solely responsible for the electrical power transmission and distribution system in the study area. The county's transmission line network includes 56 circuits covering approximately 190 miles. Although high voltage lines on the order of 500 kV and 765 kV are absent in the study area and the state as a whole, Hartford County's network is comprised of a combination of 69 kV, 115 kV, and 345 kV circuits covering a wide array of design configurations.

Unlike a number of other states that have passed legislation to regulate transmission line field limits, Connecticut follows the general electrical exposure restrictions created in 1990 by the International Radiation Protection Association (IRPA). These limits allow electric field levels of more than twice the intensity specified in some state regulations and place no limits on magnetic fields (Fitzgerald 1990). In Connecticut, the regulations on utility right-of-way width take into account only simple line characteristics and acknowledge additional policies adopted by the individual utility companies (Connecticut Power Facility Evaluation Council 1974). Corridors are set no greater than 250 feet in width even for multiple lines.

In 1991, the State of Connecticut created a task force to establish appropriate policy for EMFs. In its final recommendations, the task force did not call for implementation of EMF mitigation or the adoption of restrictive legislation but did establish a set of best management practices for EMF emissions. The practices, adopted by the Connecticut Siting Council in 1993, are primarily concerned with protecting the public from EMF hazards caused by future additions or alterations to the transmission network. Detailed electromagnetic field assessments are required for proposed projects. The methodology described in the following section produces a picture of the complex geographical relationships between existing transmission lines, their associated exposure fields, and the distribution of human populations and could be used to assess the exposure impacts of changes in the transmission network or residential development patterns.

GIS Functions

To map and analyze EMFs associated with high voltage transmission lines in Hartford County, four common spatial analytic capabilities of GIS were used: buffering, polygon overlay, areal interpolation, and point-in-polygon overlay (13). Buffering is an operation to calculate and identify the boundary of a region within a specified distance of a particular object. The procedure can be performed for any spatial entity-point, line, or area-but the output from the procedure is one or more areas outlining the particular object. In this analysis, variable buffers were created for both sides of each transmission line segment in the study area to depict the geographic extent of the 2 mG exposure field. The 2 mG level was selected based on the cited epidemiological research finding significant associations between this level of exposure and cancer incidence.

Polygon overlay is an operation to compare geographic features and their attributes across two or more data layers and to create new geographic features having particular combinations of data attributes. In this research, the polygons representing areas where EMF exposure exceed 2 mG were overlayed with the buffered street networks of the 1990 census blocks. An areal interpolation technique, the areal weighted average approach, was then used to estimate child populations within the 2 mG exposure fields based on residence. Areal interpolation involves estimation of attribute values for locations or regions for which there are no exact measurements (14). Finally, point-in-polygon overlay was used to identify school locations within the estimated exposure fields.

Analytical Procedures

Estimation of the number of children exposed to EMFs by residence or by school attendance was accomplished through 10 steps:

1. Compiling a digital database of transmission lines located in the study area using the pcARC/INFO 3.4D+GIS software and checking the positional accuracy of the lines against TM images to an accuracy of within 15 m (Figure 1).

2. Calculating winter and summer exposure fields for each transmission line segment using the FIELDS software available from Southern California Edison Company.

3. Buffering the center of each line segment right-of-way as represented in the GIS to the distance calculated for winter and summer exposure fields in excess of 2 mG and subtracting from the exposure fields for each segment the local right-of-way where no development can exist.

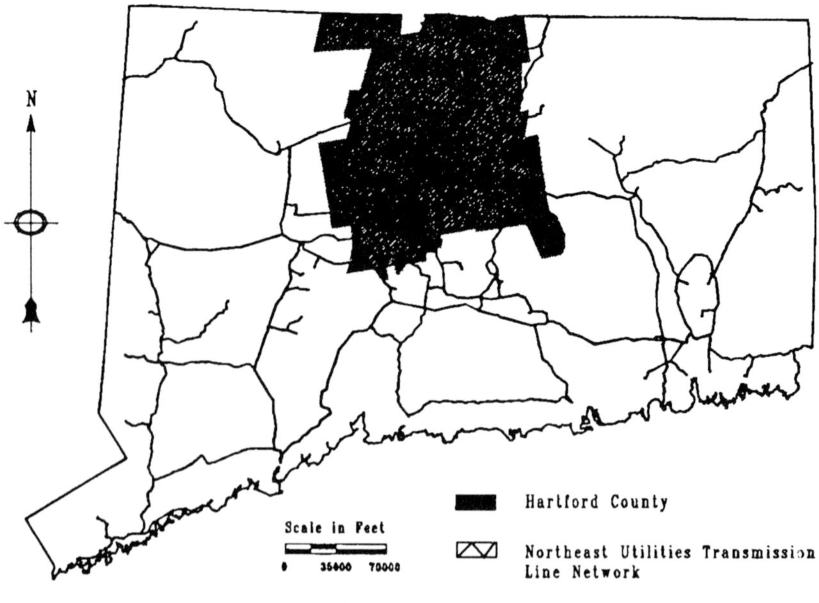

FIGURE 1. Hartford County, Connecticut, Study Area and State Transmission Line Network.

4. Creating a digital database of 1990 Census blocks (also to an accuracy of within 15 m) for Hartford County based on road, stream, and town boundary data from the Connecticut Department of Environmental Protection digital databases. An example of this database for a single town in the study area, Bloomfield, is provided (Figure 2).

5. Attaching 1990 Census data on number of children under 18 years of age to each block.

6. Creating a 300' buffer around the street network (excluding major highways) of each census block to represent the zone where residential development is most likely to occur based on residential setback and lot configurations in the study area (Figure 3).

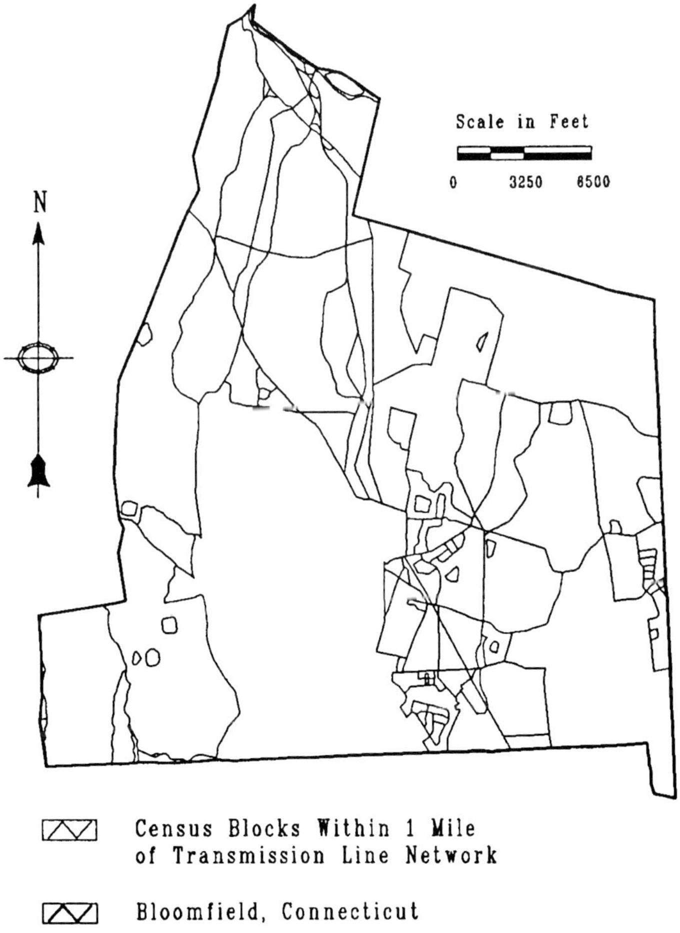

FIGURE 2. Census Block Boundaries, Bloomfield, Connecticut.

7. Overlaying the database containing the street buffer with the database containing the census blocks to localize the block population to the street network.

8. Creating a digital database of public and private school locations in the study area.

9. Overlaying the databases containing the exposure fields with the databases of localized child populations within each block and estimating the number of children by multiplying the total number of children in the block by the percentage of the buffered street region in that block that coincided with an exposure field (Figure 4).

10. Overlaying the databases containing the exposure fields with the database

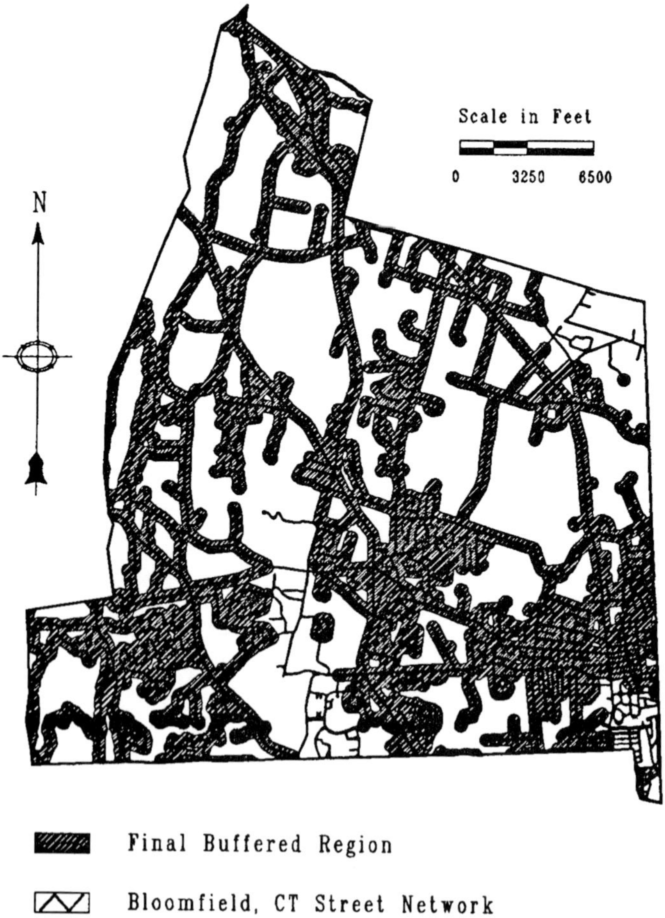

FIGURE 3. 300-Foot Buffer for All Streets within 1 Mile of High Voltage Transmission Lines in Bloomfield, Connecticut.

showing school locations and estimating the number of children in exposure fields from school enrollment figures (Figure 5).

Data Collection

The digital database for the transmission lines was created by digitizing maps of line location and type obtained from Northeast Utilities and registering these lines against a less complete but more spatially accurate database of transmission lines compiled by the State Department of Environmental Protection. The FIELDS program used to calculate the EMF field around a line requires input on

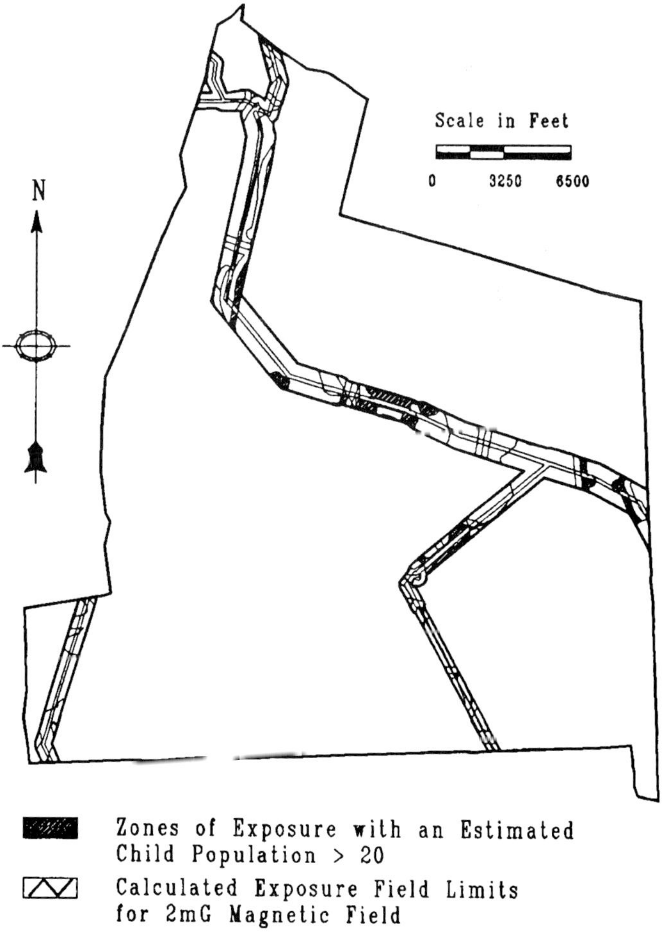

FIGURE 4. Estimation of Child Populations Living within a 2 mG Magnetic Field in Bloomfield, Connecticut.

phase coordinates (horizontal and vertical), number of subconductors per bundle, conductor diameter, bundle diameter, line kV, phase current in amps for the relevant time period, phase angle in degrees, ground wire coordinates (horizontal and vertical), ground wire diameter, ground wire current amps, and ground wire phase angle. For this research, the phase current for each transmission line segment was measured twice, first as the normal rated circuit load for a particular line in winter and second as the normal rate circuit load in summer. These data were obtained from Northeast Utilities.

Data on child population by block in Hartford County in 1990 were obtained from the 1990 Census of Population and Housing (15). School latitude/longitude data were obtained from a gazetteer and converted into State Plane Coordinates

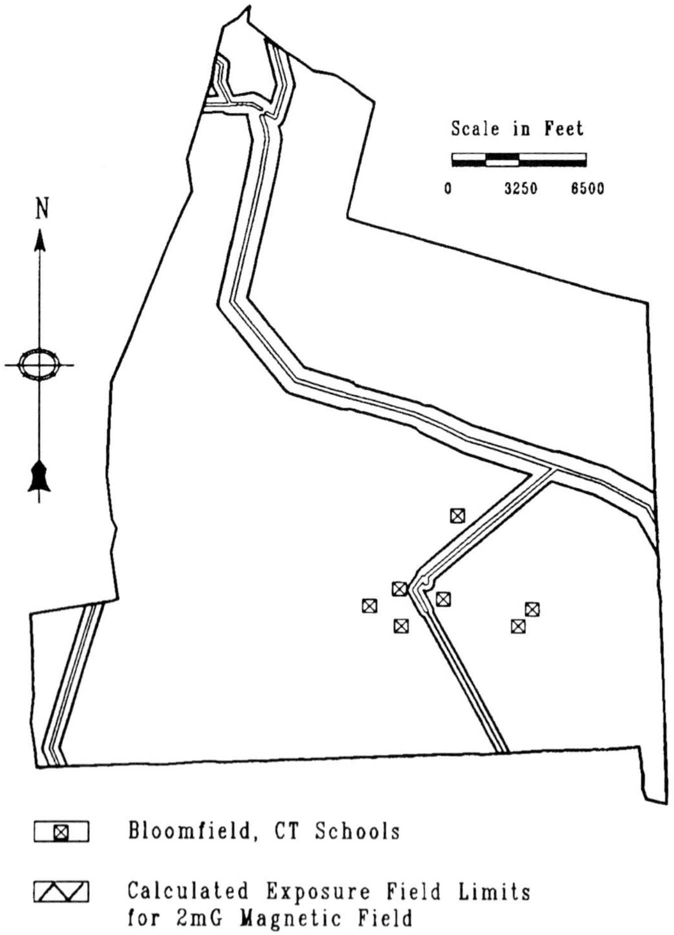

FIGURE 5. School Location in Relation to a 2 mG Magnetic Field in Bloomfield, Connecticut.

(16). Information on 1990 school enrollment was collected from the State Department of Education and from individual schools (17).

Results

This analysis indicates that, overall, the number of children exposed to EMFs in Hartford County is fairly low in relative terms. Approximately 2 percent of children under age 18 are estimated to be exposed by residential location based on the large winter exposure fields. However, in absolute terms, 4,194 children are estimated to be exposed.

Although child exposure from residential location was estimated to occur in 24 of the 29 towns in the study area, the analysis clearly identifies selected neighborhoods where exposure is much greater due to the child population of the block and the juxtaposition of the street network and transmission line exposure fields. Of the total number of blocks in Hartford County (10,599), 618 blocks had populations of children under 18 and were adjacent to a calculated winter exposure field. The average number of children per residential block adjacent to winter exposure fields was close to 7 children, although there was considerable variation across blocks. For 5 percent of the blocks in the county that had child populations and were adjacent to winter exposure fields, the estimated number of children within the 2 mG limit was more than 20 children. These blocks accounted for 46 percent of the total estimated exposed population of 4,194.

A number of the high exposure blocks were concentrated in two older suburban communities, Bloomfield and Southington, where population densities are higher and there is considerable manufacturing and commercial activity. Interestingly, there was very little residential exposure in the central city of Hartford where the transmission lines are bundled and buried. The five towns with no estimated residential exposure because no transmission lines cross into the towns were all rural or suburban communities on the periphery of the county. Small scale maps of child exposure to EMFs by residence and by school location for the entire study area could not be included in this chapter but are available in a more detailed account of this research (18).

In relation to exposure by residence, exposure to EMFs by school attendance was estimated to be much lower. Three schools with a total enrollment of 887 students (out of a countywide school enrollment of approximately 136,000) were identified as located strictly within the winter 2 mG exposure field. It should be noted, however, that many schools were located just outside the estimated exposure field and exposure based on school attendance is probably very conservatively estimated in this study.

A final and important result of this analysis is the variability in the width of the 2 mG exposure field (Figure 6) calculated based on the line attributes and the normal rated winter load. Only a small number of transmission line segment fields were within the 250' right-of-way zone. These results suggest that the common use of an arbitrary uniform buffer in epidemiological research on EMF exposure and adverse health outcomes is not appropriate. Furthermore, buffers cannot automatically be assumed to be wider for higher voltage lines because line configuration characteristics like phasing can significantly reduce EMFs.

DISCUSSION

The circumstances of an environmental contaminant like EMFs associated with electrical power transmission lines lend themselves to successful pre-emptive public health intervention to limit exposure to a possible health hazard. This is reflected in the actions of public utility companies to modify line configurations near schools where field measurements have indicated that high level electromagnetic fields are present. It is also reflected in the enactment by state

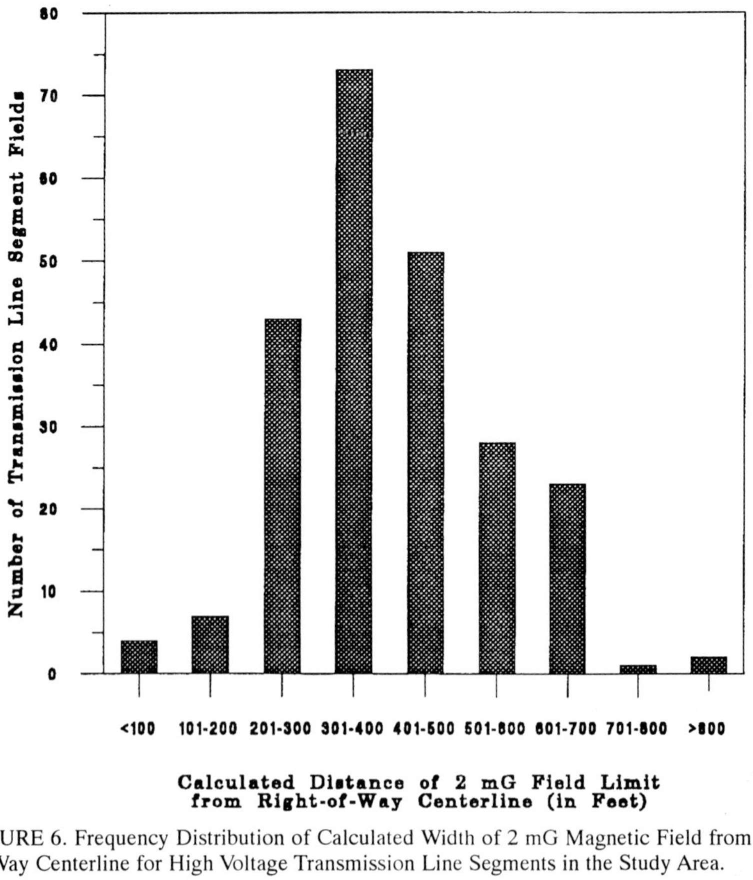

FIGURE 6. Frequency Distribution of Calculated Width of 2 mG Magnetic Field from Right-of-Way Centerline for High Voltage Transmission Line Segments in the Study Area.

legislatures of laws regulating utility rights-of-way or exposure limits.

Understanding the often complex geography of human exposure to environmental contaminants is a key element for developing and implementing effective regulations to protect human health. There is the geography of the sources of contaminants which may be fixed or mobile. There is the geography of the hazard fields surrounding these sources which are usually variable in time and space. There is also the important geography of human location and movement in the environment which describes how individuals come into contact with the environmental risk zones existing at a variety of spatial scales. A weakness of existing research on human exposure is the generalization of individuals' locations in the environment to the home location. Although the home base is clearly an important focus of the individual's activity space, daily travel and activity patterns take people to a variety of other locations where they may be exposed to environmental contaminants and individuals sharing the same residence may in fact circulate in very different parts of the larger environment. The effort to evaluate in the research described in this chapter child exposure to EMFs that might occur by school attendance is an admittedly modest effort to begin to represent exposure in relation to the entire environment of the individual.

Traditional cartographic methods for displaying spatial relationships have not been adequate to represent complex temporal and spatial patterns like those involved in human exposure to environmental contaminants. Over the last twenty years, the development of geographic information systems has created a more powerful tool for modelling spatial relationships. The capability afforded by GIS to search for specific sites with a given set of attributes (large population and high level of environmental contaminant) enables analysts and public policymakers to target intervention programs. The ability to model variables geographically is also important to improving the selection of cases and controls for epidemiological research. Although GIS has not been widely used in epidemiological research or public health to date, there are clear signs that this technology will increasingly be used to analyze environmental contaminants, ecological changes, and human health outcomes.

REFERENCES

1. Morgan, Granger M. 1985. Power-line Fields and Human Health. *IEEE Spectrum* 22;62-68.
2. Korobkova, V.P., Morozov, Y.A., Stolarov, M.D. and Y.A. Yukub. 1972. *Influence of the Electric Field in 500 and 750 kV Switchyards on Maintenance Staff and Means for Its Protection.* Technical Report 23-06. CIGRE.
3. Office of Technology Assessment, U.S. Congress. 1989. *Biological Effects of Power Frequency Electric and Magnetic Fields, Background Paper,* OTA-BP-E-53. U.S. Government Printing Office, Washington, DC.

4. Lerner, Eric J. 1984. Biological Effects of Electromagnetic Fields. *IEEE Spectrum* 21:57-69.
5. Wilson, Berry W., Stevens, Richard G. and Larry E. Anderson. 1989. Neuroendocrine Mediated Effects of Electromagnetic-Field Exposure: Possible Role of the Pineal Gland. *Life Sciences* 45:1319-1332.
6. Wertheimer, N. and E. Leeper. Electrical Wiring Configuration and Childhood Cancer. *American Journal of Epidemiology* 11:345-355.
7. Ordon, Thomas J.F. 1993. EMF Mitigation in Schools. Paper presented at the Electrical Insulation Research Center (EIRC), University of Connecticut, Tutorial Workshop on Electromagnetic Fields, Windsor Locks, Connecticut, October.
8. Kraus, John D. 1992. *Electromagnetics.* McGraw-Hill, Inc., New York.
9. Pool, Robert. 1990. Is There an EMF-Cancer Connection? Science 249:1096-1098.
10. Tomenius, L. 1986. 50Hz Electromagnetic Environment and the Incidence of Childhood Tumors in Stockholm County. Bioelectromagnetics 7:181-207.
11. Savitz, D.A., Wachtel, H.A., Barnes, F., John, E.M. and J.G. Tvrdik. 1988. Case-Controlled Study of Childhood Cancer and Exposure to Electromagnetic Fields. *American Journal of Epidemiology* 128:21-38.
12. Feychting, Maria and Anders Ahlbom. 1992. *Magnetic Fields and Cancer in People Residing Near Swedish High Voltage Power Lines.* IMM-Report 6/92. Karolinska Institutet, Institutet for Miljomedicin, Stockholm.
13. Starr, Jeff and John Estes. 1990. *Geographic Information Systems: An Introduction.* Prentice-Hall, Inc., Englewood Cliffs, New Jersey.
14. Lam, N.S. 1983. Spatial Interpolation Methods: A Review. *The American Cartographer* 25:383-397.
15. Bureau of the Census, U.S. Department of Commerce. 1991. *1990 United States Census of Population and Housing: CD90-1B-1 New England Division.* Bureau of the Census Data User Services Division, Washington, DC.
16. Abate, Frank R. 1991. *Omni Gazetteer of the United States of America, Vol. 1, New England.* Omnigraphics, Inc., Detroit, Michigan.
17. Connecticut Board of Education. 1992. *1991 Connecticut Education Directory.* State Department of Education, Hartford, Connecticut.
18. Joy, Kevin P. 1994. *Estimating Child Exposure to Electromagnetic Fields Associated with Transmission Lines in Hartford County, Connecticut.* Unpublished M.A. Thesis, Department of Geography, University of Connecticut.

Environmental Contaminants, Ecosystems and Human Health. Edited by S.K. Majumdar, E.W. Miller, and Fred J. Brenner.

Chapter Twenty-Four

CANCER MORTALITY AND RISK FACTORS IN THE LEHIGH VALLEY

ROBERT C. ZIEGENFUS
Department of Geography
Kutztown University
Kutztown, PA 19530

INTRODUCTION

In a recent publication (1), this author used cartographic and statistical techniques to analyze the spatial and temporal variation in total cancer mortality rates within the 63 minor civil divisions of Lehigh and Northampton Counties (commonly called the Lehigh Valley) in Pennsylvania for the years 1970-1989. This chapter expands on the earlier work by examining risk factors thought to be associated with cancer. Bivariate and multivariate statistical techniques are used to assess the degree, if any, of association between the mortality rates and the risk factors. The purposes of this research are twofold: 1) to assess the value of the risk factor data in understanding the male and female cancer mortality rates for the years 1970-1989; and 2) to determine the impact that small population places have on an analysis of cancer rates and risk factors.

DATA AND METHODS

It is important to state that the current study employs the minor civil division (MCD) as the geographic unit of analysis. This is quite unusual as many studies of cancer mortality, from one of the earliest (2) to one of the most recent (3), have been conducted at the county level. Indeed, the first single largest compilation of cancer mortality data that led to the establishment of etiological relationships for several cancers was based on county data (4). These studies were considered to be sound methodologically even with the obvious aggregate nature of the information (5-7). The county as the unit of analysis was considered to have the advantages of sufficient numbers of persons and deaths to produce stable mortality rates, and readily available datasets of suspected risk factors.

County data mask considerable variation. For example, Lehigh County's age adjusted total male cancer mortality rate for 1985-1989 is 233.1/100,000. Yet, within the 25 MCDs of Lehigh County the rates vary from 132.2 to 317.7. This gain in detail is accompanied by a potential sacrifice in precision. While the availability of cancer rates for all MCDs is certainly desirable, the move to a lower level of aggregation means that the numerators (deaths) and denominators (population) used to calculate mortality rates have both become smaller. This, in turn, can generate rates that are more susceptible to large fluctuations, especially in the case of very small populated places and/or low rate diseases where a difference of only one or two deaths can produce drastically different rates from one time period to another.

Cancer Data

This rate instability issue is the very reason why all of the cancer data used in this research have been aggregated into five-year totals. At the author's request, the 20 years of cancer deaths and the populations at risk were combined into four discrete time periods (1970-1974, 1975-1979, 1980-1984, 1985-1989) by the Pennsylvania Department of Health. To remove the effect of different age distributions within the MCDs, age-adjusted mortality rates were calculated using the direct method (8) and are reported as rates per 100,000. Pennsylvania's 1970 age distribution was used as the standard population. The final composition of the cancer database included an age-adjusted mortality rate for the four time periods for each of the 63 MCDs for each gender for all cancers combined.

Risk Factors

Theoretically, there should be an abundance of information related to cancer risk given all the reports and publications on the subject. Theory is rarely reality. It is well established that personal habits such as smoking, alcohol consumption and poor nutrition all create some level of risk. But these data, when they exist at all, are most often available from national or state surveys with too few observations to be applied to smaller geographic units. Environmental variables, too, have been implicated in creating excess risk. Again, the data are often absent; commonly because the environmental contaminants that may be related to cancer would have to be known from the time when the exposures might have occurred, typically 15 to 30 years ago if current cancer mortality is being examined. Another reason environmental data are difficult to assemble is that very few monitoring systems were in place to monitor toxic pollutants in air or water. Many state air monitoring systems established in the 1970's were designed to measure the levels of the criteria pollutants set forth in the Clean Air Act of 1970. Moreover, even where monitoring data are available, the spatial distribution of the monitoring network was usually so sparse that the assumption of a link between the data and actual human exposures would be questionable. The problems of different scales of measurement and lack of temporal concurrency between the measurement and the exposure make personal and environmental risk factors difficult to assemble.

Given the limitations discussed above, it is not surprising to find that researchers are forced to accept surrogates for risk factors. Usually this results in the use of U.S. Bureau of the Census data. The scale problem is immediately resolved because the Census data can be matched to the same level of aggregation as the disease data. Which decennial population census to use becomes the next issue. Since the cancer data in this study span the period 1970-1989, it makes little sense to choose any year after the start of the cancer records. Therefore, 1970 data were selected even though 1950 or 1960 might have been more appropriate considering the latency period for cancer. Neither 1950 nor 1960 data were readily available at the level of detail needed for this study.

It was difficult enough to obtain the 1970 information as the sample data for all MCDs was never published by the Census Bureau. In fact, the printed documents contained only 13 of the 63 MCDs. Compounding the dilemma, the author discovered that the 1970 Census Bureau data tapes were no longer publicly available anywhere in Pennsylvania. Four documents prepared by the planning commission that jointly serves Lehigh and Northampton Counties contained a fairly wide variety, though far from all, of the tape data (9-12). A careful perusal of these publications in conjunction with a literature review of cancer studies using census data led to the selection of the 20 variables listed in Table 1. This list is clearly a compromise compared to what was desired. While population, education, income and housing quality data are ample, the employment data are less than ideal because each refers to the total population 16 or older rather than males and females separately. Even more desirable would have been gender data by occupation instead of industry. No data were available for two other topics often addressed in risk factor studies. The mobility of the population remains unknown due to the absence of migration variables. Also missing is any indicator of the foreign born and foreign stock, sometimes used as a surrogate for dietary consumption patterns (13). In short, the risk factor analysis in this study will focus on typical socioeconomic factors.

STATISTICAL METHODS

Three statistical procedures were utilized to examine the relationships between the cancer rates and the risk factors. First, bivariate correlations were calculated for each of the 20 risk factors versus the male and female all cancer rates. Due to the skewed distributions of most of the cancer rates and the risk factors, the Spearman rank correlation coefficients will be reported. Conversion of the non-normal data by taking the logarithms was not an alternative as several of the risk factors had true values of zero.

An additional procedure employed in the correlation work was the selective reduction in the number of MCDs in the analysis. Mention was made above about the unstable results that can be generated by small numbers in the numerators and denominators of mortality rates. Even with the five year aggregation, there were a few MCDs that had no deaths in one or more time periods. As an extreme example, Chapman Borough in Northampton County had only two male cancer deaths in 1970-1974 for a rate of 548.7, the highest of the 63 MCDs in that time period. In 1975-1979, by contrast, it had zero male deaths for a rate of 0.0 and the lowest rate of any MCD. This same cycle of two deaths and no deaths occurred again in 1980-1984 and 1985-1989, respectively. For this reason, six additional rounds of calculations were conducted on each cancer type in order to determine if the rate fluctuations in the small places were influencing the correlations. The reductions occurred as follows: 1) the 6 MCDs with a 1970 population less than 1,000 were dropped; 2) the 11 MCDs less than 1,500; 3) the 18 MCDs less than 2,000; 4) the 23 MCDs less than 2,500; 5) the 26 MCDs less than 3,000; and 6) the 30 MCDs less than 3,500.

The second statistical procedure used in this study was principal components analysis. One of the criticisms of regression analysis when used with risk factors is that the so-called independent variables are not independent of each other but correlated to some degree. Therefore, principal components analysis was considered

TABLE 1
Risk Factor Variables - 1970 Census Data

1. Population density
2. Percent of population urban
3. Total population
4. Percent of total population 25+ with 8 or less years of education
5. Percent of total population 25+ with high school education
6. Percent of total population 25+ with 4 or more years of college
7. Median family income
8. Percent of families with incomes of $25,000+ in 1969
9. Percent of families with incomes less than poverty level
10. Percent of persons 65+ with incomes below poverty
11. Percent of occupied housing units lacking one or more plumbing facilities
12. Percent of occupied housing units with 1.01 or more persons per room
13. Percent of males 16+ in male civilian labor force
14. Percent of females 16+ in female civilian labor force
15. Percent of population 16+ employed in agriculture
16. Percent of population 16+ employed in mining
17. Percent of population 16+ employed in construction
18. Percent of population 16+ employed in manufacturing
19. Percent of population 16+ employed in transportation
20. Percent of population 16+ employed in education, other professional and public administration

essential to reduce the 20 risk factors, several of which are highly correlated, to a small number of uncorrelated factors. Component scores were then calculated for each MCD for each factor. In turn, the uncorrelated component scores were used as true independent variables in regression analysis (14).

Lastly, two different types of regression analyses were conducted. One type had the interval-level 1985-1989 cancer rates as the dependent variable and the component scores as the independent variables. The assumption underlying this approach was that all of the risk factors were contributing to the analysis by virtue of their placement into one of the factors generated by the principal components analysis. Moreover, the creation of component scores generated variables that were uncorrelated with each other. The second type of regression had the ranks of the 1985-1989 cancer rates as dependent variables and the ranks of the risk factors as the independent variables, but having entry into the equation a function of stepwise controls (15). This conservative technique of performing regression on the ranks served as a means of evaluating the utility of two different approaches to determine the significant risk factors for the male and female cancers. A modified scheme of reducing the number of MCDs in the analysis was instituted for both types.

RESULTS

Correlation Analysis

There are two specific issues to be addressed with the correlation results. One is the need to examine the validity of the assumption that the 1970 population data being used as risk factors will have more explanatory power as time progresses because most cancers develop long after exposure. Consequently, it is expected that the fewest statistically significant correlations will occur in the 1970-1974 period and the largest number in 1985-1989. An exception to this expectation is that previous research revealed that the 1975-1979 period to be dissimilar to the other three in its spatial pattern (1). The second issue concerns the role of the small population places and their volatile cancer rates on the interpretation of the correlations. As mentioned in the previous section, this matter will be evaluated by a selective reduction in the numbers of MCDs in the analysis.

When the ranks of male all cancers for the four time periods were correlated with the ranks of the risk factors for all 63 places, 12 coefficients were significant at $p = .05$ as shown in the first column under each time period in Table 2. Noting the number of correlations by time reveals the expected pattern of 3 for 1970-1974, 0 for 1975-1979, 4 for 1980-1984, and 5 for 1985-1989. What was not expected is the strength of the 1970-1974 correlations. Population density is the only variable very strongly positively correlated with male cancers for the three periods. Median family income and the percent of families with incomes of more than $25,000 are strongly negatively correlated in the first and last period.

This type of analysis was performed for each of the six levels of reduction, though space does not permit the number of tables necessary to show all of the intermediate steps. Therefore, a decision was made to list the results of the last step when the 30 MCDs with less than 3,500 population in 1970 were deleted. The 3,500 "rule" is not arbitrary; it represents the stage at which the correlations had become more consistent.

These final step correlations appear in the second column under each time period in Table 2. Also, factors significant only in this stage are identified below the dashed line in the Table. The pattern of an increasing number of significant correlations is repeated with 5 for 1970-1974, 6 for 1980-1984, and 7 for 1985-1989. Perhaps even more remarkable is the strength of the 1985-1989 correlations as a much larger coefficient is necessary to be significant when only 33 places are being correlated than 63. Most of the risk factors found to be significant for all 63 places were also significant for the largest 33 for 1985-1989, so there is consistency that lends credence to the existence of a true rather than spurious finding. The additional significant factors associated with the largest 33 MCDs in the 1985-1989 period reinforce the pattern. What emerges is that

TABLE 2

Correlation Coefficients and Significance Level for Male All Cancers

Risk Factor	1970-1974		1975-1979		1980-1984		1985-1989	
Population	.35	.60			.33	.63	.38	.43
Density	.006	.000			.008	.000	.002	.013
Median Family	-.25					-.37	-.27	-.45
Income	.047					.033	.031	.009
% Family Income	-.39					-.40	-.33	-.35
>$25K	.002					.022	.009	.047
% Persons 65+							.24	.52
Below Poverty							.05	.002
% Males in					-.33	-.53	-.44	-.47
Labor Force					.009	.002	.000	.006
% Employed in		-.45			-.25			
Agriculture		.009			.04			
% Employed in		-.58			-.29			
Transportation		.000			.02			

% Urban		.57				.53		
		.000				.002		
% College								-.34
Education								.05
% Families								.57
Below Poverty								.000
% Employed in		-.36				-.42		
Construction		.04				.014		
	3	5			4	6	5	7

Note: The first column under each year identifies the results when all 63 MCDs are in the analysis and the second column when only the largest 33 are in the analysis. Also the entries below the dashed line are variables that were significant only when the largest 33 MCDs are in the analysis.

population density and poverty are the primary risk factors related to male all cancers for these time periods. Employment risk factors are of lesser importance. It is interesting to note the uniform strength of the male labor force participation variable in the two 1980s periods. Without additional data, one cannot determine why communities with a smaller percentage of males in the labor force in 1970 have higher rates of male cancers in the 1980s. The enigmatic 1975-1979 period had no significant risk factors, though this period would be part of the overall pattern as population density was significant at .06 in both stages and percent urban and percent families in poverty at .07.

To summarize the male cancer findings, the expected pattern was found as more significant correlations appeared in the two 1980s time periods when all 63 MCDs were in the analysis, and across all the time periods but with more in the latter two when the smallest 30 places of less than 3,500 population had been dropped. Risk factors with the strongest positive correlations with male cancers were population density and poverty while the percent of males in the labor force was the strongest negative correlation.

The risk factor analysis for female all cancers appears in Table 3. Clearly, the

TABLE 3

Correlation Coefficients and Significance Level for Female All Cancers

Risk Factor	1970-1974		1975-1979		1980-1984		1985-1989	
Population	.39	.43			.31	.49		
Density	.002	.013			.013	.004		
% Urban	.36	.37				.35		
	.003	.033				.047		
% Less than					-.29			
8 Years Ed.					.023			
% High School					.33			
Education					.008			
% Persons 65+							.33	.48
Below Poverty							.008	.005
% H.U. Missing	-.28							
Plumbing	0.25							
% H.U. Crowded	-.30							
	.016							
% Employed Ed.	.28				.25			
Other Prof,	.026				.049			
Pub Admin								
Population			.27					
1970			.034					
Median Family		-.40						-.48
Income		.021						.005
% Family Incomes		-.39						
>$25K		.027						
% Families Below				.37				.55
Poverty				.033				.001
% Females in								.41
Labor Force								.017
	5	4	1	1	4	2	1	4

Note: This table is arranged in the same manner as Table 2.

male pattern is not replicated. Here the 1970-1974 period has the largest number of significant correlations of any period when all 63 MCDs are included while the 1980-1984 period is second. The 1985-1989 period, expected to have the most, has a mere one. Dropping the smallest 30 places improves the number of correlations for 1985-1989, though only to equal standing with 1970-1974. Table 3 also provides solid evidence of the value of comparing the 63 versus the largest 33. An examination of the risk factors for 1970-1974 reveals that the two housing variables are *negatively* correlated with female all cancer. For 1980-1984, the percent of persons who have less than eight years of education is also negative while high school education is positive. These correlations must be considered unlikely in light of the dramatic change that occurs when the smallest 30 places are deleted. Now, the analysis focuses clearly on population and poverty.

In summary, the female all pattern included some questionable correlations in the full set of 63 MCDs and an even split of significant correlations in the first and last time period. Population density and poverty are the risk factors that correlate most often.

Regression Analysis

The intent of the regression analysis is to investigate what risk factors help to explain the 1985-1989 cancer rates for males and females, as this period had more significant correlations for males than any other period and was of equal importance to the 1970-1974 period in the reduced set for females. In an applied context, the 1985-1989 period is the logical choice as there is a need to understand the risk factors of importance in at least the most recent period if this research is to have any practical value.

The first set of regressions were calculated using the cancer rates as the dependent variable and the component scores from the principal components analysis on the risk factors as the five independent variables. A base 10 logarithm was generated for the 1985-1989 male all cancer rate due to its non-normal distribution. To test the effect of the small places, and knowing from the correlation results that six steps of reduction were redundant, the regression analyses were performed on a reduced set that dropped the 11 places less than 1,500, 23 less than 2,500, and the 30 less than 3,500 (See Table 4). When the regression procedure included all 63 MCDs, component score 2 was the first to enter the analysis followed by score 4 and 3, with all three accounting for approximately one-quarter of the variance in the rate. In the next step, the order remained the same but without score 3 and with less explanatory power. The order of entry was reversed and the R-square improved to 32% when the 23 places of less than 2,500 were dropped. Finally, dropping the smallest 30 led to selection of just score 4 explaining more than a third of the variance in the rate.

An interpretation of this pattern requires an identification label for each score. To give each score a label, one must note the variables loading strongly on a factor. Accordingly, score 4 has been designated elderly and family poverty, score 2 as well educated professionals, and score 3 as rural/small town mining and construction. The poverty variables represented by score 4 have the most explanatory power for the 1985-1989 male all rate, especially when the smallest 23 and 30 MCDs are deleted. This finding corresponds well with the last column of Table 2 showing the significant correlations for male all cancers.

As the female all cancer rate for 1985-1989 was normally distributed, the regression was calculated directly on the interval-level values. Table 4 indicates that no variables entered the analysis in the first two steps. A result such as this occurs when no variable has a sufficiently large coefficient to have an F value of 3.84 (15). For the other two levels, only the poverty score enters the analysis, and it explains between 20% and 24% of the variance in the female all rate. Once again, the regression and correlation results coincide.

In summary, the first regression analysis generated results for both males and females that parallel the correlation analysis. Elderly and family poverty is the most, and sometimes sole, significant score. The percent of the variance explained exceeds 37% for males and 24% for females.

The second type of regression was calculated on the ranks of the male and female cancer rates as the dependent variables and the ranks of the risk factors as the independent variables. Results from the ranks of the male all cancer rate for 1985-1989 are listed in Table 5. It is immediately obvious that this approach has yielded a substantial amount of explanatory power with 58% of the variance accounted for at the final level. This number is also a large increase over the 38% from the first regression. In spite of the different numbers, the same risk factor, poverty, is predominant in both types of regression. Combined with poverty in a positive relationship is population density, but not population size. The interpretation, then, is that the highest male all cancer rates in 1985-1989 are related to the percent of families living in poverty in small populated places with high densities and little or no employment in mining in 1970.

Figure 1 shows the male all cancer rate pattern for 1985-1989 for the 33 MCDs in the regression. The map classes are based on the mean and standard deviation method of range-grading as follows: less than -1 SD, -1 SD to the mean

TABLE 4
Component Score Regression Analysis

		Scores					Adjusted R-square
		1	2	3	4	5	
Male	all 63		1	3	2		24.0%
	drop 11		1		2		18.1%
	drop 23		2		1		31.7%
	drop 30				1		37.6%
Female	all 63			none			
	drop 11			none			
	drop 23				1		24.3%
	drop 30				1		19.8%

TABLE 5
Regression Analysis of the Ranks

				Risk Factors	Adj. R-square
Male	all 63	-	%	Males in labor force (17.8%)	23.9%
		+	%	Females in labor force	
	drop 11	-	%	Families with incomes > $25K (13.5%)	
		+		Population density	
		+	%	Persons 65+ below poverty	31.7%
	drop 23	-	%	Males in labor force (14.9%)	
		+	%	Females in labor force	23.1%
	drop 30	+	%	Families below poverty (30.8%)	
		+		Population density	
		-		Population	
		-	%	Employed in mining	57.8%
Female	all 63	+	%	Persons 65+ below poverty	9.4%
	drop 11	+	%	Persons 65+ below poverty (14.9%)	
		+	%	Females in labor force	20.9%
	drop 23	+	%	Persons 65+ below poverty	25.0%
	drop 30	+	%	Families below poverty (27.4%)	
		+	%	Females in labor force	39.2%

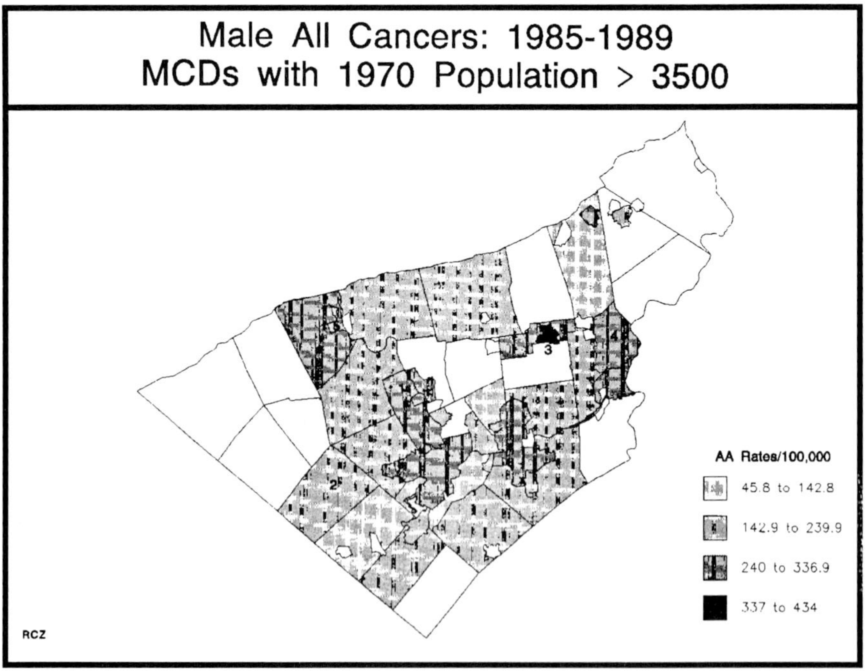

FIGURE 1.

(239.9); the mean to +1 SD, and more than +1 SD. (More extreme values do exist in the 30 MCDs, but dropped from the analysis.) While the risk factors identified in Table 5 explain 58% of the variance in these rates, there are five numbered MCDs on map 1 that had standardized residuals larger than 1.0: 1=Whitehall Township (2.3); 2=Upper Macungie Township (1.7); 3=Nazareth Borough (1.5); 4=Forks Township (1.4); and 5=Slatington Borough (1.2). This means that these places had a much higher actual cancer rate than the value predicted by the regression model. If all male cancers for 1985-1989 are quite well explained by poverty, population density and size, and lack of mining employment, it is also true that other factors may be important in certain places. The search for these other risk factors should focus on the MCDs identified by residual analysis, but that is beyond the scope of this chapter.

Table 5 also contains the numbers for the female all cancer rate. There is more internal variable consistency across the steps than was the case for the male rate, and this probably accounts for the improvement in R-square at each step. At the final step, almost two-fifths of the variance is explained, a sizable increase over the 20% in the last step of the first regression. These results suggest that the highest 1985-1989 female all cancer rates are related to those MCDs that had the largest percentage of families living below poverty and females in the labor force in 1970. It is tempting, and highly speculative, to conclude that an occupational cancer component has been identified for females when these regression results are viewed with the correlation analysis for 1985-1989.

The map pattern for female cancer rates appears in Figure 2, and the same procedures were used for creating classes and residuals as in Figure 1. Three large residuals emerged from the regression model for MCDs with rates above the median of the ranks of the rates: 1=Bangor Borough (1.8); 2=Catasaqua Borough (1.4); and 3=Forks Township (1.4). It is unknown why Bangor has the highest female cancer rate of the 33 MCDs while it had a rank of only 23.5 in percent of families below poverty, but a potential clue may be its rank of 53 in percent of females in the labor force in 1970. This suggests that an occupational cancer connection should be investigated.

DISCUSSION

All of the preceding analyses suggest that the two purposes of this research are closely intertwined. One purpose was to determine the relationship between or among socioeconomic risk factors from the 1970 Census and the male and female cancer rates for four time periods. The second purpose was to determine the influence of small population MCDs in a risk factor analysis, given the highly variable mortality experience in such places.

Male All Cancer

The correlation analysis for males revealed a consistent and expected pattern of more significant correlations in the two 1980s periods, especially 1985-1989, with population density (positive) and high family income (negative) as the key risk factors. This pattern was reinforced and expanded to additional poverty variables when the smallest 30 MCDs of less than 3,500 population were deleted from the calculations. A regression analysis on the log of the 1985-1989 rate and the component scores produced another finding that identified poverty as explaining 38% of the variance. When the regression was repeated using the ranks of the rate against the ranks of the risk factors, the percent of families below poverty emerged as the single most important risk factor. Both the explanatory power and the logic of the variables chosen by the regression improved dramatically as the number of MCDs in the analysis was reduced from 63 to 33. Based on these results for males, one can safely conclude that the 1970 risk factors had more value in helping to understand the male all cancer rate in the 1980s than the 1970s, and that a sensitivity analysis of the importance of size suggests that, at least in this research, MCD populations of less than 3,500 have substantial variation that can generate unstable results in correlation and regression analysis.

Female All Cancers

The assumption of more significant correlations in the 1980s did not hold true

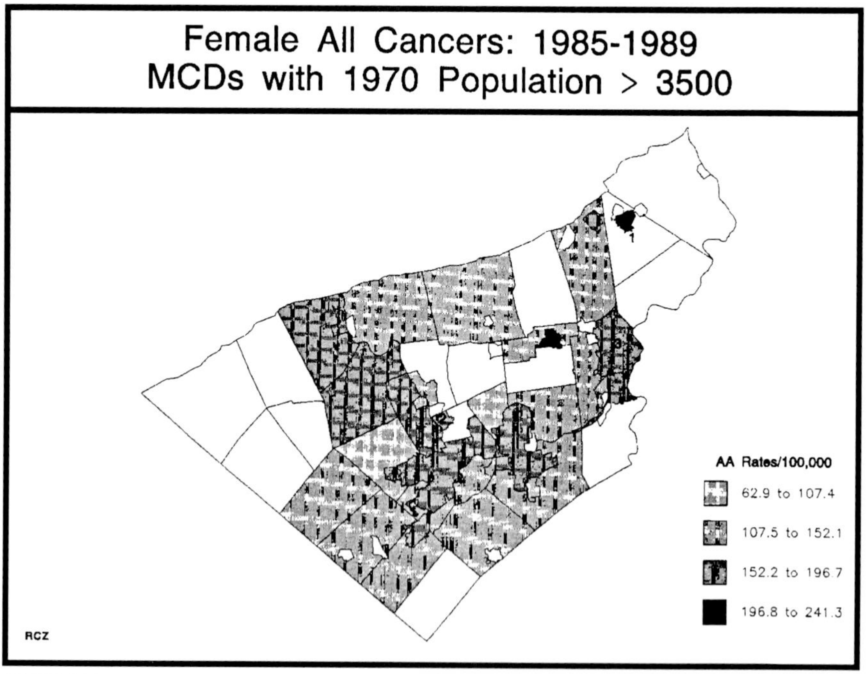

FIGURE 2.

for the female rate. For the analysis of all 63 MCDs and the reduced set of 33 there was an even split between the two decades. An otherwise numerical dominance by the 1970-1974 period for the 63 is flawed by correlations that appear highly unlikely. Dropping the smallest 33 created more reasonable results with population density and percent urban important in 1970-1974 and poverty in 1985-1989. Component score regression focused exclusively on the poverty factor, when a reduced set of MCDs were analyzed. Regression on the ranks produced a final adjusted R-square of 39% as compared to 20% in the first regression. The risk factors of importance, however, remained essentially the same. In addition to poverty, the correlations and the regression on the ranks illuminated the value of the percent of females in the labor force in 1970. One can not be assured that this is an indication of occupational exposures without additional research. To conclude, female all cancer had significant correlations scattered across time periods though the strength of the correlations for the reduced set of 33 for the 1985-1989 period were the most logical results. While not as strong as for males, the utility of the reduction technique proved its worth again.

The primary importance of poverty as a risk factor for male and female all cancers is similar to that reported in other studies (8, 13). One wonders if the stronger correlations in the 1985-1989 period represent, to some degree, the greater socioeconomic disparity in mortality from 1960 to 1986 documented recently by Pappas et al. wherein they report that the inequalities increased five times more in men than women, and more according to income than education (16).

Implications for Further Research

A perusal of the male versus female results reveals that the number of statistically significant correlations is higher, and that the correlation coefficients are larger for the males. Similarly, the component regression generated a 38% explanation of the variance for males and 20% for females while the regression on the ranks yielded 58% and 39% for males and females, respectively. Much of this differential is likely due to the choice of risk factors used in the study. Probably the major limitation is the lack of gender-based occupational data. The only gender-specific variable was the percent of labor force participation. Even this crude measure was positively correlated with female all cancer for the 1985-1989 period. It would have been highly desirable to include the percent of females in technical, managerial, professional, clerical/kindred, and operative jobs or a combination of these. No doubt the male correlation coefficients would have been strengthened by the equal availability of male-specific occupational variables. Likewise, a larger proportion of the total variance might have been explained with gender-specific education data. Any attempt to add precision to the risk factor data should also include variables for mobility/migration, foreign born/foreign stock, and environmental exposures.

Another modification that has merit for a future study is the development of an aggregation technique by which the 30 MCDs of less than 3,500 population in 1970 could be combined to create a number of new units of analysis. The objective of such an effort would be to have all the cancer deaths in the analysis at each step. It should be possible to design a clustering method that will group the small MCDs. While aggregation of individual MCDs will produce some loss of detail, the potential gain is knowledge about cancers and risk factors in small places that could potentially go beyond merely stating that their cancer rates are unstable.

A more precise risk factor database and a technique to have all the deaths accounted for in each step of reduction would afford the opportunity to expand the research beyond the male and female all cancer rates into specific body-site cancers. Such an expansion might lead to the identification of specific risk factors for specific cancers. Another contribution of this type of effort might be clues to the mystery of the 1975-1979 period that had very few correlations with the current risk factors and cancers.

ACKNOWLEDGMENTS

The author gratefully acknowledges the financial support provided by the Lehigh Valley Hospital for the purchase of the cancer mortality data used in this study. The author wishes to thank Dr. Michael Greenberg for his comments on the statistical methodology used in this paper.

DISCLAIMER

The cancer mortality data were supplied by the State Center for Health Statistics and Research, Pennsylvania Department of Health, Harrisburg, Pennsylvania. The Department specifically disclaims responsibility for any analysis, interpretations or conclusions.

REFERENCES

1. Ziegenfus, Robert C. 1993. Cancer in Lehigh Valley Communities: An Example of Applied Medical Geography. *The Pennsylvania Geographer.* XXXI:13-35.
2. Hoover, Robert et al. 1975. Geographic Patterns of Cancer Mortality in the United States, pp. 343-360. In: J.F. Fraumeni, Jr. (Ed.) *Persons at High Risk of Cancer: An Approach to Cancer Etiology and Control.* Academic Press, Inc., New York, pp. 544.
3. Harmon, Mary P. and Kathryn Coe. 1993. Cancer Mortality in U.S. Counties with Hazardous Waste Sites. *Population and Environment.* 14:463-480.
4. Mason, Thomas J. et al. 1975. *Atlas of Cancer Mortality for U.S. Counties:*

1950-1969. DHEW Publication No. (NIH) 75-780. USGPO, Washington, D.C., pp. 103.

5. Hoover, Robert et al. 1975. Cancer by County: New Resource for Etiologic Clues. *Science.* 189:1005-1007.
6. Hoover, Robert and J. Fraumeni, Jr. 1975. Cancer Mortality in U.S. Counties with Chemical Industries. *Environmental Research.* 9:196-207.
7. Blot, W. et al. 1977. Cancer Mortality in U.S. Counties with Petroleum Industries. *Science.* 198:51-53.
8. Greenberg, Michael R. 1983. *Urbanization and Cancer Mortality: The United States Experience 1950-1975.* Oxford University Press, New York, pp. 276.
9. Joint Planning Commission of Lehigh-Northampton Counties. 1971. *1970 Census Data for Minor Civil Divisions in Lehigh and Northampton Counties.* Lehigh Valley, PA n.p.
10. O'Dell, Allen R. and Barbara S. Taylor. 1972. *Population Growth Trends - 1970.* Joint Planning Commission of Lehigh-Northampton Counties, Lehigh Valley, PA pp. 52.
11. Taggart, Glenn M. 1972. *Social Profile Report.* Joint Planning Commission of Lehigh-Northampton Counties, Lehigh Valley, PA pp. 143.
12. Seitz, John C. and Joseph A. Sandova. 1973. *Housing Profile Report:1970.* Joint Planning Commission of Lehigh-Northampton Counties, Lehigh Valley, PA pp. 96.
13. Greenberg, Michael R. 1979. *The Spatial Distribution of Cancer Mortality and of High and Low Risk Factors in the New Jersey-New York-Philadelphia Metropolitan Regions, 1950-1969. Part 1.* Report to the New Jersey Department of Environmental Protection, Trenton, New Jersey, pp. 209.
14. Greenberg, Michael R. and Samy Amer. 1989. Self-interest and Direct Legislation: Public Support of a Hazardous Waste Bond Issue in New Jersey. *Political Geography Quarterly.* 8:67-78.
15. Norusis, Marija J. 1992. *SPSS/PC+ Base System User's Guide Version 5.0.* SPSS Inc., Chicago, pp. 910.
16. Pappas, Gregory et al. 1993. The Increasing Disparity in Mortality Between Socioeconomic Groups in the United States, 1960-1986. *The New England Journal of Medicine.* 329:103-109.

Environmental Contaminants, Ecosystems and Human Health. Edited by S.K. Majumdar, E.W. Miller, and Fred J. Brenner.

Chapter Twenty-Five

BLACK LUNG DISEASE

RAJA V. RAMANI and ROBERT L. FRANTZ

Department of Mineral Engineering
The Pennsylvania State University
104 Hosler Bldg.
University Park, PA 16802

"The first priority and concern of all in the coal mining industry must be the health and safety of its most precious resource-the miner."
Federal Coal Mine Health and Safety Act of 1969

INTRODUCTION

Mining is one of the oldest of human occupations. The awareness of occupational illnesses associated with mining, even in ancient times, is evident from the writings of Hippocrates (460-370 BC), Pliny and Elder (AD 23-79), and Galen (AD 131-210). Georgius Agricola (1494-1535), in his *De Re Metallica,* refers to the diseases of miners and the need for their control through good planning and operating practices and responsible management (Schwerha, 1989). *Coal workers' pneumoconiosis* (CWP), often called Black Lung disease, is an occupational disease which results from inhalation of coal mine dust which usually contains relatively small amounts of free crystalline silica.

A *pneumoconiosis* is best defined as the accumulation of dust in the lungs and the tissue's reaction to its presence (Lapp and Castranova, 1993). *Silicosis* is the name given to the fibrotic disease of the lungs caused by the inhalation of dust containing silicon dioxide, primarily in the free crystalline form. Silicosis is the older and more well known of these mineral pneumoconioses. In contrast, CWP was not recognized as different from silicosis until the pivotal work of Collis and Gilchrist (1928).

The U.S. Bureau of Mines (USBM) still regards worker exposure to respirable aerosols in mines as the most serious health problem in mining (USBM, 1994). The National Institute of Occupational Safety and Health (NIOSH) is considering recommending a reduction in the standard for occupational exposure to respirable coal mine dust from 2 mg/m^3 to 1 mg/m^3 (NIOSH, 1993). The Mine

Safety and Health Administration (MSHA) may require new criteria for sampling procedures in the mine which would evaluate the ambient dust concentration in the mine atmosphere based on a single sample, instead of the current practice of averaging five samples (MSHA, 1992; Ramani, 1993). Several high-production longwall mining faces currently have extreme difficulty meeting the existing dust standards, and some already operate at reduced production levels to achieve mandated levels (USBM, 1993). The importance of adequately protecting the miners from high concentrations of coal mine dust, as well as of reducing the incidence and severity of the Black Lung disease, cannot be overemphasized. The importance stems, first and foremost, from the humanitarian viewpoint and, second, from the perspective of the overall welfare of the coal industry.

COAL WORKERS' PNEUMOCONIOSIS

Pathological and Clinical Features

The histopathologic hallmark of CWP is the coal macule (Balaan, Weber and Banks, 1993). Autopsy or open lung biopsy reveals the coal macule to be an amorphous mass of coal particles, dust-laden pulmonary macrophages, reticulin, and collagen. No pathognomonic signs or symptoms identify workers with CWP. A CWP diagnosis requires that the worker experienced the appropriate dust-expose setting (e.g., underground coal mining) and duration (usually 10 years or more) and, has the characteristic chest radiograph. Based upon the International Labor Organization (ILO, 1980) international classification, pathologists and radiologists have established case definitions for characterizing the pathology and the appearance of CWP on chest radiography (Weeks, 1993). CWP appears in two principal forms: *simple*, with small opacities, and *complicated*, with opacities larger than 10 mm in diameter. For all practical purposes, diagnosis of CWP in the living miner is wholly dependent upon the chest radiograph interpretation.

Simple Coal Workers' Pneumoconiosis

A coal miner has simple CWP if the chest radiograph shows the presence of rounded opacities that are no larger than 10 mm in diameter. These opacities tend to occur first in the upper lung zones, but as the disease advances, the middle and lower zones can become involved. According to the ILO scheme, opacities as much as 1.5 mm in diameter are designated as *p*; 1.5 to 3 mm in diameter, as *q*; and 3 to 10 mm in diameter, as *r*. The distribution of these nodules is designated according to the major profusion categories, ranging from category 0 (no nodules) to category 3 (nodules profusely present). Each of the major categories is further subdivided into three, resulting in a total of 12 subcategories. An x-ray film classified as 1/2 indicates that the disease stage is category 1 but that serious consideration was given to categorizing it as 2. For pneumoconiosis to be present, a rating of at least 1/0 (category 1) is required.

Lung volume studies show that the residual volume (RV) is elevated in workers with CWP. Studies of gas exchange in workers with CWP show that those with predominantly pinhead or p-type opacities, as well as those with an increased profusion of irregular opacities, have lower gas transfer indices. Lapp and colleagues (1971) have observed normal pulmonary artery pressures in most patients with simple CWP and without obstructive impairment, and elevated pulmonary pressures in patients with airway obstruction, complicated CWP, or those with simple CWP of the p or pinhead opacity. Studies of ventilatory capacity in coal miners generally show that forced vital capacity (FVC) and forced expiratory volume in one second (FEV_1) tend to be reduced when compared to these parameters in non-miners. In general, reductions in these spirometric indices in coal miners with simple CWP are small, and the values are usually within the normal range or very slightly reduced. A major significance of the presence of simple CWP is its predisposition to the development of progressive massive fibrosis.

Where cigarette smoking is also involved, studies dealing with non-smoking workers alone have helped clarify the independent effects of smoking and dust. For example, Attfield (1985) has estimated the cigarette smoking effect to be 1.1 to 2.6 times the coal mine dust effect. A similar trend was demonstrated in a study of British coal miners by Love and Miller (1982).

Complicated Coal Workers' Pneumoconiosis

Complicated CWP (also called *progressive massive fibrosis*) is present when the chest radiograph demonstrates opacities exceeding 10 mm. The ILO scheme classifies the extent of these lesions into categories A, B, and C. While simple CWP is often associated with mild, if any, deficits in ventilatory capacity, severe impairment of pulmonary function may be seen in workers with complicated CWP. Dyspnea, attributable to dust exposure, occurs frequently. Shortness of breath may be mild or noticed only on exertion in the early stages, but with time and disease progression, it occurs when the individual is at rest as pulmonary function becomes further impaired. Pulmonary function testing of workers with complicated CWP reveals significant impairment of lung mechanics and gas exchange, particularly when the large opacities attain category B or C.

Epidemiology

Current knowledge of the epidemiology and clinical aspects of CWP comes mainly from studies of large populations of coal miners in Britain and in the United States (Balaan, Weber, Banks, 1993). In eight studies done in the United States between 1961 and 1970, the reported prevalence of CWP ranged from 4 to 46%, The prevalence of CWP and the rank of the coal mined seemed to be correlated. A higher prevalence of CWP occurred among workers who mined medium- to high-rank coal (e.g. anthracite, Pennsylvania), and prevalence was lower among those who mined low-rank coal (high volume bituminous, Utah).

In 1969, NIOSH initiated the National Study of Coal Workers' Pneumoconiosis (NSCWP) to study the respiratory health of coal miners in the United States. The first round of the NSCWP was performed from 1969 to 1971 and included over 9,000 coal miners from 31 mines, with a participation rate of over 90%. The reported prevalence of simple and complicated CWP was approximately 30%. Variations in the prevalence of CWP, with rank of the mined coal were again observed. Not surprisingly, progressively higher radiographic categories of simple CWP were associated with increased time spent by the miner underground.

Although the data from the first round were recently reanalyzed and reported by Attfield and colleagues (1992), to indicate lower prevalences than those reported earlier, the relationship between the CWP prevalence and years spent mining remained. The second, third and fourth rounds of the NSCWP showed the prevalence of pneumoconiosis to be decreasing significantly (Attfield and Castellan, 1992). The salient results from this study are summarized in Table 1. This general decline in the number of new cases of both simple and complicated CWP might be related to the lower participation rate in these rounds than in the first, the fact that many workers with pneumoconiosis who were recognized in the earlier rounds left the work place, and the introduction of and adherence to more stringent CWP control measures since 1969.

CONTROL OF CWP

The concern for the health of miners from respirable dust has manifested itself in four primary mechanisms of control: 1) regulatory control through the passage of mine health and safety laws that set minimum standards of performance, 2) legal and social control through workmen's compensation laws for occupation-related injuries and health deterioration, 3) medical control through periodic physical examination, wearing of personnel protective devices, etc., and 4) engineering control through design and operation of mines according to the best recommended practices (Ramani, 1988).

Regulatory Control: The 1969 Coal Mine Act

Changes in the Federal government's responsibility for mine safety developed through a series of legislative actions in 1865, 1910, 1941, 1946, 1947 and 1952, culminating with the passage of the Federal Metal and Non-Metallic Mine Safety Act of 1966 (the 1966 Metal Act), the Federal Coal Mine Health and Safety Act of 1969 (the 1969 Coal Act), and the Federal Mine Safety and Health Amendments Act of 1977 (the 1977 Mine Act).

The 1969 Coal Act represents a significant piece of legislation in the annals of mine health and safety. Setting mandatory health standards in Title II, the act stated its purpose as *"to provide to the greatest extent possible, that the working conditions in each underground coal mine are sufficiently free of respirable dust concentrations in the mine atmosphere to permit each miner the opportunity to work underground during the period of his entire adult working life without incurring any disability from pneumoconiosis or any other occupation-related disease during or at the end of such period."* In Title IV, Congress considered the plight of miners suffering from pneumoconiosis arising from employment in underground coal mines and provided for the payment of Black Lung benefits. In Title V, Congress identified the need for studies, research, experiments and demonstrations and appropriated funds to carry out these activities.

TABLE 1

Epidemiological Data on U.S. Coal Miners' Pneumoconiosis - NCSCWP Results[+]

Salient Factors	Round 1 1969-71	Round 2 1972-75	Round 3 1977-81	Round 4 1985-88
1. Number of Miners	9078	9347	5275	3534
2. Tenure Distribution of Miners (%)				
0-1 year	10.2	9.8	4.6	3.4
2-4	10.0	19.9	18.2	2.9
5-9	7.6	13.2	32.3	12.4
10-14	5.6	5.8	13.9	26.7
15-19	8.0	6.0	5.1	29.1
20-24	14.9	8.5	4.0	9.8
25-29	16.1	14.4	4.8	5.0
30+	27.5	22.3	17.1	10.7
3. Number of X-Ray Readings	9034	8385	5131	3451
4. Regional Distribution of Miners (%)[a]				
High Rank	6	3	0	0
Medium-High Rank	15	14	19	17
Medium Rank	55	58	55	51
Medium-Low Rank	25	25	27	32
5. Simple CWP Prevalence (%)[b]				
Rounded Opacities				
Category 1+	4.5	4.1	1.8	1.3
Category 2+	1.2	1.0	0.4	0.1
Combined Opacities				
Category 1+	6.6	5.1	3.6	2.3
Category 2+	1.5	1.2	0.5	0.3
6. Progressive Massive Fibrosis (%)[c]				
0-14 years tenure	0.0	0.0	0.0	0.0
15-19	0.0	0.3	0.0	0.0
20-24	1.3	1.1	0.5	0.9
25-29	2.4	2.0	1.2	0.6
30+	5.8	3.9	2.5	2.5

[+] Authors have summarized the data from Attfield and Castellan (1992). There are extensive notations and explanations to the data in the reference.

[a] Regional distribution based on number of x-ray readings.

[b] Adjusted to common-tenure distribution; total miners in U.S. in 1986 was approximately 78,000.

[c] Extracted from Figure 4 (Attfield and Castellan, 1992).

Within eight short years of the enactment of the 1969 Coal Act, Congress passed the 1977 Mine Safety Act, which combines the health and safety programs of the 1966 Metal Act and 1969 Coal Act into single legislation, and transfers the responsibility for enforcement of the provisions from the Department of the Interior to the Department of Labor. New provisions were enacted to provide mandatory health and safety training to new miners and refresher training to all miners. The 1977 Mine Act is purported to have adopted the stronger features of the Occupational Health and Safety Act of 1970, perhaps the most significant piece of legislation regarding workmen's occupational health and safety in U.S. history.

Underground Coal Mine Dust Standards

One of the major objectives of the 1969 Coal Act was the prevention of Coal Workers' Pneumoconiosis (CWP). A U.S. Bureau of Mines survey of 29 mines in the 1968-69 period found average dust concentrations in excess of 6.0 milligrams per cubic meter of air (Shepich, 1983). The 1969 Act states that *"Effective three years after the enactment of the Act, each operator shall continuously maintain the average concentration of respirable dust in the mine atmosphere during each shift to which each active miner in the active workings of such mine is exposed at or below 2.0 milligrams of respirable dust per cubic meter of air."*

In establishing, for the first time, dust standards for the nation's coal mines, Congress considered the risks of exposure to high dust levels. According to a report submitted to Congress by the Department of the Interior, the probability of developing simple pneumoconiosis decreases with decreasing dust concentrations (Table 2). The probability of developing progressive massive fibrosis (complicated pneumoconiosis) also decreases with reduced exposure. The probabilities in Table 2 were based on British medical data on dose-response relationships which regarded 70 mg/m^3-years as a critical dose. Obviously, it was the intent of

TABLE 2
Incidence Rates for Pneumoconiosis
(after Committee on Education and Labor, 1976)

	Concentration mg/m3	Incidence Rate Per 1000 Miners After 35 Years Exposure
Simple Pneumoconiosis	7.0	360
	4.5	150
	3.0	50
	2.0	20
Complicated Pneumoconiosis	7.0	130
	4.5	40
	3.0	20

Congress to ensure a life time exposure of less than 70 mg/m^3-years, (2 mg/m^3 x 35 years), and thereby reduce the risks of contacting pneumoconiosis during 35 years of working life of the progression of the disease from simple to complicated stages (Ramani, 1988).

According to Weeks (1993), the respirable dust standard is unique in several ways because it has not been subject to the kind of risk assessment by which standards proposed under regulatory agency rule-making procedures are usually analyzed, its efficacy has not been fully evaluated, due to the scarcity of exposure information that can be linked to measures of outcome, and the current assessment only considers CWP as the sole measure of outcome measured as manifestations on chest radiographs. According to Weeks (1993), at exposure levels at or below 2.0 mg/m^3, there is loss of lung function and this is important for any assessment of the efficacy of the standard.

Characterization and Measurement of Respirable Dust

During the past 40 years, the philosophy regarding the assessment of health hazards from pneumoconiosis-producing dust has changed. Prior to 1959, the number of particles below 10 μm diameter in a specified volume of air (particle standard) was used as a measure of the health hazard. In the United States, this standard was measured in terms of million particles per cubic foot (mppcf). Since 1959, the respirable mass of coal mine dust per volume of air (mass standard) was officially endorsed as the relevant health hazard measure. Historically, the British Medical Research Council (BMRC), the U.S. Atomic Energy Commission (AEC), the American Conference of Governmental Industrial Hygienists (ACGIH), the U.S. Occupational Safety and Health Administration (OSHA), the U.S. Environmental Protection Agency (EPA), and the International Standards Organization (ISO) have recognized the importance of particle-size factors in inhalation-related risks (ACGIH, 1988).

At present, with regard to underground coal mine airborne dust sampling, there are two criteria for the respirable fraction sampling instruments. The first, resulting from work performed within the U.S. Atomic Energy Commission, is defined by a sampling efficiency curve that passes through the following points: 100 percent efficiency at 2 microns and smaller, 50 percent efficiency at 3.5 microns, and zero efficiency for particles 10 microns and larger (Figure 1). Efficiency is defined as the percent of dust of a particular size retained or deposited. The other criterion for the respirable fraction of dust, recommended and adopted by the 1959 Johannesburg Pneumoconiosis Conference, is defined by a sampling efficiency curve (BMRC) that passes through the following points: 100 percent efficiency at 1 micron and smaller, 50 percent efficiency at 5 microns, and zero efficiency for particles of 7 microns and larger. Particle sizes refer to unit density having the same falling velocity as the particle in question. Comparison of the two recommended respirable size criteria with the pulmonary deposition curve is shown in Figure 1. The purpose of both criteria is to approximate dust deposition in the nonciliated portions of the lung.

Personal Sampler

In the personal respirable dust sampler, the first stage is a 10 mm nylon cyclone, with a collecting efficiency that closely approximates the criterion set by the U.S. Atomic Energy commission. This efficiency is obtained with the sampler operating at a flow rate of 2 liters/minute. At second stage, a membrane filter collects the respirable fraction of dust that passes through the cyclone. The dust collected is weighted to the nearest 0.1 mg, and the concentrations are expressed as milligrams of dust per cubic meter (mg/m^3) of air sampled. Each miner can be equipped with a personal sampler to determine his or her dust exposure during a work shift. The major advantages of the instrument are its light weight, cost, insensitivity to positioning and portability on the person. Its major disadvantage is the uncertain fluctuation of collection characteristics in the cyclone arrangement.

MRE Gravimetric Sampler

In the MRE sampling instrument, the first stage is a four-plate horizontal elutriator which has a collecting efficiency that closely approximates the criterion recommended and adopted by the Johannesburg Pneumoconiosis Conference. This efficiency is obtained with the sampler operating at a flow rate of 2.5 litres/minute. The membrane filter used as the second, or collecting stage, is

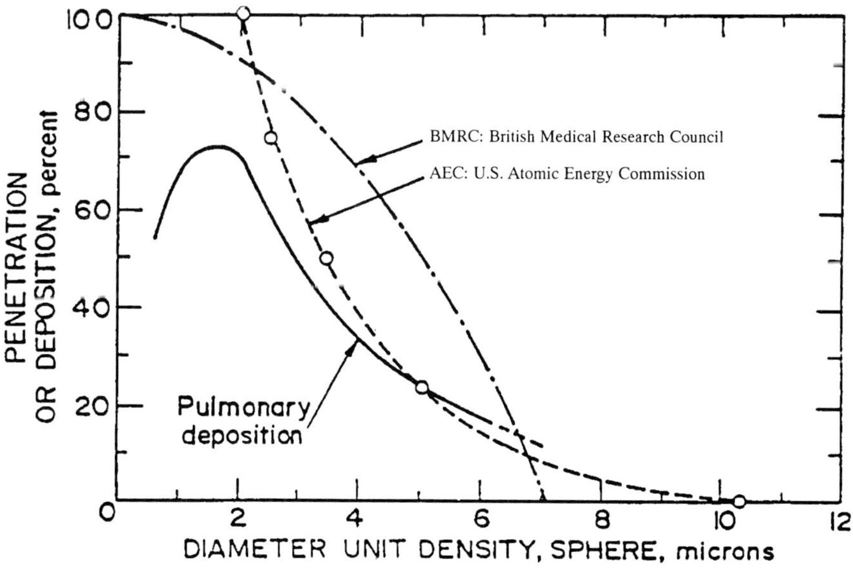

FIGURE 1. Comparison of Recommended Respirable Size Criteria With Pulmonary Deposition Curve (Jacobson, 1970).

treated in the same manner as the filters in the personal sampler. U.S. dust standards are specified in terms of this instrument. The MRE instrument provides high reliability regarding collection principles and characteristics. Its disadvantages are its size, weight and sensitivity to positioning during sampling.

The U.S. standards for airborne respirable coal mine dust, and the reduction of the standard in the presence of quartz are more stringent than the standards in many countries. The location of samples, the number of samples, and the manner of sampling are also different. For example, the respirable dust standard in Australia is 3 mg/m^3, and in U.K. significantly more than 3 mg/m^3. U.S. standards are having beneficial effects; yet there are concerns on several aspects other than CWP incidences and severity. With the recent rise of interest in airways disease, more concern is directed to larger particles that are deposited in respiratory airways. Investigations characterizing these so-called thoracic dusts have only just begun. As a result of these concerns, in recent years, increasing attention is paid to several other characteristics of the airborne dust, such as mineralogical and chemical components, surface charge, freshness and physical attributes, and their biologic influence.

Workman's Compensation Laws

Control of occupational illnesses through Workmen's Compensation laws is viewed as a deterrent since the costs are an after-the-fact penalty of potentially high magnitude. Management is thus encouraged to take many positive actions to avoid these costs. The Black Lung Benefits Program under the 1969 Coal Act is a compensation program for the afflicted miners and their dependents which is funded by a charge on every ton of coal mined. Today this charge is $1.10/ton of underground mined coal and $0.55/ton of surface mined coal. With an average figure of one billion tons of coal production, and a 60-40% surface to underground split, the annual cost to coal companies is approximately $750 million. Statistics on the number of beneficiaries of the Federal Black Lung program, and the benefits paid by the Federal Government are shown in Table 3. The total benefit paid out by both the Social Security Administration (SSA) and the Department of Labor (DOL) during the period 1970-1991 is just under $30 billion. The maximum number of beneficiaries in any one year was nearly 540,000 in 1981, and the maximum amount paid in any one single year was $1.93 billion in 1985. The number of beneficiaries and the total benefits paid have been generally decreasing since the 1980s.

Medical Control

Under the 1969 Coal Act, two activities focus on the prevention of CWP through medical control: 1) the National Coal Workers' Health Surveillance Program, and 2) the Coal Mine Respirable Dust Sampling Program (Attfield, 1984). Together they combine both primary and secondary methods for CWP prevention: the control of airborne dust in mines and the control of disease among miners. The health surveillance program has several significant purposes: 1) the protection of health of the individual miner; 2) the monitoring of prevalence and progression of CWP; and 3) the identification of miners with high incidences of CWP for greater dust exposure control. The medical control provides a useful evaluation of the individual miner's health status through early detection of abnormal conditions and prescription of preventative action. If a chest X-ray indicates CWP, then the management offers the miner the option to transfer to a

TABLE 3
Federal Black Lung Program
Number of Beneficiaries and Total Payments by the
Social Security Administration and Department of Labor, 1970-1972@

	Social Security Administration		Department of Labor+		Federal Government	
Year	Number of Beneficiaries	Total Amount ($) in Thousands	Number of Beneficiaries*	Total Amount ($) in Thousands	Number of Beneficiaries*	Total Amount ($) in Thousands
1970	111976	111000			111976	111000
1971	231729	378900			231729	378900
1972	298963	554400			298963	554400
1973	461491	1045200			461491	1045200
1974	487216	951300	*	6512	487216	957812
1975	482311	947700	*	13631	482311	961331
1976	469655	963300	*	28920	469655	992220
1977	457399	942200	*	25221	457399	967421
1978	439970	965100	*	155257	439970	1120357
1979	418948	983100	83887	622985	502835	1606085
1980	399477	1032000	139073	813206	538550	1845206
1981	376505	1081300	163401	805627	539906	1886927
1982	354569	1076000	173972	784085	528541	1860085
1983	333358	1055800	166043	859854	499401	1915654
1984	313822	1038000	163166	873932	476988	1911932
1985	294846	1025000	160437	905516	455283	1930516
1986	275783	971000	158550	629075	434333	1600075
1987	258988	940000	153289	655290	412277	1595290
1988	241626	9040000	149156	656689	390782	1560689
1989	225764	882000	144187	650123	369951	1532123
1990	210678	863400	138491	626521	349169	1489921
1991	196419	844440	133001	942428	329420	1786828
1992	182396	822500	Not Available	Not Available	Not Available	Not Available

@This data has been compiled from Social Security Administration and Department of Labor reports.

\+ The Social Security Administration was assigned initial responsibility for administering the Black Lung benefits program. The Department of Labor assumed responsibility for processing and paying claims on July 1, 1973.

* The Department of Labor approved only 6000 of the approximately 125,000 claims received between July 1973 and February1978. The remaining 119,000 claims were either denied or were still pending when the standards for approving claims were liberalized on March 1, 1978 (The American Coal Miner, The President's Commission on Coal, Washington, D.C.).

less dusty work area. In addition, the surveillance program provides a means to assess the efficacy of the dust control program by monitoring the incidence and prevalence of CWP. The results of epidemiological study of the U.S. coal miners (Table 1) have revealed a continuing decrease in the incidence and severity of the coal worker's pneumonconiosis.

Engineering Control

Engineering control encompasses the design of mines, selection of appropriate equipment, selection and training of the human resources to constitute a knowledgeable work force, and management of these resources to achieve the goals of health, safety and productivity. Mining as a process is involved with breaking in-situ materials (coal seam, orebody), and therefore generation of dust is unavoidable. However, the amount of airborne dust and the respirable fraction in the airborne dust constitute extremely small fractions of the mined volume (Figure 2). The most effective means of reducing the incidence and severity of the Black Lung disease is by controlling the exposures of miners to airborne dust. The basic engineering control methods are aimed at reduction of dust generation, dust removal to reduce entrainment, suppres-

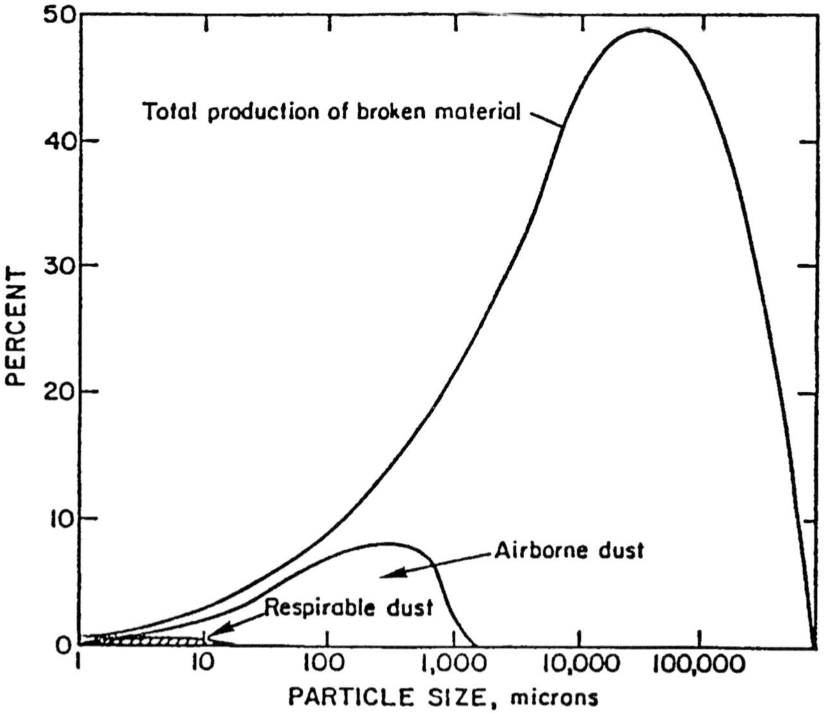

FIGURE 2. Typical Size Distribution of Breakage Product From Mining (Hamilton, 1970).

sion of entrained airborne dust, isolation of the dust or the worker from each other, and dilution of the airborne dust concentration. Proper application of engineering controls during the planning, designing and operating phases of a mining operation for reducing both ambient concentration and worker exposure is the first line of defense against the onset of CWP. The passage of the 1969 Coal Act significantly increased funds for health and safety research. A significant fraction of these funds were spent in research for improving the quality of mine air through engineering control. Significant research and development efforts were also expended by the coal companies to bring operations under compliance.

The dust control research led to methods to reduce the generation of dust during cutting, to suppress and minimize the entrainment of generated dust, to extract and collect the airborne dust, to prevent the dispersion of dust into the mine atmosphere, to dilute the airborne dust concentrations and to remove dust from work areas. The objective of reducing worker exposure to high concentrations of dust resulted in, as far as practicable, keeping the worker away from dust or the dust away from the worker. A committee of the National Academy of Sciences (1980) noted that the diligent application of existing technology resulted in procedures and practices permitting most mines to meet the 2 mg/m^3 standard. The Committee, however, concluded that *"further significant progress requires a new approach. In the future, research should be directed toward obtaining fundamental understanding of the origin, transport and characteristics of respirable coal mine dust."* At the present time, except in some high production longwall and continuous miner sections, a majority of the coal mines is in compliance with the 2 mg/m^3 standard (Figures 3 and 4). Two aspects that have become major areas of concern are quartz in airborne coal mine dust, Figure 5, and diesel particulates.

RESPIRABLE DUST RESEARCH CENTER

The prevalences of category 1 and category 2 CWP are showing continuing declining trends indicating that the 1969 Coal Mine Act actions have been effective.

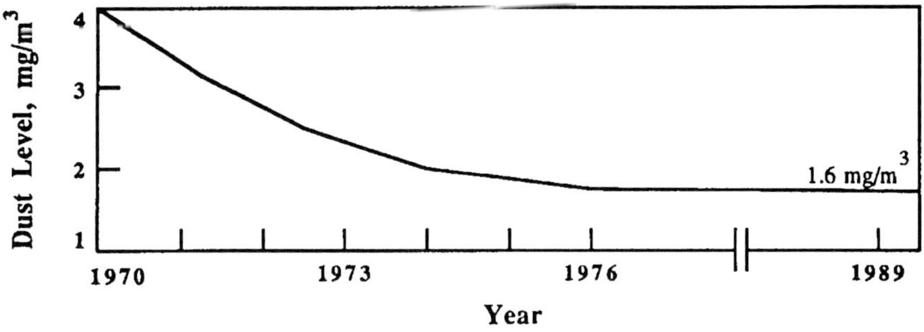

FIGURE 3. Average Respirable Dust Exposure of the Continuous Miner Operatior, 1970 Versus 1989 (after Jankowski and Frantz, 1990).

Whether all pneumoconioses will be eliminated is a question that remains to be answered. It is within this context that a comprehensive program designed to accelerate fundamental research for the control of respirable dust in mines was expanded. Early in the 1980s, the U.S. Bureau of Mines initiated the concept of the Generic Center incorporating it into its mining research program, and in 1983, the Generic Mineral Technology Center for Respirable Dust was established. The Generic Center concept was first envisioned by the U.S. Bureau of Mines as a way to bring together the various different scientific, engineering and medical research interests in the respirable dust area to effect a coordinated and efficient organization for advancing knowledge in that field. The result was a combination of the collective talents and skills of more than 100 scientists, engineers, medical personnel and graduate students, through the formation of a consortium of five major research universities, including medical facilities, to perform respirable dust research. The Generic Mineral Technology Center for Respirable Dust consortium comprises The Pennsylvania State University (with

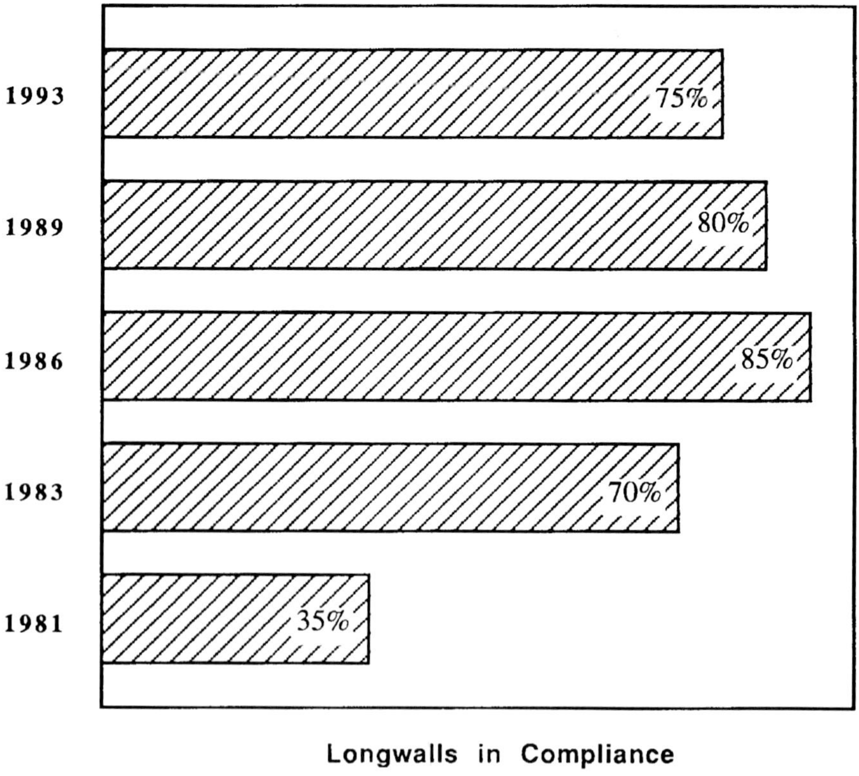

FIGURE 4 Dust Compliance Levelws Mining Sections, 1981 to 1993 (after Jankowski and Frantz, 1990).

its Hershey Medical Center) and West Virginia University (with its Medical Center) as the lead universities and the University of Minnesota, Massachusetts Institute of Technology and Michigan Technological University as affiliates (Ramani and Frantz, 1992).

The primary goal of the Center is to reduce the incidence and severity of respirable dust disease through researching and advancing the fundamental understanding of all aspects of respirable dust associated with mining and milling and the interaction of dust and lungs. The dust/lung research of the Center has several long-range objectives including finding better methods for the early diagnosis of coal workers' pneumoconiosis, (CWP), methods to modify the disease process, tests to identify toxic products, and tests to identify individuals with heightened susceptibility to pneumoconiosis. On the scientific and engineering side, the Center's objectives are to understand the fundamental aspects of dust generation, entrainment, and transportation and their impacts on mine operations planning for exposure control. A major objective of the Center is to define the mechanisms of dust-induced lung disease by determining dust characteristics that are important in causing lung disease, important interactions between dust and lungs, and earliest detectable stages of the disease. The Center's work concentrates on 1) the control of dust generation; 2) the dilution, dispersion and collection in mine airways; 3) the characterization of dust

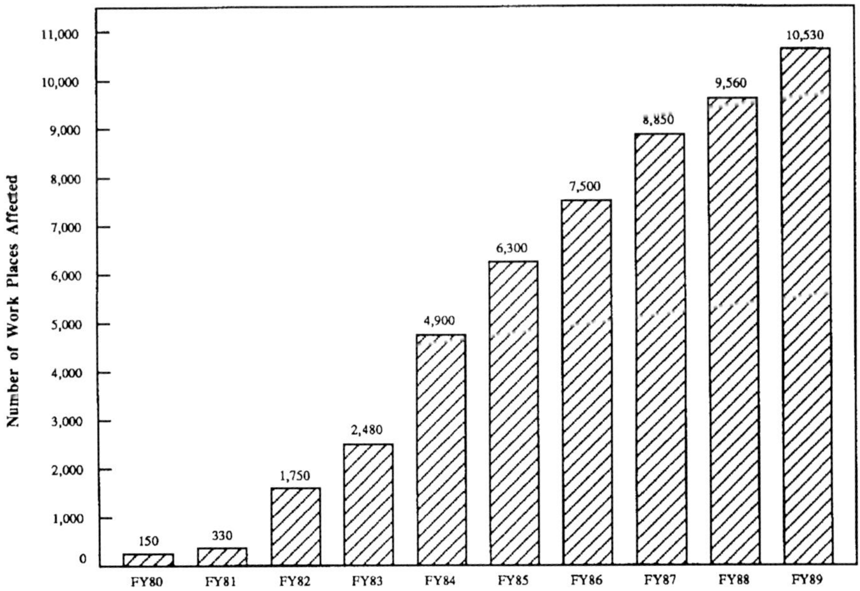

FIGURE 5 Number of Mining Sections Placed on More Stringent Dust Standards (below 2.0 mg/m3) Due to Quartz Content in Airborne Respirable Dust Samples (Jankowski and Frantz, 1990).

particles; 4) the interaction of dust and lungs; and 5) the relationship of mine environment, geology and seam characteristics to dust generation and mobility. All this work is carried out by interdisciplinary teams of scientists, engineers, and biomedical and medical researchers who interact with USBM, MSHA, NIOSH, and industry regularly on their investigations (Figure 6). A unique feature of the Center is the use of non-human primate to bridge the experimental gap between small animal research investigations and black lung victim's conditions. The fundamental aspects of the Center's work are applicable to the control of respirable dust problems in both hard rock mines and coal mines and to other dusts such as diesel-generated soot. This concerted effort has advanced the fundamental knowledge on such issues as the effects of trace elements on particle toxicity, fresh dust vs. stale dust, the role of silica, new contaminants and the responses of tissues to dust inhalation, and dust generation, entrainment, agglomeration, suppression, transport, characterization and sampling.

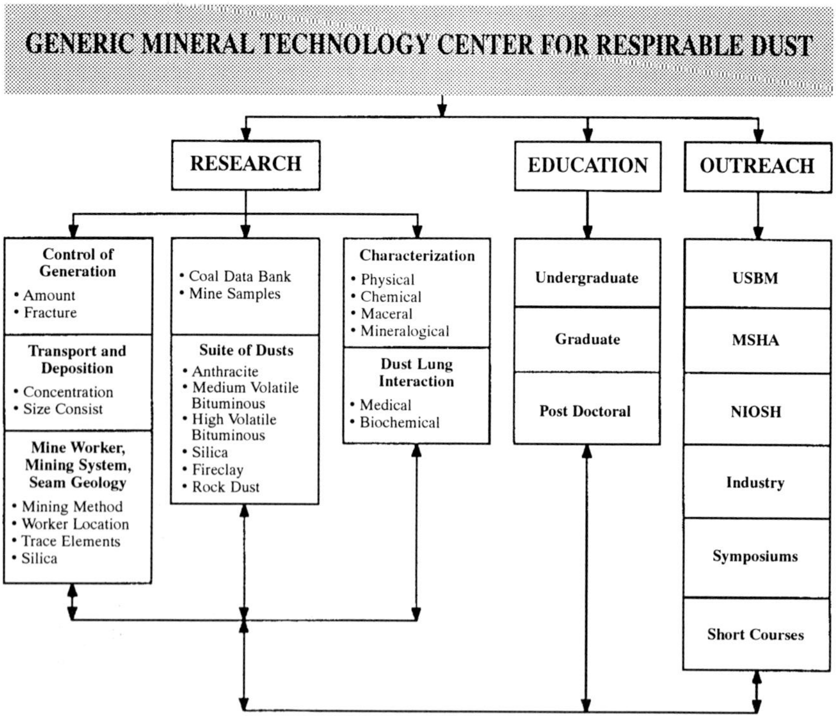

FIGURE 6. Respirable Dust Center Program Activities.

SUMMARY

The 1969 Coal Mine Act prescribed a systems approach to the eradication of CWP through new health and safety standards, more effective engineering controls, inspections, medical examinations, and black lung compensation. The act recognized the need for increased scientific, engineering, biomedical and medical research and studies to support not only the new legislation but also the development of new equipment and dust management methods. These activities drastically impacted mine production operations, and mine ventilation planning, engineering and practice, leading to both greater expectation for and fulfillment of healthier working conditions. Yet, there are outstanding issues as the high concentration of silica in airborne coal mine dust, the characterization and control of diesel particulates, and the adequacy of sampling and analysis for both airborne dust concentration and exposure determinations. These require satisfactory resolution.

There is reason to believe that there is a decline in the CWP incidence rates. However, the knowledge of the effects of coal mine dust on miners' lungs is continuing to expand beyond the effects visible in chest radiographs. (Weeks, 1993). Dust has been shown to cause obstructive lung disease, even in the absence of CWP, in the form of emphysema and bronchitis. There is a call for a broader view of occupational lung disease among coal miners which embraces the idea that chronic lung disease among miners has multiple causes, occupational and otherwise (e.g., coal mine dust and tobacco smoke). Further, there are no proven efficacious therapies that are specifically directed toward reversing or ameliorating the adverse respiratory effects of CWP or silicosis (Balaan, Weber, Banks, 1993). Therefore, current management efforts should be directed toward minimizing exposure, early identification of those workers with accelerated declines in pulmonary function and relocation of such workers from the workplace, recognition of rapid progression of radiographic lesions, prompt initiation of therapy for infection, and supportive management of complications such as bronchospasm, hypoxemia, and core pulmonale. Efforts to discourage and stop smoking are vital, as are identification, diagnosis, and therapy of patients suspected of having potentially treatable complications such as tuberculosis. Research must continue into the development of novel pharmacologic approaches to medicate or block the disease causation.

There is little doubt the underground mine air in U.S. coal mines has become cleaner. However, there are a number of questions raised particularly with regard to the complete healthful quality of the mine atmospheric environment. As the ability to monitor, analyze and characterize the mine atmospheric environment increases, these questions as well as new questions will continue to be raised. In fact, the research and development of engineering and medical solutions to the existing and emerging problems along with enlightened legislation are the only realistic avenues for the complete elimination of the health hazards from airborne dust in mine environments.

BIBLIOGRAPHY

ACGIH. 1988. *Advances in Air Sampling,* Proc. 1987 ACGIH Symposium, Lewis Publishers, Chelsea, MI 409 pp.

Attfield, M.D. 1984. "Recent Results on Forthcoming Studies in the Epidemiology of Coal Worker's Pneumoconiosis in Underground Mines," *Proceedings Coal Mine Dust Conference,* Generic Technology Center on Respirable Dust, West Virginia University, Morgantown, pp. 156-162.

Attfield, M. 1985. Longitudinal Decline in FEV1 in United States Coal Miners, *Thorax*, 40:132-137.

Attfield, M. 1992. British Data on Coal Miners' Pneumoconiosis and Relevance to U.S. Conditions, *Am. J. Public Health*, 82:978-983.

Attfield, M. and R. Althouse. 1992. Surveillance Data on U.S. Coal Miners' Pneumoconiosis, 1970 to 1986, *Am. J. Public Health*, 82:971-977.

Attfield, M. and Castellan, R. 1992. Epidemiological Data on U.S. Coal Miners' Pneumoconiosis, 1960 to 1988, *Am. J. Public Health*, 82:964-970.

Balaan, R., Weber, L. and E. Banks. 1993. "Clinical Aspects of Coal Workers' Pneumoconiosis and Silicosis," *Occupational Medicine: State of the Art Reviews,* Vol. 8, No. 1, January-March 1993, Hanley and Belfus, Inc., Philadelphia, pp. 19-34.

Collis, E.L. and J.C. Gilchrist. 1928. "Effects of Dust Upon Coal Trimmers," *J. Ind. Hyg. Toxicol.,* 10:101-109.

Committee on Education and Labor. 1970. *Legislative History of the Federal Coal Mine Health and Safety Act,* U.S. Government Printing Office, Washington, D.C., pp. 14-17.

Hamilton, R.J. 1970. "Principles of Machine Design and Operation Related to Dust Production and Control with Particular Reference to Longwall Mining in U.K.," *Proceedings of the Symposium on Respirable Coal Mine Dust,* USBM Information Circular 8458, pp. 160-178.

Jacobsen, M. 1970. "Assessing Respirable Dust in United States Coal Mines," *Proceedings of the Symposium on Respirable Coal Mine Dust,* USBM Information Circular 8458, pp. 42-54.

Jankowski, R.A. and R.L. Frantz. 1990. *U.S. Bureau of Mines Respirable Dust Program: An Overview of Accomplishments and Future Research Plans,* Department of Interior, Washington, D.C., 25 pp.

ILO. 1980. Guidelines for the Use of ILO International Classification of Radiographs of Pneumoconioses, revised ed., *International Labour Office*, Geneva.

Lapp, N.L. and V.C. Castranova. 1993. "How Silicosis and Coal Workers' Pneumoconiosis Develop-A Cellular Assessment," *Occupational Medicine: State of the Art Reviews,* Vol. 8, No. 1, January-March 1993, Hanley and Belfus, Inc., Philadelphia, pp. 35-56.

Lapp, N., Seaton, A. and K. Kaplan, et al. 1971. Pulmonary Hemodynamics in Symptomatic Coal Miners, *Am. Rev. Respir. Dis.,* 104:418-426.

Love, R. and B. Miller. 1982. Longitudinal Study of Lung Function in Coal Mines, *Thorax,* 37:193-197.

MSHA. 1992. Review of the Program to Control Respirable Dust in the United States, *U.S. Department of Labor,* Washington, D.C., 60 pp.

National Academy of Sciences. 1980. Measurement and Control of Respirable Dust in Mines, NMAB-363, Washington, D.C., 405 pp.

NIOSH. 1993. Criteria for a Recommended Standard: Occupational Exposure to Respirable Coal Mine Dust (DRAFT), *U.S. Dept. of Health and Human Services,* Washington, D.C., 346 pp.

Ramani, R.V. 1988. Air Quality in Mines: Progress and Prospects of Legal Control, *Fourth International Mine Ventilation Congress,* Australasian Institute of Mining and Metallurgy, Melbourne, pp. 13-18.

Ramani, R.V. 1993. Current Issues in U.S. Mine Ventilation, *Mining Engineering,* V. 45, No. 10, pp. 1262-1265.

Ramani, R.V. and R.L. Frantz. 1992. The Generic Mineral Technology Center for Respirable Dust: A New Multi-Disciplinary Approach to Research, *Fifth International Mine Ventilation Congress,* Mine Ventilation Society of South Africa, Johannesburg, pp. 415-425.

Schwerha, J.J. 1989. "Industry Perspective-Health Issues: Past, Present and Future," *Coal Mining Technology, Economics and Policy 1989,* American Mining Congress, Washington, D.C., pp. 419-426.

Selikoff, I.J., Key, M.M. and H.K.D. Lee. 1972. *Coal Workers' Pneumoconiosis,* Annals of New York Academy of Sciences, Vol. 200, 861 pp.

Shepich, T.J. 1983. Welcome Remarks, *Proceedings of the Symposium on Control of Respirable Coal Mine Dust,* Mine Safety and Health Administration, Washington, D.C., pp. 3-6.

USBM. 1993. Bureau of Mines Research 1993, *U.S. Dept. of the Interior,* Washington, D.C., 164 pp.

USBM. 1994. *Communications to Congressional Committee,* (Unpublished).

Weeks, L. 1993. "From Explosions to Black Lung: A History of Efforts to Control Coal Mine Dust, *Occupational Medicine: State of the Art Reviews,* Vol. 8, No. 1, January-March 1993, Hanley and Belfus, Inc., Philadelphia, pp. 1-34.

Environmental Contaminants, Ecosystems and Human Health. Edited by S.K. Majumdar, E.W. Miller, and Fred J. Brenner.

Chapter Twenty-Six

POSSIBLE LINKS BETWEEN THE PHYSICOCHEMICAL CHARACTERISTICS OF DRINKING WATER AND DEMENTIA OF THE ALZHEIMER TYPE IN A REGION OF QUEBEC, CANADA (Image Project)

ERIC GAUTHIER[1], JEAN-FRANCOIS EMARD[2] JEAN-PIERRE THOUEZ[1], and DENIS GAUVREAU[2]

[1]Université de Montréal
Département de Géographie
Montréal, Québec, H3C 3J7
[2]Projet IMAGE
Centre hospitalier Côte-des-Neiges
Montréal, Québec, H3W 1W5
4565 Queen Mary Road
Montréal, Québec, H3W 1W5

INTRODUCTION

Measures to improve public hygiene in industrialized countries have considerably reduced the incidence of infectious diseases over the last century. This was made possible largely through filtration and chlorination of drinking water, greatly reducing the risk of bacterial contamination. However, drinking water quality may also be influenced by organic and inorganic substances that can persist for a long time. These substances are both man-made (e.g. industrial, agricultural and municipal pollution) and natural (e.g. dissolution of minerals from soil and bedrock in origin. The concentrations of such substances in drinking water are rarely high enough to have acute toxic effects. However, low doses exposure over a very long period of time could have long-term effects on health. Some risks could, in the long term, be related to a deficiency or excess of minerals. For example, goitre is endemic in areas where the concentration of iodine in the water and soil is low (Hetzel and Dunn, 1989) and there is an association between the water hardness of drinking water and the risk of cardiovascular disease (Thouez, 1979).

The connection between elements in very low concentrations (trace elements) in drinking water and health risks remains difficult to prove. This is partly due to the multifactorial nature of many diseases: various factors related to exposure to trace elements, the chemical and toxic characteristics of the trace elements, the complex interaction between the trace elements and other substances, genetic and biological factors as well as individual lifestyle and the latency of the disease can all be involved.

Trace metals have become important in understanding the pathogenesis of several neurodegenerative diseases (NDD). Recent findings concerning human metabolic processes have identified several trace elements that might be associated with some NDD. Elements that are not essential to the functioning of the human organism such as lead, mercury and aluminum could play a role in the development of NDD (Verity, 1992). On the other hand, a deficiency of elements (e.g. calcium, magnesium, potassium, copper, zinc, selenium and iron) essential to the functioning of the enzymatic and metabolic systems could be involved in the etiology or pathogenesis of certain NDD (Yanagihara, 1982; Evans et al., 1992).

Neurotoxicity and the Risk of Dementia

Some metals such as lead, mercury, arsenic and aluminum could cause a toxic-type dementia (Signoret and Hauw, 1991). Lead is a potent neurotoxin, and drinking water is in many cases a major source of lead exposure (Seux and Dab, 1994). The lead comes from corrosion of lead pipes in the water supply system and from tin/lead soldered on copper pipes in houses (Subramanian and Connor, 1991). The solubility of lead in water is higher at acid pH and in soft water (Desjardins et al., 1989). Epidemiological studies carried out over the last twenty years have indicated that a variety of neurological disorders may be caused by lead (Isnard, 1994). Concentrations of lead greater than 200 μg/l have clear effects on the central and peripheral nervous system, and such effects have even been observed among individuals exposed to lower concentrations (Royal Canadian Society, 1986). Chronic exposure to lead could lead to deficiencies in certain enzymes and renal insufficiency in adults. In children, effects on neurobehavioral development and various neurophysiological changes can result from very low levels of exposure (Isnard, 1994).

Mercury exist both in organic and inorganic forms; the organic form methylmercury is particularly important as a potent neurotoxin (Rupinder and

Costa, 1992). The bioavailability of methylmercury increases with decreasing pH and depends on the catalytic action of microorganisms, on the availability of inorganic mercury, and on its binding to sediments in suspension (Carty and Malone, 1979). Mercury can accumulate in the food chain and in the human organism; the primary source of exposure is from eating fish (EPA, 1984). Organic mercury can cause serious neurological disorders as well as damage to DNA (Fan, 1987). The concentration of mercury in drinking water is usually below 100 ng/l (NAS, 1978), and it seems unlikely that such low levels can have direct toxic effects, although there are no actual studies to document this.

A large number of pesticides have been found as contaminants in drinking water (Gauthier et al., 1993). Poisoning can be acute or chronic. Arsenic-containing pesticides can cause memory problems as well as various intellectual disorders (Feldman et al., 1984). Organophosphorous and carbamate pesticides can have neurotoxic effects by inhibiting acetylcholinesterase, an enzyme fundamental to the functioning of the central nervous system (Korsak et al., 1977). There are no authoritative studies on the effects of prolonged exposure to pesticides in drinking water.

Aluminium, mostly present as aluminosilicates (compound of Al, Si, O and various other elements) is a major component of bedrock and soil. Aluminium is used in food additives, drugs, toothpaste, cutlery, antiperspirants and in water treatment. Canadian recommendations on drinking water quality do not consider aluminium as toxic (Health and Welfare Canada, 1993). However, several authors feel that aluminium is an element that could have neurotoxic properties (McLachlan et al., 1992; Perl and Good, 1991). The neurotoxicity of aluminium in humans was demonstrated by the myoclonic encephalopathy occurring in some kidney patients on haemodialysis (Alfrey et al., 1976; Will and Savory, 1983). Two sources of exposure have been identified: Al-based phosphate binders (Sedman et al., 1984) and dialysis water high aluminium (Davison et al., 1982; Milne et al., 1983). The latency period may be as short as a few months, and the toxic effects are reversible by treatment with the aluminium chleating agent desferroxamine (Ackrill et al., 1980).

It has not been excluded *a priori* that the neuropathologies considered until now as idiopathic or degenerative are not part of the group of toxic dementias. So, at least some subgroups of amyotrophic lateral sclerosis and Parkinson's disease could result from chronic poisoning from various substances (Signoret and Hauw, 1991). Exogenous toxins could also play a role in dementia of the Alzheimer type.

Amyotrophic Lateral Sclerosis

After the appearance of an unusually high number of cases of amyotrophic lateral sclerosis (ALS) in Guam and other locations in the western Pacific after the second World War, researchers studied the links between mineral deficiencies (calcium and magnesium) and toxic metals (aluminium, iron and manganese), and ALS (Garruto, 1991). The high concentration of aluminium demonstrated in the brains of patients may be caused by long-term exposure to low concentrations of calcium and magnesium in the drinking water, causing increased intestinal absorption of aluminium (Yase, 1972; Garruto, 1991). Research carried out in Guam has also associated ALS and Parkinsonism-dementia to a neurotoxin present in a species of palm tree, *Cycas circinalis* (Spencer et al., 1993). In other parts in the world epidemiological studies have indicated a link between exposure to lead and mercury and the development of ALS (Boothby et al., 1974; Brown, 1954), but others have failed conform these links (Mitchell, 1987; Gresham et al., 1986). A deficiency of zinc and selenium might play a significant role in the development of ALS (Yasui et al., 1991; Moriwaka et al., 1993).

Parkinson's Disease and Parkinsonism-Dementia

Parkinson's disease (PD) is characterized in particular by a deficiency of the dopaminergic system. The deficiency could be related to several genetic and environmental factors (Poirier et al., 1991; Oertel and Kupsch, 1993). Some authors have suggested that chronic exposure to manganese, iron or aluminium could increase the risk of developing PD, and a deficiency in calcium and magnesium could increase the absorption of the metals (Zayed, 1988; Zuber and Alperovitch, 1991). The presence of free radicals in the brain could also play a significant role in PD (Poirier et al., 1986), and zinc deficiency or high concentrations of iron could create favorable conditions for an increase in free radicals. A copper deficiency could generate a deficiency of the enzyme tyrosinase and lead to a deficiency of dopamine (Goyer, 1986). Some pesticides such as paraquat could also play a role in the development of PD (Barbeau et al., 1985). Several research have studied the different environmental factor in relation to PD, but the results have been contradictory. Rajput et al. (1987) reported a link between well water in Saskatchewan and the early development of PD. On the other hand, Semchuk et al. (1991) did not observe any significant connection between well water consumed in rural areas and PD. Cotzias et al. (1971) reported a link between manganese exposure and PD. Zayed et al. (1990) reported a relationship between PD and exposure to manganese, aluminium and iron in the workplace, but they did not find a significant link between the risk of PD and consumption of drinking water or working with pesticides.

Dementia of the Alzheimer Type

Dementia of the Alzheimer's type (DAT) is the principal form of dementia, and its etiology remains unknown. It is difficult to estimate the prevalence of DAT precisely, but there is no question that the number of people afflicted with it will increase significantly as the population ages. Estimates of the prevalence of DAT vary between 0.6 and 10.3% among people 65 years of age and over, and the prevalence increases exponentially with age (Jorm et al., 1987). In Canada, a study on health and aging has recently estimated an 8.0% prevalence rate of dementia among the population aged 65 years and older (McDowell et al., 1994). DAT may result from a genetic predisposition and exposure to environmental factors, modulated by the biological aging process (Gauvreau, 1987). Several case-control epidemiological studies have indicated links between genetic and certain environmental factors such as family history of dementia, Down's syndrome and a past history of dementia, Down's syndrome and a past history of cranial trauma and depression among afflicted subjects (Van Duijn et al., 1992). Focusing specifically on the genetic component, it appears that the early and less frequent form of DAT might be caused by: a) a gene mutation of the precursor amyloid β protein located on chromosome 21 (St. George-Hyslop et al., 1987); and/or, b) a mutation of a yet unidentified gene located on chromosome 14, which would cause yet an earlier presentation of the disease (Schellenberg et al., 1992). Recent research has shown that a protein responsible for the transport of cholesterol in blood, ApoE, whose gene is located on chromosome 19 (Pericak-Vance et al., 1991) might constitute a major risk factor in senile DAT (Corder et al., 1993). The primary environmental factors suggested in the literature are head trauma, exposure to various metals, deficiencies of various minerals, exposure to pesticides, solvents or a viral agent (Henderson, 1988; Gautrin and Gauthier, 1989; Emard et al., 1992).

Aluminium is the primary chemical element under investigation in the research into the causes of DAT (Craun, 1990). The main reason is the elevated concentrations found in the brains of afflicted subjects (McLachlan et al., 1992). Edwardson (1991) and others have suggested that people with dementia might assimilate high quantities of aluminium more easily. This vulnerability could hypothetically depend on genetic factors influencing the absorption, transport and excretion of the toxic element, and could be due to a membranous anomaly or other co-factors (Thompson et al., 1988). A nutritional deficiency in calcium or zinc could increase the level of intestinal absorption of toxic elements such as aluminium (Yase, 1972; Yasui et al., 1991; Constantinidis, 1990). Durlach (1990) suggested that magnesium deficiency could increase the intestinal absorption of aluminium in DAT. On the other hand, deposits of aluminium as well as the pathologic lesions found in DAT, differ from what has been observed for toxic dementia among dialysis patients (Signoret and Hauw, 1991). Borella et al. (1990) mention the possible role of a potassium deficiency in DAT. A recent study suggests that a high level of fluoride in drinking water could play a preventive role in DAT because it might compete with aluminium during ingestion (Forbes et al., 1992).

Nieboer et al. (1993) reviewed the literature on the possible effects of aluminium in drinking water on DAT. Epidemiological studies in England, Canada, Norway and France all indicated a connection between moderate concentrations (in the order of 0.1 mg/l) of aluminium in drinking water and the risk of DAT (Martyn et al., 1989; Flaten, 1990; Michel et al., 1991; Neri and Hewitt, 1991).

Even though the aluminium contained in drinking water only contributes a small proportion to the total amount ingested daily, it could make significant contribution to the quantity absorbed by the digestive system because the bioavailability of Al may be higher in water.

To date, most research on the possible role of chemical elements in DAT has concentrated on aluminium. Other elements in drinking water, however, could be involved in the etiology of DAT. For example, a recent study suggest that other characteristics of drinking water (e.g. calcium, fluoride, pH) might interact with aluminium (Jacqmin et al., 1994). Furthermore, no studies were able to take into consideration in a precise way the length of exposure to different sources of drinking water. Other epidemiological studies revealed higher rates of DAT in rural areas over urban areas (Rocca et al., 1990; Emard et al., 1992).

GOAL OF THIS STUDY

The goal of our research is to investigate whether exposure to the various physicochemical characteristics of drinking water can be linked to the risk of DAT. The principal aim of the present ecological study is to identify the substances that might play a role in DAT. The ecological approach is an exploratory method for understanding the interactions between each component of drinking water and their possible links with DAT. The ecological studies mentioned in the preceding section all have inherent weaknesses, and generally, the possibility of very long latency period was not taken into consideration. One of the advantages of our study is rigorous DAT diagnostics. Furthermore, we know the residential itinerary (the exact address of all individuals for a long period of time) along with the source of drinking water for each subject considered.

DATA AND METHODS

Data on Drinking Water

The area of our study is the Saguenay-Lac-Saint-Jean region of Quebec (SLSJ). The physicochemical data on drinking water used in this study were provided from the database of the urban ecosystems division of the Quebec Ministry of the Environment, drinking water samples were collected between 1978 and 1994. Data from before 1978 was not taken into consideration because the sampling and analytical protocols were different.

There are, however, sources of bias. There are several samples in which the concentrations are below the analytical detection limit. We decided to use these data, replacing data below the detection limit with values equal to half the detection limit (Newman et al., 1989). The samples were collected by local personnel, which could imply a selection bias; samples might have been collected during periods with good water quality in order not to exceed established limits, so the values for certain substances may be underestimated. Furthermore, the number of required samples varies with the number of residents in the municipality. Data from small municipalities may therefore be less representative because the number of samples is low.

We compiled more than 15,000 samples for which 33 substances have been analyzed in 72 drinking water supply sectors serving more than 268,000 inhabitants of the SLSJ region. We constructed a first analysis matrix taking into consideration the median values of substances for each water supply sector. We decided to use median values over arithmetic mean values because the physicochemical data for water is assymmetric in distribution, and extremes values may greatly influence the arithmetic mean. Medians, which are not affected by this type of distribution, give a better estimate of the annual average concentrations of the various elements.

We excluded 20 water supply sectors due to missing data. These 20 sectors cover less than 10% of the population connected to public waterworks. Under-sampled substances were also excluded from our analysis (Al, Se, As, P). This resulted in a matrix of 48 water supply sectors and 29 substances. The medians were standardized to eliminate problems with scale and unit measure heterogeneity.

Aluminium is rarely analyzed in drinking water in the SLSJ region because it is not on the list of priority substances covered by Quebec drinking water bylaws. In order to consider potential concentrations of aluminium in drinking water, we used color as an indicator. Studies done in the SLSJ region on the concentration of aluminium in lakes have shown that soluble aluminium is strongly correlated to color (r spearman = 0.72: $p \leq 0.001$) and organic carbon ($r = 0.71$: $p \leq 0.001$) (Campbell et al., 1986; Dupont, 1990).

Data on Subjects Afflicted with DAT

We based our research on the IMAGE Project case registry. The IMAGE Project has a vast network of collaborators in all of the care and social services facilities in the SLSJ region, which makes it possible to take a complete census of subjects. Seven field agents and 12 clinicians participated in the census of patients. The diagnosis of DAT was based not only on strict clinical criteria but also on neuropathological criteria and was made by a multidisciplinary team with expertise in the area (Gauvreau et al., 1989). This study involved all subjects diagnosed as "definite" or "probable" cases of DAT since January 1, 1987 who had completed a questionnaire to reconstruct a residential and occupational itinerary (ROIQ). To date, 334 ROIQ have been administered. Of this number, 50% were diagnosed as "definite" cases of DAT. The ROIQ is filled out by close relatives of subjects. Administering the ROIQ is facilitated by the marked propensity of inhabitants in the SLSJ region not change their place of residence over the course of this century. The SLSJ region has two distinct zones: the rural area around primary Lac Saint-Jean and the urban area of Chicoutimi-Jonquière-La

Baie. Of 334 patients, 13 were excluded from the study as they had lived in municipalities previously excluded.

The Ecological Approach

The ecological approach is thoroughly discussed in the literature (Legendre and Legendre, 1984; Walter, 1991; Walter, 1992). Because we wanted to better understand the relationships between the physicochemical characteristics of the water, we used principal component factorial analysis with VARIMAX rotation (Figure 1). This method is used to reduce the data matrix to a few key factors (Taylor, 1977). Principal component factorial analysis calculates loading factors for each variable (element) in the matrix. This loading indicates the strength of the connection between variable "x" and factor "y". The factorial scores applicable to each municipality are then calculated for each factor retained.

We did a cluster analysis (Ward method) on the scores for each of the 5 factors retained to group together those municipalities with the same physicochemical profiles. The clusters were selected based on the dendrogram that resulted. We used the Kruskal-Wallis method to valid the existence of a significant difference ($p = 0.05$) between clusters.

Analysis of the Distribution of Subjects with DAT in Relation to the Physicochemical Characteristics of the Drinking Water in the SLSJ Region

In order to compare the physicochemical characteristics of water and the risk of DAT, the prevalence rates based on when the very first symptoms of dementia appeared were calculated for each group of municipalities from the cluster analysis (Figure 1). The duration of the subjects' stay (time spent at each individual's addresses) was calculated for each cluster so as to identify exposure time. The prevalence rates were standardized using the direct method (Bernard and Lapointe, 1987) on the region population base at risk of presenting with symptoms significant differences ($p < 0.05$) between two standardized prevalence rates, inference tests were done on the basis of a theoretical normal distribution of subjects (z scores) as well as on confidence intervals based on Poisson's theoretical distribution (Schoenberg, 1983).

ANALYSIS RESULTS

The Physicochemical Profile of Drinking Water in the SLSJ Region

We retained the first five factors explaining 65.4% of the total variance from the factorial analysis. Table 1 shows the factorial loading of each element attributed to each factor retained. The shaded zones represent the primary elements that qualify each factor. Factor 1 explained 26.8% of the variance in the matrix and includes chloride, magnesium, orthophosphates (PO_4), potassium, total solids, sodium and conductivity. In our research, Factor 1 represents primarily magnesium and potassium as well as other related elements.

Factor 2 explains 15.6% of the variance and is made up of nitrogen, iron, color and turbidity. Factor 2, is highly related to water color and organic matter content and is a rough index of the concentration of aluminium in drinking water.

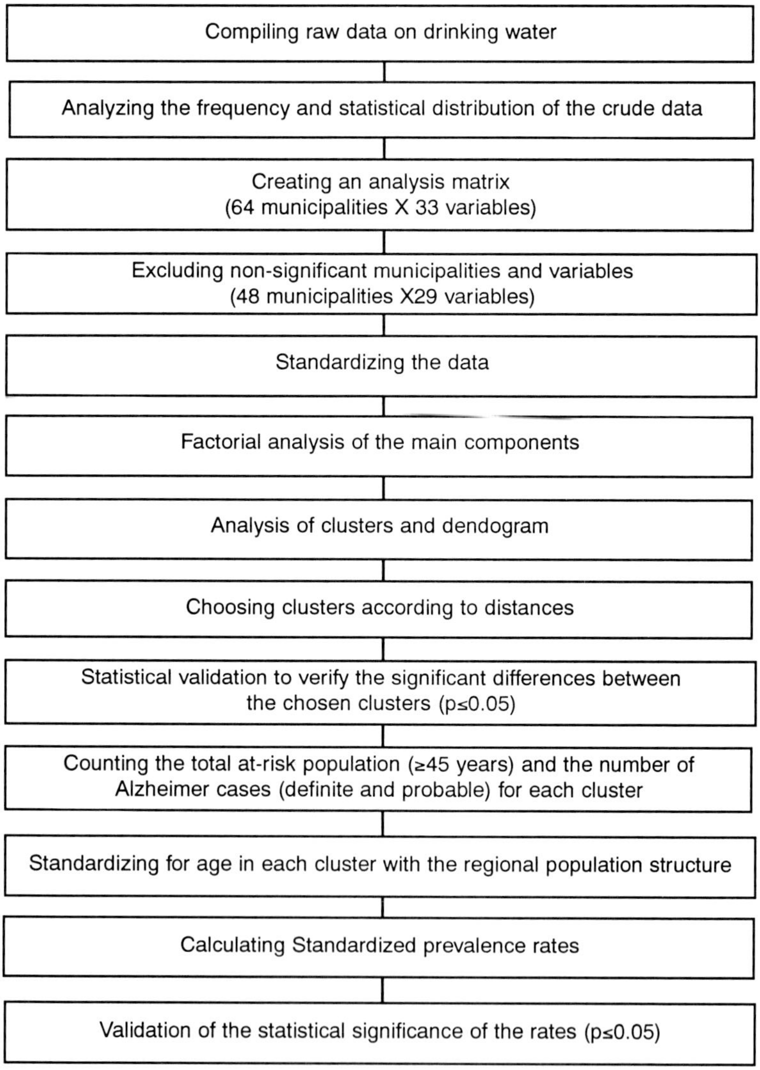

FIGURE 1. Steps in the exploratory approach

Factor 3, which explains 10.5% of the variance, indicates the presence of toxic metals such a lead, mercury, cyanide, copper and silver. Factor 4 and 5 combined explain 12.4% of the variance of the matrix. Factor 4 contains pH and fluoride, while Factor 5 contains calcium and manganese.

Three clusters of municipalities were retained based on the dendrogram of the clusters analysis (Figure 2). Statistical validation using the Kruskal-Wallis method showed significant differences ($p < 0.05$) between the three clusters. This validation highlighted the significantly different elements between the three clusters (Table 2).

Cluster 1 is characterized by very soft water, highly colored with low conductivity and a pH slightly more acidic than that of the other two clusters. The concentrations of calcium and magnesium in Cluster 1 are much lower than those in the other two clusters. Cluster 1 has concentrations of lead higher than those in Clusters 2 and 3. On the other hand, Cluster 3 shows concentrations of

TABLE 1

Principal Factors Stemming from Factorial Analysis with VARIMAX Rotation

ELEMENTS	FACTOR 1 (Non-metals)	FACTOR 2 (Coloration)	FACTOR 3 (Toxic metals)	FACTOR 4 (Fluoride-pH)	FACTOR 5 (Calcium-manganese)
Silver	-.11115	-.20970	.69535	.22814	.16628
Nitrogen	.37416	.68620	.14399	.11832	.02622
Barium	-.00750	.10260	-.29771	.03732	-.13749
Boron	-.08833	.28955	-.21320	-.65722	.01705
Cadmium	-.05439	-.00558	-.13925	-.17524	-.07300
Calcium	-.39121	-.05489	-.09785	.08461	.86274
Chloride	.84737	.11201	.00365	-.04590	.17998
Chromium	-.10457	-0.04428	.43242	.67048	-.01542
Conductivity	82711	.17704	-.16228	.08973	.25357
Colour	-.29977	.83043	-.02647	-.11223	.09404
Copper	-.24905	.33298	.44088	.06657	-.13683
Cyanide	-.01945	.03701	.79145	.08589	.13622
Iron	.15336	.85291	-.13769	-.01388	.12123
Fluoride	.30538	.23206	-.13121	.69942	-.08548
Magnesium	.91362	.07994	-.09834	.04955	.29031
Manganese	.14262	.31722	.10350	-.00465	.83978
Mercury	.01791	-.27400	.67340	.33963	-.24778
Nitrate	.05862	-.20831	.20725	-.00015	.05302
Orthophosphate	.89716	.07690	-.00425	.16692	-.15663
pH	.20296	-.19441	.07449	.71175	.28105
Lead	-.18577	.13740	.53820	-.27427	-.34157
Potassium	.87477	.10259	-.12979	.09013	.12394
Sodium	.94200	.07189	.02280	.06431	-.07959
Solid	.81618	-.04823	-.14789	.07505	.47983
Sulphate	.23376	-.38644	.08157	.22542	.08646
Turbidity	.16052	.74468	-.09215	-.31348	-00950
Percentage of variance	26.8%	15.6%	10.5%	6.9%	5.5%

fluoride much higher than those in Clusters 1 and 2. The distribution by type of drinking water source (subterranean water, lake or river) indicates that the three clusters are homogenous.

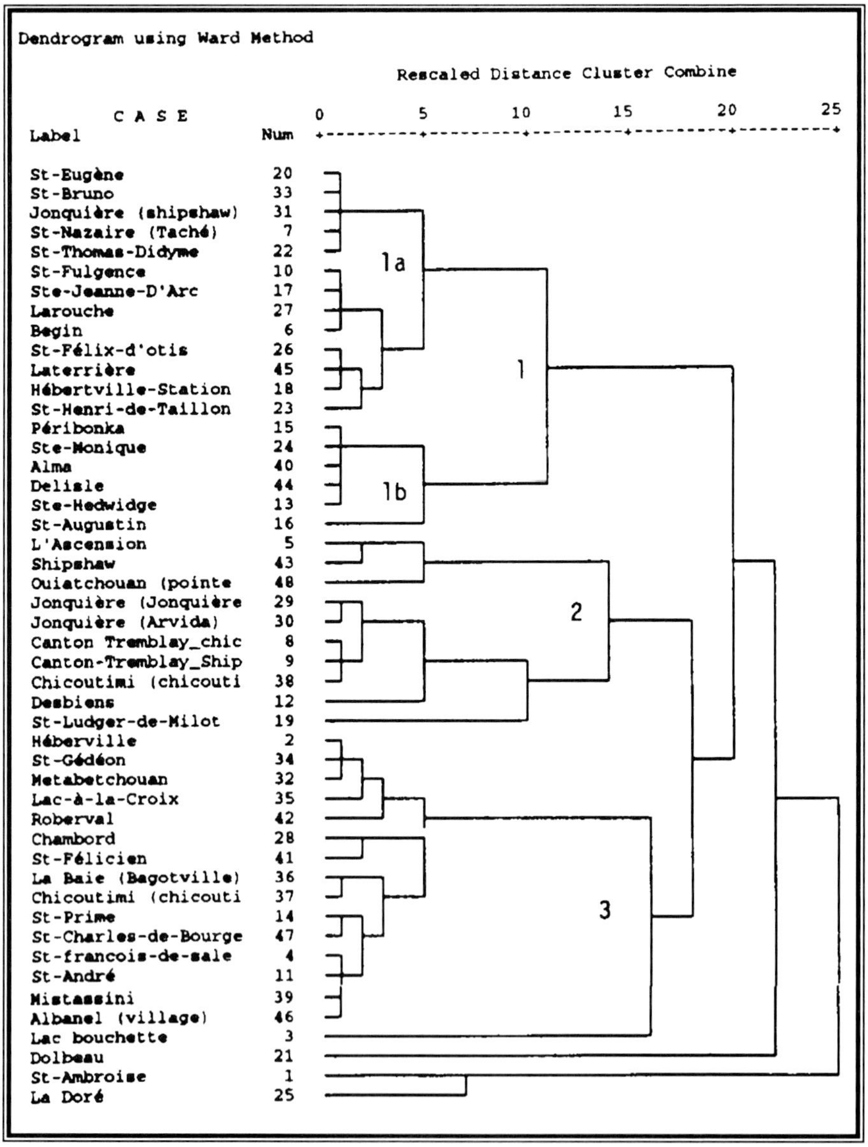

FIGURE 2. Dendrogram stemming from cluster analysis using the Ward method.

Associations Between the Distribution of DAT and the Physicochemical Profile of Drinking Water

Table 3 shows the prevalence rates of DAT for each cluster standardized for age. If we exclude the age distribution, various population characteristics show that the three clusters are homogenous. The average duration of stay for subjects who resided in the same place before the appearance of the very first symptoms of DAT shows no significant difference between clusters. Cluster 1 has the highest rate (0.006229), which is 33.9% higher than the regional rate. Prevalence rates for Clusters 2 and 3 are lower than the regional rate, but these differences are not statistically significant.

DISCUSSION

The results in Table 3 reveal a gradient in the standardized prevalence rates of DAT according to three major types of physicochemical profile of the drinking water in the SLSJ region. Cluster 1 shows the highest rate. This high rate could be related to the following observations: a) the concentrations of lead are higher in Cluster 1, probably because the water is softer and the pH more acidic than in

TABLE 2

Significantly Different Elements Among the Three Clusters Retained

		CLUSTER 1			CLUSTER 2			CLUSTER 3		
Factors	Elements*	Median	Minimum	Maximum	Median	Minimum	Maximum	Median	Minimum	Maximum
F1	Conductivity	33.50	19.00	67.00	79.50	40.50	129.50	80.75	26.00	280.00
F1	Magnesium (μg/l)	650.00	300.00	1100.00	775.00	650.00	1300.00	1100.00	500.00	3800.00
F1	Potassium (μg/l)	400.00	150.00	800.00	625.00	400.00	1250.00	600.00	250.00	2750.00
F2	Nitrogen (μg/l)	20.00	10.00	100.00	12.50	10.00	40.00	17.50	10.00	70.00
F2	Iron (μg/l)	165.00	40.00	815.00	42.50	10.00	340.00	197.50	10.00	390.00
F2	Colour (hazen)	20.00	3.50	66.00	3.00	1.00	47.00	16.00	0.00	35.00
F2	Turbidity (μg/l)	0.85	0.45	2.00	0.50	0.18	1.60	0.78	0.10	1.30
F3	Mercury (μg/l)	0.08	0.05	0.15	0.10	0.05	0.15	0.05	0.05	0.10
F3	Lead (μg/l)	5.00	1.00	17.50	2.75	2.00	5.00	1.38	0.50	5.00
F4	Fluoride (μg/l)	50.00	50.00	160.00	150.00	50.00	800.00	100.00	50.00	900.00
F4	pH	6.90	6.70	7.30	7.42	6.60	8.20	7.22	6.70	7.70
F5	Calcium (μg/l)	3600.00	1600.00	10000.00	9325.00	4850.00	22000.00	13300.00	2200.00	65000.00

*With the exception of mercury, the concentration of all of these elements was above the detection limit.

the other clusters; b) the stronger coloration of the water in Cluster 1 indicates higher concentrations of aluminium than in the other clusters; c) it is possible that a higher fraction of the mercury in Cluster 1 is the methylmercury form than in the case in the other clusters, because organic matter content is higher and the pH is lower; d) it is possible that low concentrations of calcium, magnesium and potassium observed in Cluster 1 might affect the intestinal absorption of toxic metals from the water; e) fluoride could play a protective role against DAT in Clusters 2 and 3 as it may reduce the intestinal absorption of aluminium. It is also interesting to note that high color values, which indicate high levels of organic matter, indicate that a large fraction of aluminium is found as organic species, species that might be more easily absorbed in the body than inorganic species.

On the map of standardized prevalence rates for DAT for the three different clusters (Figure 3), we found a northwest axis close to the confluence of the Saguenay River and Lac-St-Jean. This pattern is, however, difficult to understand. A more in-depth study of the duration of exposure of subjects with DAT could perhaps help us to shed light on it.

We have validated our results through analysis of the significant differences between the three physicochemical profiles using the Kruskal-Wallis test, and for the standardized prevalence rates using z scores and Poisson distribution. The high reliability of the DAT diagnosis, using both clinical and neuropathological criteria, increases the significance of our results. Furthermore, we confirmed the coherence of our results by sub-grouping the three main clusters obtained (e.g. sub-groups 1a and 1b were similar in Figure 2). The sub-groups are significantly different with $p < 0.05$ and the gradient observed on the basis of the three main clusters. For example, the prevalence rates for sub-groups 1a and 1b are 56% and 21% above the regional rate.

The concentrations of elements observed in the different clusters do not exceed the limits outlined in Quebec by-laws for drinking water (Government of Quebec, 1988). On the other hand, the median coloration of the water in Cluster 1 is higher then the level recommended by the federal government (Health and

TABLE 3
Significantly Different Elements Among the Three Clusters Retained

	Number of cases of DAT/population at risk by age group				Other characteristics of DAT subjects		Standardized prevalence rate	
Cluster	45-54	55-64	65-74	75+	% of 15 years and over who graduate University	Mean duration of stay (years)	Standardized rate (according to age)	Ratio of rate
Cluster 1	1/5245	18/4185	28/2195	27/1020	10%	24.95	0.006229	1.339
Cluster 2	1/8760	14/7670	37/4370	35/1845	12%	20.95	0.00395	0.842
Cluster 3	4/12715	31/10770	69/6280	56/3315	13%	21.46	0.00467	0.993
Region	6/28040	66/23720	136/13465	126/6525	-	-	0.0047	1.00

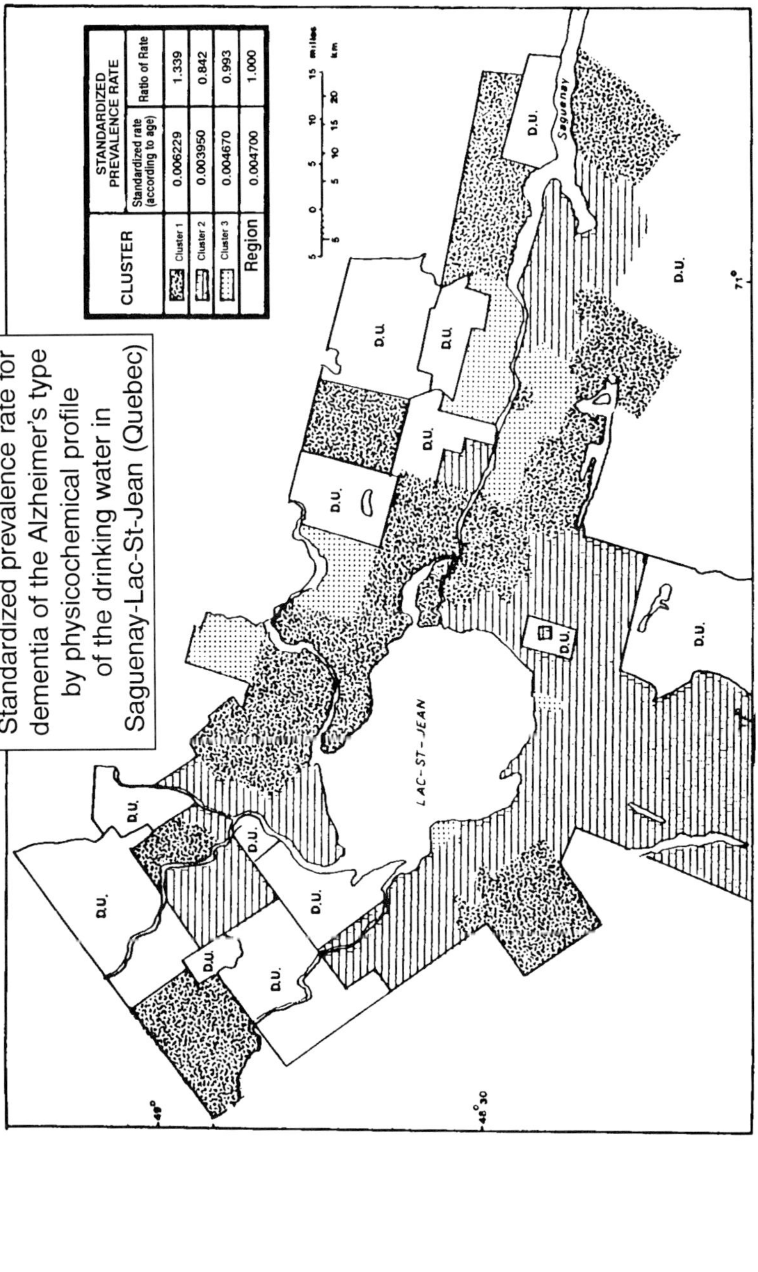
Standardized prevalence rate for dementia of the Alzheimer's type by physicochemical profile of the drinking water in Saguenay-Lac-St-Jean (Quebec)
CLUSTER
STANDARDIZED PREVALENCE RATE
Standardized rate (according to age)
Ratio of Rate
Cluster 1
0.006229
1.339
Cluster 2
0.003950
0.842
Cluster 3
0.004670
0.993
Region
0.004700
1.000
miles
km
D.U.
LAC-ST-JEAN
Saguenay
49°
48°30
71°

FIGURE 3.

Welfare Canada, 1993). The bias discussed previously probably imply an underestimation of many elements concentrations, so the links between the physicochemical profiles and DAT might in fact be higher than what we have observed.

In conclusion, several associations were found, but the ecological nature of the present study implies that cause-and-effect interpretations should be made with great care. Such ecological studies cannot alone prove any causal relationships between the diseases and drinking water constituents, but they may give rise to useful hypotheses. We will soon try to confirm these preliminary results of our analysis, taking into consideration the residential itineraries of a group of controls without DAT. These results will hopefully make it possible to identify the role that elements in the drinking water could play compared to other sources of exposure. Furthermore, univariate statistical analysis and logistic regression may contribute to a clarification of the roles and mechanism of each element in the etiology of DAT. A program of sampling for aluminium is necessary to quantify the concentrations in drinking water. Other research is currently under way within the IMAGE Project on the involvement of pesticides (organophosphates and carbamates used in forestry and agriculture) in DAT.

ACKNOWLEDGMENTS

This work was subsidized by the Human Sciences Research Council and the Alzheimer's Society of Canada. We would also like to thank the Jean-Louis Lévesque Foundation and the Berthiaume-Du Tremblay Foundation. Trond Peder Flaten, University of Trondheim, is thanked for reviewing the manuscript.

REFERENCES

Ackrill, P., Ralston, A.J., Day, J.P. and K.C. Hodge. 1980. *Successful removal of aluminium from patient with dialysis encephalopathy*. Lancet, II:692-693.

Alfrey, A.C., LeGendre, G.R. and W.D. Kaehny. 1976. *The dialysis encephalopathy syndrome: Possible aluminium intoxication.* N Engl J Med, No. 294:184-188.

Barbeau, A., Roy, M., Paris, S., Cloutier, T., Plasse, L. and J. Poirier. 1985. *Ecogenetics of Parkinson's disease: 4-hydroxylation of debriscoquine.* Lancet, II:1213-1217.

Bernard, P.M. and C. Lapointe. 1987. *Mesures statistiques en épidémiologie.* Presses de l'Université du Québec. 314 p.

Boothby, J.A., deJesus, P.V. and L.P. Rowland. 1974. *Reversible forms of motor neuron dosease: lead 'neuritis'*. Arch. Neurol., 31(1):18-23.

Borella, P., Giardino, M., Neri, M. and E. Andermarker. 1990. *Magnesium and potassium status in elderly subjects with and without dementia of the*

Alzheimer type. Magnesium Research, Vol. 3, No. 4:283-289.

Brown, I.A. 1954. *Chronic mercurialism: a cause of the clinical syndrome of amyotrophic lateral sclerosis.* Archives of Neurology and Psychiatry, 72(6):674-681.

Campbell, P.G.C., Thomassin, D. and A. Tessier. 1986. *Aluminium speciation in running water on the pre-cambrian shield: Kinetics aspects.* Water Air Soil Pollution, Vol. 30:1023-1032.

Carty, A.J. and S.F. Malone. 1979. *The biogeochemistry of mercury in the environment.* Elsevier, Amsterdam, 403-419.

Corder, E.H., Saunders, A.M., Strittmatter, W.J. and D.E. Schmechel, et al. 1993. *Gene dose of apolipoprotein E type 4 allele and the risk of Alzheimer disease in late onset families.* Science, vol. 261:921-923.

Constantinidis, J. 1990. *Démence d'Alzheimer et zinc.* Archives Suisses de Neurologie et Psychiatrie, Vol. 141, No. 6:523-556.

Cotzias, G.C., Papavisiliou, P.S. and J. et Ginos, et al. 1971. *Metabolic modification of Parkinson's disease and chronic manganese poisoning.* Ann Rev. Med., Vol. 22:305-326.

Craun, G.F. 1990. *Review of epidemiologic studies of aluminium and neurologic disorders.* Envir. Geoch. Health, 12(1-2):125-135.

Davison, A.M., Walker, G.S., Oli, H. and A.M. Lewins. 1982. *Water supply aluminium concentration, dialysis dementia, and effect of reverse-osmosis water treatment.* The Lancet, II:785-787.

Desjardins, R., Brisebois, P. and L. Potvin. 1989. *Le plomb dans l'eau potable.* Sciences et Techniques de l'Eau, Association Québécoise des Techniques de l'Eau, Vol. 22, No. 1:21-30.

Dupont, J. 1990. *État de l'acidité des lacs de la région hydrographique du Saguenay-Lac-St-Jean.* Ministère de d'environnement du Québec, Direction de la qualité du milieu aquatique, publication QEN/PA-38/1, 131 pages.

Durlach, J. 1990. *Magnesium depletion and pathogenesis of Alzheimer's disease.* Magnedium Research, Vol. 3, No. 3:217-218.

Edwardson, J.A. 1991. Aluminium and Alzheimer's disease. J Inorg Biochem, 43:885.

Emard, J.F., et al. 1992. *Répartition géographique de la maladie d'Alzheimer au Saguenay-Lac-Saint-Jean, Québec (Projet IMAGE).* Cahier de géographie du Québec, No. 36:61-75.

Emard, J.F., Thouez, J.P, and D. Gauvreau. 1992. *Maladies neurodégénératives et facteurs de risque.* Université Paul Valery, Montpellier, Cahiers GEOS, No. 24, 31 pages.

EPA. 1984. *Mercury health effects update.* Health issues assessment, Final report, Environmental Protection Agency, Washington, EPA-600/8-019f.

Evans, P.H., Yano, E., Klinowski, J. and E. Peterhans. 1992. *Oxidative damage in Alzheimer's dementia, and the potential etiopathogenic role of aluminosilicates, microglia and micronutrient interactions.* Free Radical and Aging, ed. Birkhauser Verlag Basel, Switzerland, 179-189.

Fan, A.M. 1987. *Genotoxic and Carcinogenic Metals: Environmental and Occupational Occurrence and Exposure.* Princeton Scientific Publishing, Advances in Modern Environmental Toxicology, Vol. 6, 185-210.

Feldman, R.G., Niles, C.A. and M. Kelly-Hayes. 1984. *Peripheral neuropathy in arsenic smelter workers.* Neurology, Vol. 29:939-944.

Flaten, T.P. 1990. *Geographical associations between aluminum in drinking water and death rates with dementia (including Alzheimer's disease), Parkinson's disease and amyotrophic lateral sclerosis in Norway.* Environ Geochem Health, No. 12:152-167.

Forbes, W.F., Hayward, L.M. and N. Agwani. 1992. *Geochemical Risk Factors: Functioning Based on the Ontario Longetudinal Study of Aging (LSA) I. Results from a Preliminary Investigation.* Canadian Journal on Aging, Vol. 11, No. 3:269-280.

Garruto, R.M. 1991. *Pacific Paradigms of Environmentally-induced Neurological Disorders: Clinical, Epidemiological and Molecular Perspectives.* Neurotoxicology, Vol. 12:347-378.

Gauthier, E., Dupuis, P., Banton, O. and J.P. Villeneuve. 1993. *Utilisation d'un modèle de simulation stochastique et d'un sig pour l'analyse et la représentation spatiale des risques de contamination des eaux souterraines par les pesticides.* Sciences et Technique de l'Eau, Association Québécoise des Techniques de l'Eau, Vol. 26, No. 4:273-282.

Gautrin, D. and S. Gauthier. 1989. *Alzheimer's Disease: Environmental Factors and Etiologic Hypotheses.* Canadian Journal of Neurological Sciences, Vol. 16:375-387.

Gauvreau, D. 1987. *Le paradigme de la maladie d'Alzheimer.* Interface, Vol. 8, No. 5:16-21.

Gauvreau, D., DeBraekeleer, M. and R. Bouchard, et al. 1989. *The IMAGE Projet: A multimatrix model for study of Alzheimer disease. In: Familial Alzheimer's disease, Molecular Genetics and Clinical Perspectives.* (Ed.: J.P. Blass, G. Miner, L.A. Miner, R.N. Richter, J.L.), Valentine/Dekker, New York. 330-365.

Government of Quebec. 1988. *Règlement sur l'eau potable.* Éditeur officiel du Québec, réglement (Q-2,r.4.1), 7 pages.

Goyer, R.A. 1986. *Toxic effects of metals.* In: Toxicology, McMillan Publishing Company, New York, 582-635.

Gresham, L.S., Molgaard, C.A., Golbeck, A.L. and R. Smith. 1986. *Amyotrophic lateral sclerosis and occupational heavy metal exposure: a case-control study.* NeuroEpidemiology, No. 5:29-38.

Health and Welfare Canada. 1993. *Recommendations for drinking water quality in Canada. Gourvernment of Canada,* Supply & Services Canada, Hull.

Henderson, A.S. 1988. *The risk factor for Alzheimer's disease: a review and a hypothesis.* Acta Psychiatr. Scand., Vol. 78:257-275.

Hetzel, B.S. and J.T. Dunn. 1989. *The iodine deficiency disorders: their nature and prevention.* Annual Review of Nutrition, 9:21-38.

Isnard, H. 1994. *Toxicité du plomb et saturnisme infantile.* Techniques, Sciences et Méthodes, No. 2:70-72.

Jacqin, H., Commenges, D., Letenneur, D., Barberger-Gateau, P. and J.F. Dartigues. 1994. *Components of Drinking Water and Risk of Cognitive Impairment in the Elderly.* American Journal Epidemiology, Vol. 139:48-57.

Jorm, A.F., Korten, A.E. and A.S. Henderson. 1987. *The prevalence of dementia: a quantitative integration of the literature.* Acta Psychiatr Scand, 76:465-479.

Korsac, R.J. and M.M. Sato. 1977. *Effects of chronic organophosphate pesticide exposure on central nervous system.* Clin. Toxicol, No. 11:83-95.

Legendre, L. and P. Legendre. 1984. *Écologie numérique, Tome II: La structure des données écologiques.* ed. 2., Paris, Masson et Presse de l'Université du Québec, collection d'écologie, No. 13, 335 pages.

Martyn, C.N., Osmond, C. and J.A. Edwardson, et al. 1989. *Geographical relation between Alzheimer's disease and aluminium in drinking water.* Lancet, Vol. 1:59-62.

McDowell, I., Hill, G. and J. Lindsay, et al. 1994. *Canadian Study of Health and Aging: study methods and prevalence of dementia.* Canadian Medical Association Journal, Vol. 150, No. 6:899-913.

McLachlan, D.R., Fraser, P.E. and A.J. Dalton. 1992. *Aluminium and the pathogenesis of Alzheimer's disease: a summary of evidence.* Aluminium in Biology and Medicine, Wiley, Ciba Foundation Symposium No. 169:87-108.

Michel, P., Commenges, D. and J.F. Dartigues, et al. 1991. *Study of the relationship between aluminum concentration in drinking water and risk of Alzheimer's disease.* In: Alzheimer's disease: Basic mechanisms, diagnosis and therapeutic strategies, John Wiley, New York, 387-391.

Milne, F.J., Sharf, B., Bell, P. and A.M. Meyers. 1983. *The effect of low aluminium water and desferrioxamine on the outcome of dialysis encephalopathy.* Clinical Nephrology, Vol. 20, No. 4:202-207.

Mitchell, J.D. 1987. *Heavy metals and trace elements in amyotrophic lateral sclerosis.* Neurologic Clinics, Vol. 5, No. 1:43-60.

Moriwaka, F., Satoh, H., Emima, A. and C. Watanabe, et al. 1993. *Mercury and selenium contents in amyotrophic lateral sclerosis in Hokkaido, the northernmost island of Japan.* Journal of the Neurological Sciences, Vol. 118, No. 1:38-42.

NAS. 1978. *An Assessment of Mercury in the Environment.* Panel on Mercury. Washington, DC: National Academy of Sciences.

Neri, L.C. and D. Hewitt. 1991. *Aluminum, Alzheimer's disease, and drinking water.* Lancet, No. 338:390.

Newman, M.C., Dixon, P.M., Looney, B.B. and J.E. Pinder. 1989. *Estimating mean and variance for environmental samples with below detection limit observations.* Water Resources Bulletin, American Water Resources Association, Vol. 25, No. 4:905-915.

Nieboer, E. and B.L. Gilbson, et al. 1993. *Health effects of aluminium: a critical review with emphasis on aluminium in drinking water.* Institute of

Environment and Health, University of Toronto and McMaster University, 110 pages.

Oertel, W.H. and A. Kupsch. 1993. *Pathogenesis and animal studies of Parkinson's disease*. Current Science, Neurology and Neurosurgery, Vol. 6:323-332.

Pericak-Vance, M.A., et al. 1991. American Journal of Human Genetics Vol. 48:1034-1050.

Perl, D.P. and P.F. Good. 1991. *The relationship between aluminium and neurofibrillary tangle formation.* In: Alzheimer's Disease and the Environment, Royal Society of Medicine Services, Round Table Series No. 26:60-69.

Poirier, J., Dallaire, L. and A. Barbeau. 1986. *Parkinson's disease and free radicals.* Free Radical, Cell damage and disease, London, Catherine Rice-Evans.

Poirier, J., Kogan, S. and S. Gauthier. 1991. *Environment, Genetics and Idiopathic Parkinson's Disease.* Canadian Journal of Neurological Sciences, Vol. 18:70-76.

Rajput, A.H., Uitti, R., Stern, W., Laverty, W. and K. O'Donnel, et al. 1987. *Geography, drinking water chemistry pesticides and herbices and etiology of Parkinson's disease.* Canadian Journal of Neurological Sciences, Vol. 14, No. 3:414-418.

Rocca, W.A., et al. 1990. *Prevalence of clinically diagnosed Alzheimer's disease and other dementing disorders: a door-to-door survey in Appignano, Macerata Province, Italy.* Neurology, Vol. 40:626-631.

Royal Canadian Society. 1986. *La commission d'étude du plomb dans l'environnement, Le plomb dans l'environnement au Canada, Science et réglementation: Rapport final.* Ministère de l'approvisionnement et services Canada, 469 pages.

Rupinder, K.B. and M. Costa. 1992. *In: Environmental Toxicants: Human Exposures and their Health Effects.* Morton Lippman, Van Nostrand Reinhold, New York, 575-581.

Schoenberg, B.S. 1993. *Calculating confidence intervals for rates and ratios: Simplified method utilizing tabular values based on the Poisson distribution.* Neuroepidemiology, Vol. 2:257-265.

Sedman, A.B., Wilkening, G.N. and B.A. Warady, et al. 1984. *Clinical and laboratory observations: encephalopathy in childhood secondary to aluminium toxicity*. J. Pediatr., Vol. 105:836-838.

Semchuk, K.M., Love, E.J. and R. Lee. 1991. *Parkinson's disease and exposure to rural environmental factors: a population based case-control study.* Canadian Journal of Neurological Sciences, Vol. 18:279-286.

Seux, R. and W. Dab. 1994. *Evaluation de l'exposition au plomb et stratégie de prévention.* Techniques, Sciences et Méthodes, No. 2:73-78.

Shellenberg, G.D., et al. 1992. Science, Vol. 258:668-671.

Signoret, J.L. and J.J. Hauw. 1991. *Maladie d'Alzheimer et autres démences.* Médecine-Sciences, Flammarion, Paris, 511 pages.

Spencer, P.S., Ludolph, A.C. and G.E. Kisby. 1993. *Neurologic diseases associ-*

ated with use of plant coponents with toxic potential. Environmental Research, Vol. 62, No. 1:106-113.

St. Geoge-Hyslop, P.H., et al. 1987. *The genetic defect causing familial Alzheimer's disease maps on chromosome 21.* Science, Vol. 235:885-890.

Subramanian, K.S. and J.W. Connor. 1991. *Lead contamination of drinking water: metals leaching from soldered pipes may pose health hazard.* Journal of Environmental Health, 54(2):29-32.

Taylor, P.J. 1977. *Quantitative methods in geography: An introduction to spatial analysis.* Houghton Mifflin Company, Boston, 386 p.

Thompson, C.M., et al. 1988. *Regional brain trace element studies in Alzheimer's disease.* Neurotoxicology, Vol. 9, No. 1:1-7.

Thouez, J.P. 1979. *Caractéristiques physico-chimiques de l'eau potable et la mortalité ischémique du coeur: application aux municipalités des cantons de l'Est (Québec).* The Canadian Geographer, Vol. XXIII, No. 4:308-321.

Van Duijn, C.M., Clayton, D.G., Chandra, V., Fratiglioni, L., Graves, A.B., Heyman, A., Jom, E., Kokmen, A.F., Kondo, R., Mortimer, J.A., Rocca, W.A., *Shalat, S.L., Soininen, H. and A. Hofman. 1992. Interaction between genetic and environmental factors: A re-analysis of case-control studies of Alzheimer's disease.* Symposium Proceedings: Abstracts of the third International Conference on Alzheimer's Disease. Neurobiology of Aging, Vol. 13, No. 1:S115.

Verity, M.A. 1992. *Neurotoxins and environmental poisons.* Current Opinion in Neurology & Neurosurgery, Vol. 5, No. 3:401-405.

Walter, S.D. 1991. *The ecologic method in the study of environmental health. I. Overview of the method. Environmental Health Perspectives, Vol. 94:61-65. II. Methodologic issues and feasibility.* Environmental Health Perspectives, Vol. 94:67-73.

Walter, S.D. 1992. *The analysis of regional patterns in health data II. The power to detect environmental effects.* American Journal of Epidemiology, Vol. 136, No. 6:742-759.

Wills, M.R. and J. Savory. 1983. *Aluminium poisoning: dialysis encephalopathy, osteomalacia and anaemia.* Lancet, No. 2:29-34.

Yanagihara, R. 1982. *Heavy Metals and Essential Minerals in Motor Neuron Disease.* Human Motor Neuron Disease, edited by Lewis P. Rowland. Raven Press, NY, Vol. 36:233-247.

Yase, Y. 1972. *The pathogenesis of amyotrophic lateral sclerosis.* Lancet, Vol. 2, No. 772:292-296.

Yasui, M., Kihira, T. and K. Ota, et al. 1991. *Zinc concentration in the central nervous system in case of multiple sclerosis-comparaison with other neurological diseases.* No to Shinkei-Brain & Nerve, Vol. 43, No. 10:951-955.

Zayed, J. 1988. *Maladie de Parkinson et facteurs environnementaux dans la région de Valleyfield (P.Q.).* Université de Montréal, Départment de médecine du travail et d'hygiène du milieu, Thèse de doctorat, 197 pages.

Zayed, J., Ducic, S., Campanella, J.C., Panisset, P., André, P., Masson, H. and M.

Roy. 1990. Facteurs environementaux dans l'étiologie de la maladie de Parkinson. *Canadian Journal of Neurological Sciences,* Vol. 17:286-291.

Zuber, M. and A. Alperovitch. 1991. Maladie de Parkinson et facteurs environmentaux. *Rev. Epidem.* et Santé Publ., Vol. 39:373-387.

Environmental Contaminants, Ecosystems and Human Health. Edited by S.K. Majumdar, E.W. Miller, and Fred J. Brenner.

Chapter Twenty-Seven

RADIATIONAL HEALTH HAZARDS

E. WILLARD MILLER
Professor of Geography and Associate Dean
for Resident Instruction (Emeritus)
The College of Earth and Mineral Sciences
The Pennsylvania State University
University Park, PA 16802

INTRODUCTION

There is evidence in the environment that radiation dates from the beginning of the earth. Life has thus developed with an ever present background of radiation. It was not until the late 19th century, however, that scientific investigation began to identify and understand the importance of radiation on human health.

NATURE OF RADIATION

Radiation is a process in which energy in the form of rays of light or heat is emitted from atoms and molecules as they undergo change. Radiation has four energy sources-alpha (α), beta (β), gamma (γ) and neutrons (ν). The energy and penetration ability for each of these varies greatly. Alpha particles travel only a few inches and can be stopped by skin tissue or a sheet of paper, but are very dangerous when ingested or inhaled. Beta rays move a few feet through air but can be stopped by a quarter inch of wood or a three tenths inch of aluminum. Like alpha particles, they can be harmful when emitted inside the body. In contrast, gamma rays travel hundreds of feet through the atmosphere and can penetrate a concrete block or several inches of lead. These gamma rays of electromagnetic energy can easily damage tissue. Neutrons also travel hundreds of feet in the air and can be stopped by enforced concrete and water. When neutrons collide with an atom, they can induce radioactivity by combining with the nucleus.

NATURAL SOURCES OF RADIATION

Natural sources of radiation provide, by far, the greatest proportion of radiation received by individuals. Exposure to radiation is inescapable. Radiation comes to the earth, from outer space, and from radioactive material in the earth's crust. An individual is subject to irradiation that remains outside the body as well as radiation that is inhaled in air or swallowed in food or water. Although everyone on the planet is subject to natural radiation, some people receive much more than others. If an individual lives in an area of radioactive rocks or soils, the dose of radiation is larger. The dose may also depend on life style. If an individual spends a large amount of time on a beach exposed to the sun, the radiation dose increases.

Cosmic Rays

About one-half of the radiation received by an individual comes from exposure to cosmic rays. These rays originate from the sun and from interstellar space. The rays consist of high-energy particles of which about 70 percent are protons, 20 percent alpha particles and 10 percent a variety of nuclei. The polar areas receive more cosmic rays than the equatorial region because the earth's magnetic field directs the radiation toward the polar regions. The intensity of cosmic radiation also increases with altitude. A person living at sea level will receive an effective dose equivalent to about 3,000 microsieverts of cosmic radiation each year, but at 6,000 feet the dose will be several times that amount.

Primordial Radionuclides

Primordial radionuclides are radiation that came into existence at the time of the formation of the earth. These radionuclides can be divided into two groups, those whose half-life are longer than the age of the earth and those that are the radioactive product formed after the earth came into existence. The first group has about 20 known radionuclides with half-lives ranging from 7 years, for example, for uranium 235, to about 10 years for dysprosium 156. The second group contains 40 radionuclide, all of which are decay products of uranium 235, uranium 238, or thorium 232. Each of these three radionuclides undergoes successive alpha and beta decay steps until it reaches one of the stable isotopes of lead.

The widespread distribution of uranium and thorium throughout the earth's crust leads to a significant contribution to the external irradiation dose. Because uranium 238 is unstable, the particles in its nucleus are altered as protons, and neutrons break away. Uranium 238 goes through a series of changes, becoming first thorium 234, then protactinium 234, and on to becoming radon 222 and ultimately lead 206.

SKIN CANCERS

Typing Skin Cancer

Exposure to sunlight is the most important single cause of skin cancer. The initial changes are wrinkling, looseness, and dryness of the skin. This premature aging ultimately results in cancer. The risk varies with the coloration of the skin. Caucasians because of their light colored skin are particularly susceptible, while Negroes are far less vulnerable.

The excessive exposure to ultra violet rays in the narrow band of light between 290 and 320 nanometers is thought to do the most damage. Three types of skin cancer can develop, each more serious than the initial one. The basal cell non-melanoma skin cancer is the most common among whites. This cancer rarely spreads to vital organs and is treated in a doctor's office. However, it does leave the skin blemished. Squamores cell carcenomes are fortunately less common, but can grow quickly and spread to vital organs. Malignant melanoma in its initial stage looks like a mole and is dark and irregular in outline. This skin cancer is the most dangerous for it spreads quickly to vital organs and death is common.

Trends in Skin Cancers

With growth of sun bathing in the 20th century, skin cancer has experienced a remarkable increased pace. It is estimated that 400,000 new cases of non-melanoma skin cancer occurs in the United States each year and this number is rising. (Page and Ardyce, 1985). The growth is even faster in Australia. In the 1990s Australian total number of basal cell skin cancer are about 140,000, nearly one percent of the population.

From a health viewpoint, it is estimated that there will be 32,000 cases of melanoma cancer reported in the United States in 1994. According to the American Cancer Society about 6,900 will die from this disease.

Prevention of Skin Cancer

The best protection from developing skin cancer is to limit sun exposure beginning as a child. Additional suggestions include avoiding sunbathing or visiting tanning parlors. When exposure to the sun occurs wear a hat, long sleeves, and sunglasses. Have the least exposure between 10 a.m. and 3 p.m. Sunscreens are now effective in shielding the skin from the sun's rays for they greatly reduce the amount of UV light penetrating the skin. Sunscreen should be put on at least one hour before exposure to the sun. Sunscreens that offer the best protection are those with the highest SPF (sun protection factor) number, such as 15 through 25. While most sunscreens contain PABA (para-amino-benzoic acid) which mainly block short UV rays, some newer brands (called broad spectrum sunscreens) contain the ingredient oxybenzone which blocks the longer wave lengths as well.

Cancer Cures

A number of experiments have been conducted in laboratories to find cures for skin cancers. For years scientists have known that vitamin A and its derivatives, nutrients found in green, yellow and orange fruits and vegetables, and in egg yokes and liver, are essential for healthy skin and hair. More recently epidemiologic studies have shown that people with low vitamin A intake face higher risks of certain cancers. In addition, certain chemicals have the goal of reducing cancer risks. More than 600 compounds have been identified as having some evidence of ability to interrupt the cancer process. Besides vitamin A, attention has focused on vitamin C and the trace elements selenium and calcium.

In 1994, David Berd, professor of medicine at Thomas Jefferson University in Philadelphia, announced by the American Association of Cancer Research that a vaccine to treat advanced melanoma had been found to be 70 percent effective in preventing the recurrence of this deadly skin cancer. The vaccine was made from the patients own tumors and was the first to show such a high success rate. Joseph Bertina, head of pharmacology of the J.R. Memorial Sloan-Kettering Cancer Center in New York, indicated that other anti-melanoma vaccines are being tested, but the Jefferson results have been most promising so far. In addition studies confirm the findings, the techniques used to control the melanoma or vaccine could be used to develop vaccines for other types of cancers.

RADON

The natural decay of uranium -238, the principal form of uranium found in the earth's crust, leads to the formation of radon and its daughters. When uranium occurs as an ore body the intensity of the release of radon is increased. One of the principal forms of radon is radon -222, a noble gas, which escapes from exposed rock surfaces. Radon -222 has a half-life of 3.8 days and decays by the emission of alpha (α), beta (β), and gamma (γ) rays. In the decay of radon -222 a number of short-lived nuclides (radon daughters) are produced including the isotopes polonium -214 and polonium -218. These progeny are solids and have a half-life of a few seconds.

Universal Radon Hazard

It was not until 1984 that it was recognized that health hazards from radon were extremely widely distributed over the earth. The earth began as a molten state. As a consequence, radioactive particles are found in each square foot of rock, although the concentration varies with the type of rock.

The modern recognition of the widespread nature of radon was triggered by the now famous incident near Boyertown, Pennsylvania. On a December morning in 1984, Stanley Watras set off the radiation alarm when he went to work at the Limerick nuclear power plant. It was soon proved that the radiation did not come from the plant, but his home. On testing for radon in the Watras home, 3,200 picocuries (a trillionth of a curie) were recorded. A curie is equal to 37 billion (37 x 10^9) disintegrations of radioactivity per second. This was the highest level ever discovered in a residence. Up to that time no one realized that radon would be a danger to anyone except uranium miners and people who lived close to uranium mines. A totally new concept came into existence that radon could be constantly released from the earth and find its way into buildings and accumulate to life threatening concentrations.

Location

As a response to the very high picocuries in the Watras home, the testing for radon contamination spread across the nation. As a consequence, radon was found throughout the nation with particularly high concentrations from Colorado and the Dakotas to the Northeast. The Environmental Protection Agency and the U.S. Public Health Service estimated radon poses a health problem in 8,000,000 American homes. Further, the EPA estimated that between 5,000 and 20,000 lung cancer deaths were occurring annually due to radon exposure. Only cigarette smoking causes more deaths.

It is difficult to identify places where radon will be found. Originally it was thought that radon was only associated with igneous rocks. This is now known to be incorrect. Radon is found in all types of rocks. In addition radon concentrations vary greatly from place to place. Thus every home in an area must be tested. It is now estimated that every square meter of land emits half a picocurie per second.

Health Hazard

Radon daughters either as free icons or absorbed as water droplets or dust particles, may be inhaled, deposited and retained in the respiratory system of individuals (Union of Concerned Scientists, 1975). Consequently, the respiratory epithelium, the thin layer of cells forming a tissue that covers the surface of the lungs, can be exposed to radiation doses that far exceed whole body doses from internally dissolved radon. Lung exposure from radon daughters is far more harmful than from radon itself. The bulk of the basic damage is from alpha particles, which have short ranges but high energy transfer.

Lung cancer, believed to arise from exposure to radon daughters, usually appears in the hilar region of the lung. This area is covered with a moving sheet of mucus. Since this surface layer of cells develops from a layer of basal cells, the health of the epithelium depends upon the continued integrity of the basal cells.

When radon droplets enter the epithelium they will deliver a significant fraction of energy of the alpha particles to the basal cell region. With a 30 minute half-life, the radon daughters will decay mainly on the epithelium before they can be removed by normal clearance mechanisms of mucous transport. The actual radiation dosage to the larynx depends upon a number of factors: concentrations of radon daughters in the atmosphere, respiratory volume and exchange rate, distribution of radon daughters between free ions and ions absorbed by dust particles, size of the particles and respiratory clearance rate.

Control of Radon

By the late 1980s, radon as a hazardous gas was recognized by Congress. In 1988, Title III, Indoor Radon Abatement as an amendment to the Toxic Substance Control Act was enacted. The Act established the long term goals with respect to radon levels. To implement the legislation, the Environmental Protection Agency (EPA) has set 4 picocuries of radon per liter of air as the action level at which a homeowner should consider remedial measures. The level varies with Canada setting the rate at 10 picocuries and Germany 7. Regrettably there are no precise rules to indicate a safe level. If EPA set the level too high, the agency is criticized for "letting people die." If it sets the level too low, there is criticism of over regulating.

It is estimated that there are five million homes in the U.S. where the picocuries level is at or above the 4 picocuries level. The total bill for dealing with the natural radon problem is estimated at $5 billion. The solution of the problem requires a ventilation system to remove the collected radon. This procedure can easily cost more than $1,000. Questions are now being raised concerning the EPA established level as to when the number of picocuries become harmful to health. Many epidemiologists believe that level is too low. EPA had originally estimated that about 30,000 lung cancer deaths occur annually from radon. The agency believes this number was too high and the number is now between 7,000 and 14,000.

URANIUM AND NUCLEAR ENERGY

Uranium Mining

It has long been recognized that miners in uranium mines had shortened lives. As early as the 16th century it was reported that miners in the pitchblende (radium) mines of Schneeberg in the Erz Mountains of Germany and at Joachimsthal in Czechoslovakia were plagued with pulmonary diseases. In the 19th century studies found that lung diseases accounted for 75 percent of all deaths of uranium miners. Between 1900 and 1940 a number of studies verified the earliest findings.

In spite of numerous studies indicating the health hazards of uranium mining, the danger from radon in the mines was long ignored. The first study to recognize the feasible health hazard of radon to uranium miners was prepared by the Colorado Department of Health in 1949. Other studies followed by the U.S. Public Health Service. In 1957 the USPHS, based on radon daughter surveys, predicted a significant increase in mortality from lung cancer among miners. In 1964 a survey of 3,656 white uranium miners indicated that there were 218 deaths compared with 148 expected.

The first quantitative relationships between the incidences of radiation cancer and cumulative dosage of radiation was proved by Wagoner and others in 1965. With cumulative radiation exposure, it was revealed that the incidence of cancer increased from 3.1 per 10,000 miners per year to 116 per 10,000 miners per year over six exposure categories ranging from less than 120 WLM (working level per month) to more than 3,270 WLM.

As a response to many studies it is now determined that the highest rates of lung cancer occur in miners who worked in mines in which the average concentration exposure exceeded 1,000 WLM, who were in the mines a minimum of 15 years, and who were modest to heavy cigarette smokers (Federation Radiation Council, 1967). Nevertheless, the protection of the miner was extremely slow to develop. Radon and its daughters can be removed from mines by using proper ventilation. The initial monitoring programs began in 1954. However, radon control programs were not very effective until the 1960s. Radiation protection in the mines remained limited until 1967 when uniform Federal Regulatory Commission standards came into effect.

Atomic Energy Commission failed to protect the uranium miner due to a peculiar interpretation of the Atomic Energy Act. The Act was interpreted by the AEC not to include uranium mining but to include only other aspects of the mining process. The view was held by the AEC that it could not regulate operations underground, but only after the ore was removed from the earth could AEC control begin. The AEC considered the mined ore rather than the ore body as the "place" of deposit in nature. This position was not tested by the courts and the ruling remains intact.

Uranium Ore Processing

The uranium ore requires treatment to recover the uranium. In the milling process more than 99 percent of the ore is waste. As a consequence, high piles of tailings accumulated near the mills. For many years the danger from radioactivity from the tailings was ignored. Because the extraction of a few pounds of uranium to fuel a single 1,000-megawatt light water reactor for a year requires the generation of at least 106 tons of tailings, the Environmental Protection Agency now recognizes that tailings present a massive environmental impact on the area.

By 1959 the U.S. Public Health Service recognized the seriousness of the problem, and new regulations by the Atomic Energy Commission stopped or diminished the dumping of tailings in streams. However, for many years tailings were used as fill at a number of construction sites such as in Grand Junction, Colorado. In 1966, the Colorado Public Health officials found abnormally high readings of gamma radiation and radon daughters in about 1,500 buildings built on the tailings. By 1975, more than 5,000 sites in Colorado had been surveyed to determine gamma and radon levels. Due to excessive radiation, the U.S. Congress passed regulations that will cost at least $100 million to remove tailings at hazardous sites.

ATOMIC EXPLOSIONS

The initial explosion of an atomic bomb occurred on July 10, 1945 at Los Alamos, New Mexico. Even at this early date, a target committee was planning other droppings of the bomb on Japan. After careful evaluation of potential sites, including consultation with chief of staff, George Marshall, the committee selected Hiroshima. On August 2, 1945 the first atomic bomb was dropped on Hiroshima, and on August 8, 1945 a second atomic bomb exploded over Nagasaki. Both sites were devastated. Although destruction was great, the value of the atomic bomb was still questioned. General Le May, chief of staff, stated "the atomic bomb had nothing to do with the end of the war at all."

To test the effectiveness of the atomic weapon, the Navy sought a site in the Pacific Ocean centering on Bikini and Eniwetok. Between 1946 and 1958, the U.S. Air Force conducted atomic tests on the islands. In the 1950s, studies were made of chemical reactions to the fallout, with particular attention focused on the 300 plus isotopes produced in atomic bomb detonations. Questions began to be raised as to how fission particles affected human beings. As the isotopes were detected around the world, a controversy swept through the scientific world. As evidence of the health hazards grew, it became increasingly evident that surface detonations of the atomic bomb had to stop. Finally, in 1958, President Eisenhower declared a moratorium on atomic explosions both in the United States and the South Pacific. The Soviet Union and the United Kingdom agreed to the moratorium, and since then countries have confined their testing of atomic weapons to underground locations. Only France has continued to test atomic bombs in the atmosphere in the southern Pacific Ocean.

It was hoped that the Pacific islands would become habitable a few years after the atomic explosions. In the late 1960s, plans were developed to have the inhabitants return to the islands. Houses were built and restoration of other facilities began. However, in 1972 after further studies by the AEC, it was found that the soils and water continued to have a high level of radioactivity. In 1978, the 139 people on Bikini were removed. The islands remain radioactive and health continues to be endangered. These islands will remain deserted for many more decades.

ATOMIC ELECTRIC GENERATION

When the first atomic power plant was opened at Shippingport, Pennsylvania in 1957, it was felt that the future problems of producing electric power had been solved. For more than a decade the building of atomic power plants grew rapidly throughout the industrial world. In the late 1960s, questions began to be raised about the potential problem of radioactive contamination of the environment. With the near disaster of Three Mile Island in 1979, and the explosion of the atomic plant at Chernobyl in 1982, the use of uranium as a fuel was highly questioned.

In the use of uranium to produce electricity the most controversial part of the nuclear fuel cycle is the power production. In all nuclear power plants, radioactive materials are produced in the nuclear reactor. The fissioning of the uranium and the neutron activation of the coolant produces many radioisotopes. These radioactive isotopes, particularly the gaseous ones, can escape from the fuel rods through pinhole defects to contaminate the coolant water. Constant monitoring of the atomic reactor is thus a necessity. While in the normal operation of a plant the escape of isotopes is minimal (the EPA has estimated that the health effect of one year's operation of a 1,000 MWE plant is responsible for a total of only about 0.001 M/rem/year), an explosion can send radioactive isotopes into the atmosphere for thousands of miles. For example, radioactive materials from the Chernobyl plant were deposited on the mosses of Lapland in northern Sweden and Finland destroying them as a food for the region's reindeer.

A further radiational concern in the use of uranium to produce electricity comes in the means of disposing of the radioactive waste produced by the nuclear power plant in the process of producing electricity. Attention has become so massive and the public fears have become so persuasive that the industry has been threatened. In the process of producing 800,000 kilowatts of electricity during a three-year period, 24 kilograms of uranium 238 and 25 kilograms of uranium 235 are consumed. During this period, the enriched uranium 235 has been reduced from 3.3 percent to 0.8 percent of the total. In the process of producing electricity, 35 kilograms of radioactive wastes are produced. Of this amount 8.9 kilograms of such various isotopes as plutonium, 4.6 kilograms of uranium 236, 0.5 kilograms of neptunium 237, 0.12 kilogram of Americium 243, and 0.01 kilo gram of curium 244. Since only 28 kilograms of uranium 235 are consumed and a fifth of that amount is converted into uranium 236 and neptunium 237, it is estimated that only 60 percent of the energy produced comes from uranium 235. The thirty-one percent comes from plutonium and 5 percent from the high energy neutron of uranium 238.

The annual waste material, which remains highly radioactive, from a 1,000 megawatt nuclear reactor is about two cubic feet. If an individual is within 10 miles of an unprotected radioactive waste container, he would receive a dose of 500 rems (which is about a 50 percent chance of being fatal) within 10 minutes. None of this waste has been permanently disposed of as to date. In 1984 the U.S. government proposed three sites at Hanford, Washington, Hereford, Texas, and in the isolated Yucca Mountains of Nevada (Miller and Miller, 1989). The Yucca Mountain site was chosen as a permanent depository. Its planned completion is 2003.

Although complete risk cannot be eliminated in buried radioactive waste, the danger must be placed in the proper perspective. A public fear is that the canister will ultimately leak and the groundwater will be contaminated (Shapiro,

1981). In the Yucca Mountain cave, because the canisters will be placed in impervious layers of rock the typical flow of water is less than 30 centimeters per day. At this rate it will take about a thousand years for the water to migrate 100 kilometers. During this period the decay of the wastes makes them far less dangerous. For example, cesium 137 decays to barium 137 in a period of 400 years (Cohen, 1977). During this period the total gamma ray hazard falls by more than four orders of magnitude.

There is also a fear that the waste material would reach the surface due to erosion. This is an unfounded fear. When the waste is buried more than 600 meters deep, it will require millions of years to erode this material to the earth's surface. This assumes that the waste is in a stable tectonic area not subject to earthquakes or volcanic activity.

There is also a fear that social and political conditions are not sufficiently stable to guarantee safety. Monitoring of excessive ionizing radiation would, however, require minimum personnel. Because of the radioactive effects of the waste materials, radical groups will not disturb the permanent deposits.

Although the problem is now recognized, mankind is still faced with finding a sound solution. Because of the time factor, completely rational decisions may not be possible. Nevertheless with the best advice from scientists and engineers and the recognition of social responsibility, a permanent depository for high-level nuclear waste is fundamental to the welfare of all people.

HEALTH EFFECTS OF IONIZING RADIATION

In order to assure safety health standards for the measurement of radiation, levels have been set by national and international commissions and agencies. In the United States these relevant organizations are the National Committee on Radiation Protection and Measurement (NCRPM), the Nuclear Energy Commission (NEC), and the Environmental Protection Agency (EPA).

These standards included the following limitations to exposure. For individuals who are occupationally exposed, the maximum permissible dose is 5 rems/year for a whole-body irradiation. If a dose in a single year exceeds this amount, then in succeeding years the dose must be reduced. The occupational dose to the skin in one year must not exceed 15 rems, to the hands not more than 75 rems, and to the forearms, 30 rems. The maximum permissible exposure for an individual is set at one tenth the occupational limits, that is 0.5 rems excluding natural background and medical exposures (Stolwijk, 1983).

Short Term Health Effects

After extensive research it has been discovered that the health effect of a massive exposure in a short period is quite different than the exposure of the same amount over a longer period of time. For example, if individuals are exposed to

1,000 rems over a short period, immediate severe illusions will occur and almost all persons will die within one month. In contrast, if individuals are exposed to 10 times the amount of radiation in the natural background over several decades, more will suffer acute illness, and only a few will experience genetic defects.

In the use of radiation to test some diseases, the effects vary with the size of the dose. When this radiation dose, on the entire body, is below 25 rems there is no evident chemical effect. In doses of 25 rems or more there is a decrease in the number of white blood cells and with doses between 100 and 299 rems there will be a continual reduction of the white blood cells and leukemia will develop accompanied by vomiting. At this level the human body will recover in several weeks. If the radiation dose ranges between 200 and 600 rems, vomiting will occur, accompanied by internal bleeding and complicating infections. Most hair will be lost within two weeks. Treatment will consist of blood transfusions and antibodies extending over a period of one month to nearly a year. The mortality rate will be as high as 80 percent. When the dose is between 600 and 1,000 rems the treatment is the same, but the mortality rate is between 80 and 100 percent (Stolwijk, 1983).

Long-Term Health Effects

The exposure to ionizing radiation may not become evident for many years. An understanding of long-term radiation on genetic and somatic effects is not as detailed or quantifiable as would be desired. Large human populations have not been exposed to doses of radiation over extended periods. While animal populations have been intensively studied, this extrapolation from animal behavior to humans is very difficult. The long-term effects of radiation consists of three categories - genetic effects, human growth and development, and the somatic effects (Stolwijk, 1983)

The genetic effects are most evident in a single, dominant mutation associated with a disease or abnormality. About one percent of the population has some form dominant mutation including anemia, dwarfism, and extra fingers or toes. Recessive mutations require that two recessive genes are inherited, one from each parent. These produce such diseases as sickle cell anemia, cystic fibrosis and tag sacks disease. About 0.1 percent of all births have recessive diseases. Mutations are, however, a response to complex heredity cancers.

In order to measure radiation not caused by ionizing radiation a method is used known as "doubling dose." Scientific studies indicate that a double dose of radiation is between 200 and 300 rems. Since natural radiation background is only 3 to 5 rems in a reproductive lifetime, only a very small percentage of the mutations can be attributed to this effect. The possible relationship between additional ionizing radiation and increased disease because of its mutational component is predicted to be between 0.5 and 5 percent for a general illness. This appears to be too high. Estimates indicate that the increase in the dose to which a population will be exposed by all nuclear power plants by the year 2,000 will be only 0.001 rems per year or 0.01 rems for a reproductive lifetime.

There is now scientific evidence that ionizing radiation has a major effect on the fetus in utero and on young children. From evidence collected during the period of atomic bomb explosions between 1945 and 1955, it was found that there was reduced growth, mental retardation, and microcephaly. Variations in dose levels will cause different results. When a fetus was exposed within 16 weeks after conception the individual was more sensitive to health problems than if exposed later. Individuals exposed to 25 rems were mentally retarded and if the amount was 50 rems the mental retardation was profound. If the radiation exposure occurs after birth, much higher levels are needed over a longer time period. Evidence is limited, but chronic doses of 365 rems/year or less have not produced health effects.

Except for maximum doses of ionizing radiation that produce immediate effects, the somatic effects of small doses of radiation over time may not become evident for years or even decades after exposure. The long-term effect may appear as a possible reduction in fertility, but most common developments are such diseases as leukemia cancers, and malignancies of the breast, thyroid, stomach or intestinal tract.

As a result of long-term research, it is now established that the increased risk of cancer is in proportion with the excess in radiation exposure. If a population is exposed to ionizing radiation in excess of the national background it is estimated that total number of excess deaths per million exposed per rem would be 10 from thyroid, 50 from breast, 25 to 50 from lung, 20 to 50 from leukemia and 10 to 15 from gastrointestinal cancers. To place these figures in proper perspective, however, it must be remembered that cancer causes 200,000 deaths for each one million mortality. The increase in cancer mortality due to 10 excess rems in a population of one million would be 1,000 in addition to the expected 200,000.

CONCLUSIONS

No doubt remains that excessive ionizing radiation is harmful to health. Cosmic rays are the primary cause of skin cancers. Primordial radiation is a source of potential health problems from radon and in the utilization of uranium for the production of weapons and electricity: Because of the potential health hazard, the use of radiation material requires eternal vigilance. A basic question remains, can society meet this challenge?

SELECTED REFERENCES

Anderson, Hobart, Ian. 1989. "Weather Forecasting Could Take the Sting Out of Sunbathing," *New Scientist,* 122 (May 27):30.

Boutwell, R.K. 1964. "Some Biological Aspects of Skin Carcinogenesis," *Prog Exp Tumor Res*. 41:207-250.

Brodine, Virginia, ed. 1975. *Radioactive Contamination,* New York: Harcourt Brace, Jovanovitch, 190 pp.

Brown, Phyllida. 1993. "Genetic 'Switch' to Target Skin Cancer," *New Scientist,* 127 (March 6):6.

Cohen, Bernard. 1977. "The Disposal of Radioactive Waste from Fission Reactors," *Scientific American*, 236 (June):24-31.

Colglazier, E. William, Jr., ed. 1982. *The Politics of Nuclear Waste,* New York: Pergamon, 264 pp.

Concar, David. 1992. "The Resistible Rise of Skin Cancer," *New Scientist,* 134 (May 16):23-28.

DeMasily, G., Ledonx, E., Barbreau, A. and J. Margot. 1977. "Nuclear Work Disposal: Can the Geologist Guarantee Isolation," *Science.* 197:519-27.

El-Hinnawi, E.E., ed. 1980. *Nuclear Energy and the Environment*, Elmsford, NY: Pergamon, 300 pp.

Flax, Susan Jo. 1981. "Radioactive Waste Management," *Harvard Environmental Law Review*, 5(2):259-295.

Hashimoto, Ken and Walter F. Lever. 1968. *Appendage Tumors of the Skin,* Springfield, IL: Charles C. Thomas, 179 pp.

Miller, E. Willard and Ruby M. Miller. 1990. *Environmental Hazards: Radioactive Materials and Wastes,* Santa Barbara, CA: ABC/CLIO, 298 pp.

________ and ________. 1989. *Environmental Hazards: Air Pollution,* Santa Barbara, CA: ABC/CLIO, 250 pp.

Page, Harriet and Ardyce. 1985. *Cover Rates and Risks,* Publication No. 85-691, Washington, D.C.: U.S. Department of Health and Human Service, Natural Institute of Health, pp. 111-114.

Shapiro, Fred C. 1981. *Radwaste,* New York: Random House, 288 pp.

Slaga, Thomas J., et al. Edition. 1989. *Skin Carcinogenesis: Mechanisms and Human Relevance,* New York: Alan R. Liss, 424 pp.

Stolwijk, Jan A.J. 1983. "Nuclear Waste Management and Risks to Human Health" in Walker, Charles H., Leroy C. Gould and Edward J. Woodhouse, eds., *Too Hot to Handle, Social and Political Issues in the Management of Radioactive Wastes,* New Haven, CT: Yale University Press, pp. 75-93.

Travis, John. 1992. "Closing In on Melanoma Susceptibility Gene(s)." *Science,* 258 (November 13):1080-1081.

Venkei, T. and J. Sugar. 1965. *Early Diagnosis Pathohistology and Treatment of Malignant Tumors of the Skin,* Budapest: Adademiai Kiado, 368 pp.

Environmental Contaminants, Ecosystems and Human Health. Edited by S.K. Majumdar, E.W. Miller, and Fred J. Brenner.

Chapter Twenty-Eight

DISEASE DISTRIBUTION IN A VACCINATED POPULATION: The Spatial Ecology of Measles in the United States

ANDREW D. CLIFF and PETER HAGGETT

Department of Geography
Bristol University
Bristol, England

INTRODUCTION

Since an effective measles vaccine was licensed in 1963, a vigorous vaccination program against measles has allowed the United States to reduce the morbidity of that disease by several orders of magnitude. In 1962, the Centers for Disease Control recorded 481,530 measles cases and 408 deaths; by 1983 this had been reduced to 1,714 cases and 4 deaths. Despite this success, the wild virus continues to circulate and as late as 1990 some 27,786 cases were reported.

This chapter looks at the distribution of these cases in a highly vaccinated population and explores the role of ecological factors contributing to the continuation of pockets of the disease. We look at the distribution at three spatial levels: the national, the local and the international.

GEOGRAPHICAL DISTRIBUTION OF THE DISEASE

The impact of three decades of measles vaccination on the morbidity of the disease in the United States is clearly shown in Figure 1. The decline has been a dramatic one with implications not only for measles itself, but also for other diseases linked to the measles virus (such as subacute sclerosing panencephalitis).

The changes in the geographical distribution of measles that have accompanied vaccination are dramatically illustrated in the lower part of Figure 1. These show United States counties reporting measles in any week of the year at the start

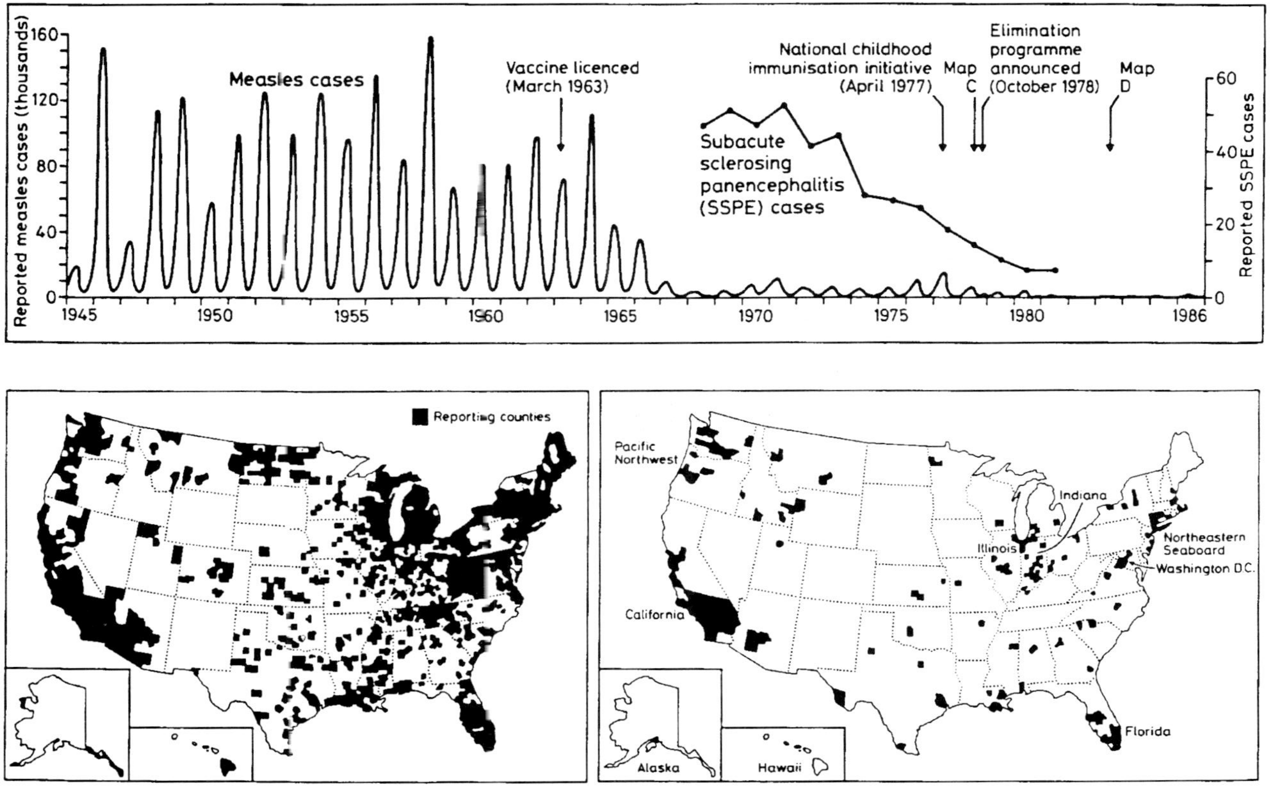

FIGURE 1. United States trends in measles cases, 1945-86. (Upper) Graphs of monthly measles cases with SSPE totals based on annual data. (Lower) Maps of United States counties repor ing measles cases during any week of the year are shown in black for 1978 (lower left) and 1983 (lower right). Source: Cliff and Haggett [1988, Figures 4.9 (A), (C) and (D), page 165]

of the Measles Elimination Program (1978) and at its end (1983). The contraction of infected areas from most of the settled parts of the United States in 1978 to only restricted areas of the Pacific Northwest, California, Florida, the northeastern seaboard and parts of the Midwest by 1983 is pronounced. The persistence of indigenous measles in many of these regions may be explained by the importation of cases from Mexico, Canada and from overseas.

In 1983, 12 states and the District of Columbia reported no measles cases, and 26 states and the district of Columbia reported no indigenous cases. Four states (Indiana, 406; Illinois, 216; California, 181; Florida, 159) accounted for 64 percent of all cases. Of the 3,139 counties, only 168 (5 percent) reported any measles cases. In contrast, measles was reported from 988 in 1978 when the Measles Elimination Program began.

Hersch and colleagues (Hersch, Markowitz, Maes, et al., 1992) have described the geographical distribution of measles cases in the United States at the county level for the 10-year period from 1980 to 1989. To do this, they analyzed the weekly county data published in the Centers for Disease Control's *Morbidity and Mortality Weekly Report,* and related the case loads recorded there to demographic data given in the 1980 and 1990 United States censuses. Over the decade, 56,775 measles cases were reported to Centers for Disease Control. But the distribution was highly skewed. Five states accounted for over half the case total: California, Texas, Illinois, New York, and Ohio. One state (South Dakota) did not report a single case of measles during the whole decade.

Over the decade, the county of occurrence could be traced for 99.2 percent of the total reported measles cases, and epidemiological and demographic profiles were produced for each of the 3,137 counties that make up the United States. Each profile included, for example, the county's mean annual incidence of measles, the number of years in which measles was reported, and its age/gender structure.

The spatial concentration was even more pronounced at the county, as compared to the state, level. As Table 1 shows, 17 counties (0.5 percent of the county total) reported measles in all ten years, while just over half the counties (53.9 percent) reported no measles cases during the decade. Just six counties accounted

TABLE 1

United States, 1980-89: Counties Reporting Measles Every Year

State	County (Mean Measles Incidence Per 100,000 Population)
California	San Diego (6.3), Los Angeles (3.1), Orange (3.0), Santa Cruz (2.2), San Mateo (1.7)
New York	Bronx (6.2), Kings (6.1), Queens (3.4), New York (2.8), Westchester (2.2)
Illinois	Cook (6.6), Du Page (4.6)
Florida	Dade (4.4)
Hawaii	Honolulu (2.2)
Texas	Tarrant (1.9)
Massachusetts	Middlesex (1.0)

Source: Hersch, Markowitz, Maes, et al. (1992, Table 3, p. 1939).

for over one fifth of all reported cases: Cook County, Illinois (3,420); Los Angeles County, California (2,543); Harris County, Texas (1,968); Milwaukee County, Wisconsin (1,872); Kings County, New York (1,383); and San Diego County, California (1,370).

Comparison of the epidemiological record with the demographic data yielded some important conclusions (Table 2). Counties reporting measles more frequently during the decade differed from non-reporting counties in four ways: they had (i) above median populations, (ii) greater population densities, (iii) higher percentage of Blacks, and (iv) larger proportions of Hispanics. Measles incidence was most strongly correlated with population size ($r = 0.44$) and population density ($r = 0.32$), and was moderately correlated with percentage of Hispanic population ($r = 0.25$) and percentage of population below the poverty level ($r = 0.24$); the weakest correlations were observed with percentage of Black population ($r = 0.10$) (See Hersch, Markowitz, Maes, et al. (1992, pp. 1937-8).

Both large populations and high population densities (see Table 2) help measles virus transmission by creating a substantial potential pool of susceptibles and an enhanced probability of contact between infected and susceptible individuals (see Anderson and May, 1984). Griffiths (1973) has estimated that, with a 50 percent measles immunization coverage level, a population of 1 million would be required to allow sustained measles transmission. Of the 17 counties that reported measles every year during the decade, 16 were part of statistical metropolitan areas (SMAs) in which the population exceeded one million. So Hersch, Markowitz, Maes, et al. (1992), p. 1940) suggests that ". . . the presence of endemic measles among most populous counties [is] sustained by the combination of crowding, large numbers of unvaccinated preschool-aged children, and imported cases."

Table 3 analyzes the characteristics of measles cases in the large number of counties that rarely reported measles (in only one to three years during the 1980s) with the 17 counties that reported every year. The table indicates that immunization levels among preschool-aged children were less than 50 percent in some areas that frequently registered measles. Further study of measles cases in counties

TABLE 2

United States, 1980-89: County Characteristics According to Years Reporting Measles

County Characteristics	Zone Years Reporting	Ten Years Reporting
Number	1,690	17
Median Population	13,514	1,382,751
Median Population Density (Persons Per Square Mile)	25.7	1,989.2
Median Hispanic (Percent)	0.7	16.7
Median Black (Percent)	0.6	11.7
Median Below Poverty Level (Percent)	16	11.3

Source: Hersch, Markowitz, Maes, et al. (1992, Table 2, p. 1938).

that recorded measles every year showed that they occurred predominately among unvaccinated preschoolers. Counties that registered measles cases less frequently exhibited a case concentration among children of school-age who had apparently been vaccinated; this presumably reflects failed vaccinations.

Both measles incidence and the number of years in which measles was reported were both strongly correlated with percentage of Hispanic population. This may be explained by lower immunization levels among Hispanics, larger numbers of international importations from counties where Hispanics live, and differences in vaccine delivery in counties with substantial numbers of Hispanics. Study of the 1988 Los Angeles epidemic indicated that Hispanics had higher attack rates than Blacks, against suggesting that immunization coverage may be lower among Hispanics (Centers for Disease Control, 1989).

Hersch's analysis is important in illustrating the highly focal nature of measles in the United States during the 1980s. His group's findings have implications for prevention. Spatially targeted strategies are needed to improve immunization levels among preschool-aged children living in large inner-city areas.

LOCAL VARIATIONS: THE PENNSYLVANIA AMISH

Religious groups exempt from immunization laws form a small but significant risk group for measles outbreaks (Centers for disease Control, 1981; Novotny, Jennings, Doran et al., 1988). Special epidemiological interest has been shown in disease outbreaks in the Amish communities. The Amish trace their religious heritage to the Anabaptist movement of sixteenth century Switzerland, migrating to the United States in 1709 to escape religious persecution (Crowley, 1978). The first settlements were founded in Pennsylvania, and today one of the largest groups (c.20,000 out of a national total of c.100,000) lives in Lancaster County, Pennsylvania. Several characteristics of the Amish predispose them to substantial measles outbreaks. They have large families (average seven live births per family), and have intensive social interaction primarily within their own communities, churches and schools. The schools usually consist of a single room that

TABLE 3

United States, 1980-89: Measles Case Characteristics According to Years Reporting Measles

Measles Case Characteristics	Counties With Years Reporting 1-3	10
Total Measles Cases	13,595	13,235
Mean Annual Measles Incidence	1.9	4.0
Median Age of Case, Years	15.3	3.0
Unvaccinated Children, <5 Years (Percent)	14.8	54.5
Vaccinated Persons, 5-19 Years (Percent)	42.4	18.1

Source: Hersch, Markowitz, Maes, et al. (1992, Table 2, p. 1938).

accommodates 30-40 students, leading to a high degree of mixing. Many Amish parents claim religious and ethical exemption from vaccination which is allowed by the Pennsylvania school immunization laws.

Sutter and colleagues (Sutter, Markowitz, Bennetch, et al., 1991) have compared the severity of measles among both primary and secondary cases in Amish households in Lebanon County, Pennsylvania. While no significant differences was found, details of the outbreak given in Sutter contain much of geographical interest.

Because interstate and intrastate travel to other Amish communities is common and vaccinations levels are low, there is a continuing risk of measles outbreaks spreading quickly amongst this religious community. Indeed, the relative epidemiological and social isolation of the Amish, except from members of their own community, means that the group responds in a similar way to the virgin soil communities of the nineteenth century as far as measles epidemics are concerned. For the outbreak considered here, which had its origin in Lawrence County, Pennsylvania, no measles cases had been reported among residents of the county from 1970 until the outbreak began on 5 December, 1987.

The first two cases were in public school students from Neshannock school district in Lawrence. The source of the infections is unknown. The subsequent sequence of spread among the Pennsylvania Amish is reconstructed in Figure 2. On 21 March 1988, the first cases were reported among Amish students in Mercer County. One virus generation later, measles had reached Indiana and Miffin Counties. On 4 April, a group of Amish from Lebanon County visited an Amish family in Miffin County and had contact with a measles patient. Of the visiting group, eight contracted measles from the patient. The disease then spread rapidly through other Amish families in Lebanon so that, by 30 June, 130 cases among 25 Amish families in the county had occurred in some five virus generations. The crude attack rate was 39 percent, and 74 percent among those susceptible to measles. Inevitably, Amish families in other Pennsylvania counties became involved so that, by 30 July 1988, cases were reported among the Amish from 13 counties.

To control the outbreak, special measles vaccination clinics were held in locations convenient to the Amish (schools and homes). In Lebanon County, two clinics were held in May 1988, but attendance was poor and only 14 Amish were vaccinated. Measles spread from the Pennsylvania Amish to other Amish communities in Kentucky (April), Michigan (June), New York (July) and Ohio (September); see Figure 2(B). All were linked epidemilogically to the Pennsylvania outbreak (Sutter, Markowitz, Bennetch, et al., 1991, p. 13).

The outbreak illustrates several themes of disease ecology: the role of schools as diffusion poles, leading to geographically contagious spread; the susceptibility to high attack rates of isolated but linked communities once infection occurs; and the impact of vaccination levels upon the spread of measles. From the viewpoint of measles elimination in the United States, it demonstrates the potential reservoir for virus survival provided by difficult-to-reach, unvaccinated groups. These may be religious sects as discussed here, mobile communities (for example, measles within gypsy travellers as discussed in Centers for Disease Control, 1983) or inner city ethnic groups (for example, the Latinos in Los Angeles studied by Ewert, Thomas, Chun et al., 1991). Accordingly, state and local health authorities have been alerted to the need to try to increase vaccination levels and raise awareness of the consequences of infection (Centers for Disease Control, 1991).

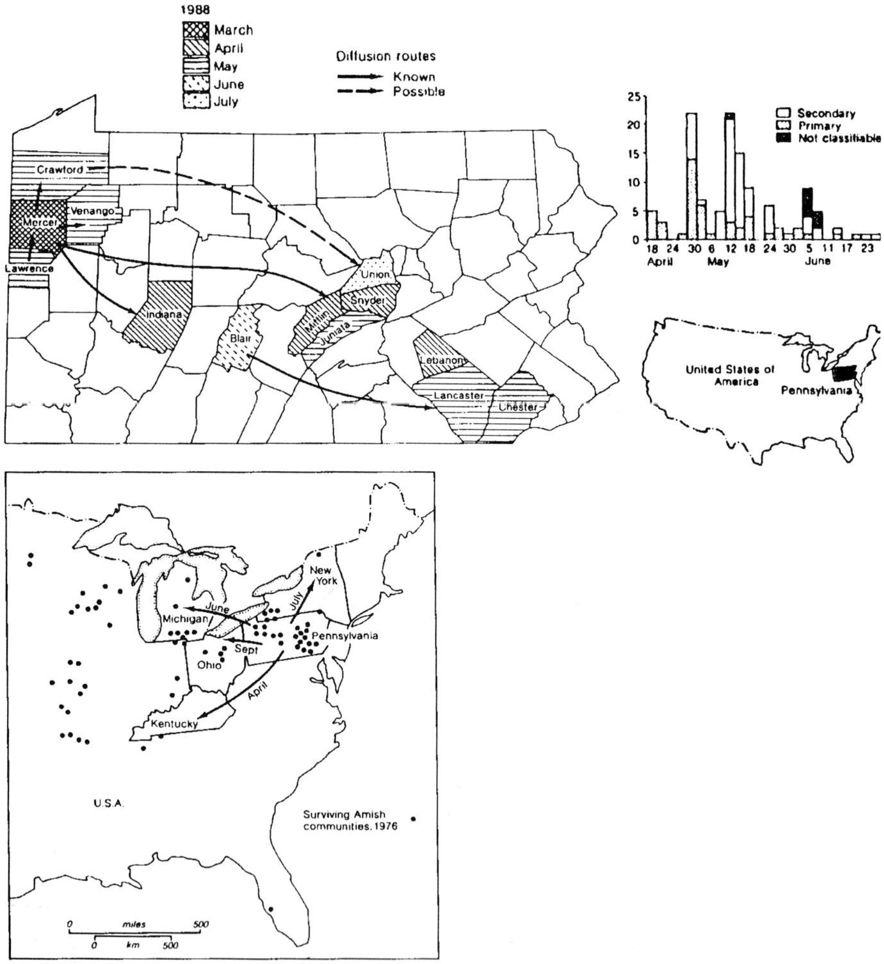

FIGURE 2. Pennsylvania measles outbreak, 1987-8: spread in the Amish community (Upper) dates of rash onset of the first Amish case in each county: vectors identify diffusion routes. Inset graph shows, by exposure status, date of rash onset among Amish patients in Lebanon County. (Lower) Spread to Amish communities of other states. Source: Cliff, Haggett and Smallman-Raynor (1992, Figure 9.15, page 240).

THE GLOBAL SETTING: MEASLES IMPORTS

International importations of measles into the United States represents a continuing problem. By the mid-1980s, annual indigenous case levels had been so reduced by vaccination that attention turned in attempts to eliminate measles, not only to the elusive residual internal sources of outbreaks discussed earlier in this section, but also to external sources of infection. Studies of measles outbreaks by the Immunization Division of the Centers for Disease Control indicated that future importation of measles into the United States will be related obviously, both to the level of measles incidence in the foreign country and to the level of exchange in terms of travel flow between that country and the United States. The specific risk in a particular part of the United States was linked to the concentration of flow from a foreign country into particular gateway cities and states where, for example, major international airports are located.

Externally-sourced measles cases can be divided into potentially preventable importations (by US citizens) and nonpreventable importations (by non-US citizens). US citizens account for the largest number of importations each year, approximately 60 percent of all importations reported from 1981 to 1984. Usually between 20 to 30 different countries, each year represent the sources for non-US citizens.

Markowitz, Tomasi, Hawkins, et al., (1988) have studied international measles importations into the United States between 1980 and 1985. In the five years, importations reported annually ranged from 89 to 121 cases, perhaps two per week overall. These comprised between 0.7 and 6.9 percent of the annual reported case total. While these measles cases were acquired in more than 70 different countries, three countries accounted for over one third: Mexico, the UK and the Philippines. Thirty-six states reported importations. However, over half were recorded in just six. California, New York, Maryland, Florida, Texas and Hawaii. Most of these importations (fluctuating annually from 71 to 84 percent of the total imported cases) did not cause secondary transmission within the US. The regional distribution of cases within the United States appears in Table 4.

The movements of infected cases into the United States may be concisely represented through flow maps. Figure 3 shows the known external sources of measles cases brought into the United States in the years 1980 and 1981. The width of the flow line from any country is proportional to its percentage share of the total number of measles cases in the United States which were attributed to external sources. In the years mapped, the immediate neighbors of the United States, Mexico and Canada, provided just over half of the externally-generated cases, while the United Kingdom and France dominated the flows emanating from Europe.

When the gateways into the United States are examined in the larger of the inset maps (lower right), then the association of the Canadian flow with the heavily populated and traditional industrial areas of the northeastern United States

TABLE 4
United States, 1980-81: Percentage of Total Foreign Arrivals in Eight US Travel Regions from 12 Principal Countries

	United States Travel Region							
Country	1	2	3	4	5	6	7	8
Canada	25.8	10.8	6.2	7.6	6.0	1.4	1.7	2.8
Mexico	0.3	4.9	1.1	1.4	0.7	4.3	0.4	0.3
UK	0..6	1.1	1.6	0.6	0.6	0.4	0.9	0.2
FRG	0.2	0.9	1.2	0.5	0.4	0.5	0.4	0.3
Japan	0.1	1.7	0.7	0.2	0.3	0.3	0.2	0.1
France	0.2	0.5	0.8	0.4	0.2	0.3	0.3	0.1
Australia	0.1	0.2	0.4	0.2	0.2	0.3	0.3	0.2
Italy	0.1	0.3	0.6	0.1	0.1	0.2	0.3	0.1
Netherlands	0.1	0.3	0.2	0.1	0.2	0.1	0.1	0.1
Venezuela	0.0	0.1	0.3	0.5	0.0	0.1	0.1	0.0
Sweden	0.1	0.2	0.3	0.1	0.2	0.1	0.1	0.0
Belgium	0.0	0.1	0.2	0.1	0.1	0.1	0.1	0.0

1 - New England; 2 - Far West; 3 - Eastern Gateway; 4 - The South; 5 - Great Lakes; 6- Frontier West; 7 - Mid Atlantic; 8 - Mountain West.
Source: Cliff, Haggett and Smallman - Raynor (1993, Table 9.10, p. 241).

and the Midwest is apparent. In the same way, the dominance of southern California as a collecting area for cases originating in Mexico is marked. This flow is made up of both illegal immigrants and seasonal agricultural and hotel workers employed in large numbers in California. It is evident that the regions within the United States, which act as the chief 'sink' areas for cases arriving from both Canada and Mexico, all have close physical proximity to the originating countries; flows collapse rapidly with increasing distance from the source areas. The main European gateway is New York, focused upon Kennedy International Airport.

CONCLUSION

Despite the massive reductions in measles morbidity in the United States over the last thirty years through vaccination, the virus continues to persist internally in small geographical pockets and through the continuing flow of measles imports from other countries. We have reviewed briefly here the distribution of the disease at state and county levels, illustrated the way it spreads within one particular Pennsylvania community, and showed how the United States is tied in to a larger international epidemiological community. Figure 4 summarizes the spatial processes involved at three different geographical scales.

Within the United States, the elimination of the measles virus will depend on the maintenance of high vaccination levels and rapid response to the remaining

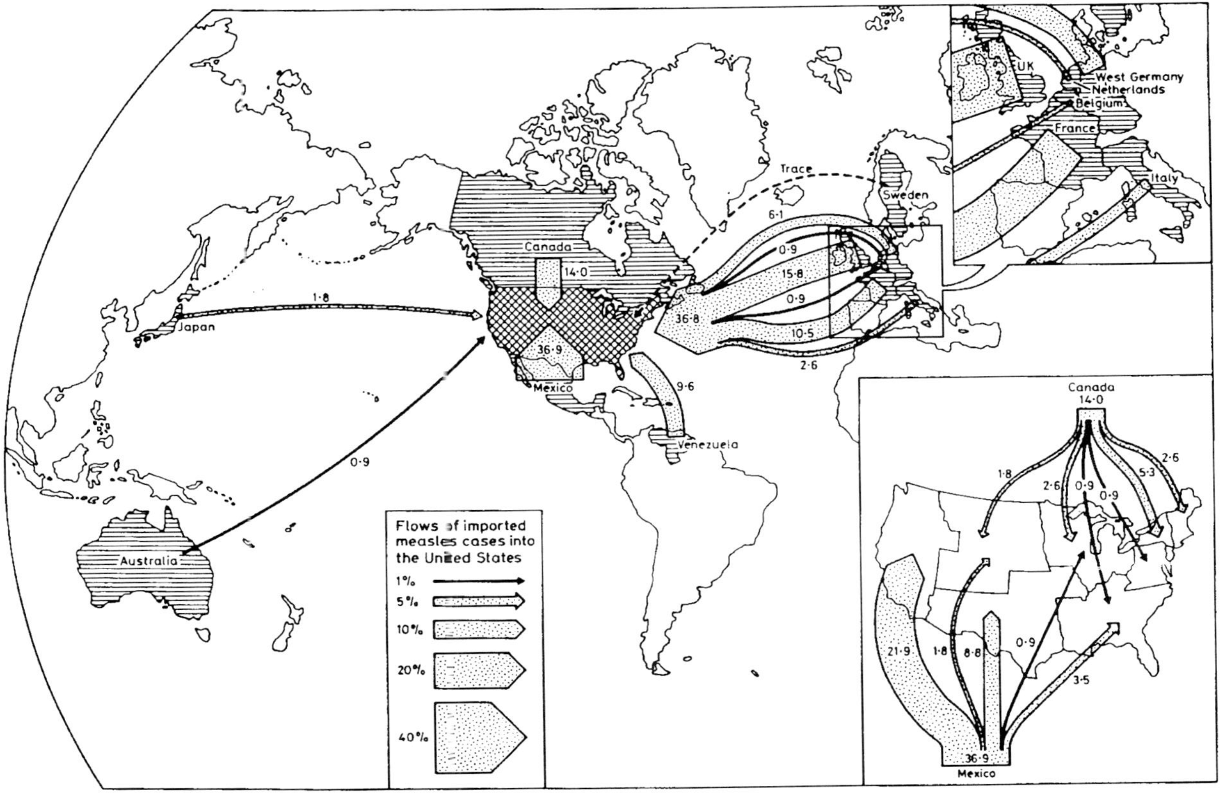

FIGURE 3. United States measles imports, 1980-5. (Main map) International flows with width of the vectors proportional to the share of all imported cases (= 100 percent). (Inset map). Regional flows within North America from the neighboring states of Canada and Mexico with vectors drawn on the same scale. Source: Cliff and Haggett [1988, Figure 3.3 (A), page 102].

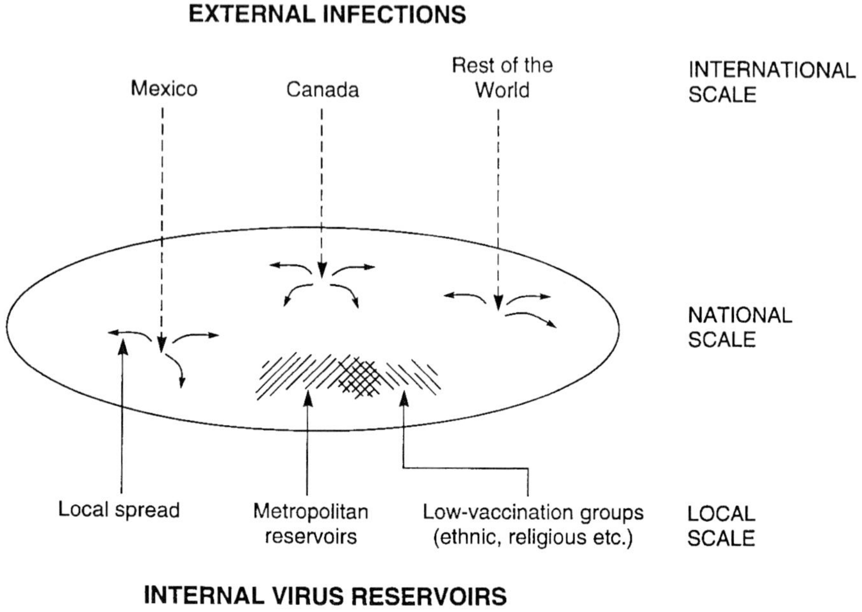

FIGURE 4. Schematic diagram of the spatial processes influencing the distribution of measles within the United States.

outbreaks in specific locations (e.g. inner parts of large metropolitan areas) and among specific groups (e.g. college freshmen). Reductions in invasions of the virus from outside sources will depend on the success of measles elimination programs such as the WHO's Expanded Program on Immunization. Given the open nature of the United States to passenger movements, it would seem likely to be several more decades before the ultimate goal of measles eradication from the United States can be achieved.

REFERENCES

Anderson, R.M. and R.M. May, 1984. 'Spatial, temporal, and genetic heterogeneity in host populations and the design of immunization programmes.' IMA Journal of Mathematics and Applied Medical Biology, 1, 233-66.

Centers for Disease Control. 1981. 'Measles among children with religious exemptions to vaccination - Massachusetts, Ohio.' Morbidity and Mortality Weekly Report, 30, 550-6.

Centers for Disease Control. 1989. 'Measles - Los Angeles County, California,

1988.' Morbidity and Mortality Weekly Report, 38, 49-52, 57.
Centers for Disease Control. 1991. 'Outbreaks of rubella among the Amish - United States, 1991.' Morbidity and Mortality Weekly Report, 40, 264-5.
Cliff, A.D. and P. Haggett. 1988. Atlas of Disease Distributions: Analytic Approaches to Epidemiological Data. Oxford: Blackwell Reference.
Cliff, A.D., Haggett, P. and M. Smallman-Raynor. 1993. Measles. An Historical Geography of a major Human Viral Disease. From Global Expansion to Local Retreat, 1840-1990. Oxford: Blackwell Reference.
Crowley, W.K. 1978. 'Old order Amish settlement: diffusion and growth.' Annals of the Association of American Geographers, 68, 249-64.
Ewert, D.P., Thomas, J.C., Chun, L.Y., Enguidanos, R.C. and S.H. Waterman. 1991. 'Measles vaccination coverage among Latino children aged 12 to 59 months in Los Angeles County: a household survey.' American Journal of Public Health, 81, 1057-9.
Griffiths, D.A. 1973. 'The effect of measles vaccination on the incidence of measles in the community.' Journal of the Royal Statistical Society A, 136, 441-9.
Hersh, B.S., Markowitz, L.E., Maes, E.F., Funkhouser, A.W., Baughman, A.L., Sirotkin, B.I. and S.C. Hadler. 1992. 'The geographic distribution of measles in the United States, 1980 through 1989.' Journal of the American Medical Association, 267, 1936-41.
Markowitz, L.E., Tomasi, A., Hawkins, C.E., Preblud, S.R., Orenstein, W.A. and A.R. Hinman. 1988. 'International measles importations United States, 1980-1985.' International Journal of Epidemiology, 17, 187-92.
Novotny, T., Jennings, C.E., Doran, M., March, C.R., Hopkins, R.S., Wassilak, S.G.F. and L.E. Narkowitz. 1988. 'Measles outbreaks in religious groups exempt from immunization laws.' Public Health Reports, 103, 49-54.
Sutter, R.W., Markowitz, S.E., Bennetch, J.M., Morris, W., Zell, E.R. and S.R. Preblud. 1991. 'Measles among the Amish: a comparative study in primary and secondary cases in households.' Journal of Infectious Diseases, 163, 12-6.

Environmental Contaminants, Ecosystems and Human Health. Edited by S.K. Majumdar, E.W. Miller, and Fred J. Brenner.

Chapter Twenty-Nine

THE SPACES OF INFECTIOUS CONTAMINANTS: AIDS On Its Hierarchical Backcloths

PETER GOULD
Evan Pugh Professor of Geography
306 Walker Building
Penn State University
University Park, PA 16802

INTRODUCTION

Contaminants, things that pollute and infect in ways that are detrimental to life, move through and over physical-natural structures. We only have to think of chemical, radioactive, electromagnetic, and particle movements through air, water and soil, with their effects often amplified by biological chains (1). Chernobyl's radionuclides carried by air (2), and DDT carried by water, are only two out of thousands of examples, and the effects of both were concentrated through food chains to levels dangerous to human and animal life. But contaminants are also carried over structures created by human beings themselves (3), through sewers and irrigation ditches, and via roads, railways, ships, and airlines. These structures form the 'spaces' shaping the movements of contaminants, and they include the socio-economic patterning of the human condition itself, the spatial arrangements of rich and poor, advantaged and disadvantaged, and how these are connected by human movements (4).

Diseases, virtually by definition, infect people in ways that are detrimental to life. They move through human (and animal) populations in ways that intimately reflect the structures of connections shaping the 'spaces' that are often far removed from the flat, and apparently smooth and homogenous space of the geographic map. Few things, diseases included, diffuse across geographic space like heat from a Bunsen burner through a copper sheet, although when the world was simpler and most travel was by foot, horse and camel, scourges like plague and cholera did appear to break out of source areas in ways that can be plotted as roughly circular isochrones, or 'time-waves,' on the traditional map.

THE STRUCTURE OF THE 'SPACE'

It follows that if we are to understand how diseases move and 'contaminate' people and animals, we must discover and explicate the structures on which they exist and move. It is not enough to think, in now-traditional epidemiological fashion, simply in the temporal domain, along the single dimension of time, focusing exclusively on the historian's 'when' and 'how many' questions (5). The answers to these, if they have any plausibility at all, are quite literally useless. In other words, nothing can be done with such predictions along the time horizon to intervene in an epidemic, for single numbers provide no guidance for intervention strategies. Only the answers to the geographer's 'where' questions have any genuine utility, for they provide means of intervening by specifying where interventions can take *place*, perhaps by intensified vaccination programs, extensions and intensification of healthcare delivery systems, and educational programs to enhance protective behaviors (6). This means we must think about the diffusion of diseases in *spatiotemporal* terms, recording cases by location, as well as in time, so we can predict the next *maps* (7), rather than useless aggregate numbers that no one can do anything with. Every scrap of information, particularly new cases, must be recorded in (x,y,t) coordinates, and our thinking must move to the structures that carry a particular disease from one person to another.

What are these structures and spaces? They are both natural and human, and clearly their relative importance will depend on the particular disease involved. In order to move from person to person, some diseases require intermediate carriers or vectors. One only has to think of malaria, yellow fever, bubonic plague, Lyme disease, etc. being carried by mosquitoes, ticks, fleas and other insects (8). Or consider the role of water in the movement of cholera, dysentery, and schizosomiasis via rivers, streams, sewers, wells, and so on. Other diseases are transmitted directly from one human to another, either by direct touch (*con-tangere*, with or by touch, hence *contagious*), or by being in very close proximity. Some require long and intimate contact, for example, Hansen's disease (leprosy); others can be transmitted in the closeness of a breath, for example tuberculosis. The focus here is upon the human immunodeficiency virus (HIV), which is used as a 'tracer' to exemplify why it is often literally vital that we think about the transmission of contaminant diseases in spatial and structural, i.e. geographic, terms.

THE SPREAD OF THE HIV

All the evidence points to Africa as the origin of the HIV-the HIV1 in East and Central Africa, and the HIV2 in West Africa (1). At some point, possibly only 50 years ago, the viruses moved in slightly mutated forms out of chimpanzee and mangabey monkey populations into human hosts. In the last ten years, both varieties have demonstrated remarkable abilities to mutate, a characteristic that makes effective vaccines even more difficult to produce. The virus is spread by intimate sexual contact, by sharing infected intravenous needles, by contaminated blood transfusions, and by infected hemophiliac injections. The question is: if, as it appears likely, the crossover from animals to humans occurred 50-100 years ago, how did the HIV 'suddenly' appear as a 'new virus' that will infect 40 million people worldwide by the year 2000?

This is not an easy question to respond to, let alone answer. Indeed, it will never be answered with absolute certainty. But this does not mean that we have to substitute mere and totally unfounded speculation. If we cannot know all the details (for example, it is impossible to know *exactly* when the virus moves from one person to another), we certainly know a great deal in more general terms about the spread of the virus across the globe. But that phrase, 'in more general terms,' takes on here specific geographical connotations. Different structures, different 'spaces,' guiding transmission of the virus exist at different geographic scales, from the global through the continental, national, regional and so on, right down to the sociographic networks of sexual encounters and needle sharing in a neighborhood or house.

For all practical purposes, and excluding viral presence in laboratory animals, the HIV requires human beings to exist, and connections between human beings to be transmitted as a contaminating traffic. If human beings are considered elements of a set, and sexual and other contact are considered as relations in a strict mathematical sense, then the connected elements form, quite literally, *structures* (9). We shall call these *backcloths,* and it is obvious that these can exist in an hierarchy of different levels of aggregation, from the individual intimacies of sexual contact to the aggregations of such people making up counties, cities, regions, nations and so on, up to the entire globe. These sets of people, at whatever level in the hierarchy of aggregation, are connected by human spatial interaction, i.e. flows of people themselves. Since HIV exists in, and moves with human beings, we ought to be able to say something sensible and important about the spread of HIV, and its manifestation as AIDS, if we focus upon such flows of people at a variety of geographic scales, i.e. levels in the hierarchy of aggregation. The fact that virtually no epidemiologists to date (1994), roughly 13-14 *years* into the pandemic, have actually done this, simply demonstrates the poverty of spatial thinking and analysis in modern epidemiology (10). Clearly, in a short exposition such as this, we cannot cover all scales everywhere. What I shall do is select six levels in the hierarchy of backcloths to indicate how such thoroughly geographic perspectives can illuminate the way HIV has spread, and how the various methodologies involved can make reasonable short term forecasts in *space* as well as time.

A GLOBETROTTING HIV

It is assumed today that relatively large-scale investments in transport infrastructure in Africa after World War II shrunk the map in 'time-space,' lowered costs and intensified human movements, and played a major part in the huge tides of rural-to-urban migration producing the still-exploding cities. Such spatial changes, resulting in easier movement and great concentration of people, provide ideal conditions for the proliferation and spread of a virus found in previously isolated and relatively immobile populations. So how did it happen that the first warning signal was received in Los Angeles in 1981 (over a year before the HIV was identified in the Pasteur Institute in Paris), as the appearance in several homosexual men of a pneumonia and cancer so rare that they were seldom seen by doctors in a lifetime of medical practice (11)? The answer, it seems reasonably clear, lies in the simple fact that few places today are more than 36 hours away from any other by air.

HIV, of course, is not the first disease to be carried by people traveling by air; warning cards are regularly given to passengers on flights from cholera-stricken areas, and attempts have been made to model the global movement of new strains of influenza in an attempt to provide sufficient warning time for inoculation programs, particularly for high-risk people (the elderly), and those who are socially important (such as firemen, policemen, or members of the Communist Party in the former Soviet Union). One model, developed by Russian and American epidemiologists (12), was based on air passenger flows between the world's 51 largest cities, and it indicated that the times at which a Hong Kong variety of influenza arrived were related to the strength of such flows connecting the set of cities into a global backcloth. The model was adopted by two French epidemiologists in Paris (13), and used in a simulation mode to demonstrate the likelihood of an African origin for HIV. Using the connectivity matrix of air flows between the 51 major cities, the virus was artificially 'injected' into the geographic backcloth and allowed to spread in a probabilistic way across the globe. When the starting points were such cities as San Francisco, Casablanca, Havana, etc., the maps simulated at successive time periods bore no relation to the global distribution of AIDS as it was then known. Only when the simulation started at Kinshasha, Zaïre, did the resulting sequence of maps resemble the pattern of global infection, providing one more piece of evidence that Central Africa was probably the origin of HIV1.

FROM GLOBE TO CONTINENT

Africa today is a continent of catastrophe. With HIV1 and HIV2 as sources in both east and west, the viruses have spread with such intensity that whole groups and, indeed, whole regions are in the process of being wiped out. Soaring AIDS cases and rates of HIV infection are roughly equal between men and women, reminding us that the capital H of HIV stands for *human,* not homosexual. The great majority of transmissions are via heterosexual contact, and ignorance and poverty are the underlying forces for the lack of protective behaviors, such as condom use on the part of most people. Few dare to guess at what the ultimate impact will be in tens of millions of lives lost.

In Africa, the spread of HIV does not appear to be especially concentrated in any particular group (unlike the situation in North America and Europe at the moment), and the major spreading mechanisms at this continental scale seem to be two-air and truck travel. Air travel by the elite groups of Africa has already ensured that the HIV1 of East Africa is thoroughly embedded in the urban populations of West Africa, as high level visitors come into sexual contact with *les femmes libres* in every city. Even by 1987, 24 percent of Abidjan's prostitutes were seropositive for HIV1, and two years later the virus had overtaken HIV2 in over half of them. By 1993, 49.5 percent were infected by HIV1 alone. Tragically, one can see it moving in the Côte d'Ivoire, through society and over geographic space, with ever greater intensity: by 1993, it was in 11.4 percent of first time male blood donors, 14.0 percent of all pregnant women, and 59.1 percent of prisoners (14). In major towns 'up country,' like Bouake and Korhogo, both with direct flights from Abidjan, rates went from zero to 18.0 percent in two years. Every air-connected city in Africa becomes a regional epicenter, a point source for the HIV to be carried by road down the regional hierarchy by truck, and from these local centers to the rural areas surrounding them.

Truckers were recognized early in Africa as major carriers of HIV, and the major transcontinental and national routes stand out on any map of HIV/AIDS, the major towns and trucks stops looking like infected beads on a string. From Djibouti to Cape Town, rates of HIV infection in 'bar girls' at truck stops often exceed 90 percent. By 1992, women in Tanzania coming for the first time to prenatal clinics were 17.9 percent infected, with rates of over 50 percent in 'high risk' individuals, but rural regions through which major truck routes passed were already reporting rates as high as 11.3 percent in all adults (15). The same tragic figures characterize the great north-south alignment all the way to South Africa. At the 'end of the line,' South Africa reported relatively low rates through the 1980's, but the great movements of truck traffic and migrant workers have proved isolation is impossible. In KwaZulu-Natal, with Durban as a major terminal, rates in the *general* African population have gone from 1.61 percent in 1990, to 9.69 percent in 1993, and other major regions are experiencing similar rates of increase. In Swaziland, HIV prevalence in all adults was 12.5 percent in 1991, and is expected to be over 27 percent by 1996, the rapid increase being due to high mobility between rural and urban areas, transit truck traffic, refugees from Mozambique, and the low status of women (16).

THE USA: DISTANCE IS NO BARRIER

It is at the national level that we can begin to see some of the details of the way the HIV, and its manifestations as AIDS, spreads in space and time. There is strong evidence that the virus was brought to major homosexual communities in the United States by the multiple sexual contacts of 'Patient Zero,' a young airline steward who engaged in literally hundreds of sexual encounters each year.

Since the median time between first infection and the onset of the cancers and pneumonias characterizing the immunodeficiency syndrome is about ten years, we can assume the virus arrived in these communities in the mid- or late-1970s. By 1982 (Figure 1), AIDS cases were already accumulating in major metropolitan areas-New York, Miami, Houston, San Francisco, Los Angeles, etc.-demonstrating almost classical hierarchical diffusion of the virus, a spreading effect strongly controlled by relations between cities in the urban hierarchy (2).

Hierarchical diffusion effects were still strong in 1984 (Figure 2), but already there is clear evidence for spatially contagious diffusion, as each urban area becomes an epicenter for its surrounding region to produce the typical 'wine stain on the tablecloth' (17). Such diffusion, described by classical distance decay effects, appears to be controlled to a high degree by the journey-to-work within the commuter field. Huge numbers of people coverage daily upon the center, only to return each evening to the surrounding communities, and such flows provide a vigorous spread mechanism for the virus. We shall examine the important details underlying such spatially contagious effects at the regional scale (below). Two years later (Figure 3), the geographic distribution of the approximately 50,000 cases begins to reflect clearly the human structuring of the geographic space by road transport, in addition to the airlines. 'Fingers' from the San Francisco and Los Angeles epicenters reach out along major interstate highways

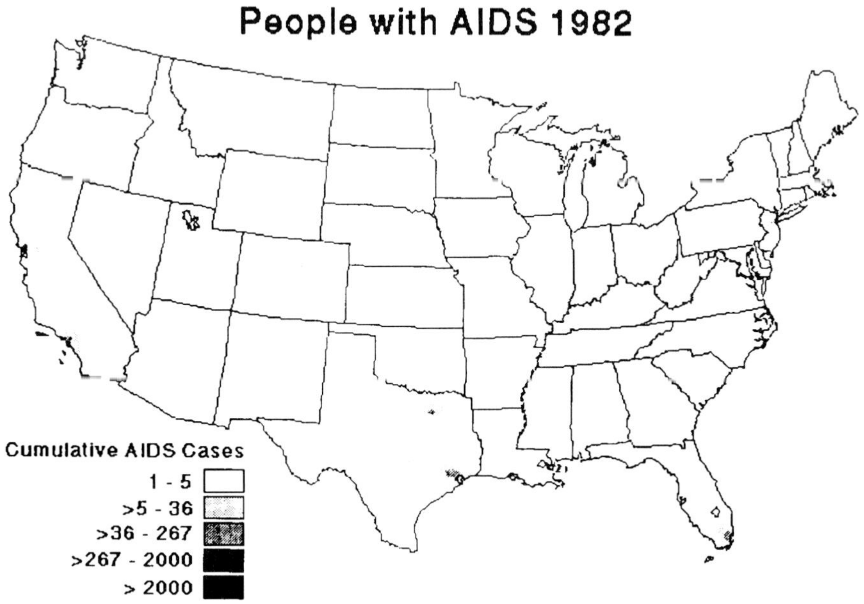

FIGURE 1. Major metropolitan centers indicate the initial preponderance of hierarchical diffusion over the airline network.

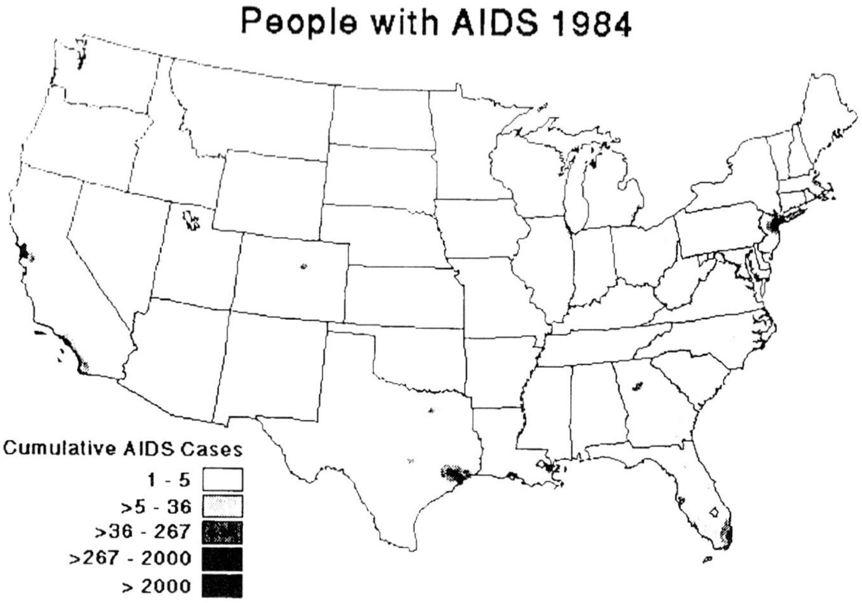

FIGURE 2. Spatially contagious diffusion around major regional epicenters.

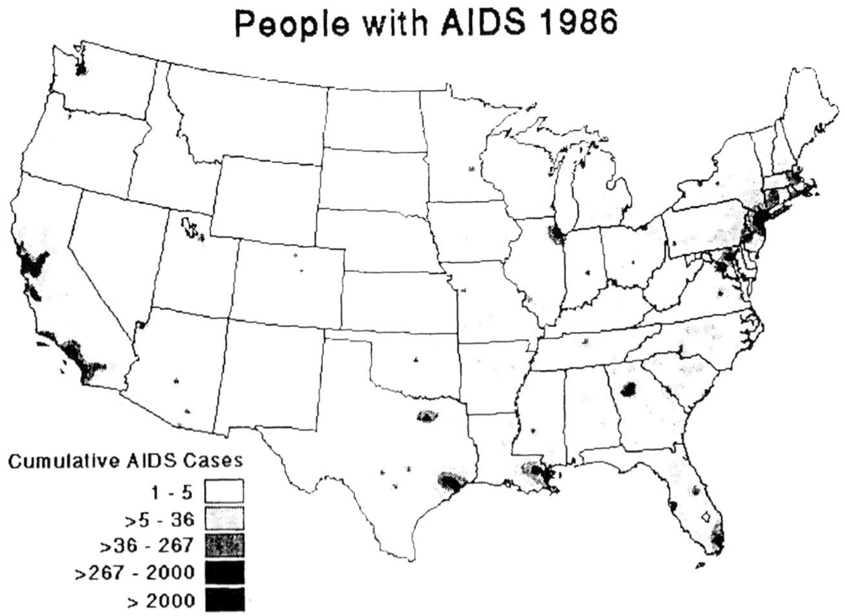

FIGURE 3. The emergence of major transportation alignments, particularly the interstate system.

(I-5 and Route 101), and on the east coast the 'I-95 effect' is already prominent, from Boston to Washington as a high ridge of infections, with 'beads on the string'-Richmond, Fayetteville, Florence, Savannah, Jacksonville-reaching south to another solid alignment reaching north from Miami.

By 1988 (Figure 4), the structuring of the national space by the interstate and other major routes is clear: a ridge of AIDS cases has joined up the west coast from San Diego to Seattle; the I-76-80 alignment has coalesced almost to Chicago, and from this the fingers of I-71 in Ohio, and I-55 in Illinois, stretch south to Cincinnati and St. Louis respectively. Such alignments are hardly surprising to geographers, even if they have not been noticed by spatially-oblivious epidemiologists. Major roads and airlines are built to facilitate human interaction, and since the HIV moves with, and is transmitted by people, the geographic pattern of infection obviously reflects such movements. By 1990 (Figure 5), there is strong interstitial filling between these major nodes and ridges of infected people, as the symptoms of the virus appear diffusing into the rural areas of America. It is worth noting that this spatio-temporal sequence provides an enormous wealth of scientific information, and it has been used to forecast the next maps with spatial adaptive filtering (18).

By 1990, approximately 160,000 AIDS cases had accumulated in the United States, and by the end of 1993, augmented by new definitions of conversion from infection to the opportunistic diseases characteristic of AIDS, the country passed

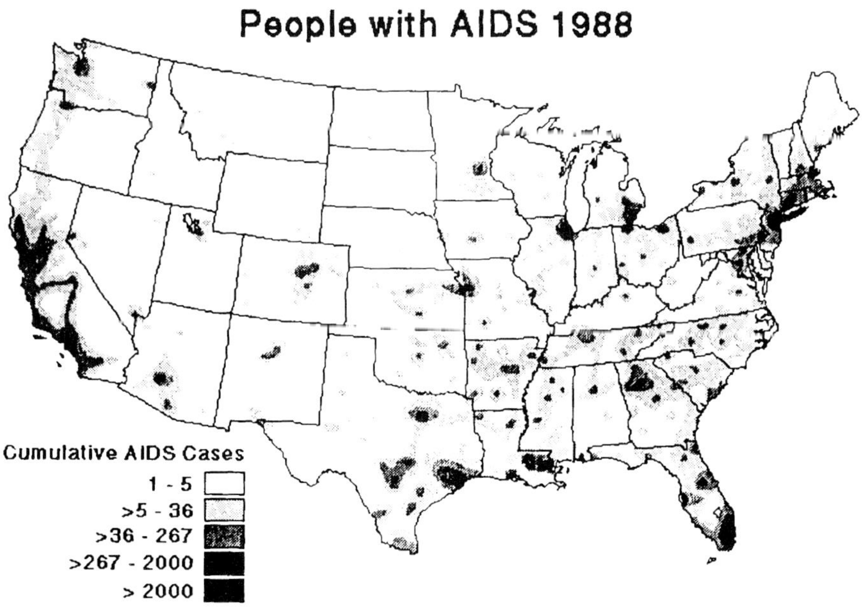

FIGURE 4. Major structures of human interaction appearing as ridges of infection.

the 300,000 mark, with possibly one million more infected. If a three-dimensional map of the 1993 distribution were made, and one hundredth of an inch (0.25 mm) were used in the vertical direction to represent each AIDS case, then the major centers of infection would appear as sharp exponential 'spikes' on the surface, the one for New York over 50 feet (15.4 meters) high. We shall examine this particular peak on the 'AIDS surface' in some detail, but our next change in geographic scale takes us to the state level, where regional details come more clearly into focus.

PENNSYLVANIA: FROM THE URBAN TO THE RURAL

There is no 'typical' state, but in focusing upon the geographic scale of Pennsylvania, we delimit a 'region' with a mixture of large metropolitan centers-Philadelphia, Pittsburgh, Harrisburg-and highly rural areas. Such a diverse human geography provides a more detailed look at the mechanisms of hierarchical and spatially contagious diffusion, distinctions made for clarifying intellectual convenience, for both operate simultaneously in any actual epidemic. Initially, and as we saw at the national level, hierarchical diffusion appears to dominate in the early years, but it continues to reinforce the epicenters later as spatially contagious diffusion begins to fill in the interstitial spaces on the map.

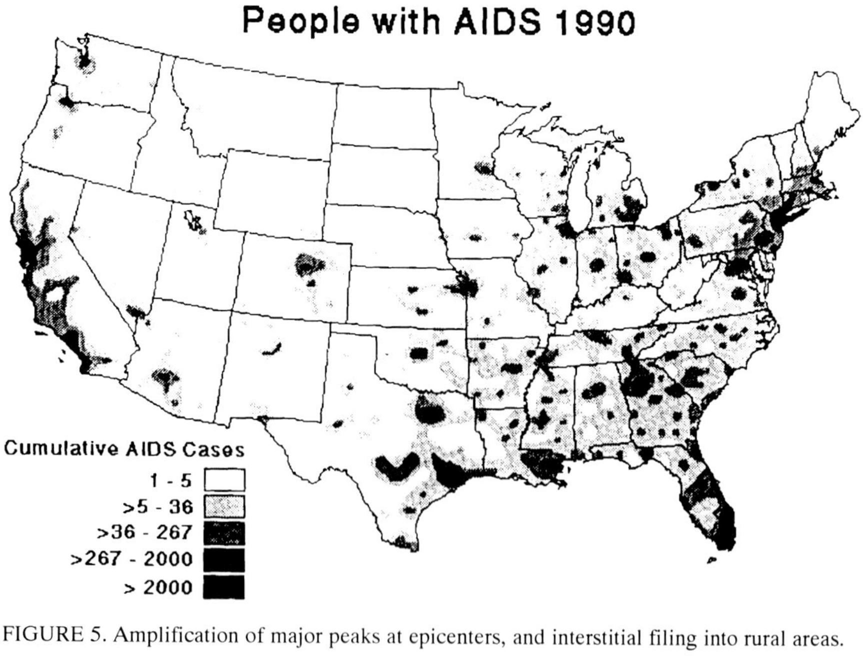

FIGURE 5. Amplification of major peaks at epicenters, and interstitial filing into rural areas.

The first cases of AIDS in Pennsylvania were 'post-reported' to 1981 in Pittsburgh and Philadelphia (Figure 6), but by 1982 there is already the beginning of spatially contagious diffusion being channeled by the major arteries, particularly I-76, the Pennsylvania Turnpike, and its northward extensions. By 1984, the

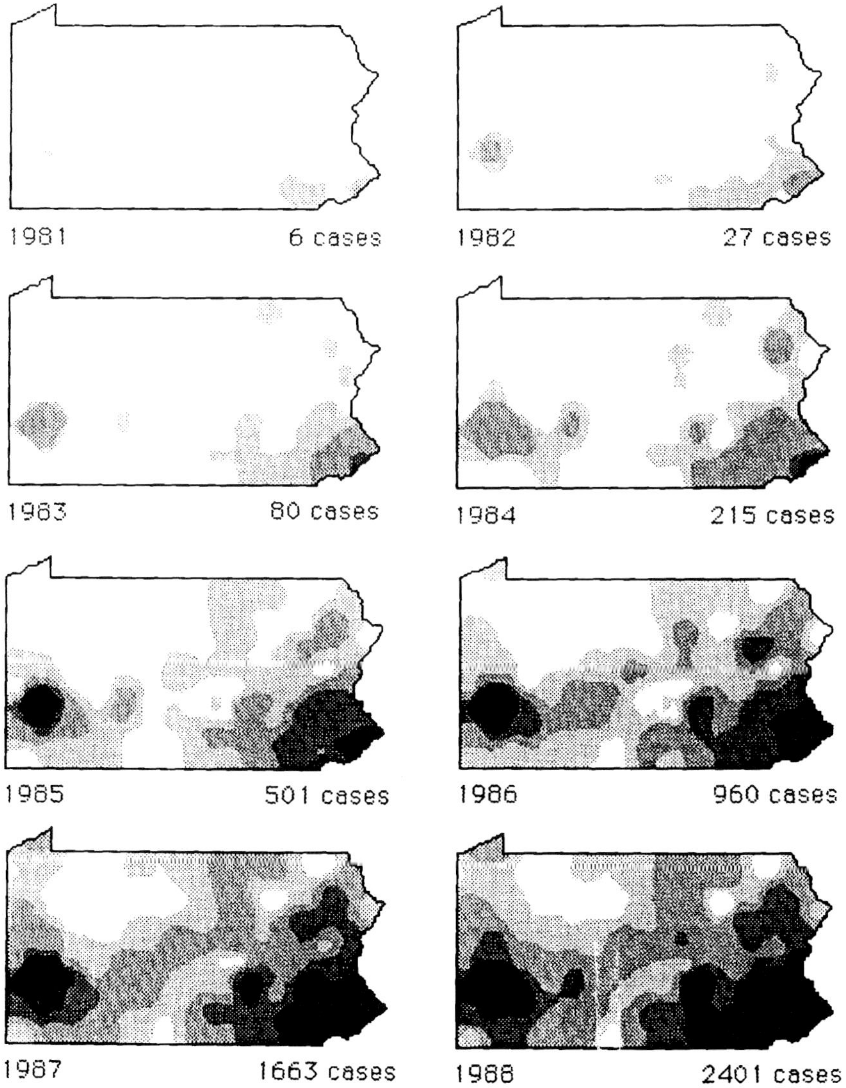

FIGURE 6. The diffusion of AIDS in Pennsylvania, 1981-1988, displaying at the regional level the details of hierarchical and spatially contagious diffusion. The initial east-west movement (I-76), and the later northwest-southeast diffusion parallel to the ridge-and-valley alignment, are prominent.

Poconos in the northeast are another source area: as weekend vacation sites for the homosexual community, the HIV has entered the local population to appear years later as early AIDS cases.

Although it is impossible to document at the moment, due to proper ethical constraints on confidentiality, it appears highly likely that some of the HIV transmission is carried by long haul truckers, a group whose mores of masculinity tend to disparage condom use. Two years later, another major transportation alignment appears, as Johnstown, Altoona, State College and Williamsport appear along the major southwest-northeast axis following the alignment of the ridges and valleys bisecting the Commonwealth. Later they join into a continuous ribbon, one example of a process of seepage into the rural areas that gradually fills in the map. By 1988 only the very sparsely settled areas of Warren, Elk, Jefferson and McKean Counties are AIDS free, although these too have infected people in the following years.

In these details of HIV/AIDS diffusion at the state level, one can see that the virus is not, and cannot, be contained in the major metropolitan centers. At the time of writing (1994), some of the highest rates of increase of AIDS cases are characteristic of rural areas (19). Transmission may be slower here, but is likely that a much higher proportion of it will be by heterosexual relationships, particularly in younger people. Nevertheless, such rural transmissions will be continually reinforced by the intense regional epicenters; by 1990, most of the counties with the highest increases of AIDS cases were adjacent to urban areas. It is time to change geographic scale once again, and examine one of the most intense nodes of infection in the United States and, indeed, in the world.

AIDS IN 'COMMUTER SPACE'

The New York Metropolitan Region can be defined as the 24 counties and boroughs within roughly 50 miles (80 kilometers) of Manhattan (Figure 7). The 18 million people form the pool for 99 percent of the daily commuter flows (20), and it is these that strongly structure the conventional geographic map to form what we might legitimately call a 'commuter space' (21). It is this structured space of human movement that strongly conditions the transmission of the HIV, transmission that ultimately appears at the national level as the spatially contagious diffusion of AIDS cases seeping out from this major epicenter.

Without all the mathematical details and notations (22), imagine a square matrix 24 x 24 recording the commuter flows between each pair of origins and destinations. For example, the number of daily commuters from Bergen County in New Jersey to Manhattan, or those commuting from Queens to Westchester, and so on for the remaining 576 possibilities. If we take the row sums, we have the total number of people *from* a particular county, say Fairfield in Connecticut, *to* all others, including itself. If the elements of each row are divided by their row sums, we have the proportion, or probability of commuters in each county going to all the others. In other words, we could take any commuter at random in, say, Somerset County, New Jersey, and ask "What is the probability that this person will commute daily to any other county or borough in the region?" We now have a representation of the commuting 'system' as a Markov process (23).

Recalling that the HIV is carried by people to people, we can imagine the virus suddenly appearing in some county or borough. At the next time interval, the virus will be 'smeared' across the region according to the distribution of probabilities from the county where it appeared to all the others. At the next time interval, these 'probablistic smears' will be redistributed in turn, and so on, until an equilibrium distribution of the virus across the region is achieved. If this model appears fanciful, recall that literally millions of people are commuting and interacting every day in all sorts of ways. For example, 70 percent of the customers of Manhattan prostitutes are from the region, and many prostitutes on crack cocaine are willing to engage in unprotected sex for as little as $1.00 (24). What becomes very clear in such a process is that no matter where the HIV might start, it quickly gets sucked into Manhattan, the daily target for millions of commuters, where it proliferates and gets spewed out over the region by the same process that focused it.

From the equilibrium values of each county or borough, it is possible to construct a commuting index which measures a county's commuting access to the entire regional system. Such commuting indices predict, to a very high degree

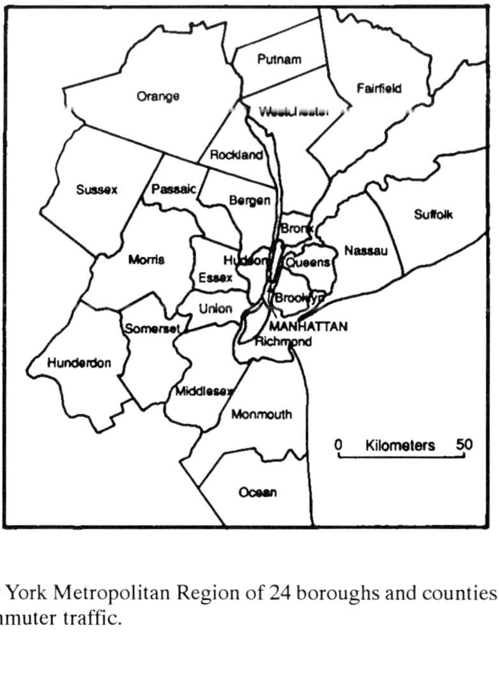

FIGURE 7. The New York Metropolitan Region of 24 boroughs and counties generating 99 percent of the daily commuter traffic.

(r = 0.85-0.92), the *rates* of AIDS throughout the pandemic (Figure 8). Rates of AIDS literally depend upon the human geographic structure of the 'commuting space' at this geographic scale. Yet within this huge metropolitan region, there is clearly another human geographic pattern which provides yet another backcloth for the HIV. The HIV and AIDS starts in the homosexual and IV drug communities, but it quickly moves into the general, heterosexual population via needle sharing and bisexual intercourse. Many of the infected are poor, disadvantaged, and often black and Hispanic. It is time to change our geographic 'lens' one more time and consider a single borough in the system-the Bronx.

WAVES OF DESPAIR: THE BRONX

The borough of the Bronx, one element in the New York region, is an impacted area of human poverty and misery, with a physical landscape that in places resembles the cities bombed during the Second World War. This is not the place to tell again the history of perfidy, and political and social irresponsibility that led to such conditions (25), but simply to note the medical and epidemiological consequences in geographical terms. During the 1950s and 60s, the south-central Bronx became the home of many poor people living in overcrowded conditions. In the 1970s, as a result of deliberate "planned shrinkage," many firc and police units were disbanded, essentially leaving the Bronx to burn. The result was a huge displacement of poor people from the heart of the Bronx to other parts of the city where they could obtain some sort of housing (Figure 9). Given the sense of helplessness in their lives, it is hardly surprising that drug abuse was common,

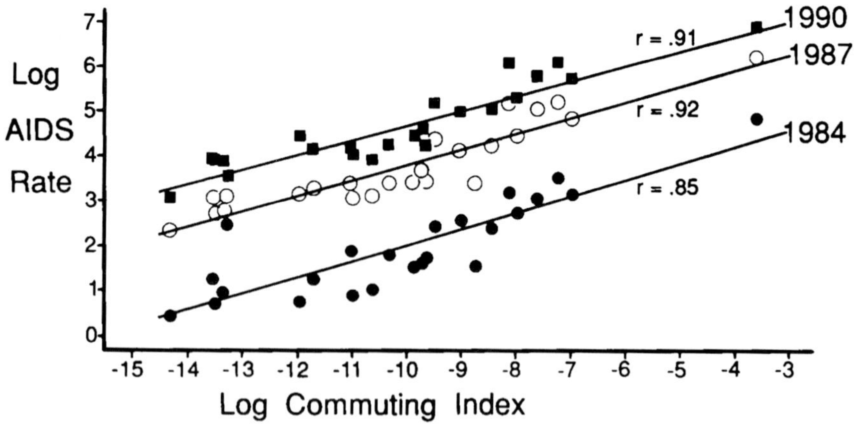

FIGURE 8. Rates of AIDS in the 24 boroughs and counties of the New York Metropolitan Region related to indices of commuting intensity.

and drug-related deaths provide a tragic index of the human and physical 'hollowing out' of the area.

Unfortunately, such an explosive, outward, wave-like movement reflected more than just drugs. Needle sharing and prostitution were common, and the HIV moved through such a population with great speed. Map sequences showing deaths from AIDS (Figure 10) show precisely the same pattern of a hollowed out core, "Burnout Valley," marked by ridges of death on either side, a pattern already visible in 1984, one that amplified by 1988, and continues to this day (1994). And now, riding on the wave of displaced people, many of them abusing drugs, and infected with HIV and other sexually transmitted diseases (STDs), there is a resurgence of tuberculosis (26). Many of the TB cases are totally resistant to any mix of antibiotics, due to mutations in the bacillus during the 1980s, when Federal

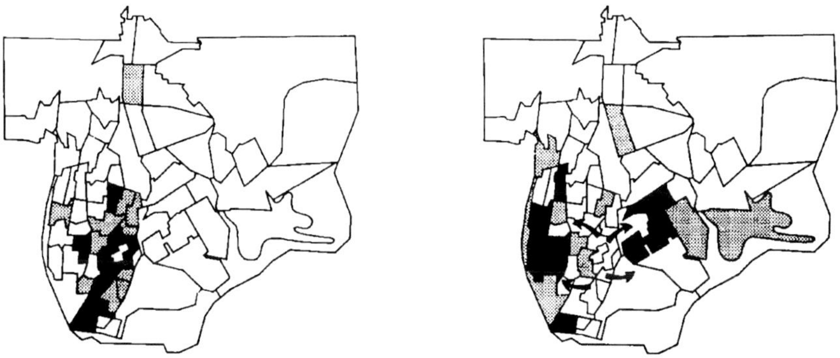

FIGURE 9. Drug-related deaths in the Bronx, indicating the hollowing out effect of the burnout driving people to seek new homes.

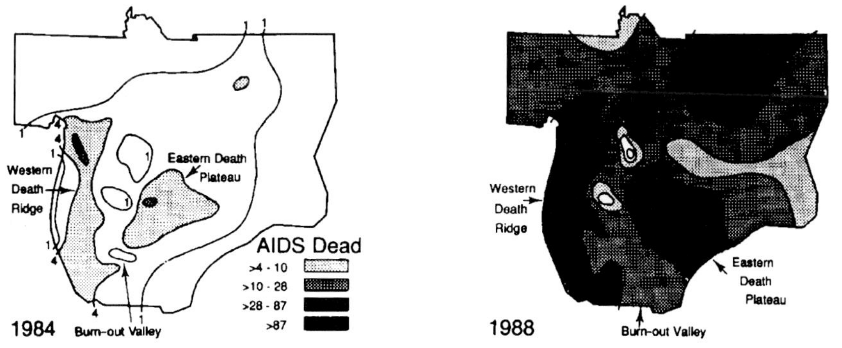

FIGURE 10. Deaths from AIDS in the Bronx, showing clearly the effects of "Burnout Valley" surrounded by high ridges of death on either side.

funds for the TB monitoring program were cut from $40 to $1 million. Tuberculosis is normally held in check by a person's immune system, but in the presence of AIDS many more active cases appear, providing still more sources of infection for the general population (27). At this geographic scale, where the sociographic connections themselves almost come into view, we have a tragic case of the human backcloth itself fragmenting and changing its geographic configuration, carrying with it the deadly mix of drugs, STDs, HIV and TB.

HUMAN GEOGRAPHIC STRUCTURES AND SPACES

All contaminating diseases in human populations require certain structures to be in place in order to exist and be transmitted. All such structures, whether cities connected by air, counties connected by commuter flows, or people joined by sexual relations, have direct geographic expression, from the scale of the globe, down a structural hierarchy to national and regional backcloths, to the social networks of individual lives. At each scale, at each level of aggregation in the hierarchical backcloth, different mechanisms may be at work diffusing diseases, including the HIV, over space and time. Every attempt to intervene constitutes a deliberate attempt to change the backcloth at some scale. The elimination of smallpox, for example, was nothing more (but nothing less!), than an effort to destroy completely the backcloth of susceptibles with a worldwide vaccination program so that the virus had no place to go, no structural backcloth on which it could exit. Measles, also requiring human hosts, is the next target of the World Health Organization. Every attempt to quarantine is equally nothing less than an attempt to disconnect, to fragment a backcloth made up of susceptible people, so that transmission across the gaps cannot occur. From the cordon sanitaire at a national scale, to clean needle programs in a city, to the latex condom at the individual level, breaking connections is the key to preventing the transmission of HIV traffic. The policy implications are immediate, particularly in view of the fact that vaccines and cures appear unlikely in the next decade. Transmission of HIV must be stopped, and in democratic societies this can only be accomplished by education, particularly of the young experiencing the first surges of sexual awakening. Approximately 20 percent of AIDS cases are under 29 years of age, implying infection occurred during the late teen or early adult years. High risk occupations must also be identified. For example, the occupation 'trucker' is appearing with increasing frequency in individual case records, suggesting intensive educational campaigns for a group of people reluctant to adopt structure-breaking condoms. In 1993, for the first time, infections in women via heterosexual transmission exceeded those of IV drug use.

It can be done: Scandinavia, with its exemplary traditions of sexual education in the schools, has been renowned for decades for cutting its gonorrhea rate to practically zero, and there are signs that HIV transmission has been brought down to the same levels. Breaking and destroying the structures on which HIV exists and travels is not difficult to understand, but education-purposeful, focused, and intense education—is all we have at the moment to produce life-saving backcloth changes at all levels in the structural and geographic hierarchy.

NOTES

1. Extensive documentation for the evidence for such statements may be found in my *The Slow Plague,* reference 5.

2. The relation between air passenger flows connecting the 102 biggest metropolitan regions of the United States, and the corresponding rates of AIDS cases, is currently (July, 1994) under investigation, replicating at the national scale a series of remarkable studies at the regional scale. See references 20, 21 and 26.

REFERENCES

1. Majumdar, S., Miller, E. and R. Schmalz (eds.). 1989. *Management of Hazardous Materials and Wastes: Treatment, Minimization and Environmental Impacts.* The Pennsylvania Academy of Science. Easton, PA, pp. 474.
2. Gould, P. 1990. *Fire In The Rain: The Democratic Consequences of Chernobyl.* Polity Press, Cambridge, pp. 163.
3. Atkin, R. 1974. *Mathematical Structure in Human Affairs.* Heinemann Educational Books. London, pp. 178.
4. Wallace, R. 1990. Urban Desertification, Public Health and Public Order: Planned Shrinkage, Violent Death, Substance Abuse and AIDS in the Bronx. *Social Science and Medicine.* 31:801-803.
5. Gould, P. 1993. *The Slow Plague: A Geography on the AIDS Pandemic.* Blackwell. Cambridge, MA, pp. 228.
6. Gould, P. 1991. Modeling the AIDS Epidemic for Educational Intervention, pp. 30-44. In: R. Ulack and W. Skinner (Eds.) *AIDS and the Social Sciences: Common Threads.* University of Kentucky Press, Lexington, KY, pp. 30-44.
7. Gould, P., Kabel, J., Gorr, H. and A. Golub. 1991. Predicting the Next Map. *Interfaces.* 21:80-92.
8. Cliff, A. and P. Haggett. 1988. *Atlas of Disease Distribution,* Blackwell, Oxford, pp. 300.
9. Gould, P. 1980. Q-Analysis, or a Language of Structure: An Introduction for Social Scientists, Geographers and Planners. *International Journal of Man-Machine Studies.* 12:169-199.
10. Gould, P. 1994. The Hidden Geography of AIDS. *Geographical: The Royal Geographical Society Magazine.* 64:18-23.
11. Auerbach, D. 1984. Cluster of Cases of the Acquired Immunodeficiency Syndrome: Patients Linked by Sexual Contact. *The American Journal of Medicine.* 76:487-492.

12. Rvachev, L. and I. Longini. 1985. A Mathematical Model for the Global Spread of Influenza. *Mathematical Biosciences.* 75:3-22.
13. Fahault, A. and A-J. Valleron. n.d. (*circa.* 1988). The Role of Air Transport in the Global Spread of HIV Infection. Available from Unité de Recherches Biomathématiques et Biostatistiques, Université de Paris VII, pp. 9.
14. Center for International Research. 1994. *HIV/AIDS Surveillance Data Base.* U.S. Bureau of the Census, Washington, DC, pp. 153.
15. _______. 1994. South African Epidemic Continues to Spread. *AIDS Analysis Africa.* 5:1.
16. Whiteside, A. 1994. Swaziland: The Epidemic Gains Momentum. *AIDS Analysis Africa.* 5:7.
17. Abler, R., Adams, J. and P. Gould. 1971. *Spatial Organizations: The Geographer's View of the World.* Prentice Hall, Englewood Cliffs, NJ, pp. 597.
18. Kabel, J. 1992. *A Geographic Perspective on AIDS in the United States: Past, Present and Future.* PhD Dissertation, The Pennsylvania State University, University Park, PA, pp. 313.
19. Siu-Ngam, N. and Kam-biu. 1994. Spread of AIDS in Rural America, 1982-1990. *Journal of Acquired Immune Deficiency Diseases.* 7:485-490.
20. Wallace, R., Wallace, D., Andrews, H., Fullilove, R. and M. Fullilove. 1994. The Spatiotemporal Dynamics of AIDS in the New York Metropolitan Region from a Sociographic Perspective: Understanding the Linkages of Central City and Suburb. (forthcoming)
21 Gould, P. and R. Wallace. 1995. Spatial Structures and Scientific Paradoxes in the AIDS Pandemic. *Geografiska Annaler*, 76B:105-116.
22. Wallace, R. and D. Wallace. 1993. The Coming Crisis of Public Health in the Suburbs. *Milbank Quarterly.* 71:943-964.
23. Kemeny, J. and J. Thompson. 1976. *Finite Markov Chains.* Springer Verlag, New York, pp. 196.
24. Nieves, E. 1992. For Better Business, Prostitutes Leave Manhattan for Jersey City. *New York Times,* September 22, p. B1.
25. Wallace, R. 1991. Expanding Coupled Shock Fronts of Urban Decay and Criminal Behavior: How US Cities are Becoming 'Hollowed Out.' *Journal of Quantitative Criminology.* 7:333-335.
26. Wallace, R. and D. Wallace. 1994. The Destruction of US Minority Urban Communities and the Resurgence of Tuberculosis: Ecological Dynamics of the White Plague in the De-developing World. PISCS Inc., New York, pp. 13.
27. Wallace, D. 1994. The Resurgence of Tuberculosis in New York City: A Mixed Hierarchically and Spatially Diffused Epidemic. *American Journal of Public Health.* 84:1000-1002.

Environmental Contaminants, Ecosystems and Human Health. Edited by S.K. Majumdar, E.W. Miller, and Fred J. Brenner.

Chapter Thirty

ENVIRONMENTAL ASSOCIATION FOR *MYCOBACTERIUM LEPRAE*

RICHARD TRUMAN

Research Scientist, Director
Laboratory Research Branch
GWL Hansen's Disease Center
Louisiana State University
Baton Rouge, LA 70894

INTRODUCTION

Leprosy, also known as Hansen's disease, is a chronic infectious disorder caused by *Mycobacterium leprae*. Once common throughout the world, it persists today mainly in the tropics and sub-tropics. Approximately 1/3 of the world's population live in areas where people are considered to be at high risk for contracting the disease, and leprosy is recognized by the World Health Organization as a major health problem for many developing countries. Because of reporting difficulties and the stigma associated with leprosy, prevalence estimates vary widely: worldwide projections range upwards to 12 million affected persons, with some 4 million cases on active therapy and approximately 500,000 new case presentations each year (24).

Mycobacterium leprae was discovered by Armauer Hansen in 1873 and was one of the first bacterial pathogens ever identified. However, the bacillus is yet to be cultivated on artificial media in the laboratory. Like all mycobacteria it has characteristic 'acid-fast' staining properties; but *M. leprae* is unique in its ability to invade nerves, the finding of which in skin biopsies is the principle diagnostic feature for leprosy. The bacillus has a highly restricted mammalian host range and was once considered an obligate intracellular parasite of man (25). Experimental infections have been reproduced in only a handful of immunocompetent animal species, and are more commonly successful with immunodeficient strains (22).

In humans, leprosy primarily affects the peripheral nervous system and secondarily involves skin and other tissues (15). It can result in severe, permanent, progressive deformities which carry serious social and economic consequences. It attacks all age groups, but is slow in onset and is believed to have an incubation period averaging 3-5 years (24). Leprosy manifests itself over a broad clinical spectrum which is differentiated by the number, type and bacillary load of the skin lesions. The tuberculoid end of the spectrum is characterized by one or a few localized lesions which contain few demonstrable bacilli; while in the lepromatous end of the spectrum the lesions are numerous, more generalized and contain abundant bacilli (26). Immunologically, persons in the tuberculoid end of the spectrum exhibit a cellular immune response with little or no detectable antibodies to *M. leprae,* while those of the lepromatous end exhibit a high antibody response to *M. leprae* with little or no cellular immunity (15, 26), *M. leprae* is antigenically similar to the other mycobacterial pathogens, *M. tuberculosis* and *M. bovis,* as well as the plethora of other environmental mycobacterial species. This antigenic cross reactivity compromises the specificity of many common diagnostic tests, and the immunological split anergy across the leprosy spectrum impacts their sensitivity (25). For the most part, epidemiologic studies in leprosy are dependent on clinical presentation data. Obviously, not all infections results in clinical illness, and, with a long incubation period it is difficult to relate risk factors for leprosy infections (8, 12).

TRANSMISSION

The transmission of leprosy is not clearly understood. The specific portal for entry or exit of bacilli is uncertain and the early events in pathogenesis are unknown. Man is considered the principle reservoir, especially lepromatous cases because of their high bacillary loads. However, the relative contribution of active leprosy cases, sub-clinical carriers or other occult reservoirs, has never been determined. Transmission may be both direct or indirect. Inanimate objects or vectors might be involved but transmission is thought to be most efficient in long term direct contact with another infected person. Lepromatous cases can shed large numbers of bacilli from ulcerative lesions on their skin or nasal mucosa, and household contacts of leprosy patients are 2-10 times more likely to develop the infection than non-contacts (8, 12). However, the majority of new cases cannot recall ever having contact with another person who had leprosy (16, 24, 31).

Transmission can occur at a variety of altitudes and climatic conditions. However, the frequency and clinical type of reported cases varies widely by region. The extent to which this is a consequence of environmental, genetic, or cultural differences between the human populations is unknown; but a number of characteristics suggest that here might be some interplay with the environment. Farming is the most common occupation of leprosy patients. Among children, incidence rates are gender neutral; but by adulthood there is commonly a 60:40

ratio of affected males to females. There is a tendency for higher incidence rates in rural areas where the disease also may manifest at a younger age. Leprosy is sometimes referred to as a disease of rural incidence and urban prevalence (8). Additionally, transmission of infection often seems to cease when endemic groups move to non-endemic areas. Though leprosy was introduced into a number of areas within the United States, transmission has persisted in only a few locales (21); and in northern Europe, there has been only one report of secondary transmission to someone other than a non-immigrant child (4, 8, 19).

Environmental factors could have some effect on the transmission of *M. leprae* organisms or they might influence the host's likelihood of developing clinical illness. Most human beings seem to be naturally immune to leprosy, and the reason so few are susceptible is still unknown. However, Abrahams has offered a theory of 'Original Antigenic Sin', and proposes that continuous exposure to cross reactive antigens found on non-pathogenic mycobacteria in the environment tends to suppress the cellular immune response, or promote antibody productions, that masks antigens whose recognition is important for limiting pathogenic mycobacterial infections (1). In mice, pre-exposure to different environmental mycobacteria can dramatically change their hypersensitivity response to mycobacterial antigens (27); and the varying success or failure of BCG vaccination trials to protect against human tuberculosis in different regions has been related to the people's exposures to differing mycobacteria in their environment (30).

If the bacilli shed by an infected host do not die in the environment immediately, they may survive independently or in conjunction with another host for short periods. Or, they may be taken up and amplified by a reservoir host. Some early leprologists speculated that *M. leprae* might exist in nature, but laboratory studies have shown that the bacilli will survive desiccation in sputum or moist soil (6, 10). Organisms sharing certain staining and culture characteristics of *M. leprae* have been isolated from soil, mosses, and pond water in India and Norway (3, 17). However, without other defining characteristics or quantitative studies it is difficult to conclude that these organisms are naturally occurring *M. leprae* or what relevance they might have in disease transmission.

The possibility for transmission of *M. leprae* by anthropod vectors has not been wholly discounted. Flying insects can become heavily contaminated with *M. leprae* when feeding on nasal secretions or ulcerated skin of leprosy patients, and they could contribute to the mechanical dispersal of *M. leprae.* Though the geographic distribution of leprosy does not conform to a single biting insect vector, the particular insect species might vary by region, or could involve some cosmopolitan pests like lice, fleas, or mites. Laboratory bred mosquitoes (*Culex fatigens, Aedes aegypti*) and bed bugs (*Cimex hemiterus*) allowed to feed on leprosy patients have been shown to contain acid fast bacilli (AFB); and AFB have been found in these pests taken from leprosy households. However, the rate was low and not significantly different between infected and non-infected households. Biting arthropods can transfer the infection from infected humans to laboratory mice, and *M. leprae* can remain viable in arthropods for up to 6 days. However, there is no evidence for sustained growth of *M. leprae* within arthropods, and insect vectors would require constant replenishment from other reservoirs in order to be involved in transmission (29).

Many investigators have sought to find animals that might act as reservoirs of *M. leprae* in nature or that would support experimental infection in the laboratory. Granulomas which histopathologically appear to be similar to leprosy have been found in tree frogs and some water buffaloes; but the causative agents of these rather bizarre manifestations have not been confirmed to be *M. leprae* (3). Attempts to experimentally infect animals have met with only limited success, and no susceptible species is common to all endemic areas (22). Of the numerous early attempts to infect primates with *M. leprae,* there was only one successful infection of a gibbon ape. On the other hand, during the last 20 years 3 chimpanzees, 2 African monkeys (Mangabys), and 1 Asian monkey (Macaque fasiculotis) have spontaneously developed leprosy while in captivity. However, leprosy has not been found among free ranging primates (22).

ARMADILLOS

Other than man, nine-banded armadillos (*Dasypus novemcinctus*) are the only known self perpetuating reservoir for *M. leprae.* Studies which began in 1968 demonstrated that armadillos are susceptible to experimental infection with *M. leprae*. More than 90% of the animals develop a lepromatous type of disease and up to 10^{12} leprosy bacilli can be recovered from their internal tissues (18). Armadillos cannot be bred in captivity and must be obtained from the wild for investigative purposes. A leprosy-like disease was first detected among wild armadillos in 1975 (39). Since that time the etiologic agent has been confirmed to be *M. leprae,* and today leprosy is viewed as an indigenous infection of nine-banded armadillos (38).

Armadillos are not native to the U.S.; but began expanding their range north from Mexico in about 1880. Today they are found from Argentina to Colorado and southeast in the U.S. to Florida panhandle. Small groups have been accidentally introduced into the state of Florida since 1925. They established a second U.S. population that now ranges north into the Carolinas and which merged with the main group in the vicinity of Applichacola, Florida in 1985 (9). Leprosy is found only among armadillos from the southern migration.

Early investitures detected the naturally occurring leprosy infection when they examined histopathological sections of armadillo ear tissues and found AFB in dermal nerves. Using this test, between 2-5% of the animals obtained from parts of Texas and Louisiana appear to be infected (39). Though relatively few armadillos have been examined in other countries, 2 leprosy infected animals were found in Argentina (20) and 1 in Mexico (2). The seeming preponderance of infection in the U.S., however, fueled early speculation among scientists that leprosy was geographically restricted to this country and that the animals may have acquired *M. leprae* through some sort of inadvertent contamination of the environment on the part of leprosy researchers (39).

Histopathological examination of armadillo ear tissues has been shown to have low sensitivity for detecting *M. leprae* infection. Bacilli disseminate and become detectable in ear tissues only shortly before the animal succumbs to leprosy, and serology is much more sensitive and convenient (37). The phenolic glycolipid-1 (PGL-1) was the first major species specific antigen discovered for *M. leprae* and it has been used extensively in serological surveys (13). PGL-1 provokes primarily an IgM antibody response, which, in armadillos, arises in about a third the time required for *M. leprae* to become manifest in ear tissues. Nearly all (98%) of the experimentally infected armadillos which develop detectable PGL-1 IgM antibodies eventually die from complications due to their leprosy. PGL-1 IgM antibodies have high positive predictive value in armadillos, they mainly identify well established active infections and are not confounded by non-fulminating or self-healing infections (34).

The advent of serological methods has helped bring a better understanding to the natural history of *M. leprae* in armadillo populations. A retrospective study in Louisiana found that 14% of the wild armadillos sampled in 1960-64, 10-15 years before leprosy researchers began using armadillos, already had IgM antibodies to PGL-1. Therefore *M. leprae* infections must have occurred naturally and could not have originated through any inadvertent contamination of the environment (38). Geographically, more than 3,000 armadillos from across the southern U.S. have been examined today; but no evidence for the infection has been found among animals in the states of Alabama, Georgia (11), Florida (39) or western Oklahoma (35). There are no published reports concerning leprosy among armadillos from Arkansas, and only 4 infected armadillos have been found within the state of Mississippi (35, 39). *M. leprae* infection occurs principally among armadillos in the states of Texas and Louisiana, where the prevalence rates average 4% histopathologically and 16% serologically. However, even within these two states, there is marked variation in leprosy prevalence, with significantly higher rates found among animals in low living poorly drained areas than in other environments (36).

ENVIRONMENTAL FEATURES

Epidemiologic studies on leprosy have been inconsistent in regards to their findings about environmental factors. Rainfall or relative humidity seem to have high correlation with disease rates in some locales, but none in others (14, 23). Verbal descriptions or discrete quantitative variables are rarely adequate to express the dynamic character of natural environments. Recently however, remotely sensed data from earth orbiting satellites have become available. The digital data of satellite images can be manipulated with a variety of standard formula to determine quantitative characteristics of the environment or to enhance for specific qualitative traits. Satellite images are relatively inexpensive, capture the entire environment as opposed to a single variable, and can be applied with uniformity across the globe. This technology is likely to bring important new understanding about the relationships between disease and environment in the future.

Satellite images readily differentiate the environment of the lower Mississippi valley in the southern U.S. (Figure 1 - Plate). When enhanced as a vegetation index the bottom-land hardwoods of the Mississippi and Red rivers combine as a single system of surface aquifers with the coastal marshes and plains in the states of Texas and Louisiana. Soil moisture in this area is high. Along the Texas coast it changes only gradually for a distance of about 50 miles inland; but it ends abruptly at the coastal margin in the state of Mississippi. There is little moist soil habitat east of the Mississippi river. Antibody prevalence rates reported from within these low lying areas of Texas and Louisiana range from 18-25%. However, significantly lower prevalence rates (0-8%) are found in the rolling woodlands of Louisiana or Mississippi (36), and only 3 leprosy infected armadillos have been found at a distance of more than 50 miles inland from the coast of Texas (39). The distribution of armadillo leprosy seems dependent on environmental characteristics.

Just as an animal's morphology may be thought to be the result of natural selection so too may population characteristics be selected in different environments. Armadillos are insectivorous and are found most plentifully in low lying areas. In Louisiana, however, there is no correlation between animal density and the leprosy prevalence rate. The uniformly increased rates of low lying areas, therefore, is not likely to be the result of increased transmission consequent to direct contact (46). The susceptibility of armadillos to *M. leprae* infection, however, may vary in different environments. Reported lepromin skin test results and experimental infection data suggest that armadillos in Florida are markedly more resistant to infection than those from Louisiana. Additional observations on animals from elsewhere in the armadillo range are needed, though, before any firm conclusions can be drawn.

In addition to possible influences on disease transmission or susceptibility, environmental factors can also affect the detectability of disease in animal populations. The prevalence of armadillo leprosy seems to be relatively similar from year to year within a given environment. Repeat observations on one site in the bottom-land hardwoods of Louisiana found that the average yearly prevalence rate remained similar over a 6 year period, and had not varied significantly from what was observed there some 30 years before. Long term steady maintenance of high rates of disease over a large geographic area suggests that armadillos have harbored *M. leprae* for several generations. However, prevalence rates were significantly lower each Spring before returning to higher levels by Summer. Leprosy is a slow chronic disease which manifests itself mainly in adult armadillos. Adult female armadillos sequester themselves each spring for parturition and nurturing of their newborns, and this behavior temporarily removes a large

FIGURE 1. Plate: NOAH-Trios-9 satellite image of Louisiana and the southern Mississippi river valley during November and enhanced as a vegetation index. Light colored areas highlight the surface aquifer system of the Mississippi and Red rivers in Louisiana joining with the coastal marshes and plains in Louisiana and Texas.

proportion of the affected individuals from the population available for sampling thereby influencing the detectable rate of disease. It seems likely that age structure differences between armadillo groups in different locations have probably contributed to some of the unsubstantiated geographic trends purported in the distribution of armadillo leprosy. Interestingly, if the ovulatory status of a population is determined, the observed prevalence rates can be corrected by the frequency of adult females present in the sample and the rate is then seen to remain similar throughout the year and within similar environments. However, the difference in prevalence rates seen between low lying and better drained areas remains significant (36).

HUMAN LEPROSY

Armadillos are the only proven non-human reservoir for *M. leprae* and there has been considerable interest in determining their impact on humans. The U.S. is not an endemic country but leprosy cases have been reported among people from every state. The majority of these are among immigrants or nationals who may have acquired leprosy while living or working in endemic areas abroad (4). However, there is sporadic transmission within the U.S. and a persistent locus of infection within the states of Texas and Louisiana producing some 35 new indigenous cases each year (16). Leprosy did not occur among native American Indians and it was probably brought to the new world by European settlers and Africans being traded as slaves. It has occurred in the western gulf region of the U.S. since at least 1748 and, clearly, could not have been introduced thereby armadillos (33). The incidence of indigenous human infection in the region declined throughout the first part of this century, but it has remained at a low constant rate since the 1950's. Though once principally a disease of families, leprosy is now being increasingly acquired from other sources in the community. During the last 17 years, none of the indigenous Louisiana cases could recall ever coming in contact with another person with leprosy. Geographically, the indigenous cases tend to cluster within low lying environments in a distribution pattern similar to that seen for leprosy in armadillos (Figure 2 - Map). The historical incidence for indigenous human infection in Louisiana has been 4 times higher among people in these low-lying areas than in other parts of the state; but the total incidence of leprosy in the U.S. remains quite low (35).

Large numbers of infected armadillos are present in the U.S. and there has been considerable speculation about the prospects of zoonotic transmission. The literature contains case reports on at least 14 individuals with leprosy whose only exposure to *M. leprae* may have been in association with armadillos (28, 39). Significantly increased risks for lepromatous leprosy also are reported among Mexican born patients who lived in areas where armadillos were common (32). Armadillos are frequently taken for food and about a third of all indigenous leprosy patients in the U.S. acknowledge having had contact with the animals.

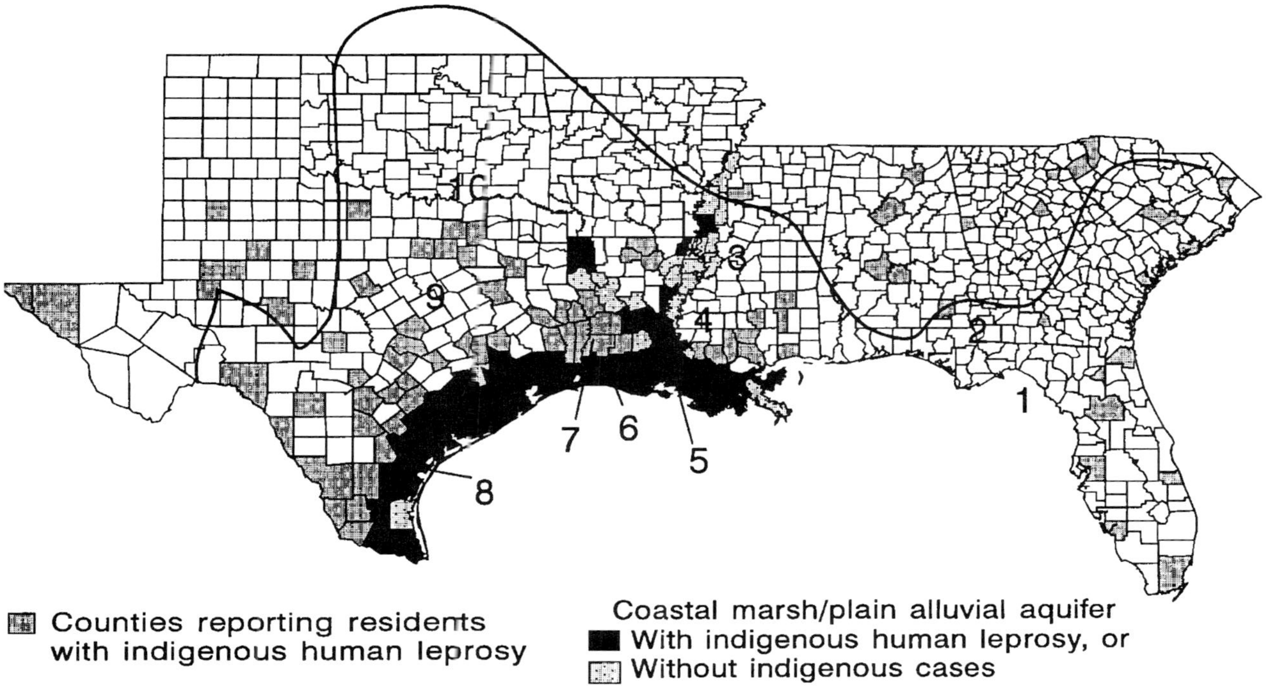

FIGURE 2. Map: Counties of the southern United States within the range occupied by nine-banded armadillos (*Dasypus novemcinctus*). Dark line marks estimated armadillo range extent. The coastal-alluvial aquifer system is highlighted. Residence counties of individuals presenting with indigenous human leprosy since 1894 are shaded. Numbers represent areas where armadillos have been surveyed for leprosy. Respective prevalence results were (ratios are histopathological results, percents are antibody prevalence rates): 1=0/987 infected found, 2=0/843 infected found, 3=24% infected, 4=4/242 infected found, 5=18% infected, 6=19% infected, 7=8% infected, 8=18% infected, 9=3/342 infected found, 10=0% infected.

Patients presenting from the states of Louisiana and Texas are significantly more likely (55%) to report contact with armadillos than patients presenting from other states (35). However, an early case control study found that Louisiana leprosy patients were no more likely to have contact with armadillos than other people in the population (7). While those observations might be biased by the generally low rate of susceptibility to leprosy among humans, a recent study among indigenous leprosy patients presenting from areas known to harbor high or low rates of armadillo leprosy also failed to show that contact with armadillos was an important risk factor for their infection (35). Certainly the size of the armadillo reservoir alone suggests that the animals might contribute to some cases of human leprosy infection. However, the incidence of disease in the U.S. remains quite low and it has not increased even in the face of growing numbers of armadillos. Though armadillos might play some indirect role in helping to perpetuate leprosy, they are probably of only secondary significance in the epidemilogy of disease in this country.

CONCLUSIONS

The similar environmental tendencies in the geographical distribution of leprosy infections in both humans and armadillos in the southern United States are intriguing. Both moved into the area from elsewhere but have established nearly identical geographical patterns for their leprosy infections. What it is about these environments that so much favor leprosy remains unclear. With the diminishing prospects of zoonotic transmission, the data might also suggest mutually enhanced susceptibility to leprosy for both man and armadillos in certain environments, or the possibility for transmission of leprosy to both man and armadillos from third reservoirs in those environments. Certainly a better understanding of how environmental factors interact to affect the distribution of leprosy is needed. Armadillos appear to be a unique parallel population which supports autochthonous transmission of leprosy. The animals can be useful population models for evaluating strategies used in control of this disease and potential modes of transmission. Contined studies in this area are likely to reveal important new basic information that will significantly benefit our abilities to finally eradicate leprosy.

REFERENCES

1. Abrahams, E.W. 1970. Original Mycobacterial Sin. Tubercle 51:316-321.
2. Amezcua, M.E., Escobar-Guitierrez, A., Storrs, E.E., Dhople, A.M. and H.P. Burchfield. 1984. Wild Mexican armadillo with leprosy-like infection. Int. J. Lepr. 52(2):254.

3. Blake, L.A., West, B.C., Lary, C.H. and J.R. Todd. 1987. Environmental non-human sources of leprosy. Reviews of Infectious Diseases 9:562-577.
4. Browne, S.G. 1985. The history of leprosy, pp. 1-14. *In* R.C. Hastings (ed.), Leprosy, 1st ed. Churchill Livingstone, New York.
5. Davey, T.F. and R.J.W. Rees. 1974. The nasal discharge in leprosy: clinical and bacteriological aspects. Lepr. Rev. 45:121-134.
6. Desikan, K.V. 1977. Viability of *Mycobacterium leprae* outside the human body. Lepr. Rev. 48:231-235.
7. Filice, G.A., Greenberg, R.N. and D.W. Fraser. 1977. Lack of Observed Association Between Armadillo Contact and Leprosy in Humans. Am. J. Trop. Med. Hyg. 26(1):137-139.
8. Fine, P.E.M. 1982. Leprosy: the epidemiology of a slow bacterium. Epidemilogic Reviews 4:161-187.
9. Hall, E.R. (ed.) 1981. The mammals of North America, 2nd ed., vol. 1. John Wiley & Sons, New York.
10. Harris, E.B., Landry, M.T., Sanchez, R.M. and C.K. Job. 1990. Viability of Mycobacterium leprae in soil, water and degenerating armadillo tissue. Microbios Letters 44:73-76.
11. Howerth, E.W., Stallknecht, D.E., Davidson, W.R. and E.J. Wentworth. 1990. Survey for leprosy in nine-banded armadillos (*Dasypus novemcinctus*) from the southeastern United States. J. Wildl. Dis. 26:112-115.
12. Huang, C.L.H. 1980. The transmission of leprosy in man. Int. J. Lepr. 48(3):309-318.
13. Hunter, S.W., Fujiwara, T. and P.J. Brennan. 1982. Structure and antigenicity of the major specific glyco lipid antigen of *Mycobacterium leprae*. J. Biol. Chem. 257(24):15072-15078.
14. Irgens, L.M. 1980. Leprosy in Norway. Lepr. Rev. 51(Supp 1):1-130.
15. Jopling, W.H. and A.C. McDougall (eds.). 1988. Handbook of leprosy, 4th ed. Heinemann, Oxford.
16. Joseph, B.Z., Yoder, L.J. and R.R. Jacobson. 1985. Hansen's Disease in native-born citizens of the United States. Pub. Hlth. Rpts. 100(6):666-671.
17. Kazda, J., Irgens, L.M. and K. Miller. 1980. Isolation of non-cultivable acid fast bacilli in sphagnum moss and vegetation by foot pad technique in mice. Int. J. Lepr. 48(1):1-6.
18. Kirchheimer, W.F., Storrs, E.E. and C.H. Binford. 1972. Attempts to establish the armadillo *Dasypus novemcinctus* as a model for the study of leprosy, Part 2 histopathologic and bacteriologic post mortem findings in lepromatoid leprosy in the armadillo. Int. J. Lepr. 40(3):229-242.
19. Leiker, D.L. 1977. On the mode of transmission of *Mycobacterium leprae*. Lepr. Rev. 48:9-16.
20. Martinez, A.R., Resoagli, E.H., deMillan, S.G., Resoagli, J.P., Rameriez, M.M., Cicuta, M.E., Rott, M.I.O. and A. Sandoval. 1984. Lepra salvje en *Dasypus novemcinctus* (linneo 1758). Arch. Argent. Dermat. 34:21-30.

21. Mastro, T.D., S.C. Redd and R.F. Brieman. 1992. Imported leprosy in the United States 1978-1988, an epidemic without secondary transmission. Am. J. Public Health 82(8):1127-1130.
22. Meyers, W.M., Gormus, B.J. and G.P. Walsh. 1994. Experimental leprosy, pp. 385-408. *In* R.C. Hastings (ed.), Leprosy, 2nd ed. Churchill Livingstone, New York.
23. Motta, C.P. 1981. The epidemiological situation in the Americas. Leprosy Review 52:61-68.
24. Noordeen, S.K. 1994. The Epidemilogy of Leprosy, pp. 29-45. In R.C. Hastings (ed.), Leprosy, 2nd ed. Churchill Livingstone, New York.
25. Rees, R.J.W. 1985. The microbiology of leprosy, pp. 31-52. In R.C. Hastings (ed.), Leprosy, 1st ed. Churchill Livingstone, New York.
26. Ridlye, D.S. and W.J. Jopling. 1966. Classification of leprosy according to immunity. Int. J. Lepr. 34(2):255-273.
27. Rook, G.A.W., Bahr, G.M. and J.L. Stanford. 1981. Hypothesis 2: The effect of two distinct forms of cell-mediated response to mycobacteria on the protective efficacy of BCG. Tubercle 62:63-68.
28. Rushing, D., Brannon, H.E., King, J.W., Fowler, M.R. and B.C. West. 1986. Autochthonous transmission of leprosy in north Louisiana USA associated with armadillos. Clin. Res. 34(1):193A.
29. Sreevatsa. 1993. Leprosy in arthropods. Indian J. Lepr. 65(2):189-200.
30. Standford, J.L., Shield, M.J. and G.A.W. Rook. 1981. Hypothesis 1: How environmental mycobacteria may predetermine the protective efficacy of BCG. Tubercle 62:55-62.
31. Taylor, C.E., Elliston, E.P. and H. Gideon. 1965. Asymptomatic infections in leprosy. Int. J. Lepr. 33(3):716-731.
32. Thomas, D.A., Mines, J.S., Mack, T.M., Thomas, D.C. and T.H. Rea. 1987. Armadillo exposure among Mexican-born patients with lepromatous leprosy. J. Infect. Dis. 156(6):990-993.
33. Trautman, J.R. 1933. The History of Leprosy, pp. 21-31. In R.C. Hastings (ed.), Leprosy, 2nd ed. Churchill-Livingstone, New York.
34. Truman, R.W., Job, C.K. and R.C. Hastings. 1990. Antibodies to the phenolic glycolipid-1 antigen of *Mycobacterium leprae* for epidemilogic investigations of leprosy in armadillos (Dasypus novemcinctus). Lepr. Rev. 61:19-24.
35. Truman, R.W., Kumaresan, J.A. and M.E. Hugh-Jones. 1994. An analysis of the origins and risks of *Mycobacterium leprae* infections in wild armadillos. in press.
36. Truman, R.W., Kumaresan, J.A., McDonough, C.M., Job, C.K. and R.C. Hastings. 1991. Seasonal and spatial trends in the detectability of leprosy in wild armadillos. Epidem. & Infect. 106(3):549-560.
37. Truman, R.W., Morales, M.J., Shannon, E.J. and R.C. Hastings. 1986. Evaluation of monitoring antibodies to PGL-1 in armadillos experimentally infected with M. leprae. Int. J. Lepr. 54(4):556-559.

38. Truman, R.W., Shannon, E.J., Hagstad, H.V., Hugh-Jones, M.E., Wolff, A. and R.C. Hastings. 1986. Evaluation of the origin of *Mycobacterium leprae* infections in the wild armadillo dasypus-novemcinctus. Am. J. Trop. Med. Hyg. 35(3):588-593.
39. Walsh, G.P., Meyers, W.M. and C.H. Binford. 1986. Naturally acquired leprosy in the nine-banded armadillo *Dasypus novemcinctus* a decade of experience 1975-1985. J. Leukocyte. Biol. 40(5):645-656.

Environmental Contaminants, Ecosystems and Human Health. Edited by S.K. Majumdar, E.W. Miller, and Fred J. Brenner. © 1995, The Pennsylvania Academy of Science.

Chapter Thirty-One

POVERTY VERSUS POLITICS IN POLLUTION ABATEMENT

DWIGHT R. LEE
Ramsey Professor of Economics and Private Enterprise
Economics Department
University of Georgia
Athens, GA 30602

INTRODUCTION

It is widely believed that the use of private property in the pursuit of private advantage has led to excessive pollution. It is as widely accepted that the solution to this problem requires more government control over market decisions. Both views are wrong. Pollution is excessive not because of too much reliance on private property but because of too little. Furthermore, attempts to control pollution with direct government control over polluting activities are badly flawed because they lack both the information and motivation to protect the environment effectively. In fact, the problem with government attempts to reduce pollution is completely analogous to the problem that leads to excessive pollution in the first place.

POLLUTION AS AN ECONOMIC PROBLEM

The problem of environmental pollution cannot be adequately understood until it is recognized as fundamentally an economic problem, that is, a problem of scarcity. Our resources simply are not sufficient to satisfy all of our demands. We want clean air, pure water, and unspoiled landscapes to sustain life and health, and indulge our aesthetic sensitivities. But equally important for maintaining life and enhancing its quality is the use of our environment as a waste-sink. Every productive act we engage in-whether breathing in and out, growing wheat, or generating electricity-creates unwanted by-products that have to be discharged into the environment. One of the costs of producing more goods and services is the sacrifice of some environmental quality. Similarly, the cost of a cleaner environment is measured in terms of the resulting sacrifice of some goods and services.

To say that we all value a cleaner environment gives no guidance in deciding whether or not the environment should be a little bit cleaner, or a little bit dirtier. Although we want a cleaner environment, we also want more housing, entertainment, medical research, schooling, fine wines, fast-acting detergent, warmth in the winter, air conditioning in the summer, convenient and safe transportation, attractive clothing, etc. The relevant question is, what is the appropriate combination of environmental purity and all the other desirable things we enjoy, given that we can have more of one only at the cost of sacrificing some of the others?

It is not easy to answer this question. But it is possible to say something about the combination of desirable things that can be characterized as appropriate. This combination will not contain as much of anything as we would ideally like. Because of the sacrifice involved in expanding our consumption of clean air, gourmet food, or whatever item you care to consider, it will never make sense to continue consuming it until the additional unit consumed is worth zero. So even if we were enjoying the appropriate, or optimal, amount of environmental quality, there would remain, at least at times, unpleasant and unhealthful levels of environmental pollution.

This raises the question: just because we have more pollution than we would ideally like (i.e., more than we would like if pollution could be reduced at no cost), is there any reason for believing that we have too much? Isn't it possible that we are suffering from not enough pollution, in that the additional production possible with more pollution would be worth more to those in the society than the benefits sacrificed in environmental quality?

This is a serious question, and it cannot be answered with appeals to horror stories about the environmental and health hazards that come from existing levels of pollution. As acute and as horrible as such stories may be, they consider only one side of the coin, the side that tells us about the benefits from reducing pollution. In order to make an informed judgment, we have to turn the coin over and ask, what are the costs of reducing pollution? Unfortunately, costs and benefits are very difficult to measure, for they are ultimately based on subjective judgments of individuals.

This does not mean that we cannot reach an informed judgment on the question as to whether or not we are experiencing too much pollution. Economics is fundamentally a study of scarcity, and has long been concerned with how individuals can best interact to allocate limited resources among their many alternative uses. This allocative process, if it is to work well, requires institutional arrangements to facilitate the ability of people to communicate their feelings on the costs and benefits of different resource decisions. Ideally, this communication process will provide both the information and the incentives required to direct resources into their highest value activities—or, equivalently, into their lowest cost activities.

Communication is never perfect, and we are not likely to achieve the perfect allocation of our resources. But some institutional arrangements are better than others in this regard. But recognizing certain deficiencies in the economic institutions that guide our decisions on the use of important environmental resources, economists can argue compellingly that pollution is *excessive*: environmental resources are over-used in providing man-made goods and under-used in providing environmental quality.

It is accurate, then, to characterize most economists as environmentalists. But economists generally part company with other environmentalists when it comes to recommendations for reducing pollution. The environmental lobby has traditionally seen the solution to environmental problems in more detailed government control over the actions of polluters. In this view, if we have too much pollution, it is the responsibility of government to pass a law, not only requiring polluters to reduce pollution but, in many cases, telling them how to do it. This would be a fine approach if government institutions worked in some ideal way. But this clearly is not the case. Indeed, the types of institutional deficiencies in the market place that lead to excessive pollution in the absence of government regulation are also prevalent in the political process. Because they recognize this fact, economists are highly critical of "solutions" to environmental problems that rely on extensive government control and regulation. When such governmental "solutions" do manage to reduce pollution, they invariably do so at much higher cost than is necessary. And it is unfortunately the case that relying on political control to reduce pollution occasionally results in both high pollution control costs and more, rather than less, pollution.

In order to understand best how institutional deficiencies 1) lead to excessive pollution and 2) limit the ability of government to respond efficiently to this problem, it will be helpful to consider the institutional features of a system of communication which, when working perfectly, directs our scarce resources into the pattern of use that maximizes their value.

COMMUNICATION, HONESTY, AND CONCERN FOR OTHERS

In an ideal world (ideal except for scarcity), every resource decision would be guided by information on the value of the resources in all possible employments. The way of realizing this ideal, at least conceptually, would be to first arrange for everyone to be able to communicate with everyone else regarding the value we place on different resource uses. In addition, we would have to insure that people communicate with each other honestly and that no one would give greater weight to his or her personal benefits than to the benefits of others. Under these circumstances, resources would be directed into their highest valued employments. If someone communicated to you that you had resource that is worth more to him than it is to you, you would know this is an honest statement and, in your concern for his interest, you would transfer the resource to him.

Our discussion here may appear to be nothing more than a utopian dream. It would be hard even to imagine the technological advances necessary to allow all users and potential users of resources to communicate simultaneously with each other. And even if the technological problems of communication were solved, we would still be left with the moral challenge of motivating people to communicate honestly and to give the interests of others the same consideration they give their own. If avoiding the horrors of Hell and achieving the ecstasy of Heaven, have not been sufficient motivation to insure honesty and brotherly love, there is little reason to believe that the mundane goal of efficient resource allocation will inspire the realization of these noble virtues.

But as utopian as it may appear, we should not dismiss the idea that resources can be allocated efficiently as a consequence of people communicating their values to each other, both honestly and with apparent regard for the interests of others. Indeed, this is a quite accurate description of exactly what happens with most of the resources we use. For a wide range of our activities, this system of information and consideration is coordinating our actions and interests with the actions and interests of literally millions of others in such a way that an amazingly efficient pattern of resource use emerges. In addition, this system of communication and consideration works so smoothly that few have either an awareness of, or appreciation for, the truly astonishing task that is being performed.

Interestingly, the social institution upon which the system of communication and consideration is based is private property, an institution that has long been criticized by people who claim great concern with promoting social communication and cooperation. The fact that private property can be the source of enormous private wealth is surely seen by some as a major problem. To the economist, however, it is the fact that individuals can profit from private property that explains one of its major advantages. When individuals own resources and are thus able to capture most, if not all, of the value from their use, they have a strong motivation to use them carefully and to direct them into their most productive employments. People use resources they own more wisely than resources that are commonly owned. If you want to find graffiti on a bathroom wall, you are well-advised to look in public restrooms, not privately owned bathrooms.

But if you are interested in efficient resource allocation, it is not enough to have individuals simply putting the resources they own into those uses they value most. The best use the owner had for a resource may be less valuable than the best use someone else has for it. How can each individual be informed as to the value of his resources to others, and then be motivated to take this value fully into consideration? The answer is provided by the voluntary exchange that is made possible by the private ownership of property.

Only when people have well-defined and enforced ownership rights in resources is exchange possible. No one is going to pay you for a resource unless you have a transferable property right entitling you to control and use that resource. Not until transferable property rights exist in a resource will a market develop in which that resource can be bought and sold, and a market price for that resource emerge. The price will reflect the fact that if you want to purchase the resource you will have to pay at least as much as it is worth to the owner, and this can be no less than the value others place on the resource and are therefore willing to pay the owner. In other words, market exchange results in resource prices that honestly communicate to everyone the value of those resources to others. This is a communication process that not only provides an enormous amount of information, but provides that information in such a way that all relevant decisionmakers are motivated to take it fully into consideration.

Whether you already own a resource or will have to purchase it, its market price reflects its cost to you. If you already own it and retain it, the market price reflects the value you have to forego in not selling it, just as it would if you had to purchase the resource. People will be motivated to use resources that are privately owned and easily exchanged only as long as these resources are worth at least as much to them as they are to others, as reflected in the market prices of those resources. In other words, peole are motivated to make economic decisions as if they considered the interests of others as important as their own.

This discussion of resource allocation through a system of communication and consideration, and the examination of the institutions of property rights and market exchange necessary for this system to function properly may seem something of a digression from our main focus on the problem of environmental pollution. But having looked carefully at the allocative efficiencies that result from a properly functioning system of communication and consideration, it is possible to understand better the problems that arise when such a system is not working properly. This leads us back to the pollution problem.

TOO LITTLE PRIVATE PROPERTY MEAN TOO MUCH POLLUTION

We can now explain why pollution is excessive. Those environmental resources, such as water-ways and airsheds that, among other things, necessarily serve as repositories for pollutants, are not easily parceled out to individuals as privately owned and controlled property. Because these receptor resources are not privately owned, market exchange does not generate market prices to govern the uses to which they are put. All of us make use of commonly owned receptor resources for dumping our waste, but we do so in the absence of the market information and consideration that efficiently guide most of our decisions on resource use.

Consider, for example, your decision to drive your car to work. Driving to work requires the use of resources that are valued by others, such as oil, gasoline, and the labor required for automotive upkeep. But because these resources are privately owned and easily exchanged on markets, you give full consideration to the value of these resources to others in the prices you pay. This is not the case with the clean air into which you vent your exhaust as you drive. The value that others place on the clean air you foul is not made known to you, because clean air is not rationed by market exchange. There is no price on clean air that informs you of its value to others. And even if somehow you did know this value, in the absence of a market price you have to pay for the use of clean air, there would be little motivation to take its value into consideration. The harmful effects generated by your polluting activity are imposed almost entirely on others. Not surprisingly, it can be expected that you will put much greater weight on the benefits you derive from polluting than on the costs.

With privately owned resources, people have the information and motivation to employ more of each resource only so long as the value it provides to them is at least as great as the value it would otherwise provide to others. With commonly owned environmental resources, people are motivated to employ these resources as waste-sinks as long as the value to them of doing so is positive, regardless of the value that others forego as a consequence. For this reason our environmental resources are over used as waste-sinks relative to their use for providing environmental quality. Pollution is excessive.

FIGHTING POLLUTION WITH PRIVATE PROPERTY

Before discussing the approach that most economists recommend for solving the problem of excessive pollution, it will be useful to identify just what it is that the ideal solution would accomplish. First, and most obviously, we want pollution reduced to that level consistent with the efficient allocation of our resources, the allocation which maximizes the value generated by our resources. Put differently, we want to reduce pollution as long as reducing pollution by one more unit provides more value in improved environmental quality than it costs in terms of the value of sacrificed alternative goods and services.

A second, and related, objective is to reduce pollution as cheaply as possible. There are two separate considerations here. Each pollution source has to be abated at minimum cost. There are many ways to cut back a given amount on pollution, but in general there will be only one least-cost way. But even if all polluters are abating pollution as cheaply as possible, it is not necessarily the case that the least-cost pattern of pollution is being realized. The least-cost pattern of pollution abatement will find the cost of reducing pollution by one more unit (the marginal abatement costs) the same for all polluters. If this is not true then it would be possible to reduce abatement cost without increasing pollution by having the polluter with the lowest marginal abatement cost reduce pollution by one unit while the polluter with the highest abatement cost increases pollution by one unit. Since polluters will vary in terms of the cost of pollution reduction, the least-cost pattern of pollution abatement will require a different abatement level for different polluters.

A third objective of a pollution management program is to establish incentives that will motivate advances in pollution abatement technology. This dynamic aspect of pollution control is as important as the point-in-time, or static, considerations discussed in the previous two paragraphs.

These three objectives—1) achieving the efficient level of pollution, 2) achieving the least-cost pattern of pollution reduction, and 3) motivating advances in the technology of pollution control-probably will never be fully realized. This is particularly true of the first objective. Not being able to own and control identifiable and isolated portions of the atmosphere, for example, no one is in a position to let a polluter foul his, and only his, clean air in exchange for a claim on other desirable things (this claim generally being money). Without such exchange possibilities, prices do not spring up that reflect the value people place on clean air, and without this information there is no way of determining the efficient level of air pollution.

Likewise, private ownership of identifiable and isolated portions of water in our lakes, rivers, and oceans is not possible, and thus there is no accurate way of determining the efficient level of water pollution. With it impossible to determine the efficient level of pollution through the information provided by market exchange, we have to rely upon the political process for this determination. In a democratic political order, there is the presumption that the information provided by voting and lobbying will keep the political process responsive to the preferences of the citizens. While we will soon see that there are problems with the political process that call this presumption into question, we have to rely on political decisions to determine an acceptable, if not efficient, level of pollution.

Assuming that the politically acceptable level of pollution has been determined, the last two objectives of a pollution control program are considered. It is possible to go a long way in realizing these objectives by having the government create and enforce a system of property rights in the use of the environment as a waste-sink. The essential idea is simply to have the government issue transferable pollution rights which give the holder the right to discharge one unit of pollution each week, for example. The total number of rights issued would conform to the targeted level of pollution as determined by the political process. This scheme, assuming adequate enforcement, would serve to limit pollution to the acceptable level. Also each polluter, having to reduce pollution to the level allowed by the number of rights held, will be motivated to do so at the minimum cost. But the crucial advantage in the pollution rights approach comes from the fact that the rights are transferable.

Because the rights are transferable, a market would develop for them, and the resulting exchanges would determine a market price for rights. The cost of discharging another unit of pollution per week would then equal the price of a pollution right-the value others place on the ability to increase their pollution. People will be motivated to increase their use of the environment as a waste-sink only if the additional pollution benefits them at least as much as it would benefit others. A pattern of polluting activities results that maximizes the value realized from the allowable level of pollution. Another way of stating this is that the reduction in pollution necessary to realize the acceptable level is achieved with the least cost (minimum sacrifice in valuable alternatives) pattern of abatement.

Faced with a positive price for pollution rights, each polluter has a strong motivation to discover the cheapest way to reduce pollution and to apply it. Each polluter will also be motivated to reduce pollution as long as the coast of reducing one more unit is less than the price of a pollution right. With all polluters facing the same market price for pollution rights, the cost of abating one more unit of pollution will be the same for all polluters. This is another way of stating the previously discussed requirement for the least-cost abatement pattern. The information and incentives generated by private ownership and market exchange automatically lead to the desirable pattern of pollution abatement.

The pollution rights approach also creates an incentive for polluters to develop improved abatement technologies. History is full of examples of technological development allowing more man-made output to be produced with less land and labor. Conspicuously absent have been technological improvements designed to conserve on the use of the environment as a waste-sink. Market prices on land and labor have always provided a strong incentive to conserve on these resources. The absence of prices for the use of our atmosphere and waterways, however, made it privately unprofitable to worry about conserving on their use. Marketable pollution rights would remedy this neglect.

There is another advantage of the pollution rights approach, although it is not as big an advantage as it may first appear to be. The discussion has proceeded to this point as if the only reason for purchasing a pollution right were to support a polluting activity. But pollution rights could be purchased also for the purpose of keeping them out of the hands of polluters, thereby reducing pollution. Groups interested in protecting the natural environment, such as the Sierra Club, Friends of the Earth, the National Rifle Association, and the Audubon Society, have an opportunity with pollution rights to put some of their funds to direct use in reducing pollution. They could do this by purchasing and hoarding pollution rights. If everyone who valued clean air could be depended upon to buy pollution rights for this purpose, then the price of these rights would reflect more than just their value in polluting activities; it would reflect also their value in providing a cleaner environment. In this case, the pollution rights approach could result in the efficient *level* of pollution as well as the least-cost *pattern* of abatement.

Unfortunately, there is little reason to expect people, whether or not they express concern over the environment, to buy rights to protect the environment. Buying a pollution right to reduce air pollution, for example, reduces air pollution for everyone in the area, not just for the person who pays for the right. So the total value of the reduced pollution is much greater than realized by the one individual who buys the right. Consequently, the value of keeping rights out of the hands of polluters could easily be greater than their market price, yet no one individual would be willing to buy one for that purpose. Of course, individuals could band together for the collective purchase of pollution rights. But enticing people to join such a cooperative effort would not be easy. Each person would realize that unless others joined the effort in large numbers, their individual contribution would do little good. On the other hand, each individual realizes also that, if others contribute in large number, he will be able to benefit as a free rider from the clean air they purchase. Either ways, the individual sees the personal advantage in not contributing to the common cause: he has little motivation to communicate honestly his preferences on clean air through market prices. This, of course, is just another way of looking at the general problem that exists when important resources cannot be parceled out and controlled as private property. Without property rights that are completely private, the communication needed to use our resources wisely cannot be generated.

The above argument is not meant to imply that the desire to reduce pollution will activate no purchases of pollution rights. Environmental organizations exist that have to some degree overcome the free rider problem, and one could expect such groups to purchase and hoard some pollution rights. But it would be naive to believe that such purchases would come close to reflecting the full social value of clean air if the government issued pollution rights far in excess of that number which corresponds to the efficient level of pollution. The best we could hope for is that the number of pollution rights issued will allow something close to the efficient level of pollution when they are purchased only for the purpose of polluting. But even without pollution reduction purchases motivating the efficient level of pollution, the minimum abatement cost feature of the pollution rights approach should still make it a very attractive means for controlling pollution. This is particularly true given the clumsy alternative of having the government directly regulate and control pollution sources.

Briefly stated, the direct regulation and control approach has the government determine an acceptable level of pollution and then attempt to achieve this level by requiring individual polluters to reduce their discharges by specified amounts and/or mandating the use of particular abatement technologies. In both cases, the political process will determine the acceptable level. But direct regulation and control of pollution sources cannot be expected to achieve the required abatement as cheaply as will a market in pollution rights. Not knowing the least-cost abatement approach for each pollution source, the government agency charged with pollution control will generally require a uniform approach across each class of polluters, despite the fact that the most appropriate approach will vary from source to source. Neither will the government have the information necessary to determine the least-cost abatement pattern. In the absence of market exchange, this information is unattainable. Furthermore, requiring that polluters use the "best available control technology" is vague, and, if anything, discourages technological improvements (since "better" technologies are commonly more expensive to employ). Government commands and control designed to reduce pollution is being exercised in what comes close to an information vacuum.

The result is that the current approach to pollution control cost far more than would be the case under the private property (transferable pollution rights) approach. The economic case for the private property approach goes beyond armchair theorizing by academic economists. Several simulation studies on air pollution abatement have estimated the likely cost differences between the uniform command-and-control approach and the pollution permit approach. These studies commonly find that the command-and-control approach is four to six (and in one case, 22) times more costly than the market oriented pollution rights approach (Tietenberg, 1992; pp. 402-405). This means we could achieve both enormous savings (reducing air pollution currently costs tens of billions of dollars) and a cleaner environment by substituting market incentives for government controls in the fight against pollution. There are serious obstacles, however, in making this substitution. The government's command and control approach to pollution abatement may be operating in an informational vacuum, but it is not being exercised in a political vacuum. Powerful political considerations that have little to do with protecting the environmental, but a great deal to do with protecting organized interests, favor the current command and control approach to environmental policy.

POLITICAL POLLUTION

Despite the advantages of a pollution rights approach, the political response to our environmental concerns has been to embrace direct government regulation and controls in the effort to reduce pollution. There are reasons for the political popularity of directly regulating pollution sources that have nothing to do with environmental concerns. In some cases, environmental concerns are simply a convenient vehicle for promoting hidden agendas that can actually result in a reduction in environment quality. Before turning to a particular example, it will be helpful in providing a general understanding of such nefarious political practices to examine a deficiency in the political decision-making process. This deficiency is closely analogous to the problem that explains excess pollution in private market settings.

Government is the means whereby we are able to do things collectively that would not be done if we had to rely on the volunteer actions of individuals. Government is able to go beyond what is possible though private action through the exercise of coercive power. The existence of this power is to each of us both a threat and an opportunity. On the one hand, government can impose a cost on us without providing us with offsetting benefits. On the other hand, it is possible for us to use government power to acquire benefits for which others will have to pay. The hope of the democratic process is to make government responsive to all the citizens by providing only those public services that are worth at least as much as they cost and by preventing some groups from exploiting other groups. In comparison with alternative political arrangements, democratic institutions perform this task tolerably well. However, narrowly motivated special interests find it relatively easy to use the democratic process to acquire advantages at the expense of the general public.

Government programs, expenditures, and protections often convey benefits that are largely focused on particular industries, occupational associations, or social groups. Agricultural price supports, import restrictions of shoes, and protecting the Maritime Union against competition are a few of a large number of examples. Benefiting groups have a strong motivation to become involved politically for the purpose of protecting and, if possible, expanding their particular programs, subsidies, and privileges.

These groups will each be organized to one degree or another quite independently of their political activity and will therefore find it relatively easy to collectively confront and influence key political decisionmakers. Of course, these special interest programs will impose costs on the general public in the form of higher taxes and prices. But organizing the general public for the purpose of generating political opposition to these programs will face the same problems encountered by an environmental group attempting to get those suffering from pollution to contribute toward the purchase of pollution rights. If others are successful in controlling a special interest program, your taxes will be lowered or your benefits raised whether or not you contributed to the effort. So when a program is being considered that benefits the few at the expense of the many, our political representatives can expect to hear from the few but not from the many. The resulting bias in political results is not hard to predict.

We suffer from excessive *environmental* pollution because the private benefits each of us receives from polluting activities are paid for, in large part, by a defenseless public. We suffer from *political* pollution, or excessive government involvement in a whole host of activities, because the various private benefits received from the government programs also are paid for, in large measure, by a defenseless public. Of course, each of us also suffers from the pollution and pays for the government programs of others. Most of us would be willing to reduce our pollution and the programs we favor if everyone else would do the same. Unfortunately, this does not happen because the political process is flawed for the same reason that the market for a clean environment is flawed: in the absence of private property and exchange, people are not able to communicate their preferences to each other in such a way as to insure honesty and reciprocity.

SPECIAL INTERESTS VERSUS THE ENVIRONMENT

The power that government has assumed in order to impose direct regulation and control over polluting activities has not been ignored by special interests intent on realizing private advantages completely unrelated to environmental concerns: An important case in point is the Clean Air Act, as amended in 1977. The following discussion of this case is based on Navarro (1980). The original Clean Air Act of 1970 was passed in response to widespread concern over a serious environmental problem. While we may quibble over whether this act was the best policy response to our air pollution problems, the evidence indicates that it did have a favorable impact on air quality. It also had other effects.

The 1970 Clean Air Act imposed much tighter pollution controls in areas that were heavily polluted-primarily the heavily industrialized East-than in those areas that were relatively unpolluted-the sparsely populated West. This gave the West a significant advantage over the East in the competition for industry, jobs, and tax revenues. The 1970 act imposed tight standards on the amount of sulfur-dioxide new, or newly modified, electric power plants could discharge. How these emission standards were to be met, however, was not specified by the 1970 act. Electric power plants had two alternatives: 1) burn high-sulfur coal and remove the sulfur by installing expensive, and hard to maintain, stack scrubbers, or 2) burn low-sulfur coal, in which case the standards could be met without scrubbers. Unfortunately for the Eastern coal industry and the United Mine Workers (which has had little success organizing Western miners), low sulfur coal is found in the West. Rather than take a chance on expensive scrubbers, many utilities, some as far East as West Virginia, found importing coal from Wyoming and Montana the most attractive way of satisfying the pollution control requirements.

The beneficiaries of the 1970 Clean Air Act were the general public, who enjoyed cleaner air and, as consumers of electricity, benefited from the flexibility utilities had in meeting the Clean Air Act standards as cheaply as possible. But the general interests of the unorganized public are much more difficult to communicate through the political process than are the focused and organized interests of the United Mine Workers and established Eastern coal producers. A blatant example of political pollution, or the dominance of special over general interests in the political process, came when Senator Howard Metzenbaum of Ohio successfully engineered through Congress the "local coal amendment" as one of the 1977 amendments to the Clean Air Act. This amendment made it illegal to import Western coal into Ohio or Illinois if high-sulfur coal mining jobs are threatened in these states. In order to shelter fewer than 10,000 people from competition, consumers of electricity had to spend millions of extra dollars and breathe dirtier air.

Other major provisions of the 1977 amendments to the Clean Air Act were less transparently motivated by special-interest concerns, but were no less effective at protecting Eastern interests against competition from the West, and doing so at the expense of the environment Environmentalists (many of whom wanted to halt development in the West), Eastern coal producers, the UMW, and Eastern industrialists found common cause in lobbying for passage of several of the 1977 amendments to the Clean Air Act. As one result of the amendments, the Environmental Protection Agency was given formal authority to pursue a "prevention of significant deterioration" (PSD) policy. The PSD policy greatly added to the cost of acquiring the political permission necessary to shift industrial operations out of the East and into the West. Despite the fact that the environmental costs of increased pollution in the sparsely populated and largely pollution-free areas of the West are significantly less than the environmental benefits to be derived from reducing pollution in the congested and heavily polluted areas of the East, the policy of PSD discouraged such pollution transfers.

The 1977 amendments also imposed a "best available control technology" (BACT) requirement on all new sources of potentially significant pollution, regardless of whether they were located in clean areas or dirty areas. This provision forced all new coal fueled power plants to use scrubbers regardless of the sulfur content of the coal they burned. Not being able to satisfy the pollution control authorities by substituting Western coal for the expensive scrubber technology, utilities in the East have little reason to purchase the more expensive Western Coal (Western coal has a lower BTU equivalent than Eastern coal and tends to be more expensive per BTU, particularly in the East).

Unfortunately, the price the public pays for the political privileges of the Eastern coal interests is not just higher utility bills. In many cases the price is more pollution. For example, sulfur emissions from burning Eastern coal with the scrubber technology are often greater than sulfur emissions when Western coal is burned without a scrubber. Some observers argue that the overall effect of the scrubber requirement has been to increase the sulfate air pollution in the Eastern half of the United States. Interestingly, this may be true also in the Western States, even though low sulfur Western coal continues to be burned. The scrubber technology requires some minimum level of sulfur in the coal that is being burned. Since the sulfur content of much Western coal is less than this minimum amount, sulfur often has to be added to the coal before it is burned?

Also, a by-product of using the scrubber technology is a "sludge" that is generally disposed of in land fills. This sludge is not biodegradable and is not firm enough to support buildings, so its disposal effectively removes productive land from use. It has been estimated that over the thirty-year productive life of an 1,800 megawatt electric generating plant, approximately 1,400 acres of land will be required for sludge disposal if the plant uses stack scrubbers. Finally, the scrubbing technology requires water in large quantities. The result is increased chemical and thermal water pollution that presents serious problems even when water is abundant. The problem is, of course, worse in the arid West.

There are other examples of private interests dominating the public interest through the political process, and doing so in ways that are harmful to the environment. For a discussion of some of these examples, see Baden (1986). And it should not go unmentioned that one of the most powerful interest groups that stands to gain at the expense of the environment is the Environmental Protection Agency itself. The EPA has persistently resisted market oriented approaches to environmental protection despite the tremendous environmental benefits that could be realized by those approaches. The most plausible explanation for this resistance is that the centralized command and control policies favored by the EPA can be used to justify far larger budgetary appropriations than market oriented environmental policies. As The *Economist* recently observed, "The EPA exists to regulate things, not to see the market do the job for it." (*The Economist*, 1991).

CONCLUSION

Concern over environmental quality is often used as a convenient rationale for the exercise of political power designed to favor the organized interests of the few at the expense of the unorganized interests of the many. This political pollution takes the form of higher prices, less efficient allocation of our resources, and far less pollution abatement than we are paying for. The best way of abating this political pollution, and providing the information and incentives needed to seriously combat environmental pollution, is by increasing our reliance on private property and free market exchange.

In the absence of private property rights in key environmental resources, market activity will lead to excessive pollution, and a properly structured government response can protect the environment to the benefit of us all. But caution is advisable before embracing the most obvious government solution to the problem of pollution: granting a government agency (or agencies) the power to centrally issue commands and controls. Our bureaucracies are not any better at socialism than were the Soviet bureaucracies. Political power is exercised in the absence of the type of information and motivation that characterizes market exchange based on private property. The consequence is that those who are able to position themselves politically can capture private benefits by imposing cost on others, much the same as polluters can gain private benefits by imposing costs on others. And once political power is created for the purpose of achieving some objective, no matter how noble that objective may be, it is naive in the extreme to imagine that those with political influence will refrain from using this power for their own narrow purposes.

REFERENCES

1. Tietenberg, Tom. 1992. *Environmental and Natural Resource Economics.* Harper Collins Publishers, New York, NY.
2. Navarro, Peter. 1980. The Politics of Air Pollution. *Public Interest*: 36-44.
3. Baden, John. 1986. Destroying the Environment: Government Mismanagement of Our Natural Resources. National Center for Policy Analysis Policy Report. Dallas, TX. No. 24, October.
4. *The Economist.* 1991. William Reilly's Green Precision Weapons. March 30, 1991: p. 28.

Environmental Contaminants, Ecosystems and Human Health. Edited by S.K. Majumdar, E.W. Miller, and Fred J. Brenner.

Chapter Thirty-Two

SEPARATE AND NOT EQUAL: Health-Environmental Risk and Economic-Social Impacts in Remediating Hazardous Waste Sites

MICHAEL GREENBERG

Department of Urban Studies and Community Health
Livingston Campus
Rutgers University
Lucy Stone Hall
New Brunswick, NJ 08903

INTRODUCTION

Spokespersons for African, Latino, and Native-American groups have accused the United States Environmental Protection Agency (EPA), some environmental organizations and industries of environmental racism. These organizations are charged with concentrating locally unwanted land uses (LULUs) in minority areas, bogging down the remediation of contaminants in minority neighborhoods, and allowing minorities to be exposed to dangerous levels of lead, pesticides, asbestos, and other contaminants (1-11). Using hazardous waste site remediation as an example, this paper argues that economic and social considerations should be explicitly included in risk studies. Their inclusion, I believe will help reduce controversies, such as environmental inequity, caused by too narrow a focus on health and environmental risk. I chose hazardous waste sites because their remediation is by far the most expensive environmental health program in the United States and has been featured in the environmental racism contentions.

ENVIRONMENTAL INEQUITY AND HAZARDOUS WASTE SITES

Evidence for inequity consists of case studies in urban and rural areas of Alabama, Illinois, Michigan, Mississippi, New Jersey, New York, North Carolina, Texas, other states, and a few national studies. The following provides a brief reporting of three of the national studies. *The National Law Journal* reported that the EPA was less likely to place toxic waste sites located in minority communities on the National Priority List (or Superfund) and was more likely to propose the waste be contained on site rather than removed (1).

As part of a Greenpeace challenge to incineration, Costner and Thornton compared the proportion of nonwhite residents and home owners, as well as average income, home values, and rent of zip code areas with 16 existing commercial hazardous waste incinerators and 24 proposed incinerators with U.S. average values. They reported that in zip code areas with facilities, the average percent minority population was 89 percent higher (2).

Without question, the strongest and most widely cited evidence of environmental inequity was a study done for the Commission for Racial Justice of the United Church of Christ (3). Using univariate and multivariate statistical methods, the analysts, which included this writer as the technical consultant, showed that zip code areas containing one hazardous waste site had an average of 24 percent nonwhites compared to 12 percent for zip code areas without one. Areas with two or more facilities or one of the nation's five largest hazardous waste landfills had an average of 38 percent minorities.

In 1984, in a book *Hazardous Waste Sites: the Credibility Gap,* I classified racial inequity as one potential type of social impact of a hazardous waste site (Table 1) (12). The other three impacts were health effects ranging from immediate death to emotional illness; environmental damage including loss of an ecosystem and loss of a water source; and economic degradation, such as severe property damage and need to raise taxes to pay for security and cleanup. These four broad categories are the focus of this paper.

TABLE 1
Potential Impacts Near Hazardous Waste Sites

Health	Environment
immediate death	ecosystem elimination
life-shortening exposures	elimination of species
acute illness	reduction of diversity
severe disability	reduction of species
chronic illness	reduction of biomass
chronic disability	loss of water resource
minor or temporary illness	
emotional illness	
Economic	Social
severe damage of structures	disruption of neighborhoods
severe property devaluation	distributional effects:
reduced property appreciation	on poor, specific
loss of land productivity	racial/ethnic groups,
local taxpayers pay for site	areas, future generations
cleanup and security	

Greenberg and Anderson, 1984.

INCLUDED: HEALTH AND ENVIRONMENTAL RISK AT HAZARDOUS WASTE SITES

Since 1981, risk analysis has been the methodological foundation for much of the United State's environmental health policy (13-16). The following structural equation summarizes the variables contributing to health and environmental risk at a hazardous waste site:

(1) Risk = (probability of hazardous waste escaping from the site) x (quantity of the release) x (1/dispersion of the released materials) x (population exposed to contaminants) x (uptake by the population) x (health response of the exposed population).

The federal agencies and nearly all of the state agencies have used quantitative methods embodying the elements of the equation to prioritize sites for remediation. For example, the EPA has the primary responsibility of remediating sites. Charged by Congress with taking into account health, environment, and welfare in its actions, the EPA and its consultants produced a structured-value scoring model that could select at least 400 sites for a National Priority List (NPL) from among the tens of thousands of eligible sites. Using data about waste characteristics, the likelihood of releases through the soil, air, ground and surface waters, and number of people and sensitive environments that have or could be exposed, the Hazard Ranking System (HRS) scores each waste site from 0 to 100 (17).

The HRS is not a quantitative risk assessment. For example, it produces a scaled risk score, not an absolute estimate of risk; it focuses on the worst contaminants, not all contaminants; and it uses mathematical heuristics to estimate the fate and transport of contaminants, not actual fate and transport models. In other words, it uses short-cuts to rate sites, rather than estimating the aggregate risk as morbidity, mortality, or environmental damage.

The HRS model has been applied to thousands of sites and has been the major source of objective information used by the EPA to place over 1,200 sites on the NPL. Although the HRS takes into account population distribution in its calculations, it contains no explicit component(s) for economic damage, neighborhood disruption, and the location of particular racial/ethnic, age, economic or other groups. In other words, it is devoid of socio-economic and community-based criteria.

When the HRS was first applied to selecting 400 NPL sites, a score of 28.5 out of 100 was the threshold for inclusion on the NPL. With allowance for challenges, EPA has retained the 28.5 threshold. A high score for a single contaminant on any single pathway will trigger a score of 28.5 or more. An astute state or local official has an excellent chance of getting a hazardous waste site on the NPL by instituting a concerted effort to find contamination. Some states and local governments have spend considerable amounts of money to look for signs of contamination in or near a well. Other states have not. For example, writing about the limited number of Superfund sites in Louisiana, Schleifstein and O'Byrne concluded: "We used to say there weren't any hazardous waste sites in this state, so when Superfund came along New Jersey got all the money" (18).

Some states provided so little data that the agency models use coefficients to estimate "potential" health and environmental impacts.

In short, two sites posing identical risks to public health and the environment might not get a similar HRS score. One located in an aggressive state is likely to get a high-priority NPL listing, whereas the second, located in a recalcitrant state, may not be listed on the NPL. The difference could mean millions of cleanup dollars for one and no remediation of the second.

Other federal agencies have their own prioritizing models. The United States Department of Defense (DOD) has a site ranking model, the Defense Priority Model (DPM) (19). Structured along the lines of the EPA's HRS model, the DPM also does not have an explicit component for economic and social impact.

The Agency for Toxic Substances and Disease Registry (ATSDR) was charged by Congress with providing a second opinion on EPA's Superfund sites. It was to consider the impact of waste sites on "quality-of-life" in its analyses. ATSDR has written an internal document on quality of life, but has not yet systematically examined the impact of sites and their remediation on the economy and social functions of neighborhoods (20).

The United States Department of Energy (DOE) is the only major agency that to the knowledge of this writer has an explicit economic/social component in its site ranking approach (21-22). The DOE developed the Environmental Restoration Priority System (ERPS), a multiattribute model, which includes a socioeconomic impact factor that has a weight of 9.5 out of 100. By comparison, health risk has a weight of 36, uncertainty reduction a weight of 32, environmental risk a weight of 13, and regulatory compliance a weight of 9.5. There may be questions about the merits of 9.5 out of 100 assigned to socioeconomic impacts. But more important is the fact that the DOE has decided not to use ERPS for the foreseeable future. Approximately 70 percent of its funds for remediation have been committed by legal agreements with states. Other states have no agreements and are not assured of remediation funds.

In addition to these site-ranking methods used by the federal agencies, a more formal form of quantitative risk assessment is used at sites which the agencies judge to be among the most hazardous. None of these include economic and social components (23). Full risk assessments are expensive, have been done at relatively few sites, and are limited by the generic problems associated with risk assessment. A National Research Council Committee studied all of the published scientific literature on health effects that could be linked with hazardous waste sites (24). They concluded that "a decade after implementation of Superfund, and despite Congressional efforts to redirect the program, substantial public health concerns remain, and critical information on the distribution of exposures and health effects associated with hazardous waste sites is still lacking" (p. 1). The committee could nether confirm nor deny health effects. There were insufficient data, and there is uncertainty about the ability of existing epidemiological methods to find health effects problems when they existed.

Risk assessment has generic limitations. Talcott, a research analyst at the EPA Office of Policy Analysis, describes a "gallery of uncertainties" in risk analysis at every stage. In the case of hazardous waste sites, the probability of a release is uncertain, the quantity of the release is uncertain, the dispersion is difficult to model, the ability to estimate the uptake of toxins by the population-at-risk is problematic, and the response of the population exposed to a wide variety of substances in low doses is unknown. Talcott concluded that uncertainties are "large, ubiquitous, inevitable, and only partly reducible" (14). Even though risk assessments have limitations, they are preferable to the site-ranking models which depend on short-cuts. Hence it is distressing to report that Doty and Travis found a poor correlation between HRS scores and potential danger at sites as measured by a full risk assessment (25).

NOT INCLUDED: ECONOMIC AND SOCIAL IMPACTS AT HAZARDOUS WASTE SITES

In comparison to health and environmental risks, economic and social impacts listed in Table 1 can be estimated with far greater certainty and less opportunity for blatant gaming. Methods exist for separating out economically impacted areas and for estimating the benefits of remediation. For example, Greenberg and Hughes used New Jersey's housing sales data base to compare changes in housing prices in towns with and without Superfund waste sites. Rural townships and areas experiencing rapid price increases that also had an NPL were found to have a statistically significant tendency to have lower rates of increase in housing prices (26).

Looking for a second screening method, Greenberg and Hughes surveyed tax assessors in New Jersey to estimate how many towns with hazardous waste sites suffered property value and land use impacts (27). Tax assessors representing about 20 percent of municipalities with a hazardous waste site reported that proximity to the site lowered the appreciation of property values, deterred land uses, and affected land uses within 1/2 mile of the site. But they reported many sites had no impact because they were in remote locations, contamination had already been remediated, and the confounding impacts of an economic recession, high taxes, and planning regulations made it impossible to isolate the effect of the waste site.

We suspect that some tax assessor's responses were not completely accurate, and we know that housing value data obscure actual impacts because housing turnover and sales do not occur immediately. Yet these studies show that it is feasible to determine the kinds of neighborhoods and municipalities that suffer economic damage without spending a massive amount of money and time.

More detailed analyses can be done in neighborhoods where severe impacts are expected. For example, using a housing sales data in Kitchener (Ontario), Skaburskis found that property devaluation was about 15 percent adjacent to the site and attenuated at an average of $37 per foot (28). No impact was observed at a distance beyond 354 feet. McClelland et al. reported an average increase in value of $5,000 per home (almost $20 million for all homes) in a Los Angeles neighborhood after a nearby landfill was remediated. Psychometric methods can be used to determine the underlying feelings that motivate property devaluation (29).

Methods also exist to measure the extent of inequity by race, age, income and other demographic characteristics. But before summarizing these methods, I note that the existing hazardous waste remediation program rests on the ethical principle that resources should be used to minimize the overall risk to society. It is not the only relevant ethical principle. Locally unwanted land uses (LULUs) should not be disproportionately located in areas occupied by economically and politically disadvantaged populations, especially since the disadvantaged probably use less resources. Nor should sewage treatment plants be allowed to smell, hazardous waste sites be allowed to leak off-site, and other environmental problems be neglected in these areas simply because the people have insufficient economic power and political connections to insist on remediation. In other words, there should be some balance between the benefits to a community and the costs it bears.

An initial list of disadvantaged populations includes African-American, Latino-American, and Native-Americans; poor; very young and elderly; and the infirm. The U.S. decennial census counts suffer from underreporting of the poor and minorities. With this caveat in mind, data for these populations are reported at the census block, tract, municipality, and county scales, all of which are potentially appropriate as areas of burden or impact. Similar data are available for larger units, such as groups of counties or municipalities, metropolitan areas, states, and state groupings and the United States. These could represent areas of benefit. Researchers need to determine the closest approximation to the areas of potential benefit and burden. In addition, multiple statistical methods are needed to be assured that the results are independent of the method of evaluation (30-31).

NOT EQUAL: ECONOMIC AND SOCIAL IMPACTS IN THE RISK MANAGEMENT OF NPL SITES

The American federalist system has permitted states to play a major role in remediating hazardous waste sites. Some have, others have not. In 1984, I examined the initial list of 546 sites on the NPL (12). New Jersey had 85, Michigan 48, Pennsylvania 39, and New York 29 sites. In fact, 10 states had sixty percent of the sites. Striking because of their absence from the list of 10 were Louisiana, Texas, Tennessee, and North Carolina - states with large industrial complexes. In order for states to have many sites on the list they must have hazardous waste management programs that search for sites and contamination.

When an industry is responsible for the contamination it can be made responsible for 100 percent of the cleanup costs. But it may legally challenge the designation of the responsible party. Some states will pay the cost and try to recover the money later. Others do not. When a site is not attributable to a responsible party, government directly becomes the responsible party. States must contribute ten percent of the remediation cost to receive federal funds. Some do not. Finally states are responsible for 100 percent of operating and maintenance costs of Superfund sites. Some states cannot afford O&M payments. Some states have funds or fees to remediate sites that are not covered by federal money. Others do not. The net result is that states like California, Michigan, New Jersey, New York, and a few others have aggressively moved to remediate sites and obtain as many dollars as possible from federal and responsible parties. Other states have chosen to do almost nothing. The huge discrepancy between states is underscored by two realities that exist: some states were doing so little that EPA permitted every state to designate one site for the NPL - any site; other states were moving so rapidly that they passed their own bond issues and threatened to sue EPA to obtain money from the federal Superfund. In short, the limitations of the risk-ranking models and risk management processes lead me to conclude that funds for remediation are not being distributed in proportion to health, environmental, economic or social impacts.

It is abundantly clear from the loud protests of communities that local viewpoints have not been adequately represented in decision-making. EPA's own management review of Superfund criticized the agency for not reaching out to communities (32). More recently, in testimony to Congress, EPA's efforts to involve citizens was labeled "community relations" rather than participation (33). In principle, federal and state agencies should make every effort to communicate with local officials about remediation, especially local health officers (LHOs) who are likely to be familiar with existing local environmental health problems. Yet a recent study of New Jersey health officers found considerable anger on the part of LHOs who felt that government officials, especially federal ones, were not keeping them informed about plans for local sites nor soliciting their views on site-related health issues. The LHOs, many of whom have advanced degrees in public health and are familiar with risk assessment, indicated that they were learning about major decisions after the fact. LHOs who felt that federal agencies actively solicited their views about monitoring, environmental impacts, and site remediation were more likely to trust the federal representative and to believe that agency plans were consistent with the health officer's goals for the town as whole (34).

IMPLICATIONS OF ADDING EXPLICIT ECONOMIC AND SOCIAL VARIABLES TO THE PRIORITIZING PROCESS

Adding economic and social variable will create some problems for officials. For example, adding property damage could lead to less priority for poor neighborhoods because poor Americans tend to live in less valuable housing. Some analysts trained in existing site-ranking and risk assessment methods and not familiar with economic and social analyses will be troubled. Districts that already have a high remediation priority may fear that their remediation will be deferred. Some scientists refer to economic and social impacts as "soft" indicators used by people with irrational phantom fears. Doubtless, many of these scientists will consider economic and social analyses to be a waste of time and money.

These and other concerns are balanced against two major advantages. Advocates of remediation find it difficult to justify the expenditure of $750 billion or more over the next three decades to remediate inactive and active sites, especially since it has been difficult to find evidence of health effects (35-37). Estimating local economic benefits of remediation sites will show that at least some cleanups will more than pay for themselves in expected increases in property values, jobs, and land use redevelopment opportunities.

Many senior administrators, in my opinion, were blind-sided by the environmental racism charge-that is, they were neither forewarned nor prepared. Explicit recognition of the importance of economic and social impacts will require officials to contend with concerns that trouble many Americans, such as the concerns that led to the charge of environmental racism. I use the State of New Jersey to illustrate this point. No state has made a larger commitment to remediating hazardous waste sites than has New Jersey. New Jersey had 113 sites on the NPL at the time this paper was written, almost 10 percent of the total in the United States. In addition the New Jersey Department of Environmental Protection and Energy (NJDEPE) manages thousands of sites with its own hazardous waste site, landfill, leaking underground storage tank, water pollution, and related programs. In fiscal year 1992, the NJDEPE hazardous waste program had 750 positions and a budget of $48 million (38).

Are NPL ratings in New Jersey associated with race and ethnicity? That is, are NPL sites in New Jersey with a high priority for cleanup located in municipalities with few or many black and Hispanic-Americans? Using 1989 Superfund data, the 90 municipalities with Superfund sites were divided into quintiles based on their HRS score. Municipalities hosting the first quintile of sites (NPL ratings 1-168) had an average of 9.3 percent blacks and Hispanics. With the caveat that some blacks also identify as Hispanics so that there is some double-counting, the remaining four quintiles averaged 10.1, 11.9, 15.8, and 16.7 percent blacks and Hispanics. In other words, the sites at the top of the remediation list had the smallest proportion minority, and the proportion of minorities increased as the priority decreased. T-tests of difference of proportions show that these differences are not statistically significant ($P<0.05$), yet they are troublesome.

Has the government conspired to rank sites with significant minority populations as lower priority in New Jersey? I think not. When I compared my initial lists of hazardous waste sites in 1981 to the 1989 Superfund list, I noticed that many sites located in cities with large minority populations were remediated before the Superfund process began or shortly thereafter because these sites contained explosive and flammable materials that threatened the community. Once these hazards were removed, the site no longer needed to be on the priority list or was a low-priority because the chronic threat was limited. Second, in New Jersey and most other places in the United States, groundwater contamination has been the reason that most sites have been placed on the NPL and received a high priority. For example, New Jersey had 10 of the 50 highest rated NPL sites in the United States. Six were former landfills, four were industrial sites; none were in the northern part of the state where most of the population, including minorities live. All ten sites had one characteristic in common: they threatened or had already contaminated the domestic underground drinking water supplies. Disproportionately Black, Hispanic, and city residents live in municipalities that rely on surface water not on groundwater. Should New Jersey's priority ratings be changed as a result of these data? It is not the purpose of this paper to answer that question. It is the purpose of this paper, however, to point out that it is better for the agencies to generate these data and think about the ethical and practical implications of their choices than it is to have a nonprofit organization or newspaper reporter generate the data and then confront and embarrass the agency with the data.

DISCUSSION

On February 10, 1994, the *New York Times* reported that President Clinton was going to order all federal agencies to make sure that their programs do not unfairly burden the poor and minority populations (39). The order is a step in the direction of no longer treating economic and social impacts as separate and unequal impacts. Yet I fear that some officials will treat environmental equity as another hurdle to clear rather than as an opportunity to think about the goals of the hazardous waste remediation program. They will measure the geographical distribution of race and income, determine if there is any evidence of inequity, and make adjustments in priorities and cleanup schedules, when required. They will, I am afraid, not recognize that the outrage encapsulated in the environmental inequity charge is an opportunity to broaden the scope of the remediation program from a narrow focus on health and environmental risk to a much more realistic view of the concerns of people who live near waste sites.

I have argued in this paper that present funding for remediation is not driven by hard scientific facts nor does it assure equity among regions, states, or populations. The credibility gap I pointed to over a decade ago (12) will grow as more people recognize the veneer of science guiding the expenditure of hundreds of billions of dollars. A far better alternative is to include economic and social impacts in the risk-based ranking models and make the entire decision-making process as transparent as possible. For example, the agencies could borrow the idea of weighting impacts from the DOE's ERPS model. The federal government could create a set of national weights that are defaults. State governments could be permitted to modify the weights based on their priorities. In the short run, if my suggestions are accepted, the federal government, states, and local governments are likely to become embroiled in arguments over the importance of health, environmental, economic, and social impacts. In the long-run the outcome will reflect a consensus and have greater credibility than a process that is largely hidden from the view of interested parties.

I view President Clinton's executive order and the upcoming reauthorization of the Superfund legislation as genuine opportunities to reconsider issues such as how clean is clean, and who should pay for cleanups. But it is also an opportunity to consider how the economic benefits of remediation should be weighed, how the interests of the poor, minorities, and future generations should be protected, and how recalcitrant states should be treated. This does not mean that health and environmental risk-based analyses are not important. Nor does it mean that we need complex economic and social analyses to match the complex health and environmental risk-based ones. It does mean that adding new economic and social concerns is more important than continually refining the variables and parameters of the risk-based models, and it does mean that we need a broad context for a program that will cost hundreds of billions of dollars. To call for the explicit inclusion of economic and social variables does not elevate phantom risks propagated by uneducated and irrational people to prominence. The economic and social impacts in some locations are real and can be measured. To continue to pretend that they do not exist and cannot be measured is irrational.

ACKNOWLEDGMENTS

I would like to thank my colleagues Caron Chess, Mark Lapping, Frank Popper, and Dona Schneider for their helpful comments on an earlier draft of this paper.

REFERENCES

1. *National Law Journal.* 1992. Unequal Protection: the Racial Divide in Environmental Law. September 21.
2. Costner, Pat and Joe Thornton. 1990. *Playing With Fire: Hazardous Waste Incineration.* Greenpeace, Washington, D.C.
3. Public Data Access. 1987. *Toxic Waste and Race in the United States.* Commission or Racial Justice, United Church of Christ, New York City (NY).
4. Environmental Protection - Has It Been Fair. 1992. *EPA Journal.* 18, March/April, 64 pp.
5. Goldberg, Daniel. 1992. The Quest for Environmental Equity. MSW *Management.* March/April, 25-33.
6. Bullard, Robert. 1992. *Dumping on Dixie: Race, Class, and Environmental Quality.* Westview Press, Boulder (CO).
7. Roberts, Sam. 1992. In My Backyard? Where New York City Puts Its Problems. *New York Times.* December 6, L54.
8. Bryant, Bunyan and Paul Mohai (eds.). 1992. *Race and the Incidence of Environmental Hazards: A Time for Discourse.* Westview Press, Boulder, CO.

9. Cushman, John Jr. 1992. Environmental Hazards to Poor Gain New Focus at E.P.A. *New York Times.* January 21, C4.
10. Collin, Robert. 1992. Environmental Racism: a Law and Planning Approach to Environmental Racism. *Virginia Environmental Law Journal.* 11, 495.
11. Schneider, Keith. 1993. The Regulatory Thickets of Environmental Racism. *New York Times.* E5.
12. Greenberg, Michael and Richard Anderson. 1984. *Hazardous Waste Sites: the Credibility Gap.* Center for Urban Policy Research, New Brunswick, NJ.
13. Silbergeld, Ellen. 1993. Risk Assessment: the Perspective and Experience of U.S. Environmentalists. *Environmental Health Perspectives.* 1010:100-104.
14. Talcott, Frederick. 1992. How Certain Is That Environmental Risk Estimate? *Resources.* 107:10-15.
15. Harris, Robert and David Burmaster. 1992. Restoring Science to Superfund Risk Assessment. *Environ Report.* 6:6-11.
16. National Research Council. 1993. *Issues in Risk Assessment.* National Academy Press, Washington, D.C.
17. United States Environmental Protection Agency. 1988. *Review of the Superfund Hazard Ranking System.* U.S. EPA, Washington, D.C.
18. Schleifstein, Mark and James O'Byrne. 1991. The Wasteland. *The Times-Picayunne.* March 24, 1.
19. United States Department of Defense. 1991. Defense *Priority Model, User's Manual, FY 92 Version.* U.S. Department of Defense, Washington, D.C.
20. Von Allem, Steven, Greenwell, Mike, Hansen, Hugh, Perdue, Susan, Price, Patricia and Chris Schmidt. 1992. *Quality of Life and ATSDR's Mission.* 1992.
21. Keystone Center. 1991. Keystone National Dialogue on Federal Facility Environmental Management: First Plenary Session Summary, June 12-13.
22. U.S. Department of Energy. 1991. *Preliminary Design Report: DOE Environmental Restoration Priority System.* U.S. Department of Energy, Washington, D.C.
23. United States Environmental Protection Agency. 1989. *Risk Assessment Guidance for Superfund.* 2 volumes, U.S. EPA, Washington, D.C.
24. Committee on Environmental Epidemiology, National Research Council. 1991. *Environmental Epidemiology: Public Health and Hazardous Wastes.* National Academy Press, Washington, D.C.
25. Doty, Carolyn and Curtis Travis. 1990. Is EPA's National Priorities List Correct? *Environmental Science and Technology.* 24:1778-1780.
26. Greenberg, Michael and James Hughes. 1992. The Impact of Hazardous Waste Superfund Sites on the Value of Homes Sold in New Jersey. *Annals of Regional* Science, 26:147-153.
27. Greenberg, Michael and James Hughes. 1993. Impact of Hazardous Waste Sites on Property Values and Land Use: Tax Assessors' Appraisal. *The Appraisal Journal.* January, 42-51.

28. Skaburskis, Andrejs. 1989. Impact Attenuation in Conflict Situations: the Price Effects of a Nuisance Land-Use. *Environmental and Planning A.* 21:375-383.
29. McClelland, Gary, Schulze, William and Brian Hurd. 1990. The Effect of Risk Beliefs on Property Values: A Case Study of a Hazardous Waste Site. *Risk Analysis.* 10:485-497.
30. Zimmerman, Rae. 1993. Social Equity and Environmental Risk. *Risk Analysis.* 13:649-666.
31. Greenberg, Michael. 1993. Proving Environmental Inequity in Siting Locally Unwanted Land Uses. *Risk: Issues in Health and Safety.* 4:235-252.
32. U.S. Environmental Protection Agency. 1989. *A Management Review of Superfund.* U.S. Environmental Protection Agency, Washington, D.C.
33. EPA Efforts to Involve Citizens Called More P.R. Than Substance. 1993. *Hazardous Waste News.* October 19, 327-328.
34. Greenberg, Michael. 1993. Local Health Officers' Views on Hazardous Waste Remediation. *Amer. J. Public Health.* 83:752-754.
35. Russell, Milton, Colglazier, E. William and Mary English. 1991. *Hazardous Waste Remediation: the Task Ahead.* Waste Management Research and Education Institute, Knoxville (TN).
36. Williams, Ted. 1993. The Sabotage of Superfund. *Audubon.* July/August, 30-37.
37. Van Voorst, Bruce. 1993. Toxic Dumps: the Lawyers' Money Pit. *Time.* September 13, 63-64.
38. Conversations with Lance Miller, Deputy Commissioner, New Jersey Department of Environmental Protection and Energy, 1992-1993.
39. Cushman, John Jr. 1994. Clinton to Order Pollution Policy Cleared of Bias. *New York Times* February 10, A1, D20.

Environmental Contaminants, Ecosystems and Human Health. Edited by S.K. Majumdar, E.W. Miller, and Fred J. Brenner. © 1995, The Pennsylvania Academy of Science.

Chapter Thirty-Three

RATIONALE FOR CONSIDERING AROCLOR 1260 AS A REPRESENTATIVE PCB IN RISK ASSESSMENT

D. MUKERJEE[1,3], SHANE QUE HEE[2] and O. WASSERMANN[3]

[1]U.S. Environmental Protection Agency
Environmental Criteria and Assessment Office
Cincinnati, OH 45268
[2]Department of Environmental Health Sciences
and the UCLA Center for Occupational and Environmental Health
School of Public Health
University of California at Los Angeles
Los Angeles, CA 90095-1772
[3]Institut Für Toxikologie
Klinikum Der Christian-Albrechts-
Universität Zu Kiel
Kiel, Germany

ABSTRACT

Epidemiologic data are inadequate to prove the carcinogenicity of PCBs in humans. Chronic animal bioassay data for commercial PCB mixtures such as Aroclor 1260, Aroclor 1254, Kanechlor 500, Clophen A-60 and Clophen A-30, suggest carcinogenicity. Only the data for Aroclor 1260 have requisite quality for cancer risk estimations. Since two strains of rats showed dose responses that were statistically significant for hepatocellular carcinoma in chronic feeding studies. Kanechlor 500 exposure in the diet produced liver tumors in mice as did

The opinions and assertions expressed herein are those of the authors and not to be construed as reflecting the opinion of US EPA. Mention of trade names or commercial products does not constitute endorsement or recommendation for use.

Clophen A-30, and Clophen A-60 in rats, Aroclor 1254 in the diet induced dose response incidences of liver tumors in mice and rats but were not statistically significant. Thus PCB mixtures are classified as probable human carcinogens with the same potency as Aroclor 1260.

INTRODUCTION

Polychlorinated biphenyls (PCBs) are a family of compounds based on biphenyl as the parent compound. There are 209 possible PCB isomers and congeners of which some 187 have been chromatographically separated on the one column, and all have been synthesized (Mullin et al., 1984).

Commercial PCBs are complex mixtures of PCB congeners and isomers. The US manufacturer designated a 4-digit code for its PCBs marketed under the trade name Aroclor. In Germany the trade name is Clophen; in France, Phenoclor or Pyralene; in Japan, Kanechlor or Santotherm; in Italy, Fenclor; and in USSR, Sovol. For Aroclors the first two digits, 12, indicate that the preparation is a biphenyl (C_{12}) mixture and the second two digits indicate the chlorine content percentage by mass. Thus, Aroclor 1242 is a biphenyl mixture with an average chlorine content of 42%. The US manufacturer also developed a mixture that was primarily mono-, di-, and tri-chloro isomers, Aroclor 1016. The chromatographic pattern and the actual percent chlorine (41%) is very similar to Aroclor 1242.

The composition of an Aroclor in the environment changes with time because of selective adsorption, evaporation, solubility and biodegradation of the specific PCBs. Water solubilities and vapor pressures of PCBs increase with decreasing chlorination (Hutzinger et al., 1974). Lower chlorinated PCBs are mostly removed from water bodies by biodegradation, volatilization and solubilization (Hutzinger et al., 1974), although sedimentation removal also occurs. Higher molecular weight PCBs adsorb to sediments and are absorbed by biota, although some volatilization may occur.

PCB residues have been detected in all three primary (soil, air, water) and secondary (aquatic and terrestrial plant and animal food systems; consumer products etc.) compartments of the environment. PCBs volatilize from hazardous waste sites and ambient waters with half-lives ranging between 2 months to more than 150 years. PCBs are highly lipophilic and relatively stable, and bioaccumulate easily in biota and in humans. PCBs are often detected in fish, wildlife and human milk, adipose tissue and blood serum (Risebrough et al., 1968; Buckley, 1982; Ballschmiter et al., 1981; Wasserman et al., 1979; Higuchi, 1976).

EXPOSURE

The major exposure routes to humans are through food and by inhalation. Exposures may also occur from contact with industrial products, accidental contamination of foodstuffs, or from association with contaminated environmental components. Dermal exposure is also important in occupational exposures, in cleaning up PCB spills or on hazardous waste sites containing PCBs. Total PCB concentrations are based on specific congener analysis or direct perchlorination rather than in terms of Aroclors because the congener patterns in environmental media and biological tissues usually do not match those in Aroclors unless massive contamination has occurred. Consequently, predictive models based on specific congener data must also be utilized.

Incineration is one of the major sources of PCB emissions to the air compartment. PCBs form polychlorinated dibenzofurans (PCDFs) when heated. Consequently, a route with public health implications is thermal production of PCDFs from PCBs. There are at least four major pathways: loss of two *ortho* chlorines, loss of *ortho* hydrogen as well as chlorine, loss of an *ortho* hydrogen as well as chlorine but involving a chlorine shift from the 2- to the 3-position, and loss of two ortho hydrogens. The less chlorinated congeners predominate in air samples from known contaminated areas and in water and wet deposition samples.Temperature and the amount of sediment in river and water samples are important covariables.

Dietary intake of PCBs has been reported for adults, infants, and toddlers, in the Food and Drug Administration [FDA's] Market Basket Studies (FDA, 1982a, b). Bioaccumulation of PCBs in fish and other aquatic life is a major source of exposure to humans. PCBs have also been measured in the subcutaneous fat and breast muscle of 55 waterfowl collected in New York State along the polluted Hudson River and near Long Island during 1981 and 1982 (Kim et al., 1985).

Total dietary intakes of PCBs for adults in FY79 was 0.0133 μg/kg/day (FDA 1982a). A 0.0053 μg/kg/day was from dairy products, 0.0075 μg/kg/day from meat, fish and poultry, and 0.0005 μg/kg/day from oils and fats. Total dietary intakes of PCBs for infants and toddlers in FY78, as reported by FDA, 1982b, were 11.3 and 98.5 ng/kg/day, respectively. The FDA calculation assumed that the average infant weighs 8.2 kg (6-month-old) and the average toddler (2-year-old) 13.7 kg.

Even though PCBs have low vapor pressures they have been detected in ambient air, indoor air and in occupational environments. The congeners of highest vapor pressure are enriched in the air so that PCB profiles in air differ from the original PCB.

PCBs degrade at a 2 sec residence time at 1200C in 3% excess oxygen and at 1.5 sec residence time at 1600C with 2% excess oxygen (Johnston, 1985).

HAZARD ASSESSMENT

PCBs are highly lipophilic and relatively stable. Higher chlorinated PCB mixtures rapidly bioaccumulate in the biota, and are detected in fish, wildlife and human milk, adipose tissue and blood.

Chloracne is the most commonly encountered dermatologic symptom. Other reported dermatologic symptoms include rash, burning sensation, pigmentation (darkening), thickening, and discoloration of the fingernails. It is not clear whether PCB mixtures are solely responsible for chloracne or whether PCDF contaminants are also important in chloracne and other adverse health effects. In the Yusho and Yu-cheng poisoning incidents, the correlation of toxic effects is to the PCDFs rather than the PCBs in the contaminated rice oil, in the liver, and in other tissues of victims. Hepatic effects associated with PCB exposure were hepatomegaly, hepatic enzyme induction with accelerated rate of metabolism of drugs and hormones, and hepatic dysfunction indicated by increases in serum hepatic enzyme activities. A decrease in pulmonary function associated with cough, wheezing, tightness in chest, and upper respiratory and eye irritation were also reported in PCB-exposed capacitor manufacturing workers.

In evaluating the toxicity of PCBs in animals, it is important to consider the congener and isomer composition, both of the PCBs and of the potential impurities, the length of exposure, and the species under investigation. In general, PCB mixtures have low to moderate acute toxicity to mammals. Single dose oral LD_{50} values of commercial PCB mixtures in rats range from 1.0-11.3 g/kg bw. Limited data also suggest that the acute toxic potency of PCB mixtures is similar in other species following oral, dermal or i.p. exposure. Data on purified PCB isomers have established that the toxic, metabolic and toxicokinetic behaviors of the isomers and congeners varies not only with the degree of chlorination but also with the position of the chlorine atoms. The most toxic and persistent *sym* -PCB was 3,4,5-*sym*-hexa-CB (LD_{50}=19 mg/kg bw/day) isomer, followed by 2,4,6-*sym*-hexa-CB; 2,4,5-*sym*-hexa-CB; and 2,3,6-*sym*-hexa-CB. Although structure-activity relationships are very interesting, the highly toxic, coplanar PCB isomers, such as 3,4,5-*sym*-hex-CB, have only been detected as very minor constituents of commercial PCB formulations.

PCB rapidly bioaccumulate to toxic levels following long-term, low-level exposure. A major target organ is the liver. An increase in the liver-to-body weight ratio is one of the most sensitive exposure indicators. Hepatomegaly has been observed in rats ingesting diets containing PCB's. Hepatomegaly results from liver cell hypertrophy, which is caused by fatty infiltration and proliferation of the smooth endoplasmic reticulum. The latter response is associated with the induction of hepatic enzymes, particularly the microsomal mixed function oxidases. Urinary fluorescence, which precedes porphyria has also been reported after exposure of rats for 16 weeks to 10 ppm of Aroclor 1254 (0.5 mg/kg bw/day). Focal necrosis and iron-containing deposits in the Kupffer cells occur at higher levels of exposure. PCBs also produce immunosuppression, which may be associated with thymic atrophy, lymphocytopenia and splenomegaly. Other PCB-related toxic effects include reduction in food and water intake, wasting syndrome (though this has been associated with PCDF toxicity), and decreased body temperature.

Another sensitive indicator of PCB exposure is an enlarged thyroid. Ultrastructural evidence suggestive of increased thyroid gland activity has been reported in rats maintained on diets containing as little as 5 ppm Aroclor 1254 (0.25 mg/kg bw/day) for 4 weeks.

Tests of the acute lethality of PCBs in laboratory animals, with the possible exception of guinea pig, suggest that, in general, commercial products of PCBs have similar acute toxicity regardless of route of exposure, species or age of the animal. The single dose oral LD_{50} values for commercial formulations in rats, rabbits, mice and mink range from 0.5-11.3 g/kg bw (Grant and Phillips, 1974; Bruckner et al., 1973; Kimbrough et al., 1978; Garthoff et al., 1981; Aulerich and Ringer, 1977). Route of exposure also has little effect (<1 order of magnitude) on lethality with the lethal dose for dermal administration in rabbits ranging from 0.8-3.2 g/kg bw (Nelson et al., 1972); while the lethal dose in mice administered PCBs by i.p injection, ranged from 0.9-1.2 g/kg bw (Lewin et al., 1972).

In a chronic study that defined a NOAEL, BALB/CJ mice were maintained for 9 months on diets containing 0, 3.75, 37.5 or 375 ppm of the Aroclors 1221, 1242 or 1254 (0.45, 4.57 or 45.7 mg/kg bw/day). The Aroclor with the lowest chlorine content (Aroclor 1221) produced no liver lesions, while exposure to Aroclor 1242 resulted in increased liver weight in the high dose group. Mice exposed to Aroclor 1254 showed increased mortality in the high-dose group, mild hepatopathology in the mid-dose group, and no liver lesions in the low-dose group. The NOEL observed in the study, using mice of 0.45 mg/kg bw/day, is nearly identical to the LOELs of 0.5 mg/kg bw/day associated with porphyria in rats and 0.25 mg/kg bw/day associated with enlarged thyroid. Monkeys are highly sensitive to the toxic effect of PCBs. Signs of toxicity with chronic exposure to Aroclor 1248 include severe facial and subcutaneous edema, comedones and cysts of the meibomian glands, gastric lesions, body weight loss and reduced hemoglobin and leukocytes. The lesions are unique to the monkey and resemble chloracne in man. In the monkey, chronic exposure to as low as 0.1 mg/kg bw/day of Aroclor 1248 produces frank toxic effects; no studies have been conducted from which a NOAEL can be derived or to indicate how close 0.1 mg/kg bw/day is to the NOAEL for monkeys. There are reports of a slight increase in the incidence of cleft palate in the progeny of the mice exposed to PCBs during gestation. PCBs can cause fetal toxicity including resorptions, abortions, reduced birth weight, and decreased postnatal survival in several species. The mink and monkey are the most sensitive species tested relative to reproductive toxicity. Complete reproductive failure was observed in mink maintained on a diet containing 5 ppm Aroclor 1242 (0.75 mg/kg bw/day) for 18 months. Rhesus monkeys maintained on diets containing 2.5 of 5 ppm Aroclor 1248 (0.1 or 0.3 mg/kg bw/day) for 18 months had increased abortions at the low dose and maternal toxicity with no live births at the high dose.

HIGH RISK POPULATIONS

Two separate groups of high-risk include those persons with the potential for frequent or high exposure, namely, occupationally-exposed workers and breast-fed infants. PCBs are excreted in the breast milk of lactating humans. Fetuses and neonates (2-3 months old) have a limited ability to metabolize and excrete PCBs. Infants born to women exposed to PCB during pregnancy, as in the Yusho incident, are generally small for gestation age and exhibited dark brown pigmentation on the skin and mucous membranes, gingival hyperplasia, or early eruption of teeth, and facial edema. The PCDF impurities and their components may also contribute to the severity of these effects. The WHO/EURO working group recommended that "despite the presence of PCBs, PCDDs and PCDFs in human milk, breast feeding should be encouraged and promoted on the basis of convincing evidence of the benefits of human milk to the overall health and development of the infant" (WHO, 1988).

RISK ASSESSMENT

PCBs (Aroclor 1260, Kanechlor 500, Aroclor 1254, Clophen A-30 and Clophen A-60) have been evaluated for carcinogenicity in several animal bioassays. Aroclor 1260 induced a statistically significant increase of hepatocellular carcinomas in two rat (Sherman, Sprague-Dawley) feeding studies. Kanechlor 500 produced a statistically significant liver tumor response in dd mice after dietary ingestion for 32 weeks. Aroclor 1254 fed in the diet to mice (Balb/cj) and rats (Fischer 344) induced liver tumors that were dose-related but not statistically significant. Clophen A-30 and A-60 induced hepatocellular carcinomas in rats after 832 days of feeding 100 ppm in the diet. This level of carcinogenic evidence in rats and mice constitutes sufficient evidence for carcinogenicity of these commercial PCBs in animals.

Significant risk of liver cancer has been reported among victims of the Yusho accident in Japan, which involved ingestion of contaminated rice oil but there is some uncertainty regarding concurrent exposure to other possible carcinogens like PCDFs. At present, the human epidemiologic evidence is suggestive, but not conclusive.

The positive carcinogenic evidence in rats and mice, together with inadequate evidence in humans, places Aroclor 1254 and 1260, Kanechlor 500, and Clophen A-30 and A-60 in the weight-of-evidence category B2, as a probable human carcinogen. PCB mixtures with significant amounts of common components present also as in Aroclor 1254 and 1260, Kanechlor 500, and Clophen A-30 and A-60 are likely to present a carcinogenic hazard to the humans. Recognizing the variety and variability to PCB mixtures, all PCB mixtures are currently considered to have a carcinogenic potential (category B2) similar to that of Aroclor 1260 (EPA 1990). The human carcinogenic potency of Aroclor 1260 is estimated to be 7.7/mg/kg/day from the multistage model using the hepatocellular carcinoma data from a Sprague-Dawley rat chronic Aroclor 1260 feeding study (Norback and Weltman, 1985). Given a lack of adequate information about which pure constituents of Aroclor 1260 or any other PCB mixture are carcinogenic, Aroclor 1260 is assumed to be representative of other PCB mixtures. This is a prudent public health measure for which changes could be made if additional statistically significant evidence is forthcoming.

REFERENCES

1. U.S. Environmental Protection Agency. 1988. Drinking Water Criteria Document for Polychlorinated Biphenyls (PCBs). Final. ECAO-CIN-414.
2. Norback, D.H. and R.H. Weltman. Polychlorinated biphenyl induction of hepatocellular carcinoma in Sprague-Dawley rat. Environ. Health Perspect. 60:97-105.
3. Ballschmiter, K., Buckert, H. and S. Bihler. 1981. Baseline studies - the global pollution. Z. Anal. Chem. 306:323-339.
4. Buckley, E.H. 1982. Accumulation of airborne polychlorinated biphenyls in foliage. Science 216:520-522.
5. FDA (Food and Drug Administration). 1980a. Compliance program report of findings. FY77 total diet studies - Adult (7320.73). Food and Drug Administration, U.S. Department of Health, Education and Welfare, Washington, DC.
6. FDA (Food and Drug Administration). 1980b. Compliance program report of findings. FY77 total diet studies - Infant and Toddlers (7320.74). Food and Drug Administration, U.S. Department of Health, Education and Welfare, Washington, DC.
7. Higuchi, K. (Editor). 1976. PCB Poisoning and Pollution. Kodansha Ltd., Tokyo and Academic Press, Tokyo.
8. Hutzinger, O., Safe, S. and V. Zitko. 1974. The Chemistry of PCBs. CRC Press, Boca Raton, FL. pp. 9, 13, 113, 119-131, 143-144, 151-159.
9. Kim, H.T., Kim, K.S., Kim, J.S. and W.B. Stone. 1985. Levels of polychlorinated biphenyls (PCBs), DDE and Mirex in waterfowl collected in New York State, 1981-1982. Arch. Environ. Contam. Toxicol. 14:13-18.
10. Risebrough, R.W., Reiche, P., Peakall, D.B., Herman, S.G. and M.N. Kriven. 1968. Polychlorinated biphenyls in the global ecosystem. Nature. 220:1098-1102.
11. Wasserman, M., Wasserman, D., Cucos, S. and H.J. Miller. 1979. World PCBs map: Storage and effects in man and his biologic environment in the 1970s. Ann. N.Y. Acad. Sc. 320:69-124.

NOTE

The full citations to the other references in the text are to be found in U.S. EPA, ECAO-CIN-414.

Environmental Contaminants, Ecosystems and Human Health. Edited by S.K. Majumdar, E.W. Miller, and Fred J. Brenner.

Chapter Thirty-Four

A COMPUTER ASSISTED RISK ASSESSMENT MODEL FOR UNTESTED AIR POLLUTANTS USING NEWLY DEVELOPED PARAMETERS RESIDUAL INFORMATION FLOW (RIF), RESIDUAL ELECTRONIC FORCE (REF) AND PARTIAL CHARGE FACTOR (PCF)

S.K. RAY*, M.V.S. MURTHY, K. RAJESH, S. SRISUNDER, N. RANGARAJ and P.S.R. MURTHY

*Department of Biological Sciences
Department of Pharmacy
Birla Institute of Technology and Science
Pilani (Raj), 333031, India

CLASSIFICATION OF AIR CONTAMINANTS

One of the major sources of air pollution is the low molecular weight chemicals. Seventy eight industrial chemical have been identified which are toxic and cause injury to different organs of the human body. They are classified on the basis of their mode of action (1). See the Appendix - List of compounds.

Irritants

Most of the irritants are active in the nasal tract. The mucus surface recognizes the particular type of compounds, and the interaction between the surface membrane and these compound cause discomfort. The corrosive and vesicant action of the

APPENDIX

List of Compounds Under the Different Classes of Air Pollutants

I. Irritants

1. Acetaldehyde
2. Acetic acid
3. Acrolein
4. Methyl Formate
5. Ethyl Formate

II. Asphyxiants

1. Acrylonitrile
2. Aniline
3. Butylamine
4. Diethylamine
5. Dimethylaniline
6. Ethylamine
7. Ethylene diamine
8. Isopropylamine
9. Nitroglycerin
10. Tetranitro methane
11. Chloropicrin
12. Dichloro-initroethane
13. Propylene imine

III. Anesthetics & Narcotics

1. Acetone
2. Amyl alcohol
3. Butadiene
4. Butanone
5. Butyl alcohol
6. Butyl cellosolve
7. Cellosolve
8. Diacetone alcohol
9. Dichloro ethyl ether
10. Diisobutyl ketone
11. Ethyl alcohol
12. Ethyl ether
13. Hexanone
14. Hexone
15. Methyl cellosole
16. Methyl alcohol
17. Propyl alcohol
18. Propyl ether
19. Methyl isobutyl carbinol
20. Pentanone
21. Pentane

IV. Visceral Systemic Poisons

1. Ethyl chloride
2. Ethylene dibromide
3. Florotrichloromethane
4. Methyl chloride
5. Methyl chloroform
6. Methylene chloride
7. Propylene dichloride
8. Perchloroethylene
9. Tetrachloroethane
10. Trichloroethylene
11. Trifloromonobromomethane
12. Vinylchloride
13. Allylchloride
14. Chloroform
15. Chloropirene
16. Dichlorofloromethane
17. Dichloroethane
18. Dichlorotetrafloroetane
19. Diflorodibromomethane
20. Ethylbromide

V. Hematopoietic Systemic Poisons

1. Phenylethylene
2. Teritiarybutyltoluene
3. Toluene
4. Xylene
5. Benzene
6. Chlorobenzene
7. Cresol
8. Dichlorbezene
9. Ethylbenzene

VI. Non-Specific Systemic Poisons

1. Amylacetate
2. Butylacetate
3. Cellosolveacetate
4. Ethylacetate
5. Ethylacrylate
6. Methylacrylate
7. Methylcelloslve acetate
8. Methylacetate
9. Propylacetate
10. Tetrahydrofiran

irritants are in effect a reaction between the irritating materials and the membrane components like lipids, proteins, glycolipids, glycoproteins, etc. These biochemical reactions may cause reversible or irreversible changes on the mucus surfaces. Irreversible reactions are type of biochemical interactions which cause molecular change of surface active proteins whose activity thereby is inhibited; e.g., aldehydes, acetic acid, nitro compounds.

Asphyxiants

Asphyxiants are more harmful in the biological system than ordinary irritants. They affect the tissue system in the lungs and restrict the oxidation process of the tissues. They cause dilution of the atmospheric oxygen below partial pressure which is required to maintain an oxygen saturation of the blood, sufficient for the normal tissue respiration; e.g., nitriles, amines, aliphatic compounds.

Anesthetics & Narcotics

This group exerts its principle action as simple anesthesia without serious effects. This group has a depressant action on central nervous system governed by their partial pressure in the blood supply to the brain; e.g., acetylene hydrocarbons, aliphatic ketones.

Visceral Systemic Poisons

Viscerals act on tissues of the chest and abdomen like heart, lungs, liver, etc. These compounds damage one or more of these organs; e.g., halogenated hydrocarbons.

Hematopoietic Systemic Poisons

These compounds mainly affect the circulatory system damaging the hematopoetic system and connected organs; e.g., tolune, xylene.

Non-specific Systemic Poisons

They are non-specific in action. They might damage any part of the body; e.g., acetates, tetrahydrofuran.

BIOLOGICAL SYSTEM INVOLVED

The respiratory system in higher animals is of high complexity. There are special organs to carry out exchange process. This gaseous exchange process takes place between air and blood which is called external respiration. In this case no chemical process is involved. Gaseous exchange between blood and active cells of the organism involves internal respiration involving chemical process. During normal respiration man takes in about 500 cc of air with each act of inspiration and expiration which is known as tidal air. The lungs are never deflated during expiration. About 2500 cc of air is left out in the deflated lungs known as stationary air. The blood is continuously getting oxidized at the lungs by passive diffusion. Cells of all living organism are known to contain metals, especially copper and iron. These metal ions, either as inorganic ions as such or combined with organic residues as in carbonized animal blood, catalyzes the oxidation of number of organic compounds such as aminoacids, thiols, etc. (2).

PARAMETERIZATION OF TOXIC COMPOUNDS

The parameter is a measure of a chemical structure or the potential contribution of its group to a particular property of the parent compound. In a typical procedure, a series of related compounds are examined, and the relevant parameters of their structure or substituent groups are compared with the toxic activities of the compounds. This quantification method relating structure and toxicity is known as quantitative structure-toxicity-relationship studies (QSTR). The following parameters are calculated and used in QSTR studies of the air pollutant (3-6).

Molecular weight (MW)

This is the summation of all the individual atomic weights of the atoms present in the molecule. Logarithm of molecular weight is taken to simplify the calculation.

logP

This is the hydrophobic parameter taken for octanol-water system which represents the biological system. Most of the experimental values are detailed by Hansch and logP for other compounds is calculated by fragmentation method (3).

$2X^v$

The valence molecular connectivity index $2X^v$ is calculated from a hydrogen suppressed graph of a molecule (4). A valence connectivity value δ_i^v is assigned on each atom on the basis of the following expression:

$$\delta_i^v = Z_i^v - h_i$$

where Z_i is the number of valence electron of atom I and h_i is the number of hydrogen atoms attached to atom I. Thus the use of δ_i^v permits the calculation of valance X term of first or higher order.

$$1\chi^v = \sum (\delta_i^v \cdot \delta_j^v)^{-1/2} ; 2\chi^v = \sum (\delta_i^v \cdot \delta_j^v \cdot \delta_k^v)^{-1/2}$$

where I, j, k are atoms bonded in sequence in a path and the sum is one, all distinct sets of one edge, two edge and three edge paths respectively.

Information Content (IC)

The information content calculations are based on the assignment of coordinates to an atom in a molecular graph (5, 6).

The mathematical treatments is given below:

Let A = $a_1, a_2 \ldots, a_n$ be a collection of n elements of a finite discrete system, G, representing a molecule. The set A is called the atom set of G. By a chemical graph G based on the atom set A, we mean a multigraph without self loop with "vertex set" A (G) and "edge set" D(G) such that degree of each alphanumeric symbol a_i, I = 1 . . . n is equal to or less than its corresponding valency value.

A probability distribution is then assigned to the set of equivalence classes:

$A_1, A_2 \ldots A_n$
$P_1, P_2 \ldots P_n$

where $P_i = n_i/n$ is the probability that a randomly selected element of A will occur in the ith subset. The mean information content is defined by Shannon's relation (5):

$$IC = - \sum_{i=1}^{n} (p_i \ln p_i)$$

Structural Information Content, SIC = IC/ln(n)

Complementary Information Content, CIC = ln(n) - IC

Bonding Information Content, BIC = IC/B (Number of Bonds)

Residual Electronic Force (REF)

(i) The net atomic charge on-each atom is calculated using Jaffe-Mulliken electro negativity values in Sanderson's formula (7).

$$\text{Formula: } \delta_{A \leftarrow B} = \frac{a_B - a_A}{b_B + b_A}$$

(ii) The non-bonded interatomic distances in the molecule are calculated. For this, standard bond angles and bond distances are used.

(iii) Inter atomic affinity is calculated by

$$a_{ij} = \frac{d_{ij}^2}{q_i\, q_j}$$

where q_i, q_j are the partial charges and d_{ij} is the distance between them.

(iv) For each fragment or collectively, these forces are classified as positive and negative clusters and by using information theory on the positive and negative clusters, the positive and negative electronic information flows (p_{eif}, n_{eif}) are obtained.

$$P_{eif} = - \sum_i \frac{f_i}{t_{cef}} \left(\log_2 \frac{f_i}{t_{cef}}\right)$$

where f_i is the individual force t_{cef} and is the total positive cluster electronic force. Similarly N_{eif} is calculated.

(v) REF of the fragment is given by the difference of the above information flows: REF = P_{eif} - N_{eif}

Finally the Global REF of the molecule is obtained by REF_M = REF (1) + REF (2) + . . . + REF (n)

Where REF_M: REF of the total molecular structure.

$$REF_M^2 = \frac{1}{REF(1)} + \frac{1}{REF(2)} + \_\_\_\_ + \frac{1}{REF(n)}$$

Residual Information Flow (RIF)

Partial charge on an atom is defined as the sum of all the partial charges due to charges on all atoms attached to it.

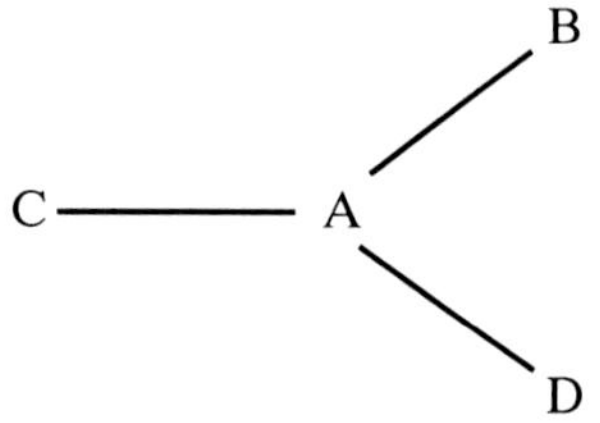

$$\delta_{A \leftarrow B} = \frac{a_B - a_A}{b_B + b_A}$$

$$\delta_A = \delta_{A \leftarrow B} + \delta_{A \leftarrow C} + \delta_{A \leftarrow D}$$

The individual partial charges of the atoms are summed according to the groups made as in information content (6). The positive cluster formed and information flow is given by Shannon's formula.

$$P_{IF} = - \sum \frac{gpc_i}{tpc^+} \log_2 \frac{gpc_i}{tpc^+}$$

Similarly,

$$N_{IF} = - \sum \frac{gpc_i}{tpc^-} \log_2 \frac{gpc_i}{tpc^-}$$

$$RIF = P_{IF} - N_{IF}$$

where, gpc_i = ith group partial charge
tpc = total partial charge

Partial Charge Factor (PCF)

It utilizes the partial charge of atoms as in RIF. The negative cluster and the positive cluster are formed by grouping the partial charges of all atoms in the compound. The PCF is calculated using the following Shannon's formula.

$$P_{IC} = - \sum \frac{ipc^+}{tpc} \log_2 \frac{ipc^+}{tpc} \; ; \; NIC = - \sum \frac{ipc^-}{tpc} \log_2 \frac{ipc^-}{tpc}$$

$$PCF = P_{IC} - N_{IC}$$

tpc = total partial charge ipc = individual partial charge

COMPUTER ASSISTED BEST FIT RESULTS

No chemical is a poison by itself. It is the dose that makes a substance poison at the toxic dose. The poisonous effects of the air pollutants depend on the concentration of the compound in the air and the time of exposure of the subject. The toxic values (tx) of the 78 industrial chemicals were experimentally determined for an eight hour exposure on human beings under definite standard safe parameters (1).

The log of toxicity (l(tx)) of these compounds have been rationalized using various parameters. The parametric values of different class of toxicants are presented in Tables 1-6. The average van der Waals volume of halogens (Hvw) in the molecules is a relevant parameter in toxicology. The best fit equations that surfaced from computer search for each class of compounds are given as below. The equations are significant at above 97.5% level.

TABLE 1
Parametric Values for Irritants

S.NO.	l(mw)	logP.	ref.	rif.	2Xv.	ic.	pcf.	cic.	sic.	bic.	1(tx).
1	1.644	-0.300	0.102	0.301	0.236	1.665	1.179	0.143	0.593	0.593	2.301
2	1.782	-0.310	0.433	0.108	0.568	1.750	-0.407	1.250	0.583	0.583	1.000
3	1.749	0.480	0.150	-0.342	0.372	2.000	0.725	1.000	0.667	0.631	-0.301
4	1.778	1.040	-0.403	0.425	0.332	2.000	0.844	1.000	0.667	0.667	2.000
5	1.870	0.180	-0.042	0.104	0.552	2.050	0.785	1.410	0.460	0.592	2.000

TABLE 2
Parametric Values for Asphyxiants

S.NO.	l(mw)	logP.	ref.	rif.	2Xv.	ic.	pcf.	cic.	sic.	bic.	1(tx).
1	1.725	-0.920	-1.100	0.095	0.333	2.128	0.877	0.679	0.758	0.709	1.301
2	1.969	0.900	-5.210	0.451	1.411	2.006	0.892	1.802	0.527	0.491	0.699
3	1.864	0.680	-4.570	0.125	1.142	1.967	1.756	2.033	0.492	0.503	0.699
4	1.864	0.430	-0.710	0.583	0.957	1.674	2.211	2.326	0.418	0.428	1.398
5	2.083	2.310	10.720	0.061	2.230	1.739	1.615	2.583	0.402	0.384	0.699
6	1.654	-0.130	3.780	0.670	0.408	1.961	1.931	1.361	0.590	0.619	1.398
7	1.893	-0.650	6.750	1.150	0.577	1.918	2.004	1.667	0.535	0.554	1.000
8	1.772	0.480	2.700	0.566	1.244	1.881	1.742	1.820	0.508	0.525	0.699
9	2.356	1.870	-8.010	-0.145	1.981	2.409	-0.946	1.913	0.557	0.519	-0.301
10	2.292	-0.030	9.120	0.941	1.628	1.239	-0.776	2.462	0.335	0.287	0.000
11	2.216	1.520	12.130	0.076	3.239	1.842	-1.114	0.965	0.656	0.614	0.000
12	2.158	2.950	3.500	0.166	2.948	2.446	-0.406	0.876	0.736	0.707	1.000
13	1.764	0.030	0.610	0.320	0.612	2.050	1.490	1.410	0.592	0.617	1.398

TABLE 3

Parametric Values for Anesthetics and Narcotics

S.NO.	L(mw)	2Xv	IC.	PCF.	RIF.	CIC.	SIC.	BIC.	logP.	ref.	1(tx).
1	1.740	0.908	1.571	0.984	0.098	1.751	0.473	0.473	-0.240	-0.363	3.000
2	1.945	1.431	1.792	1.401	0.251	2.379	0.430	0.438	0.890	1.380	2.000
3	1.733	0.471	1.371	0.667	-0.918	1.951	0.413	0.396	1.990	0.462	3.000
4	1.858	1.056	1.700	1.035	-0.245	2.000	0.459	0.459	0.260	-0.429	2.398
5	1.870	1.077	1.872	1.437	0.361	2.129	0.468	0.491	0.880	0.736	2.000
6	2.072	1.690	1.966	1.702	-0.154	2.494	0.441	0.448	0.830	0.283	1.699
7	1.955	0.902	1.920	1.786	0.171	2.080	0.480	0.491	-0.540	-0.583	1.301
8	2.065	2.691	2.182	1.066	0.301	2.105	0.505	0.489	1.880	-0.819	1.699
9	2.155	1.816	1.640	1.788	0.268	2.267	0.420	0.431	2.250	-0.775	1.176
10	2.153	3.735	1.698	1.111	-0.775	3.110	0.353	0.357	2.770	0.479	1.699
11	1.663	0.316	1.880	1.530	0.912	1.290	0.593	0.627	-0.300	1.064	3.000
12	1.870	0.781	1.426	1.959	0.133	2.482	0.365	0.239	0.830	-0.465	2.602
13	2.001	1.806	1.628	1.028	-0.581	2.620	0.383	0.383	1.780	0.050	2.000
14	2.001	2.300	1.734	1.067	-0.543	2.515	0.408	0.408	2.160	0.002	2.000
15	2.009	2.479	1.875	1.304	0.087	2.518	0.427	0.434	1.480	0.070	1.398
16	1.935	1.452	1.674	1.038	-0.477	2.326	0.418	0.418	0.910	0.309	2.301
17	1.779	0.724	1.947	1.479	0.534	1.638	0.543	0.563	0.300	-0.592	2.602
18	2.009	2.234	1.378	1.577	0.007	3.015	0.314	0.314	2.030	0.397	2.699
19	1.881	0.716	2.035	1.927	0.932	1.666	0.550	0.568	-0.770	0.250	1.398
20	1.506	0.000	1.792	1.756	1.455	0.530	0.772	0.772	-0.820	0.240	2.301
21	1.858	1.354	1.160	1.292	-1.000	3.804	0.284	0.290	1.980	-0.909	3.000

TABLE 4
Parametric Values for Visceral Systemic Poisons

S.NO.	l(mw)	logp.	ref.	rif.	2xv	ic.	pcf.	cic.	sic.	bic.	pcv.	1(tx).
1	1.810	1.430	-8.680	-0.949	0.851	1.299	1.888	1.701	0.433	0.463	0.567	3.000
2	2.274	3.530	-5.910	-1.581	2.067	1.750	0.120	1.250	0.383	0.623	0.329	1.398
3	2.135	2.530	2.570	-0.991	1.770	1.371	-1.870	0.951	0.590	0.685	0.869	3.000
4	1.977	1.190	-2.760	-0.720	0.000	1.371	0.865	0.951	0.509	0.689	0.433	1.301
5	1.708	0.910	-9.030	-0.963	0.000	1.371	0.622	0.951	0.509	0.689	0.473	2.000
6	2.125	2.490	0.910	0.994	3.380	1.561	0.468	1.439	0.520	0.576	0.646	2.699
7	1.929	1.250	-12.640	0.994	1.025	1.522	0.539	0.800	0.655	0.761	0.432	2.699
8	2.053	2.000	-0.110	-0.460	2.086	1.686	1.136	1.774	0.487	0.570	0.507	1.875
9	2.220	2.600	-6.210	-0.900	2.652	0.981	1.000	1.667	0.355	0.355	0.209	2.301
10	2.225	3.230	-4.030	1.572	3.279	1.750	-0.155	1.250	0.585	0.623	0.479	0.699
11	2.119	2.290	-6.740	0.580	1.767	1.459	-0.979	1.126	0.564	0.564	0.183	2.301
12	2.173	2.370	-1.100	-0.256	-0.618	1.371	-1.773	0.951	0.590	0.685	0.645	3.000
13	1.796	0.910	-6.840	-0.554	0.491	1.459	0.346	1.126	0.564	0.564	1.703	2.699
14	1.884	0.170	-0.520	0.392	0.471	2.161	1.778	1.161	0.650	0.650	0.819	0.699
15	2.077	0.910	4.220	-0.963	0.000	1.371	0.622	0.951	0.590	0.685	0.473	2.000
16	1.947	2.700	-1.220	-0.914	1.181	1.761	0.743	1.561	0.530	0.509	0.791	1.398
17	2.082	2.160	-1.450	-0.873	0.211	1.522	-1.873	0.710	0.682	0.761	1.252	3.000
18	1.995	1.790	-0.720	0.407	2.229	1.500	0.627	1.500	0.500	0.534	0.364	2.000
19	1.986	1.480	6.290	-1.576	0.803	1.918	-0.963	0.667	0.742	0.742	0.381	2.000
20	2.013	1.550	-10.030	-0.455	0.526	1.922	-0.908	0.400	0.828	0.961	0.875	3.000
21	2.233	2.820	-6.140	0.338	-0.110	1.750	-1.472	1.189	0.604	0.645	1.433	3.000
22	2.322	3.090	-3.060	-0.527	1.237	1.522	-1.527	0.800	0.655	0.761	0.899	2.000
23	2.037	1.610	-7.670	-0.528	1.462	1.299	0.999	1.701	0.433	0.463	0.532	2.301

TABLE 5

Parametric Values for Haemotopoietic Systemic Poisons

S.NO.	l(mw)	logp.	ref.	rif.	2xv.	ic.	pcf.	cic.	sic.	bic.	1(tx).
1	2.018	2.650	-6.130	-1.299	1.618	1.774	0.250	2.226	0.444	0.434	2.000
2	2.171	2.650	-10.030	0.175	3.750	1.871	1.014	2.885	0.393	0.381	1.000
3	1.964	2.730	-12.560	0.062	1.250	1.688	0.242	2.222	0.431	0.404	2.301
4	2.028	3.150	-0.200	0.062	1.750	1.842	0.565	2.328	0.442	0.407	2.301
5	1.893	2.130	0.000	0.000	1.155	1.000	0.000	2.585	0.279	0.256	1.398
6	2.051	2.840	-8.140	-0.393	1.772	1.325	0.000	2.260	0.369	0.339	1.875
7	2.167	3.380	-0.120	-0.709	2.304	1.459	0.020	0.126	0.407	0.373	1.699
8	2.034	1.960	-2.470	0.542	1.474	2.522	0.592	1.478	0.630	0.594	0.699
9	2.026	3.150	7.400	-1.383	1.839	1.679	0.320	2.491	0.421	0.382	2.301

TABLE 6

Parametric Values for Non Specific Systemic Poisons

S.NO.	l(mw)	logp.	ref.	rif.	2xv.	ic.	pcf.	cic.	sic.	bic.	1(tx).
1	2.115	2.490	0.094	-0.348	1.694	1.971	0.723	2.553	0.436	-0.442	2.301
2	2.065	2.270	0.054	0.357	1.537	1.927	1.180	2.395	0.446	0.454	2.301
3	2.121	0.190	1.386	0.647	1.523	2.153	1.249	2.239	0.490	0.490	2.000
4	1.945	0.660	0.857	-0.003	0.925	1.950	0.684	1.857	0.512	0.527	2.602
5	2.000	0.340	0.629	-0.508	0.956	2.307	0.496	1.600	0.590	0.577	1.398
6	1.944	0.740	0.440	-0.265	0.727	2.292	0.512	1.293	0.639	0.619	1.000
7	2.072	0.190	0.909	0.858	1.333	1.102	1.243	3.068	02.64	0.264	1.398
8	1.897	0.180	1.027	0.101	0.650	2.050	0.725	1.410	0.592	0.592	0.301
9	2.009	1.210	0.417	-0.244	1.340	2.016	0.705	2.072	0.493	0.493	2.301
10	1.858	0.460	-0.802	0.238	1.319	1.547	1.874	2.154	0.418	0.431	2.301

IRRITANTS: aldehydes, acrolein, methyl formate, etc.
1) l (tx) = 3.319 RIF - 4.431 SIC + 3.651;
n = 5; r = 0.968; F = 15.07
ASPHYXIANTS: nitriles, amines, etc.
2) l (tx) = -2.117 l (mw) + 4.938;
n = 13; r = 0.870; F = 28.38
ANESTHETICS AND NARCOTICS: acetylene hydrocarbons, etc.
3) l (tx) = -2.189 l (mx) -1.280 IC + 8.540
n = 21; r = 0.832; F = 20.25
VISCERAL SYSTEMIC POISONS: halogenated hydrocarbons.
4) l (tx) = 0.006 REF2 -18.296 Hvw-0.354 IC2+6.571;
n = 20%; r + 0.875; F = 18.67
HEMATOPOIETIC SYSTEMIC POISONS: benzene, phenol, etc.
5) l (tx) = 1.315 logP -5086 l (mw) +8.502;
n = 9; r = 0.968; F = 29.65
NON-SPECIFIC TOXIC COMPOUNDS: acetates, acrylates, etc.
6) l (tx) = 6.258 l (mw) -1.575 RIF+2.018 PCF-12.679;
n = 9; r = 0.925; F = 12.52

DISCUSSION

Irritants

Defining irritants from the pharmacological point of view, most of them are active in the nasal tract (1). New methods of characterizing the irritant molecules have been developed. The characterization method may be applicable to other toxic compounds, drugs and micromolecules (8). A relationship has been developed between the structural characteristics or indices of a compound and the log of its toxic dose where the toxic dose is in parts per million (ppm). The best fit model which surfaced from all the possible combinations of physicochemical and structural parameters is given in equation no. 1.

The biphasic equations consisting RIF+'X', where 'X' = REF, 2Xv IC, do not increase the correlation coefficient, r = 0.90 and F-values, 4.33, 4.47, 4.462 respectively. The equations are less significant compared to equation no. 1.

The biphasic equations consisting RIF+'X2', where 'X2' = PCF and BIC increase the correlation coefficient considerably, r = 0.93 and F-values, 6.16 and 6.57 respectively. The significance level is approx. 95% and 97.5%. This implies the significance of electronic parameter in the toxic effects of irritants.

Irritants are recognized on the mucus surface of the nasal tract mainly by the electronic factor of the toxicant structure, quantified by RIF (r = 0.90). Although the electronic parameter plays the major role, identification of an irritant and its effective risk is assessed by its coupled property of electronic and normalized steric factor of the compound. The steric factor here is represented by structural information content which is a normalized form of information content of the chemical structure (6).

Asphyxiants

They are more harmful because they affect the tissue system in lungs by restricting the oxidation process (1). The other possible mechanism is that even though the tissues are exposed to oxygen the presence of asphyxiants causes the inactivation of oxygen absorption system of the tissue due to a definite interaction between cell system in tissue and the toxic compounds. The computer results show that the molecular weight of this group of compounds constitute the best fit equation. The molecular weight is negatively related with toxicity of compound. So, if the molecular weight increases toxicity decreases. Another set of equations involving the electronic parameter, PCF, also gives highly significant equation above 99% level. The multiparametric equation combining PCF and X (cic, bic, sic) correlates with 1 (tx) of the asphyxiants, $r = 0.865$. Absorption or dispersion in the cell system is one factor which is represented by mol wet., SIC, BIC, CIC, logP where as recognition and interaction at the molecular level within the cell after absorption or dispersion is the second factor as represented by the strong influence of PCF in the resolution of toxicity of the compounds. This interaction may be reversible or irreversible causing inactivation, at molecular levels, of electron transport chain in the respiratory system. The moment electron transport chain is intercepted by PCF of the molecule, the oxygen absorption capability of the tissue decreases. Experimentation in this light is highly suggestive.

Anesthetics and Narcotics

There are many narcotics that inhibit the respiratory system but not all anesthetics and narcotics are inhibitors of respiration (1). Being light at the molecular level and most of them being polar, during exchange process they are cotransported by passive diffusion along with oxygen into blood where they remain in the dissolved state. Many compounds form stable complexes with iron and copper in blood and are thereby neutralized, making them free from poisonous and narcotic effects. But there are certain sets of compounds which are not involved in the oxidation and reduction process, and manage to remain in the blood system. Most of the compounds do not reach the brain due to the protective Blood Brain Barrier (BBB) and the compounds which slip through BBB cause the anesthetic or narcotic activity. The anesthetic and narcotic activity of the set of compounds is rationalized by a set of parameters. Molecular weight of the toxic compounds emerged as primary factor in showing such toxic effects. The molecular weight alone gives a correlation coefficient of $r = 0.646$. The total number of compounds in this class taken for analysis is twenty one. Electronic properties of these molecules are not relevant for explaining the biological response (rif: $r = 0.159$; ref: $r = 0.085$). The partition parameter logP was also put to examination and correlation coefficient r = 0.084 was obtained. The mode of action of this class of compounds is clearly by a different mechanism. The mechanism of anesthetic and narcotics from the computer study is a clear indication that the transportation phenomena is not mimicked by logP of the co-transported gaseous molecules. The electronic parameter plays no role in the equation, indicating that any recognition factor or interaction leading to blockade of electron transport is ruled out. The pharmacodynamic aspect is also ruled out from the fact that, the triparametric equations comprising of steric, electronic, and hydrophobic parameters are not significant in rationalizing the toxicity.

For anesthetics and narcotics the transport phenomena is not solely guided by l(mw). The best represented equation for this mechanism is the biphasic equation no. 3 with common parameter l(mw). In the process of transportation, more likely at the level of the molecular structures, intrinsic property of a chemical graph becomes more relevant. The other equations are not considered because of inter-correlation factor and lower correlation values. The common factor being mol wt and the other factor is the normalized form of the information theoretic topological index SIC, BIC. The coefficient of molecular weight in the regression equation is positive which indicates that higher the molecular weight, lesser is the possibility of the compound crossing the BBB. The coefficient of topological indices SIC, BIC in the equation are also positive. This also shows that higher the topological factor and heterogeneity of the molecule, lower is the transportation rate. So in the screening of compounds, in the light of risk assessment, it is necessary to investigate the limiting value of l(mw) as well as SIC or BIC.

Hematopoietic Systemic Poisons

In the hematopoietic systems, the best fit model is the biparametric equation 5 involving logP and l(mw). In the equation, the coefficient of logP is positive and the coefficient of l(mw) is negative. So, higher the molecular weight, lesser the toxicity and higher the logP, higher the toxicity. The main difference in the mode of action of asphyxiants and this class is that in case of asphyxiants, l(mw) is a major determining factor, whereas in hematopoietic poisons logP is a major determining factor (r = 0.68). It is likely that the hematopoietic systemic compounds are co-transported by diffusion into the tissues where these compounds are absorbed easily. The absorption and transport are mainly governed by the partition parameter logP. In the risk assessment studies absorption factor coupled with molecular weight of the hematopoietic compounds determines its toxic effect in the human circulatory system.

Other Toxic Compounds

This class of non-specific compounds can be sub-classified into visceral systemic poisons and non-specific systemic poisons. Visceral systemic poisons affect tissues in the chest or abdomen. All attempts to correlate toxicity of this class of compounds show that standard deviations of the equations are very high and correlation coefficients are low (about 0.6=r). Careful observation of the compounds reveal that most of them are halogenated compounds. Incorporating a factor for halogen (van der Waals) volume of halogens, HV_W in the multiparametric equations, gives rise to a very significant equation, no. 4. The equation has REF_2, IC and HV_W as independent variables. This search indicates that the percentage volume of halogen in a compound or halogen content per unit volume of the compound, is a relevant factor mimicking toxicity of the compounds. The best fit equation has $r = 0.875$. Non-Specific toxic compounds do not act specifically. Most of these systemic poisons are acetates, which on hydrolysis give rise to alcohol and acid. In the rationalization of toxicity of these compounds, independent parameters that emerged from computer search are l(mw), RIF & PCF. Both RIF and PCF were developed from the concept of electronegetivity of elements in a structure and are electronic parameters. So, complex influence of electronic aspect of the toxic compounds along with the molecular weight is the guiding factor for toxicity of the non-specific organic compounds. The coefficient of the l(mw) is positive. Interestingly, the coefficient of RIF is negative whereas the coefficient of PCF is positive. Both RIF and PCF being electronic parameters, the role of electronic aspect of the molecular structure is complex which imparts non-specific poisoning effect on the human system.

REFERENCES

1. Patty, Frank A. 1958. Mode of Entry And Action of Toxic Materials, pp. 143-172. In: F.A. Patty (Ed.) *Industrial Hygiene and Toxicology: General Principles. Second Ed.* Interscience Publishers, Inc., New York, pp. 830.
2. Chatterjee, C.C. 1987. Respiratory System, pp. 365-395. In: C.C. Chatterjee (Ed.) *Human Physiology Vol. 1.* Ashutosh Lithographic Co., Calcutta, pp. 759.
3. Leo, Albert J. 1990. Methods of Calculating Partition Coefficients, pp. 295-319. In: Corwin Hansch, Peter G. Sammes, John B. Taylor and Christopher A. Ramsden (Ed.) *Comprehensive Medicinal Chemistry Vol. 4.* Pergamon Press, Oxford, pp. 766.
4. Subhash, C. Basak, Bertelsen, Sharon and Gregory D. Grunwald. 1994. Application of Graph Theoretical Parameters in Quantifying Molecular Similarity and Structure Activity Relationships. *J. Chem. Inf. Comput. Sci.* 34:270-276.
5. Danail, Bonchev. 1983. Information Indices For Molecules, pp. 55-180. In: D. Bonchev (Ed.) *Information Theoretic Indices for Characterization of Chemical Structure.* Research Studies Press, Chichester, New York, pp. 249.

6. Subhash, C. Basak. 1987. Use of Molecular Connectivity Indices in Predictive Pharmacology and Toxicology: A QSAR Approach. *Med. Sci. Res.* 15:605-609.
7. Huheey, James E. 1983. Bonding Models In Inorganic Chemistry, pp. 53-183. In: J.E. Huheey (Ed.). *Inorganic Chemistry: Principles of Structure and Reactivity,* Harper International SI Edition. Cambridge, Philadelphia, San Francisco, pp. 936.
8. Ray, S.K., Premkanth, K.L., Chandra, R., Sekar, M.V.S. Murthy, Sridhar, V., Lakshmi, Narasimhan S. and S. Rudra. 1991. Design of Potent Antifungal Cinnamic Acid Derivatives: A Comparative Multiparametric QSAR Study Involving Newly Developed Electronic Parameters, Residual Information Flow and Residual Electronic Force. *Med. Sci. Res.* 19:759-763.

Subject Index

A

B

C

D

E

F

G

M

N

O

P

R

S

T

U

V

W